Lecture Notes in Physics

Editorial Board

R. Beig, Wien, Austria
W. Beiglböck, Heidelberg, Germany
W. Domcke, Garching, Germany
B.-G. Englert, Singapore
U. Frisch, Nice, France
P. Hänggi, Augsburg, Germany
G. Hasinger, Garching, Germany
K. Hepp, Zürich, Switzerland
W. Hillebrandt, Garching, Germany
D. Imboden, Zürich, Switzerland
R. L. Jaffe, Cambridge, MA, USA
R. Lipowsky, Golm, Germany
H. v. Löhneysen, Karlsruhe, Germany
I. Ojima, Kyoto, Japan
D. Sornette, Nice, France, and Los Angeles, CA, USA
S. Theisen, Golm, Germany
W. Weise, Garching, Germany
J. Wess, München, Germany
J. Zittartz, Köln, Germany

The Editorial Policy for Edited Volumes

The series *Lecture Notes in Physics* (LNP), founded in 1969, reports new developments in physics research and teaching - quickly, informally but with a high degree of quality. Manuscripts to be considered for publication are topical volumes consisting of a limited number of contributions, carefully edited and closely related to each other. Each contribution should contain at least partly original and previously unpublished material, be written in a clear, pedagogical style and aimed at a broader readership, especially graduate students and nonspecialist researchers wishing to familiarize themselves with the topic concerned. For this reason, traditional proceedings cannot be considered for this series though volumes to appear in this series are often based on material presented at conferences, workshops and schools.

Acceptance

A project can only be accepted tentatively for publication, by both the editorial board and the publisher, following thorough examination of the material submitted. The book proposal sent to the publisher should consist at least of a preliminary table of contents outlining the structure of the book together with abstracts of all contributions to be included. Final acceptance is issued by the series editor in charge, in consultation with the publisher, only after receiving the complete manuscript. Final acceptance, possibly requiring minor corrections, usually follows the tentative acceptance unless the final manuscript differs significantly from expectations (project outline). In particular, the series editors are entitled to reject individual contributions if they do not meet the high quality standards of this series. The final manuscript must be ready to print, and should include both an informative introduction and a sufficiently detailed subject index.

Contractual Aspects

Publication in LNP is free of charge. There is no formal contract, no royalties are paid, and no bulk orders are required, although special discounts are offered in this case. The volume editors receive jointly 30 free copies for their personal use and are entitled, as are the contributing authors, to purchase Springer books at a reduced rate. The publisher secures the copyright for each volume. As a rule, no reprints of individual contributions can be supplied.

Manuscript Submission

The manuscript in its final and approved version must be submitted in ready to print form. The corresponding electronic source files are also required for the production process, in particular the online version. Technical assistance in compiling the final manuscript can be provided by the publisher's production editor(s), especially with regard to the publisher's own LaTeX macro package which has been specially designed for this series.

LNP Homepage (springerlink.com)

On the LNP homepage you will find:
- The LNP online archive. It contains the full texts (PDF) of all volumes published since 2000. Abstracts, table of contents and prefaces are accessible free of charge to everyone. Information about the availability of printed volumes can be obtained.
- The subscription information. The online archive is free of charge to all subscribers of the printed volumes.
- The editorial contacts, with respect to both scientific and technical matters.
- The author's / editor's instructions.

A. Dinklage T. Klinger G. Marx L. Schweikhard
(Editors)

Plasma Physics

Confinement, Transport and Collective Effects

Springer

Editors

Priv.-Doz. Dr. Andreas Dinklage
Professor Dr. Thomas Klinger
MPI Plasmaschutz
EURATOM Association
Wendelsteinstr. 1
17491 Greifswald, Germany
andreas.dinklage@ipp.mpg.de
thomas.klinger@ipp.mpg.de

Dr. Gerrit Marx
Professor Dr. Lutz Schweikhard
Ernst-Moritz-Arndt Universität
Institut für Physik
Domstr. 10a
17489 Greifswald, Germany
marx @physik.uni-greifswald.de
lutz.schweikhard@physik.uni-greifswald.de

Andreas Dinklage et al., *Plasma Physics*,
Lect. Notes Phys. 670 (Springer, Berlin Heidelberg 2005), DOI 10.1007/b103882

Library of Congress Control Number: 2005923687

ISSN 0075-8450

ISBN-10 3-540-25274-6 Springer Berlin Heidelberg New York
ISBN-13 978-3-540-25274-0 Springer Berlin Heidelberg New York

This work is subject to copyright. All rights are reserved, whether the whole or part of the material is concerned, specifically the rights of translation, reprinting, reuse of illustrations, recitation, broadcasting, reproduction on microfilm or in any other way, and storage in data banks. Duplication of this publication or parts thereof is permitted only under the provisions of the German Copyright Law of September 9, 1965, in its current version, and permission for use must always be obtained from Springer. Violations are liable for prosecution under the German Copyright Law.

Springer is a part of Springer Science+Business Media
springeronline.com
© Springer-Verlag Berlin Heidelberg 2005
Printed in The Netherlands

The use of general descriptive names, registered names, trademarks, etc. in this publication does not imply, even in the absence of a specific statement, that such names are exempt from the relevant protective laws and regulations and therefore free for general use.

Typesetting: by the authors and TechBooks using a Springer LaTeX macro package
Cover production: *design &production* GmbH, Heidelberg

Printed on acid-free paper   SPIN: 11360360   57/3141/jl   5 4 3 2 1 0

Dedicated to Billa, Frauke and Johanna

Preface

Plasma, sometimes called the fourth state of matter, is a multifaceted substance which poses a variety of challenges. Plasma physics deals with the complex interaction of many charged particles with external or self-generated electromagnetic fields. It is this unique entanglement which makes plasma physics a fascinating field for basic research. At the same time, plasma plays an essential role in many applications, ranging, e.g., from advanced lighting devices and surface treatments for semiconductor applications or surface layer generation to the efforts to tame nuclear fusion as an energy source for our future harnessing the nuclear processes which fuel our sun.

Modern plasma research is a multidisciplinary endeavor which includes aspects of electrodynamics, many-particle physics, quantum effects and nonlinear dynamics. But even though the spatial extension, the density, the ionization degree and the plasma temperature may vary by many orders of magnitude, the physical similarities – or the plasma properties – of, e.g., the solar corona, non-neutral plasmas in ion-traps, the electron gas of metals or planetary interiors lead to similarities of these systems.

Plasmas on earth are evanescent. The confinement of plasmas for extended times is a very difficult task and one of the central keys for plasma research and applications. Consequently, transport phenomena which go far beyond classical transport are highly relevant. This also leads to the ultimate challenge of many-particle physics: the understanding of turbulence. In addition, a variety of "ordered" collective effects can be studied in unique clarity, for example, phase transitions in "dusty" plasmas or the large variety of plasma waves. The corresponding investigations are at the forefront of current research and development.

This volume of Springer Lecture Notes in Physics provides an overview of modern plasma research with a special focus on confinement and related issues. Beginning with a broad introduction, the book leads graduate students and researchers – including those not specialized in plasma research – to the state of the art of modern plasma physics. The book also presents a methodological cross section ranging from plasma applications and plasma diagnostics to numerical simulations, an important link between theory and experiment which is gaining more and more importance. The references are chosen to guide the reader from basic concepts to current research. Exercises in computational plasma physics are supplied on a Web site (see Chap. 16 in Part III of this book).

The contributions are structured in three parts: After a broad introduction to *Fundamental Plasma Physics*, the focus of this volume on *Confinement, Transport and Collective Effects* is covered. Modern plasma physics is also applied science and has many methodological branches as described in the third part on *Methods and Applications*.

The chapters have been written by prominent experts in their respective fields. The book is based on a series of lectures for graduate students in the framework of a W.E.–Heraeus Summer School.

We would like to thank the W.E.–Heraeus Foundation for funding and the International Max Planck Research School "Bounded Plasmas" for supporting the 50th Heraeus Summer School "Plasma Physics: Confinement, Transport and Collective Effects" held in Greifswald during October 2003. We are indebted to those speakers who contributed; this book has benefitted from their encouragement and support.

We thank Dr. Angela Lahee from Springer Heidelberg for her friendly collaboration throughout this project. We also appreciate the professional and friendly support from Ms. Jaqueline Lenz, Ms. Gabriele Hakuba, Ms. Elke Sauer and Ms. Shanya Rehman during the editorial and technical realization of this book.

And last – but certainly not least – we are deeply grateful to Ms. Andrea Pulss, for whom it must have been much more than a "challenging effort" to do the technical editorial work.

Greifswald,
April 2005

Andreas Dinklage
Thomas Klinger
Gerrit Marx
Lutz Schweikhard

Contents

Part I Fundamental Plasma Physics

1 Basics of Plasma Physics
U. Schumacher 3
1.1 Definition, Occurrence and Typical Parameters
    of Plasmas 3
1.2 Ideal Plasmas 5
1.3 Important Plasma Properties 7
    1.3.1 Debye Shielding 7
    1.3.2 The Plasma Parameter 8
    1.3.3 Landau Length 8
    1.3.4 Plasma Frequency 9
1.4 Single Particle Behavior in Plasmas 10
    1.4.1 Coulomb Collisions, Collision Times and Lengths 11
    1.4.2 Electrical Conductivity of Plasmas 14
    1.4.3 Single Charged Particle Motion in Electric
          and Magnetic Fields 15
1.5 Kinetic Description 19
References 20

2 Waves in Plasmas
A. Piel 21
2.1 Introduction 21
2.2 Dispersion Relation for Waves in a Fluid Plasma 22
    2.2.1 Maxwell's Equations 22
    2.2.2 The Equation of Motion 22
    2.2.3 Normal Modes 23
    2.2.4 The Dielectric Tensor 23
    2.2.5 Phase and Group Velocity 24
2.3 Waves in Unmagnetized Plasmas 24
    2.3.1 Transverse Waves 25
    2.3.2 Longitudinal Waves 31
    2.3.3 Electron Beam Driven Waves 35
2.4 Waves in Magnetized Plasmas 37
    2.4.1 Propagation Along the Magnetic Field 39

	2.4.2	Cut-Offs and Resonances	40
	2.4.3	Propagation Across the Magnetic Field	43
2.5	Concluding Remarks		47
References			48

3 An Introduction to Magnetohydrodynamics (MHD), or Magnetic Fluid Dynamics

B.D. Scott 51

- 3.1 What MHD Is 51
- 3.2 The Ideas of Fluid Dynamics 52
  - 3.2.1 The Density in a Changing Flow Field – Conservation of Particles 52
  - 3.2.2 The Advective Derivative and the Co-moving Reference Frame 54
  - 3.2.3 Forces on the Fluid – How the Velocity Changes 55
  - 3.2.4 Thermodynamics of an Ideal Fluid – How the Temperature Changes 57
  - 3.2.5 The Composite Fluid Plasma System 58
- 3.3 From Many to One – the MHD System 59
  - 3.3.1 The MHD Force Equation 60
  - 3.3.2 Treating Several Ion Species 60
  - 3.3.3 The MHD Kinematic Equation 61
  - 3.3.4 MHD at a Glance 62
- 3.4 The Flux Conservation Theorem of Ideal MHD 62
  - 3.4.1 Proving Flux Conservation 62
  - 3.4.2 Magnetic Flux Tubes 64
- 3.5 Dynamics, or the Wires-in-Molasses Picture of MHD 64
  - 3.5.1 Magnetic Pressure Waves 65
  - 3.5.2 Alfvén Waves: Magnetic Tension Waves 67
- 3.6 The Validity of MHD 68
  - 3.6.1 Characteristic Time Scales of MHD 68
  - 3.6.2 Checking the Assumptions 69
  - 3.6.3 A Comment on the Plasma Beta 70
- 3.7 Parallel Dynamics and Resistivity, or Relaxing the Ideal Assumption 71
- 3.8 Towards Multi-Fluid MHD 73
- 3.9 Further Reading 73
- References 74

4 Physics of "Hot" Plasmas

H. Zohm 75

- 4.1 What is a Hot Plasma? 75
- 4.2 Kinetic Description of Plasmas 77
  - 4.2.1 The Kinetic Equation 77

		4.2.2 Landau Damping	78
4.3	Fluid Description of Plasmas		79
	4.3.1	The MHD Equations	79
	4.3.2	Consequences of the MHD Equations	82
4.4	MHD Instabilities		86
	4.4.1	Classification of MHD Instabilities	86
	4.4.2	Examples of MHD Instabilities	88
4.5	Summary		92
References			93

5 Low Temperature Plasmas
J. Meichsner 95

5.1	Introduction	95
5.2	Gas Discharges and Low Temperature Plasmas: Basic Mechanisms and Characteristics	98
	5.2.1 Classical Townsend Mechanism and Electric Breakdown in Gases	98
	5.2.2 Townsend and Glow Discharge	100
	5.2.3 Arc Discharge	102
	5.2.4 Streamer Mechanism and Micro-Discharges, Dielectric Barrier and Corona Discharge	102
	5.2.5 Glow Discharge at Alternating Electric Field, RF and Microwave Discharge	104
5.3	Plasma Surface Transition	106
	5.3.1 Plasma Boundary Sheath, Bohm Criterion	106
	5.3.2 RF Plasma Sheath	108
	5.3.3 Electric Probes	110
5.4	Reactive Plasmas and Plasma Surface Interaction	114
References		116

6 Strongly Coupled Plasmas
R. Redmer 117

6.1	Introduction	117
6.2	Many-Particle Effects and Plasma Properties	118
	6.2.1 Green's Function Technique: Spectral Function	119
	6.2.2 Cluster Decomposition of the Self-Energy	122
6.3	Composition of Strongly Coupled Plasmas	125
6.4	Electrical Conductivity	127
6.5	Conclusion	130
References		131

Part II Confinement, Transport and Collective Effects

7 Magnetic Confinement
F. Wagner and H. Wobig 137
7.1 Conditions for Fusion 137
7.2 The Need for Magnetic Confinement 138
7.3 Particle Motion in Electro-Magnetic Fields 139
7.4 Constants of Motion 144
    7.4.1 Exact Invariants 144
    7.4.2 Adiabatic Invariants 144
7.5 Concepts of Magnetic Confinement 147
    7.5.1 Introduction 147
    7.5.2 The Mirror Machine 147
    7.5.3 Toroidal Confinement 148
    7.5.4 Magnetic Surfaces and Toroidal Equilibrium 149
    7.5.5 Confinement in Tokamaks 151
    7.5.6 Coil System of Tokamaks 152
    7.5.7 Theory of Tokamak Equilibria 153
    7.5.8 Cylindrical Approximation 154
    7.5.9 Confinement in Stellarators 156
7.6 Transport in Plasmas 163
    7.6.1 Collisional Losses 164
    7.6.2 Particle Picture of Classical Diffusion 165
    7.6.3 Neoclassical Transport 166
    7.6.4 Turbulent Transport 168
    7.6.5 Empirical Scaling Laws 169
References 171

8 Introduction to Turbulence in Magnetized Plasmas
B.D. Scott 173
8.1 Part A – Statistical Nonlinearity
    and Cascade Dynamics 173
8.2 Eddy Mitosis and the Cascade Model 176
8.3 The Statistical Nature of Turbulence 179
8.4 Quadratic Nonlinearity and Three Wave Coupling
    for Small Disturbances 180
8.5 Incompressible Hydrodynamic Turbulence – Energy and Enstrophy 182
8.6 MHD Turbulence 187
8.7 Part B – Gradient Driven Turbulence
    in Magnetized Plasmas 190
8.8 Passive Scalar Dynamics 192
8.9 Dissipative Coupling and the Adiabatic Response 193
8.10 Computations in the Dissipative Coupling Model for Drift Wave
    Turbulence 198

8.11	No Coupling – the Hydrodynamic Limit	201
8.12	The Effects of Dissipative Coupling	205
8.13	Summary	208
8.14	Further Reading	210
References		211

9 Transport in Toroidal Plasmas
U. Stroth 213

- 9.1 Experimental Confinement Times and Diffusion Coefficients ... 214
  - 9.1.1 Global Confinement Times ... 214
  - 9.1.2 Diffusion Coefficients ... 218
  - 9.1.3 The Collisional Transport Matrix ... 221
  - 9.1.4 Diffusion as Random-Walk ... 223
- 9.2 Particle Orbits in Toroidal Magnetic Fields ... 225
  - 9.2.1 Particles in a Toroidal Magnetic Mirror ... 225
  - 9.2.2 Passing Particles ... 226
  - 9.2.3 Trapped Particles and Banana Orbits ... 227
  - 9.2.4 Trajectories in Stellarator Fields ... 228
  - 9.2.5 Influence of a Radial Electric Field ... 229
- 9.3 Collisional Transport ... 231
  - 9.3.1 Classical Transport in the Particle Picture ... 231
  - 9.3.2 Classical Transport in the Fluid Picture ... 232
  - 9.3.3 Pfirsch–Schlüter Transport in the Particle Picture ... 234
  - 9.3.4 Pfirsch–Schlüter Transport in the Fluid Picture ... 235
  - 9.3.5 The Toroidal Resonance ... 236
  - 9.3.6 Neoclassical Transport in the Particle Picture ... 237
  - 9.3.7 Elements of Stellarator Transport ... 239
  - 9.3.8 Neoclassical Transport in the Fluid Picture ... 240
  - 9.3.9 The Ambipolar Electric Field ... 242
- 9.4 Turbulent Transport ... 245
  - 9.4.1 Fluid Turbulence ... 245
  - 9.4.2 Phenomenology of Turbulent Plasma Transport ... 249
  - 9.4.3 Two Fundamental Linear Instabilities ... 252
  - 9.4.4 Elements of a Drift Wave Model ... 255
  - 9.4.5 Experimental Results ... 257
  - 9.4.6 Transport Barriers ... 260

References 264

10 Non-Neutral Plasmas and Collective Phenomena in Ion Traps
G. Werth 269

- 10.1 Introduction ... 269
  - 10.1.1 Basics of Ion Traps ... 269
- 10.2 Ion Cloud as Non-Neutral Plasma ... 278

XIV   Contents

10.3   Weakly Coupled Non-Neutral Plasmas 279
       10.3.1 Plasma Oscillations 280
       10.3.2 Rotating Walls 282
10.4   Collective Effects 284
       10.4.1 Individual and Center-of-Mass Oscillations 284
       10.4.2 Instabilities in the Ion Motion 287
10.5   Strongly Coupled Non-Neutral Plasmas 287
10.6   Summary 293
References 294

11 Collective Effects in Dusty Plasmas
A. Melzer 297
11.1   Introduction 297
11.2   Particle Charging 298
       11.2.1 Orbital Motion Limit Currents 298
       11.2.2 Other Charging Currents 300
       11.2.3 Particles as Floating Probes 300
       11.2.4 Charging in the RF Sheath 302
11.3   Forces on Particles 302
       11.3.1 Gravity 302
       11.3.2 Electric Field Force 302
       11.3.3 Ion Drag Force 303
       11.3.4 Neutral Drag Force 304
       11.3.5 Thermophoresis 304
       11.3.6 Dust Levitation and Trapping 305
       11.3.7 Vertical Oscillations and Dust Charges 305
11.4   Particle–Particle Interaction 309
       11.4.1 Strongly Coupled Systems and Plasma Crystals 309
       11.4.2 Horizontal Interaction 311
       11.4.3 Vertical Interaction 311
       11.4.4 Phase Transitions 312
11.5   Waves in Weakly Coupled Dusty Plasmas 313
       11.5.1 Dust-Acoustic Waves 313
       11.5.2 Dust Ion-Acoustic Wave 316
11.6   Waves in Strongly Coupled Dusty Plasmas 316
       11.6.1 Compressional Mode in 1D 317
       11.6.2 Compressional Dust Lattice Waves 319
       11.6.3 Shear Dust Lattice Waves 320
       11.6.4 Mach Cones 320
       11.6.5 Transverse Dust Lattice Waves 322
       11.6.6 Normal Modes in Finite Clusters 324
11.7   Summary 327
References 327

12 Plasmas in Planetary Interiors
R. Redmer 331
12.1 Introduction 331
12.2 Solar System 332
12.3 Extrasolar Planets 334
12.4 Equation of State for Partially Ionized Plasmas 337
    12.4.1 Dense Hydrogen and Helium 337
    12.4.2 Free Energy 337
    12.4.3 Fluid Variational Theory 338
    12.4.4 Plasma Component 339
    12.4.5 Hugoniot Curves 341
12.5 Electrical and Thermal Conductivity 343
12.6 Conclusion 345
References 346

Part III Methods and Applications

13 Plasma Diagnostics
H.-J. Kunze 351
13.1 Introduction 351
13.2 Scattering of Laser Radiation
    by Plasma Electrons 352
    13.2.1 Laser-aided Diagnostics 352
    13.2.2 Incoherent Thomson Scattering 353
    13.2.3 Collective Thomson Scattering 357
    13.2.4 X-ray Scattering 361
13.3 Plasma Spectroscopy 361
    13.3.1 Overview 361
    13.3.2 Charge State Distribution 364
    13.3.3 Line Emission 366
    13.3.4 Line Profiles 370
    13.3.5 Continuum Radiation 372
References 372

14 Observation of Plasma Fluctuations
O. Grulke and T. Klinger 375
14.1 Introduction 375
14.2 Basics 376
14.3 Fluctuation Diagnostics 383
    14.3.1 Invasive Fluctuation Diagnostics 384
    14.3.2 Non-invasive Fluctuation Diagnostics 390
    14.3.3 Electron Cyclotron Emission 392
    14.3.4 Beam Emission Spectroscopy 393
    14.3.5 Heavy Ion Beam Probe 394

        14.3.6 Laser-induced Fluorescence 395
14.4 Concluding Remarks 396
References 396

15 Research on Modern Gas Discharge Light Sources
M. Born and T. Markus 399
15.1 Introduction to Light Sources 399
        15.1.1 The Lighting Market 399
        15.1.2 Overview of Discharge Lamps and Applications 401
        15.1.3 Aspects of Lamp Research 404
15.2 High Intensity Discharge Lamps 404
        15.2.1 Construction and Working Principle 404
        15.2.2 Light Technical Properties 406
15.3 Modelling of High Intensity Discharge Lamps 407
        15.3.1 Physical Modelling 407
        15.3.2 Thermochemical Modelling 412
15.4 Thermochemical Experiments 414
        15.4.1 Knudsen Effusion Mass Spectrometry (KEMS) 414
        15.4.2 Corrosion Analysis 418
15.5 Conclusions 421
References 422

16 Computational Plasma Physics
R. Schneider and R. Kleiber 425
16.1 Introduction 425
16.2 Plasma Edge Physics 426
        16.2.1 Models 428
16.3 Turbulence 434
        16.3.1 Gyro-kinetic Theory 435
        16.3.2 The PIC Method 437
16.4 Outlook 441
16.5 Seminars 441
References 441

17 Nuclear Fusion
H.-S. Bosch 445
17.1 Introduction 445
17.2 Energy Production in the Sun 448
17.3 Fusion on Earth 449
17.4 Conditions for Nuclear Fusion 453
17.5 Power Balances 455
17.6 Development of a Fusion Power Plant 458
17.7 Muon-catalyzed Fusion 458
References 459

18 The Possible Role of Nuclear Fusion in the 21st Century

T. Hamacher 461
18.1 Introduction 461
18.2 The Challenges 462
      18.2.1 Energy Demand and Lifestyle 463
      18.2.2 Efficient Use of Energy and Energy Saving 464
      18.2.3 Energy Resources 464
      18.2.4 Geopolitical Frictions 465
      18.2.5 Environmental Damages 465
      18.2.6 Possible Supply Options 466
18.3 Characteristics of Nuclear Fusion as Power Source 467
      18.3.1 Overall Design of a Fusion Power Plant 467
      18.3.2 Resources 469
      18.3.3 Environmental and Safety Characteristics, External Costs . 470
      18.3.4 Economic Consideration 473
18.4 The Possible Role of Fusion in a Future Energy System 475
      18.4.1 The Global Dimension 475
      18.4.2 Fusion in Western Europe 476
      18.4.3 Fusion in India 477
18.5 Conclusion and Outlook 480
References 481

Abbreviations 483

Index 489

List of Contributors

Matthias Born
Philips Research Laboratories
Aachen
Weisshausstr. 2
D-52066 Aachen, *Germany*
matthias.born@philips.com

Hans-Stephan Bosch
Max-Planck-Inst. für Plasmaphysik
IPP-EURATOM Association
Wendelsteinstr. 1
D-17491 Greifswald, *Germany*
bosch@ipp.mpg.de

Olaf Grulke
Max-Planck-Inst. für Plasmaphysik
IPP-EURATOM Association
Wendelsteinstr. 1
D-17491 Greifswald, *Germany*
grulke@ipp.mpg.de

Thomas Hamacher
Max-Planck-Inst. für Plasmaphysik
IPP-EURATOM Association
Boltzmannstr. 2
D-85748 Garching, *Germany*
hamacher@ipp.mpg.de

Ralf Kleiber
Max-Planck-Inst. für Plasmaphysik
IPP-EURATOM Association
Wendelsteinstr. 1
D-17491 Greifswald, *Germany*
Ralf.Kleiber@ipp.mpg.de

Thomas Klinger
Max-Planck-Inst. für Plasmaphysik
IPP-EURATOM Association
Wendelsteinstr. 1
D-17491 Greifswald, *Germany*
thomas.klinger@ipp.mpg.de

Hans-Joachim Kunze
Ruhr-Universität Bochum
Universitätsstr. 150
D-44780 Bochum, *Germany*
Hans-Joachim.Kunze
  @ruhr-uni-bochum.de

Torsten Markus
Forschungszentrum Jülich
Institut für Werkstoffe und
Verfahren in der Energietechnik
D-52425 Jülich, *Germany*
t.markus@fz-juelich.de

Jürgen Meichsner
E.-M.-Arndt-Universität Greifswald
Institut für Physik
Domstraße 10a
D-17489 Greifswald, *Germany*
meichsner@physik.
  uni-greifswald.de

André Melzer
E.-M.-Arndt-Universität Greifswald
Institut für Physik
Domstraße 10a
D-17489 Greifswald, *Germany*
melzer@physik.uni-greifswald.de

Alexander Piel
Christian-Albrechts-Univ. zu Kiel
Institut für Experimentelle und
Angewandte Physik
Olshausenstr. 40
D-24098 Kiel, *Germany*
piel@physik.uni-kiel.de

Ronald Redmer
Universität Rostock
Fachbereich Physik
D-18051 Rostock, *Germany*
ronald.redmer
 @physik.uni-rostock.de

Ralf Schneider
Max-Planck-Inst. für Plasmaphysik
IPP-EURATOM Association
Wendelsteinstr. 1
D-17491 Greifswald, *Germany*
Ralf.Schneider@ipp.mpg.de

Uwe Schumacher
Universität Stuttgart
Institut für Plasmaforschung
Pfaffenwaldring 31
D-70569 Stuttgart,*Germany*
schumach@ipf.uni-stuttgart.de

Bruce D. Scott
Max-Planck-Inst. für Plasmaphysik
IPP-EURATOM Association
Boltzmannstr. 2
D-85748 Garching, *Germany*
bds@ipp.mpg.de

Ulrich Stroth
Christian-Albrechts-Univ. zu Kiel
Institut für Experimentelle und
Angewandte Physik
Olshausenstr. 40
D-24098 Kiel, *Germany*
stroth@physik.uni-kiel.de

Friedrich Wagner
Max-Planck-Inst. für Plasmaphysik
IPP-EURATOM Association
Wendelsteinstr. 1
D-17491 Greifswald, *Germany*
fritz.wagner@ipp.mpg.de

Günter Werth
Johannes Gutenberg Universität
Institut für Physik
D-55099 Mainz, *Germany*
werth@uni-mainz.de

Horst Wobig
Max-Planck-Inst. für Plasmaphysik
IPP-EURATOM Association
Boltzmannstr. 2
D-85748 Garching, *Germany*
how@ipp.mpg.de

Hartmut Zohm
Max-Planck-Inst. für Plasmaphysik
IPP-EURATOM Association
Boltzmannstr. 2
D-85748 Garching, *Germany*
zohme@ipp.mpg.de

Part I

Fundamental Plasma Physics

1 Basics of Plasma Physics

U. Schumacher

Abstract. Basic properties of plasmas are introduced, which are valid for an extremely wide range of plasma parameters. Plasmas are classified by different physical behaviour. The motion of charged particles in electromagnetic fields is revised with respect to drift motions. Adiabatic invariants are discussed and the kinetic description of plasmas is briefly presented.

This Chapter is also a guideline connecting the subsequent Chapters.

1.1 Definition, Occurrence and Typical Parameters of Plasmas

A plasma (Greek $\pi\lambda\alpha\sigma\mu\alpha$) is an ionised gas, consisting of free electrons, ions and atoms or molecules. It is characterised by its collective behaviour. Plasmas are many-particles ensembles; the charged particles are coupled by electric and magnetic self-generated and self-consistent fields.

The majority of the matter of our visible universe is in the plasma state. The fascinating fact of these plasmas is their description by the same physical mechanisms and the same formulae, even if their parameters range e.g. from extremely low charged particle densities (a few particles per cubic meter) as in intercluster gases, which are the plasmas between the clusters of galaxies in the universe, up to plasmas with 45 orders of magnitude higher electron densities as in neutron stars. Several examples of these plasmas will be treated in the different Chapters throughout this book.

The plasmas in nature as well as the man-made plasmas on Earth cover an extremely wide range of their parameters like temperatures, particle densities and plasma generated magnetic field strengths, as expressed in Table 1.1.

Examples of plasmas in nature are all stars like the Sun, which has a central temperature of about 17 million degrees. Its surface, the photosphere, radiates at a temperature of 5 700 K, and the corona has a temperature of more than one million degrees. Also the outer parts of the Earth's atmosphere consist of plasmas as, e.g. the ionosphere, the plasmosphere, and the radiation belts in the magnetosphere. Terrestric plasmas are found in gas discharges as in lightnings, in sparks, arcs, fluorescent lamps, energy saving lamps, arc lamps, plasma displays, plasma thrusters and plasma torches. These plasmas

Table 1.1. Parameter range of plasmas in the universe and on Earth

	Temperature T(K)	Density n(m^{-3})	Magnetic Field B(T)
Intergalactic gas	10^8	1	10^{-10}
Interstellar medium	10^4	10^6	10^{-10}
gas clouds in galaxies	$10^4 \ldots 10^6$	10^{12}	$10^{-8} \ldots 10^{-7}$
fusion plasmas	10^8	$10^{20} \ldots 10^{32}$	$10^{-3} \ldots 10$
technical plasmas	$10^3 \ldots 10^5$	$10^{15} \ldots 10^{25}$	10^{-2}
electron cloud in metals	10^5	8×10^{28}	
surface of stars	10^4	10^{22}	$10^{-4} \ldots 10^{-1}$
star center	$10^7 \ldots 10^8$	10^{30}	1
white dwarf	10^4	10^{36}	10^4
neutron star	10^4	10^{45}	10^8

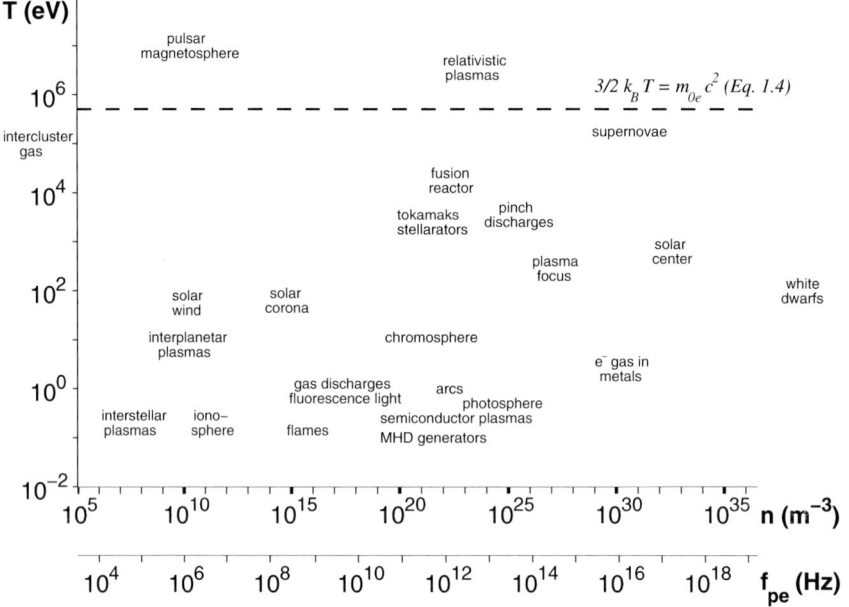

Fig. 1.1. Temperature (T) versus density (n) and plasma frequency (f_{pe}), respectively, diagram of typical natural and man-made plasmas. f_{pe} will be defined in Sect. 1.3.4

can be presented in a plot of their temperatures versus their particle densities, as given in Fig. 1.1. The temperatures are usually given in electron volts (eV) with $1\,\text{eV} \equiv 11\,605.4\,\text{K}$ representing the one degree of freedom energy equivalent eV/k_B.

1.2 Ideal Plasmas

Matter is in the plasma state, if it is ionized to a certain degree. In thermodynamic equilibrium the ionization degree is given by the ratio of the particle densities $n_{z+1,1}$ and $n_{z,1}$ of ions in the ground states of the ionization stages $z+1$ and z, respectively, multiplied by the electron density n_e, which is expressed by Saha's equation

$$\frac{n_{z+1,1}\, n_e}{n_{z,1}} = \frac{g_{z+1,1}}{g_{z,1}} \frac{2\,(2\pi m_e k_B T_e)^{3/2}}{h^3} \exp\left(-\frac{\chi_z}{k_B T_e}\right), \qquad (1.1)$$

where $g_{z+1,1}$ and $g_{z,1}$ are the statistical weights of the ground states of the ionization stages $z+1$ and z, respectively, T_e is the electron temperature, and χ_z is the ionization energy of the ion in stage z. The transition from neutral gases to plasmas may be represented by the lower boundary line of plasmas given by about 50% ionization of hydrogen and plotted in Fig. 1.2, the plasma boundary diagram.

The majority of the plasmas, which occur in nature, are in the ideal state. A plasma is *called ideal*, if the mean thermal energy $E_{th} = 3/2\ k_B T$ of the particles exceeds the mean electrostatic interaction energy, which for

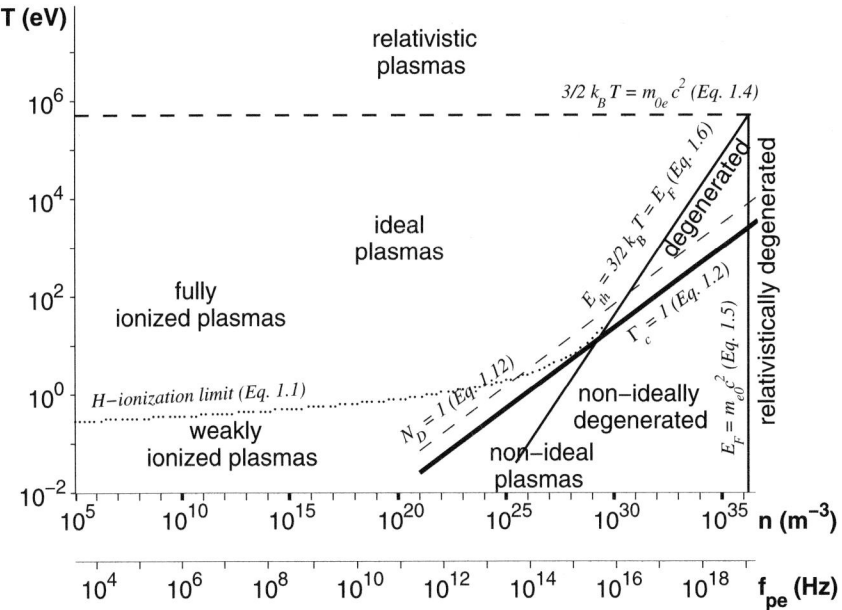

Fig. 1.2. Plasma boundary diagram

hydrogen as an example of equal electron density n_e and ion density n_i ($n_e = n_i = n$) is given by $E_e = 1/(4\pi\varepsilon_0)\, e^2\, n^{1/3}$, where k_B is Boltzmann's constant. In this case the ratio of the mean electrostatic interaction energy and the mean thermal energy, the coupling parameter

$$\Gamma_c = \frac{e^2}{4\pi\varepsilon_0\, a} \frac{1}{k_B T} \quad (1.2)$$

with the mean particle distance (Wigner–Seitz radius) $a = (4\pi n/3)^{-1/3}$ is smaller than 1.

The boundary line between ideal and non-ideal plasmas hence is given by

$$\frac{3}{2} k_B T \approx \frac{e^2}{4\pi\varepsilon_0} n^{1/3} \quad \text{or} \quad \Gamma_c \approx 1 \,. \quad (1.3)$$

Numerically this line can be expressed by $T \approx 10^{-9}\, (n)^{1/3}$, T in eV and n in m^{-3}.

The boundaries of the ideal plasmas, which mainly occur in nature, are given by the ionization limit and by the coupling parameter, which gives the boundary to the non-ideal plasmas, that recently – as well as the ultra-cold plasmas – gained increasing interest (see also Chap. 6 in Part I and Chap. 10 in Part II).

The boundary line between non-relativistic and relativistic plasmas is given by the upper horizontal line in Fig. 1.2, for which the mean thermal energy $E_{th} = 3/2\, k_B T$ equals the rest energy $m_{0e} c^2$ of the electron:

$$\frac{3}{2} k_B T = m_{0e} c^2 = 511 \,\text{keV} \,. \quad (1.4)$$

Non-relativistically degenerated and relativistically degenerated plasmas are separated by the vertical line at the electron density of about 3.1×10^{36} m^{-3}, which is given by the equality of the electron rest energy and the Fermi energy E_F:

$$m_{0e} c^2 = E_F = \frac{p_F^2}{2 m_e} = \frac{\hbar^2 \left(3\pi^2 n_e\right)^{2/3}}{2 m_e} \,, \quad (1.5)$$

where $p_F = \hbar\left(3\pi^2 n_e\right)^{1/3}$ is the Fermi momentum. The dividing line between non-degenerated and non-relativistically degenerated plasmas results from the equation of the mean thermal electron energy $E_{th} = 3/2\, k_B T_e$ and the Fermi energy E_F:

$$\frac{3}{2} k_B T_e = \frac{\hbar^2 \left(3\pi^2 n_e\right)^{2/3}}{2 m_e} \,, \quad T_e = 2.4 \times 10^{-19}\, (n_e)^{2/3} \,. \quad (1.6)$$

1.3 Important Plasma Properties

1.3.1 Debye Shielding

One of the most important properties of a plasma is the shielding of every charge in the plasma by a cloud of oppositely charged particles, the Debye shielding. Its typical spatial scale, the Debye length λ_D, is estimated – in one dimension (x) – by equating the potential energy of charge separation $E_p = e\phi(\lambda_D)$ over this distance λ_D with the kinetic particle energy $1/2\, k_B T$. In this approximation the electric field $E(x)$ in a hydrogen plasma, e.g. with $n_e = n_i = n$, is obtained from $\mathrm{div} E = \varrho/\varepsilon_0 = ne/\varepsilon_0 \simeq E(x)/x$. The potential energy hence results in

$$E_p = e\phi(\lambda_D) = e \int_0^{\lambda_D} E(x)\, dx = ne^2 \lambda_D^2/(2\varepsilon_0),$$

which gives

$$\lambda_D = \left(\frac{\varepsilon_0 k_B T}{ne^2}\right)^{1/2}. \qquad (1.7)$$

The solution of Poisson's equation

$$\triangle \phi = \frac{1}{r^2} \frac{d}{dr}\left(r^2 \frac{d\phi}{dr}\right) = \lambda_D^{-2} \phi \qquad (1.8)$$

gives the total Debye length $\lambda_D^{-2} = \lambda_{De}^{-2} + \lambda_{Di}^{-2}$, which consists of the Debye lengths of electrons (index e) and ions (index i):

$$\lambda_{De,i} = \left(\frac{\varepsilon_0 k_B T_{e,\,i}}{n_{e,i} e^2}\right)^{1/2} \qquad (1.9)$$

and the potential distribution

$$\phi(r) = \frac{q}{4\pi\varepsilon_0} \frac{1}{r} \exp\left(-\frac{r}{\lambda_D}\right), \qquad (1.10)$$

which is plotted in Fig. 1.3 as dotted line depicting the screening of the Coulomb potential of every charge q of a particle in a plasma. Hence plasmas are quasi-neutral, i.e., on the macroscopic scale of the plasma extension L, with $L \gg \lambda_D$, the plasma appears to be neutral. So-called non-neutral plasmas – charged particle ensembles confined by electromagnetic fields – lack of quasi-neutrality. Non-neutral plasmas will be addressed in Chap. 10 of Part II. The solid line represents the familiar potential distribution of a charged particle in vacuum.

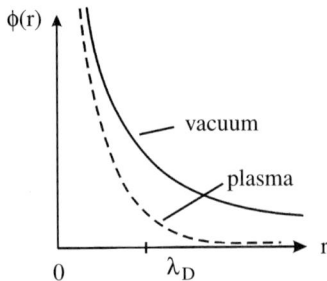

Fig. 1.3. Potential distribution of a charged particle in vacuum (*solid line*) and in a plasma (*dotted line*)

1.3.2 The Plasma Parameter

The plasma parameter N_D describes the number of particles in the Debye sphere. For a plasma with singly charged ions like a hydrogen plasma ($n_e = n_i = n$) N_D is given by

$$N_D = n \frac{4}{3}\pi \lambda_D^3 \quad \text{with} \quad \lambda_{De,i} = \left(\frac{\varepsilon_0 k_B T_{e,i}}{ne^2}\right)^{1/2} . \quad (1.11)$$

With the mean particle distance (Wigner–Seitz radius) $a = (4\pi n/3)^{-1/3}$ the plasma parameter takes the simple form

$$N_D = \left(\frac{\lambda_D}{a}\right)^3 . \quad (1.12)$$

In the parameter diagram of typical plasmas (Fig. 1.4) the lines of constant values of the Debye radius λ_D and of the plasma parameter N_D, respectively, are plotted.

1.3.3 Landau Length

The Landau length λ_L is a typical scale for Coulomb collisions to be addressed in Sect. 1.4.1. It is the critical distance of two charged particles, for which the potential energy $E_p = (Z\,e^2)/(4\,\pi\,\varepsilon_0\,\lambda_L)$ equals the kinetic energy $E_i = k_B T$ and is equivalent to the backscattering criterion of Rutherford scattering (central force scattering), resulting in

$$\lambda_L = \frac{Ze^2}{4\pi\varepsilon_0\, k_B T} . \quad (1.13)$$

For the ion charge $Z = 1$ the plasma parameter can also be written as

$$N_D = \frac{\lambda_D}{3\lambda_L} . \quad (1.14)$$

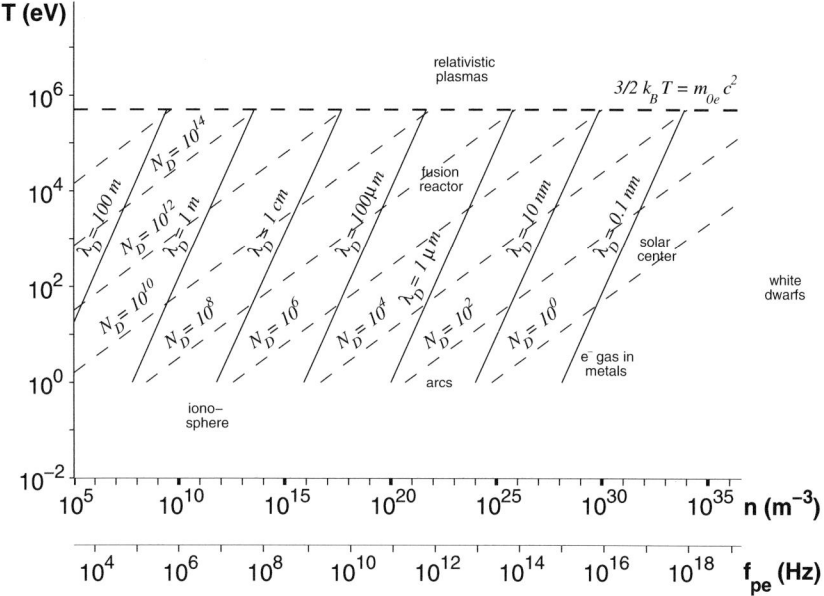

Fig. 1.4. Lines of constant Debye length λ_D and plasma parameter N_D, respectively, in the diagram of typical plasmas

1.3.4 Plasma Frequency

Microscopic deviations from quasi-neutrality in a plasma result in plasma (or Langmuir) oscillations. They represent the most simple form of oscillations in a plasma, and they are an obvious example for the collective behavior of the plasmas (see Chap. 2 in Part I on waves in plasmas). The equation of motion

$$m_e \frac{\mathrm{d}^2 x}{\mathrm{d}t^2} = -eE = -\frac{n_e e^2 x}{\varepsilon_0} \qquad (1.15)$$

of a plane plasma sheath in linear approximation relates the space charge electric field E to the separation x of the electrons from the ions and hence describes non-damped Langmuir oscillations with their electron plasma frequency

$$\omega_{pe} = \left(\frac{n_e e^2}{\varepsilon_0 m_e}\right)^{1/2}. \qquad (1.16)$$

The total plasma frequency is given by

$$\omega_p^2 = \omega_{pe}^2 + \omega_{pi}^2 \qquad (1.17)$$

with the ion plasma frequency

$$\omega_{pi} = \left(\frac{n_i Z^2 e^2}{\varepsilon_0 m_i}\right)^{1/2}, \tag{1.18}$$

which can be approximated by the electron plasma frequency because of the large mass ratio of ions to electrons. Due to the mere square root dependence on the electron density n_e the plasma frequency ω_p can replace the electron density as abscissa in all the plasma diagrams, i.e. $f_{pe}(\mathrm{s}^{-1}) = \omega_{pe}/2\pi = 8.98\,(n_e)^{1/2}$.

With these expressions some important plasma properties can be given:

- A plasma is quasi-neutral: $n = n_e = n_i$ (for ion charge $Z = 1$); generally $n_e = \sum_j Z_j n_{i,j}$.
- The product of Debye length λ_{D_e} (λ_{D_i}) and ω_{pe} (ω_{pi}) approximately equals the electron (ion) thermal speed $v_{th,e} \approx \sqrt{3}\lambda_{D_e}\omega_{pe}$ or $v_{th,i} \approx \sqrt{3}\lambda_{D_i}\omega_{pi}$, respectively.
- Ideal plasmas are characterized by the following relations:

$$\lambda_D \ll L \tag{1.19}$$
$$N_D \gg 1 \quad \text{or} \quad \lambda_L \ll \lambda_D \quad \text{or} \quad \Gamma_c \ll 1 \tag{1.20}$$
$$\omega_p \tau > 1, \tag{1.21}$$

where L is the plasma dimension, and τ stands for the collision time of the charged plasma particles with neutrals.

The first condition reflects the plasma being a many-particle ensemble, the particle charges of which are shielded outside their Debye sphere. The last condition expresses the Coulomb interaction as the dominant interaction mechanism in a plasma, as compared to collisions with neutral particles.

1.4 Single Particle Behavior in Plasmas

The understanding of plasmas is based on the properties of the single particle behavior with respect to Coulomb collisions as well as of the charged particle behavior in magnetic fields. The single particle collision times, their mean free paths and the Coulomb logarithm determine the particle distribution functions, which can be calculated from the kinetic theory (Vlasov and Boltzmann equations). Gyration around the magnetic field lines, particle drift motions and adiabatic invariants characterize the charged particle behavior in magnetic fields and thus form the basis for the magnetic confinement of high temperature plasmas.

1.4.1 Coulomb Collisions, Collision Times and Lengths

If the collision parameter p, the lateral distance of two charged particles approaching to each other (e.g. an electron interacting with another electron or a negative ion, see Fig. 1.5), is equal to the Landau length λ_L (see Sect. 1.3.3), it is called critical collision parameter, because then the electron is deflected by 90°.

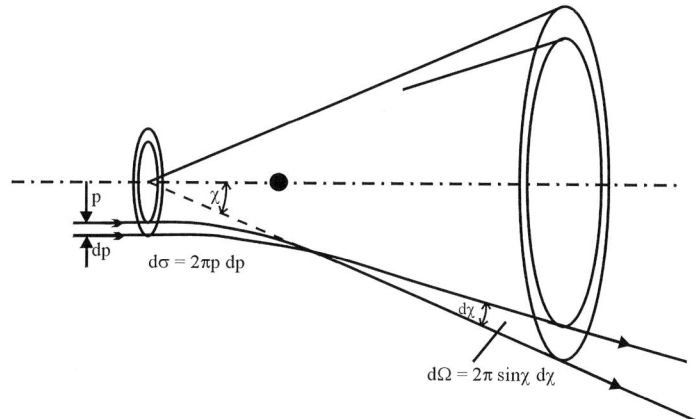

Fig. 1.5. Scheme of Coulomb interaction

The cross section for deflections larger than 90° (backscattering cross section) in a single interaction is given by

$$\sigma_{90°} = \pi \lambda_L^2 = \pi \left(\frac{Ze^2}{4\pi\varepsilon_0 k_B T} \right)^2 , \qquad (1.22)$$

and the corresponding mean free path length hence is $\lambda_{90°} = 1/(\sigma_{90°} n)$, which results in the following relation:

$$\lambda_{90°} \lambda_L = \lambda_D^2 \quad \text{or} \quad \frac{\lambda_{90°}}{\lambda_D} = \frac{\lambda_D}{\lambda_L} . \qquad (1.23)$$

From this relation it is obvious, that for ideal plasmas the mean free path length always exceeds the Debye length (and the Landau length as well).

For Coulomb interaction with the screened potential

$$\phi(r) = \frac{q}{4\pi\varepsilon_0} \frac{1}{r} \exp\left(-\frac{r}{\lambda_D}\right) \qquad (1.24)$$

[by Debye shielding, see (1.10)] the differential cross section in Born approximation is given by

$$\frac{d\sigma}{d\Omega} = \left[\frac{Ze^2/(4\pi\varepsilon_0)}{\hbar^2/(2m_r\lambda_D^2) + 2m_r v^2 \sin^2(\chi/2)}\right]^2. \quad (1.25)$$

Due to the cumulative action of many small angle collisions this results in the effective total cross section of

$$\sigma_{eff} = \int_\Omega \frac{d\sigma}{d\Omega} d\Omega = \sigma_{90°} \cdot \ln \Lambda \quad (1.26)$$

and thus gives the definition of the Coulomb logarithm $\ln \Lambda$, which is used for collision times calculation subsequently. The Coulomb logarithm is a weak function of temperature and density, as can be seen in the diagram of temperature versus density of typical plasmas in Fig. 1.6. Since Λ can be approximated by $\Lambda \approx 9N_D$ the lines in Fig. 1.6 are directly related to the dashed lines in Fig. 1.4.

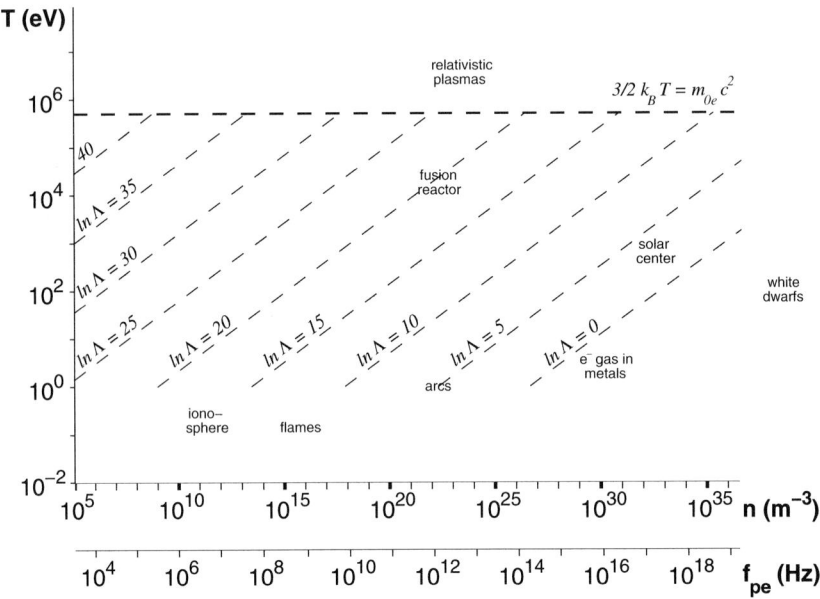

Fig. 1.6. The Coulomb logarithm $\ln \Lambda$ in the diagram of typical plasmas

The total Coulomb collision cross section

$$\sigma = \sigma_{eff} = \ln \Lambda \, \pi \left(\frac{Ze^2}{4\pi\varepsilon_0 k_B T}\right)^2, \quad (1.27)$$

the density n and the thermal speed $v_{th} = [(3k_B T)/m]^{1/2}$, result in the typical collision times $\tau = 1/(\sigma n v_{th})$, which lead to 90° deflection by the cumulative action of many small angle collisions. They are the electron–electron collision time τ_{ee}, which is about equal to the electron–ion collision time τ_{ei}, the ion–ion collision time τ_{ii}, and the ion–electron collision time τ_{ie}. If we assume test particles (first index in collision times) penetrating into thermal ensembles, we arrive with Maxwellian velocity distributions for the collision times at:

$$\tau_{ee} \cong \frac{3^{-1/2}\,16\pi\,\varepsilon_0^2\,m_e^{1/2}\,(k_B T_e)^{3/2}}{n_e\,e^4\,\ln\Lambda} \equiv \tau_{ei}\,, \qquad (1.28)$$

$$\tau_{ii} \cong \frac{3^{-1/2}\,16\pi\,\varepsilon_0^2\,m_i^{1/2}\,(k_B T_i)^{3/2}}{Z^4\,n_i\,e^4\,\ln\Lambda}\,, \qquad (1.29)$$

$$\tau_{ie} \cong \frac{3^{-1/2}\,8\pi\,\varepsilon_0^2\,m_i\,m_e^{-1/2}\,(k_B T_e)^{3/2}}{Z^2\,n_i\,e^4\,\ln\Lambda}\,. \qquad (1.30)$$

Keeping quasi-neutrality $n_e = Z\,n_i$ in mind, the ratios of these Coulomb collision times result in

$$\tau_{ee} : \tau_{ii} : \tau_{ie} \cong 1 : \frac{1}{Z^3}\left(\frac{m_i}{m_e}\right)^{1/2}\left(\frac{T_i}{T_e}\right)^{3/2} : \frac{1}{2Z}\frac{m_i}{m_e}\,. \qquad (1.31)$$

Figure 1.7 gives the lines of constant Coulomb collision times in the temperature versus density plot of some typical plasma parameters.

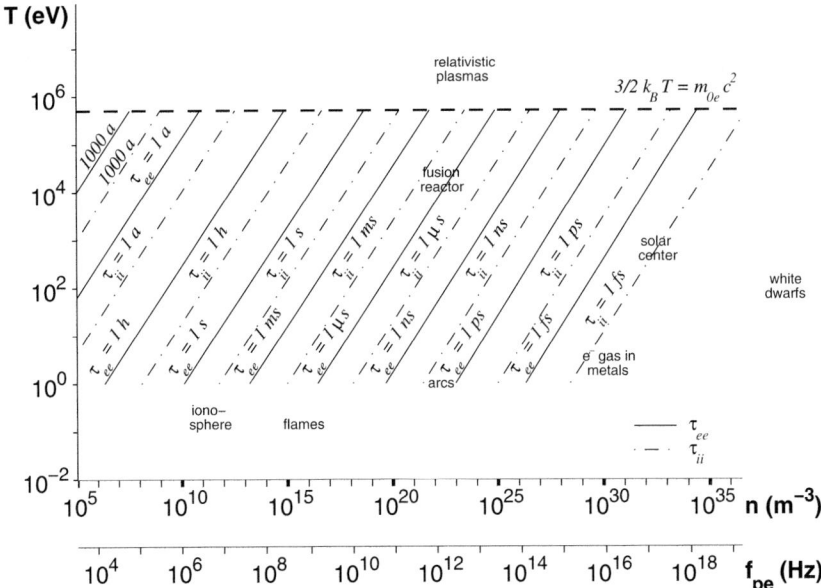

Fig. 1.7. The lines of constant Coulomb collision times τ_{ee} for electron–electron and τ_{ii} for ion–ion collisions, respectively, in the diagram of typical plasma parameters

Multiplying the collision times τ_{ee} and τ_{ii} with the corresponding mean thermal speeds, respectively, we obtain the mean free paths λ_e and λ_i for electrons and ions, respectively:

$$\lambda_e \cong \frac{16\pi \, \varepsilon_0^2 (k_B T_e)^2}{n_e e^4 \ln \Lambda} \quad \text{and} \quad \lambda_i \cong \frac{16\pi \, \varepsilon_0^2 (k_B T_i)^2}{Z^4 n_i e^4 \ln \Lambda} \,. \tag{1.32}$$

For singly charged ions ($Z = 1$) – due to quasi-neutrality – both mean free paths are equal. They are plotted in the diagram of Fig. 1.8.

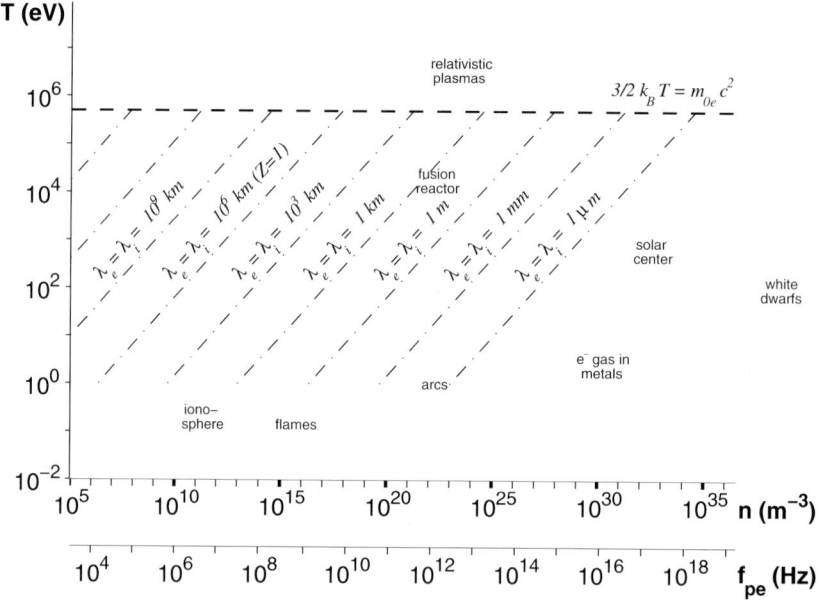

Fig. 1.8. The lines of constant mean free paths in the diagram of typical plasma parameters

1.4.2 Electrical Conductivity of Plasmas

If an electric field **E** is applied to the plasma, it accelerates the charged plasma particles, which transfer the gained energy by Coulomb collisions to the other plasma particles. From the balance of the energy gain in the electric field and the Coulomb collisional energy loss of the electrons, that play the dominant role, to the other charged particles we obtain the specific electric resistance $\eta_\parallel$ of a plasma:

$$\eta_\parallel = \frac{m_e^{1/2} \, Z \, e^2 \ln \Lambda}{3 \, (2\pi)^{3/2} \, \varepsilon_0^2 (k_B T_e)^{3/2}} \,, \tag{1.33}$$

which expresses the strong dependence of the specific electrical conductivity of the plasma mainly on the electron temperature T_e due to the insensitivity of the Coulomb logarithm (see Fig. 1.6) on density. In the diagram of typical plasma parameters in Fig. 1.9 the lines of constant specific electrical conductivity are plotted.

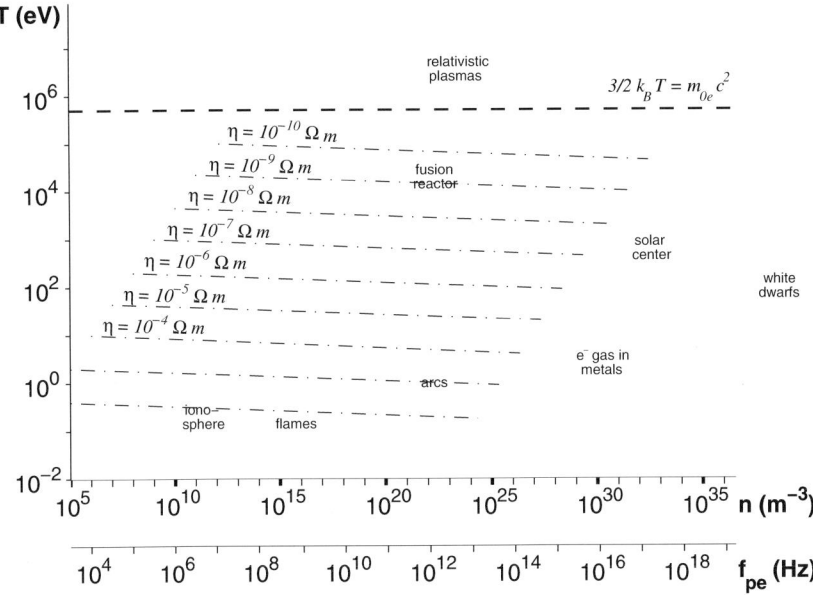

Fig. 1.9. Lines of constant specific electrical conductivity in the diagram of typical plasma parameters

1.4.3 Single Charged Particle Motion in Electric and Magnetic Fields

Charged Particle Motion in Magnetic Fields and Diamagnetism of Plasmas

Particles of charge q_α and mass $m_\alpha = \gamma m_{\alpha 0}$, where γ is the relativistic mass factor and $m_{\alpha 0}$ is the rest mass, are accelerated in electric fields **E** and magnetic fields **B** according to the (relativistic) equation of motion

$$\frac{d(m_\alpha \mathbf{v})}{dt} = q_\alpha \left(\mathbf{E} + \mathbf{v} \times \mathbf{B} \right) . \tag{1.34}$$

In magnetic fields the Lorentz force balanced by the centrifugal force leads to the spiral particle motion around the magnetic field lines with the gyration radius of $r_g = \gamma m_{\alpha 0} v/(q_\alpha B)$, also known as Larmor radius, and the cyclotron frequency $\omega_c = v/r_g = q_\alpha B/(\gamma m_{\alpha 0})$ including the relativistic mass factor $\gamma = \left[1 - (v/c)^2\right]^{-1/2}$. The charged particles hence spiral around the magnetic field lines. Without collisions they can not move across the field lines. They are free, however, to move along the field lines. This is the basic property that leads to the magnetic confinement of plasmas in closed configurations (see Chap. 7 on magnetic confinement in Part II).

Since ions and electrons gyrate around the magnetic field lines (in opposite directions, see Fig. 1.10) such, that the magnetic field generated by this rotation of charges is oppositely directed to the applied magnetic field, it is obvious, that plasmas are diamagnetic.

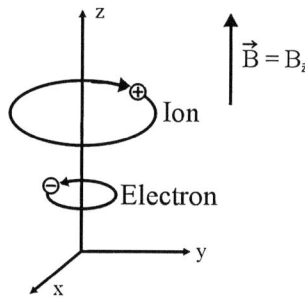

Fig. 1.10. The charged particle rotation around the magnetic field lines expresses the diamagnetism of the plasma

Charged Particle Drifts

If a homogeneous electric field is superimposed to a homogeneous magnetic field the charged particles move at a constant drift velocity, the $\mathbf{E} \times \mathbf{B}$–drift velocity

$$\mathbf{v}_E = \frac{\mathbf{E} \times \mathbf{B}}{B^2}, \tag{1.35}$$

perpendicular to the magnetic and to the electric field as well (see Fig. 1.11). This can easily be obtained from vector multiplication of the equation of motion [see (1.34)] with **B** from right and averaging over the motion of gyration. This drift velocity in crossed homogeneous electric and magnetic fields is the same for all charged particles irrespective of their velocity, mass, charge quantity and sign. Therefore, the $\mathbf{E} \times \mathbf{B}$–drift does not lead to charge separation. These specific properties do not hold for all other drift motions.

Replacing the electric field **E** by the corresponding force **F** divided by the particle charge q we obtain the general drift velocity

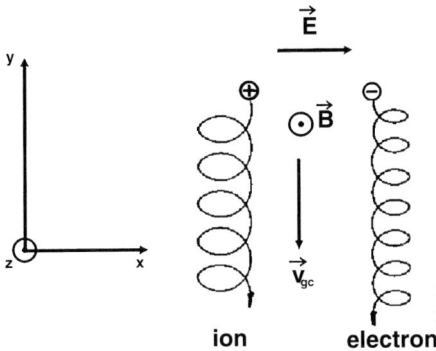

Fig. 1.11. $\mathbf{E} \times \mathbf{B}$–drift motion in crossed homogeneous electric and magnetic fields

$$\mathbf{v}_F = \frac{1}{q} \frac{\mathbf{F} \times \mathbf{B}}{B^2} . \qquad (1.36)$$

The gravitational force $\mathbf{F} = m\mathbf{g}$ immediately leads to the gravitational drift velocity

$$\mathbf{v}_g = \frac{m}{q} \frac{\mathbf{g} \times \mathbf{B}}{B^2} , \qquad (1.37)$$

which is schematically drawn in Fig. 1.12.

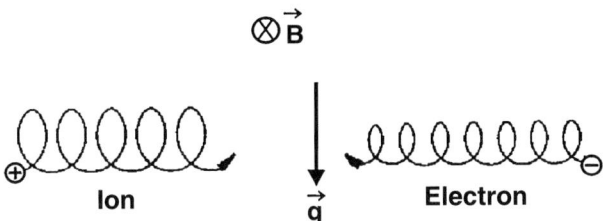

Fig. 1.12. Gravitational drift with opposite drift motions of electrons and ions

Although being negligible for the Earth's low gravitation, the gravitational drift with its opposite drift motions of electrons and ions directly demonstrates, that a net current density $\mathbf{j} = n_e (m_i + m_e)(\mathbf{g} \times \mathbf{B})/B^2$ is generated. Moreover, it is a simple model for particle drift motions under the action of other forces like the centrifugal force for charged particles in curved magnetic fields. In a curved magnetic field in vacuum with a curvature radius $\mathbf{R_c}$ due to $\nabla \times \mathbf{B} = 0$ the curvature is inherently related to a magnetic field gradient $(\nabla B)/B = -\mathbf{R_c}/R_c^2$. Hence the total drift velocity in a curved magnetic field is the superposition of the drift velocities of the curvature drift $\mathbf{v}_R$ and of the gradient–B drift $\mathbf{v}_{\nabla B}$:

$$\mathbf{v}_R + \mathbf{v}_{\nabla B} = \frac{m}{q} \frac{\mathbf{R}_c \times \mathbf{B}}{R_c^2 B^2} \left(v_\|^2 + v_\perp^2/2 \right) . \tag{1.38}$$

These drifts characterize the charged particle motion in inhomogeneous magnetic fields like in the Earth's magnetosphere. They explain the capture of charged high energy particles in the Van Allen belts of the Earth, their mirroring between areas of higher magnetic field strength near the poles, which is an example for the second adiabatic invariant J [see (1.41) in the following subsection], and the generation of the so called electron current around the equator of the Earth, which is related to the third adiabatic invariant [see (1.42) in the following subsection as well].

These drift motions, moreover, are the reason for pure toroidal magnetic fields (configurations with magnetic field lines, that are closed just after one circumference around the symmetry axis) not being suitable for plasma confinement, because in such magnetic fields the curvature and gradient–B drifts lead to charge separation in axis-parallel direction and hence to an electric field perpendicular to the magnetic field direction, which results in an $\mathbf{E} \times \mathbf{B}$–drift in radial direction for all charged particles (see Chap. 7 on magnetic confinement in Part II). There are several additional drift velocities, that occur, e.g. in inhomogeneous electric fields and in time dependent electric fields (polarization drift).

Adiabatic Invariants

The description of the single particle motion is appreciably simplified for periodic or quasi-periodic motions by the constancy of the adiabatic invariants. These are based on the line integral

$$\oint p \, dq , \tag{1.39}$$

taken over one period for the canonically conjugated quantities p and q, being a constant of motion, an adiabatic invariant.

The first adiabatic invariant is related to the gyration motion of the charged particles around the magnetic field lines. This is the magnetic moment

$$\mu = \frac{mv_\perp^2}{2B} , \tag{1.40}$$

which can directly be obtained from the product of the current related to the charged particle rotation and the area the particle encircles during its gyration.

The second adiabatic invariant J is related to the periodic bounce motion of the particle between two mirror regions in which the particle is captured. For the second (or longitudinal) adiabatic invariant J the line integral is taken over the particle's velocity component $v_\parallel$ parallel to the magnetic field **B** over the co-ordinate s:

$$J = m \oint v_\parallel \, ds \, . \tag{1.41}$$

The third adiabatic invariant is related to the constancy of the magnetic flux through the drift orbits:

$$\Phi = \int_F \mathbf{B}(f) \cdot \mathrm{d}\mathbf{f} \, , \tag{1.42}$$

where F is the area enclosed by the particles' drift orbit. The third adiabatic invariant, however, in contrast to the first one can easily be violated by fluctuations or other temporal changes of the plasma.

1.5 Kinetic Description

Although the single particle behavior already gives much detailed information, the plasma as a many-particle system is determined by collective motions as well. The characteristic properties of collective motions in a plasma determine the propagation, the damping and the absorption of waves in plasmas and the wave–plasma interaction, too (see the following Chapter on waves in plasmas). Their knowledge allows to develop efficient wave heating systems and specific new plasma diagnostics, too (see also Chap. 2 on waves in plasmas in Part I and the Chap. 13 in Part III).

The plasma is described by distribution functions $f_\alpha(\mathbf{p}, \mathbf{r}, t)$ for each species α, which depend on the momentum $\mathbf{p}$, the radius vector $\mathbf{r}$ and the time t. The distribution function is obtained from the solution of a kinetic equation, which in the limit of negligible particle interaction represents – analogous to the Liouville equation – the continuity equation in phase space. From the equation of motion for the particle species α

$$\frac{\mathrm{d}\mathbf{p}}{\mathrm{d}t} = e_\alpha \left(\mathbf{E} + \mathbf{v} \times \mathbf{B} \right) \quad \text{and} \quad \frac{\mathrm{d}\mathbf{r}}{\mathrm{d}t} = \mathbf{v} \tag{1.43}$$

we obtain the Vlasov equation (collision-less Boltzmann equation), which contains self-consistent fields **E** and **B**:

$$\frac{\partial f_\alpha}{\partial t} + \mathbf{v} \cdot \frac{\partial f_\alpha}{\partial \mathbf{r}} + e_\alpha \left(\mathbf{E} + \mathbf{v} \times \mathbf{B} \right) \cdot \frac{\partial f_\alpha}{\partial \mathbf{p}} = 0 \, . \tag{1.44}$$

If collisions are taken into account the Boltzmann equation

$$\frac{\partial f_\alpha}{\partial t} + \mathbf{v} \cdot \frac{\partial f_\alpha}{\partial \mathbf{r}} + e_\alpha \left(\mathbf{E} + \mathbf{v} \times \mathbf{B} \right) \cdot \frac{\partial f_\alpha}{\partial \mathbf{p}} = \left(\frac{\partial f_\alpha}{\partial t} \right)_c , \tag{1.45}$$

has to be solved to obtain the distribution function $f_\alpha(\mathbf{p}, \mathbf{r}, t)$, which contains the Boltzmann collision term (or collision integral) on the right side. Combined with Maxwell's equations this system forms the basis of magnetohydrodynamics (MHD, see also Chaps. 3 and 4 in Part I).

For further reading see [1, 2, 3, 4, 5, 6, 7].

References

1. F.F. Chen: *Introduction to Plasma Physics* (Plenum Press, New York 1988)
2. K. Miyamoto: *Fundamentals of Plasma Physics and Nuclear Fusion* (Iwanami Book Service Center, Toyko 1997)
3. J. Raeder: *Kontrollierte Kernfusion* (B.G. Teubner, Stuttgart 1981)
4. U. Schumacher: *Fusionsforschung – Eine Einführung* (Wissenschaftliche Buchgesellschaft Darmstadt 1993)
5. R.J. Goldston and P.H. Rutherford: *Introduction to Plasma Physics* (Institute of Physics Publishing, Bristol and Philadelphia 1995)
6. M. Kaufmann: *Plasmaphysik und Fusionsforschung* (B.G. Teubner, Stuttgart 2003)
7. G. Franz: *Oberflächentechnologie mit Niederdruckplasmen* (Springer–Verlag, Berlin Heidelberg New York 1994)

2 Waves in Plasmas

A. Piel

Abstract. Waves are basic manifestation of collective effects in plasmas. Wave types occurring in the plasma state are introduced and discussed with experimental applications, e.g. for the diagnostics of plasmas.

This Chapter is fundamental to Part II and covers many aspect introductory to Chapters on applications (see Part III), e.g. on plasma diagnostics.

2.1 Introduction

Wave phenomena are ubiquitous in nature and immediately affect human life. Sound waves in air let us hear, light waves give us vision, and vibrations of solids can form music. To the physicist, the study of wave phenomena gives an immediate insight into the elastic properties of matter, which define the wave speed. Sound waves in air and light waves in vacuum have the property that perturbations of different frequencies propagate with a unique velocity, the sound speed or the speed of light, respectively. This is quite different from surface waves on a pond, where sinusoidal waves of different frequency have different propagation speed and are therefore called dispersive. They resemble light in a glass prism, which is dispersed into its various colors.

Light waves and sound waves differ in their way of oscillation, which is transverse to the direction of propagation in light and along the propagation for sound. In solid matter, transverse "shear" waves and longitudinal "compressional" waves can even coexist at the same frequency. However, the different propagation speed of these two "modes" gives the two waves different wavelength at the same frequency. Ordinary gases can only support longitudinal compressional waves, because there is no restoring shear force.

Plasmas have a much higher variety of wave modes than ordinary matter because plasmas combine the aspects of a gas with electromagnetic forces. Moreover, the entanglement of particle motion with magnetic fields leads to wave types unknown in other fields of physics. This chapter attempts to give a survey of the various wave types in unmagnetized and magnetized plasmas. The mathematical apparatus is kept as simple as possible. The chosen topics of this tutorial, which emphasize diagnostic applications (see also Chap. 13 in Part III), reflect the preferences of the author, who is a devoted experimentalist. Important aspects, like Landau damping and Alfvén waves (see Chap. 4 in Part I) or nonlinear waves (see Chap. 8 in Part II on turbulence), had to be omitted as a tribute to the limited space. The new field of dusty plasmas, which leads to many more modes, like dust-acoustic waves, lattice waves or Mach cones, is left to a companion article (Chapter 11 in Part II).

There is quite a number of excellent textbooks on plasma waves that give a general survey [1, 2, 3], focus on cold plasmas [4] or on kinetic effects [5], which are recommended to the reader for more thorough studies.

2.2 Dispersion Relation for Waves in a Fluid Plasma

2.2.1 Maxwell's Equations

Electromagnetic waves are derived from Maxwell's equations, which define the relationship between the electric field $\mathbf{E}$, the magnetic field $\mathbf{B}$ and the electric current density $\mathbf{j}$.

$$\nabla \times \mathbf{E} = -\frac{\partial \mathbf{B}}{\partial t} \tag{2.1}$$

$$\nabla \times \mathbf{B} = \mu_0 \left(\mathbf{j} + \epsilon_0 \frac{\partial \mathbf{E}}{\partial t} \right) \tag{2.2}$$

Throughout this chapter SI-units will be used.

2.2.2 The Equation of Motion

The properties of the plasma medium appear in the dynamic response of the various plasma species to these fields. Because of their different mass and sign of charge, this response is quite different for electrons and (positive) ions. For each species, the periodic motion of the charged particles represents an alternating electric current. The superposition of the alternating currents of all species is the required information about the plasma medium to close Maxwell's equations.

The simplest approach to calculate the alternating current is to start from a single-particle model. For a plasma consisting of electrons and one species of singly charged positive ions the current density is

$$\mathbf{j} = ne(\mathbf{v}_i - \mathbf{v}_e) \ . \tag{2.3}$$

The velocities $\mathbf{v}_i$ and $\mathbf{v}_e$ of ions and electrons are the solution of the equation of motion

$$m\dot{\mathbf{v}} = q(\mathbf{E} + \mathbf{v} \times \mathbf{B}) \ . \tag{2.4}$$

The single-particle model is a suitable description of cold plasmas. This simple model can be extended by including temperature effects, which exert an electron pressure force per particle $-\nabla p_e/n_e$, as we will see in Sect. 2.3.2.

2.2.3 Normal Modes

Because we are interested in diagnostic applications, it is sufficient to consider small wave amplitudes. Therefore we may assume a linear relationship between the electric current at a certain (angular) frequency $\mathbf{j}(\omega)$ and the ac electric field $\mathbf{E}(\omega)$.

$$\mathbf{j}(\omega) = \sigma(\omega) \cdot \mathbf{E}(\omega) \tag{2.5}$$

Here $\sigma(\omega)$ is the conductivity tensor. As in optics of anisotropic crystals, the orientation of the current vector needs not to be parallel to the electric field vector. In magnetized plasmas, we will even find non-diagonal elements. For studying the dielectric properties of a plasma, we assume the propagation of plane waves

$$\mathbf{E}(\mathbf{r},t) = \hat{\mathbf{E}}(\mathbf{k},\omega) \exp(\mathrm{i}\mathbf{k}\cdot\mathbf{r} - \omega t) \,. \tag{2.6}$$

Here $\mathbf{k}$ is the wave vector which defines the propagation direction of the wave and is related to the wavelength λ by $|\mathbf{k}| = 2\pi/\lambda$. We convert the set of Maxwell's equations to the wave differential equation by taking the curl in the induction law (2.1) and using (2.2)

$$\nabla \times (\nabla \times \mathbf{E}) = -\frac{1}{c^2}\frac{\partial^2 \mathbf{E}}{\partial t^2} - \mu_0 \frac{\partial \mathbf{j}}{\partial t} \tag{2.7}$$

with $\mu_0 \epsilon_0 = 1/c^2$. Using plane wave expressions like (2.6) for the field quantities and the current, we transform Maxwell's equations into a set of algebraic equations for the amplitude factors. These amplitudes are in general complex numbers, which expresses the phase shift between the electric field and the current density.

$$\mathrm{i}\mathbf{k} \times \hat{\mathbf{E}} = \mathrm{i}\omega \hat{\mathbf{B}} \tag{2.8}$$

$$\mathrm{i}\mathbf{k} \times \hat{\mathbf{B}} = -\mathrm{i}\omega\epsilon_0\mu_0\hat{\mathbf{E}} + \mu_0\hat{\mathbf{j}} \tag{2.9}$$

Combining these two expressions and using (2.5), we obtain the algebraic equivalent of the wave equation

$$\mathbf{k} \times (\mathbf{k} \times \hat{\mathbf{E}}) + \frac{\omega^2}{c^2}\hat{\mathbf{E}} + \mathrm{i}\omega\mu_0\,\sigma(\omega)\cdot\hat{\mathbf{E}} = 0 \,. \tag{2.10}$$

2.2.4 The Dielectric Tensor

Although we have introduced a conductivity σ this does not imply that field and current are in phase. Rather, the complex quantity σ gives a smooth transition from dielectric behavior of the plasma, where the current lags behind the field by $\pi/2$, to a conductor. In the first case, the plasma is better described by a dielectric tensor $\epsilon(\omega)$, which has real components

$$\epsilon(\omega) = \mathbf{1} + \frac{\mathrm{i}}{\omega\epsilon_0}\sigma(\omega) \,. \tag{2.11}$$

With the dielectric tensor $\epsilon(\omega)$ the algebraic wave equation takes the form

$$\mathbf{k} \times (\mathbf{k} \times \hat{\mathbf{E}}) + \frac{\omega^2}{c^2} \epsilon(\omega) \cdot \hat{\mathbf{E}} = 0 \, . \tag{2.12}$$

This is a set of three homogeneous linear equations, which can be written in matrix form using the identity $\mathbf{k} \times (\mathbf{k} \times \hat{\mathbf{E}}) = (\mathbf{k}\mathbf{k} - k^2 \mathbf{1}) \cdot \hat{\mathbf{E}}$ as

$$\begin{pmatrix} k_x k_x - k^2 + \frac{\omega^2}{c^2}\epsilon_{xx} & k_x k_y + \frac{\omega^2}{c^2}\epsilon_{xy} & k_x k_z + \frac{\omega^2}{c^2}\epsilon_{xz} \\ k_y k_x + \frac{\omega^2}{c^2}\epsilon_{yx} & k_y k_y - k^2 + \frac{\omega^2}{c^2}\epsilon_{yy} & k_y k_z + \frac{\omega^2}{c^2}\epsilon_{yz} \\ k_z k_x + \frac{\omega^2}{c^2}\epsilon_{zx} & k_z k_y + \frac{\omega^2}{c^2}\epsilon_{zy} & k_z k_z - k^2 + \frac{\omega^2}{c^2}\epsilon_{zz} \end{pmatrix} \cdot \begin{pmatrix} \hat{E}_x \\ \hat{E}_y \\ \hat{E}_z \end{pmatrix} = 0 \, . \tag{2.13}$$

The necessary condition for this set of equations to have solutions with a non-zero electric field is the vanishing of the determinant $D(\mathbf{k}, \omega)$

$$0 = D(\mathbf{k}, \omega) = \det \left(\mathbf{k}\mathbf{k} - k^2 \mathbf{1} + \frac{\omega^2}{c^2} \epsilon(\omega) \right) \, . \tag{2.14}$$

This is the general dispersion relation for waves in a dielectric medium. For a given orientation of the electric field with respect to the wave vector it defines the relationship $\omega(\mathbf{k})$ for this mode. There can be simultaneously more than one mode at a given frequency with different polarization. When only one of these modes is considered, the explicit form $\omega(\mathbf{k})$ is called the dispersion relation of this mode.

2.2.5 Phase and Group Velocity

From (2.14) the phase velocity

$$\mathbf{v}_\varphi = \frac{\omega}{k^2} \mathbf{k} \tag{2.15}$$

and group velocity

$$\mathbf{v}_g = \left(\frac{\partial \omega}{\partial k_x}, \frac{\partial \omega}{\partial k_y}, \frac{\partial \omega}{\partial k_z} \right) \tag{2.16}$$

can easily be derived. The phase velocity is the speed of the wave fronts, whereas the group velocity describes the propagation of the envelope of a wave packet and is connected to the energy flow. While the direction of the phase velocity is given by the wave vector, the group velocity can, e.g. in a magnetized plasma, point into a different direction, a typical feature of anisotropic media.

2.3 Waves in Unmagnetized Plasmas

In this Section we are interested in transverse electromagnetic waves, which exist at high frequencies and are determined by the electron contribution to the dielectric tensor, and in longitudinal acoustic waves. Because of the absence of a static magnetic field, in (2.4) the Lorentz force $\mathbf{F}_L = q(\mathbf{v} \times \mathbf{B})$ contains two small quantities, the oscillation velocity $\mathbf{v}$ and the wave magnetic field $\mathbf{B}$. This product is therefore neglected as a second-order perturbation compared to the electric field force $\mathbf{F}_E = q\mathbf{E}$, which is of first-order.

2.3.1 Transverse Waves

The response of electrons and ions to an alternating electric field at high frequencies is quite different, because the oscillation velocity depends on the different masses. Neglecting electron pressure effects for the moment, we have

$$\hat{\mathbf{v}} = \mathrm{i}\frac{q}{m\omega}\hat{\mathbf{E}} \ . \tag{2.17}$$

Hence, for high frequencies we can neglect the ion current, which is by a factor m_e/m_i smaller than the electron current. At high frequencies the ions only form an immobile neutralizing background. The resulting current density is then

$$\hat{\mathbf{j}} = nq\hat{\mathbf{v}} = \mathrm{i}\frac{ne^2}{\omega m_e}\hat{\mathbf{E}} \ . \tag{2.18}$$

Because of the absence of the Lorentz force the plasma is isotropic and the current vector is parallel to the electric field. Hence the conductivity tensor is diagonal with three identical elements

$$\sigma_{xx} = \sigma_{yy} = \sigma_{zz} = \mathrm{i}\frac{ne^2}{\omega m_e} \ . \tag{2.19}$$

Likewise the dielectric tensor is also diagonal

$$\epsilon_{xx} = \epsilon_{yy} = \epsilon_{zz} = 1 - \frac{\omega_{pe}^2}{\omega^2} \tag{2.20}$$

where the electron plasma frequency ω_{pe} has been introduced (see also Chap. 1 in Part I)

$$\omega_{pe} = \left(\frac{ne^2}{\epsilon_0 m_e}\right)^{1/2} \ . \tag{2.21}$$

When we now consider wave propagation in the x-direction, the matrix wave equation (2.13) yields the following set of equations

$$\frac{\omega^2}{c^2}\left(1 - \frac{\omega_{pe}^2}{\omega^2}\right)\hat{E}_x = 0 \tag{2.22}$$

$$\left[-k^2 + \frac{\omega^2}{c^2}\left(1 - \frac{\omega_{pe}^2}{\omega^2}\right)\right]\hat{E}_y = 0 \tag{2.23}$$

$$\left[-k^2 + \frac{\omega^2}{c^2}\left(1 - \frac{\omega_{pe}^2}{\omega^2}\right)\right]\hat{E}_z = 0 \ . \tag{2.24}$$

For the existence of longitudinal electric fields $\hat{E}_x$ the condition $1-\omega_{pe}^2/\omega^2 = 0$ must be fulfilled, which is equivalent to $\omega = \omega_{pe}$. Hence, there are longitudinal oscillations at the electron plasma frequency for arbitrary wavelength.

From (2.23) and (2.24) we see that there are two identical but independent transverse modes corresponding to the two orthogonal transverse polarizations $\hat{E}_y$ and $\hat{E}_z$. From the vanishing of the square bracket, we see that these transverse modes have the dispersion relation

$$\omega = \left(\omega_{pe}^2 + k^2 c^2\right)^{1/2} . \tag{2.25}$$

These waves are the transverse electromagnetic waves in an unmagnetized plasma and have frequencies greater than the electron plasma frequency. In other words, a plasma is only transparent for electromagnetic waves at $\omega > \omega_{pe}$. At lower frequencies, the wave is reflected by the plasma.

Microwave and Laser Interferometry

Interferometry is based on comparing the phase of a wave that traverses a medium of refractive index N with a reference path that is unaffected by the medium. By definition, the refractive index is the ratio of the speed of light in vacuum to the phase velocity in the medium. The refractive index for the waves described by (2.25) is

$$N = \frac{kc}{\omega} = \left(1 - \frac{\omega_{pe}^2}{\omega^2}\right)^{1/2} = \left(1 - \frac{n}{n_{co}}\right)^{1/2} . \tag{2.26}$$

The refractive index is real for $\omega > \omega_{pe}$ and imaginary else, marking the change from a dielectric medium to a perfect conductor. The plasma frequency is therefore called the *cut-off frequency*, which is defined by $N = 0$. Likewise for a given wave frequency, the critical electron density, which gives $\omega = \omega_{pe}$ is called the *cut-off density* n_{co}. Note that the refractive index of an unmagnetized plasma is smaller than unity thus making the waves longer than in vacuum. A typical *Mach-Zehnder* arrangement is often used for microwave or laser interferometry of plasmas (Fig. 2.1). The original wave is split into a probe branch that traverses the plasma and a reference branch, which are eventually recombined to give a temporal interference pattern at a detector. For a plasma of length L the phase shift between the two branches is

$$\Delta\varphi = 2\pi \frac{(N_{plasma} - 1)L}{\lambda} . \tag{2.27}$$

If the wave frequency is chosen sufficiently higher than the cut-off frequency, the refractive index can be expanded into a Taylor series and the phase shift takes the simple form

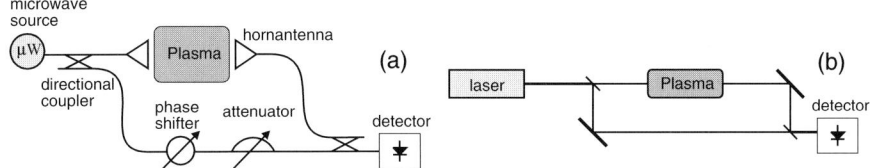

Fig. 2.1. Mach-Zehnder arrangement for (**a**) a microwave interferometer and (**b**) a laser interferometer. Splitting and recombining of the two branches is accomplished by directional couplers resp. semitransparent mirrors

$$\Delta\varphi \simeq -\pi \frac{L}{\lambda} \frac{n}{n_{co}} \propto \lambda L n_e \,. \tag{2.28}$$

The last step follows from $n_{co} \propto \lambda^{-2}$. Hence the sensitivity of an interferometer is proportional to the product of plasma length and wavelength. Microwaves are used in medium-density gas discharges while laser interferometers are required for high-density plasmas.

A typical signal from a microwave interferometer is shown in Fig. 2.2 together with the resulting electron density decay. The plasma is produced by a strong current pulse, which generates an electron density much higher than the cut-off density of the interferometer. After the current pulse the plasma decays and reaches the cut-off density at 1200 μs and the interferometer shows the typical oscillations ("fringes"). The reduced amplitude of the fringes near the cut-off and the invisibility of fringes during start-up of the current pulse are due to the limited time response of the detector circuit. The electron

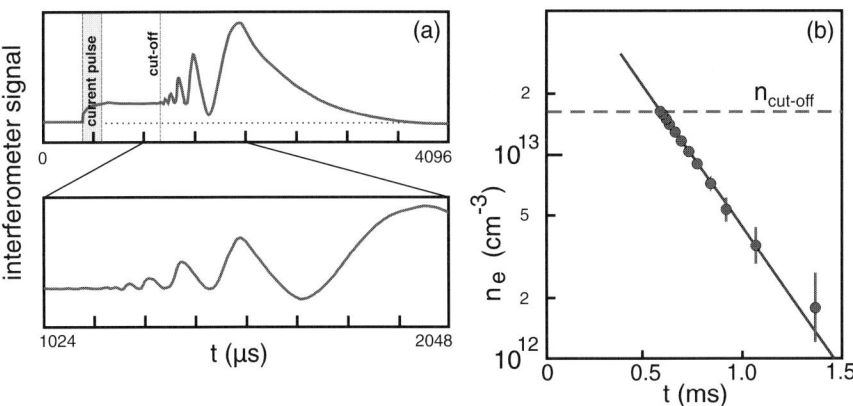

Fig. 2.2. Microwave interferometry of a pulsed low-pressure arc. (**a**) Plasma is produced with a current pulse of 250 μs duration. Thereafter the plasma decays to reach the cut-off density at about 1200 μs. (**b**) A semi-logarithmic plot of the plasma density shows the exponential decay of the plasma. The evolution corresponds to the magnified part of the interferometer signal in (**a**)

density is recovered by counting the interference fringes from the end of the discharge backwards until the cut-off density is reached. For the evaluation of the fringes near the cut-off the full expression (2.27) is used. The resulting density decay is exponential with a time constant $\tau = 315\,\mu\text{s}$.

Second Harmonic Interferometer

A special kind of two-wavelength interferometer is the second-harmonic interferometer, which was introduced by Hopf and coworkers [6]. This technique is presently used, e.g., for the diagnostics of the Alcator C-mode tokamak [7, 8]. Second-harmonic interferometry has the advantage that probe and reference beam take the same path, which reduces the sensitivity of the interferometer to mechanical vibrations that affect conventional interferometers with a separate reference branch. The original signal from a Nd:YAG laser at 1064 nm wavelength is used as the probe beam (see Fig. 2.3). The frequency doubled wave at 532 nm also traverses the plasma but experiences a different phase shift. Behind the plasma the strong probe wave is frequency doubled with a second crystal. Both waves at 532 nm produce interference fringes on the detector.

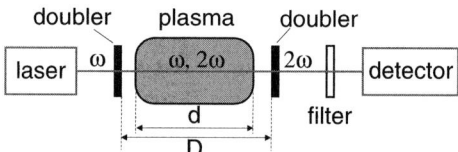

Fig. 2.3. Principle of a second-harmonic interferometer. The beam of a Nd:YAG laser at the fundamental wavelength 1064 nm is frequency doubled with a LiB_3O_5 crystal. Both the fundamental and frequency doubled wave traverse the plasma. Behind the plasma the remaining fundamental wave is doubled and the two signals at 532 nm wavelength interfere at the detector

The fundamental laser wave at ω experiences between the first and second doubler crystal a phase shift $\varphi_p(\omega)$ by the plasma and $\varphi_{air}(\omega)$ in the air gaps.

$$\varphi_p(\omega) = \frac{\omega}{c}\left(d - \frac{1}{2}\frac{\bar{n}_e d e^2}{\omega^2 \epsilon_0 m_e}\right) \tag{2.29}$$

$$\varphi_{air}(\omega) = \frac{\omega}{c}(D - d) \tag{2.30}$$

Behind the second doubler the phase of the frequency doubled signal is

$$\varphi_1(2\omega) = 2\frac{\omega}{c}\left(D - \frac{1}{2}d\frac{\bar{n}_e}{\omega^2}a\right), \tag{2.31}$$

where $a = e^2/\epsilon_0 m_e$. The frequency doubled signal has a total phase shift between the doublers of

$$\varphi_2(2\omega) = \frac{2\omega}{c}\left[D - \frac{1}{2}d\frac{\bar{n}_e}{(2\omega)^2}a\right] . \tag{2.32}$$

The phase difference of these two signals becomes

$$\varphi_2(2\omega) - \varphi_1(2\omega) = \frac{3}{2}d\frac{\bar{n}_e}{2\omega}a . \tag{2.33}$$

We see that the contribution from the air gap cancels when we neglect the different refractive indices of air at the two wavelengths. The second-harmonic interferometer is by a factor of 1.5 more sensitive than a conventional interferometer operating at the fundamental frequency ω.

Radio Beacon Technique

Radio beacons are used on satellites to probe the ionospheric plasma along the line of sight between the satellite and a ground receiver. The beacon radiates two waves with frequencies that exceed the maximum plasma frequency in the ionosphere ($\omega_{pe} \approx 20\,\text{MHz}$). The two waves are phase locked. The total phase shift is proportional to the line integrated electron density $\Delta\varphi \propto \int n_e(s)\mathrm{d}s$. This technique has also been used on a rocket with transmitter frequencies of 40 MHz and 400 MHz (see Fig. 2.4). The higher frequency is practically unaffected by the ionospheric plasma and serves as phase reference. The phase

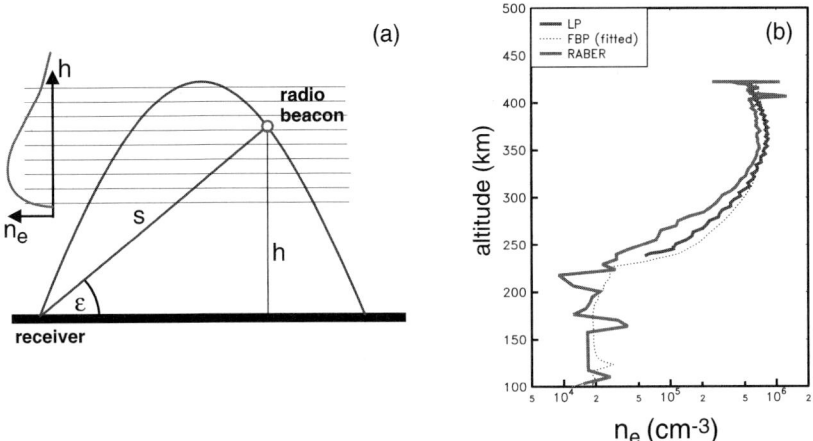

Fig. 2.4. Radio beacon on a rocket. (a) Rocket trajectory and ray path through the stratified ionosphere. (b) Resulting density profile from radio beacon (RABER: **R**adio **B**eacon on **R**ocket) compared with Langmuir probe (LP) and fixed bias probe (FBP) results (after [9])

shift between the 40 MHz signal and the reference signal is then measured as a function of rocket position along the trajectory. The density profile of the ionosphere can be reconstructed in the following manner.

The total phase difference the wave experiences on its way from transmitter to receiver is ($\theta = \pi/2 - \epsilon$)

$$\Delta\varphi = 2\pi \int \frac{[N(s)-1]\mathrm{d}s}{\lambda} = \frac{2\pi}{\lambda\cos\theta} \int [N(h)-1]\mathrm{d}h \ . \tag{2.34}$$

The integral $\int [N(s)-1]\mathrm{d}s \propto \int n_e(s)\mathrm{d}s$ is proportional to the *total electron content* (TEC). By taking the derivative with respect to time, we obtain the *differential TEC signal*

$$\frac{\mathrm{d}(\Delta\varphi)}{\mathrm{d}t} = \frac{2\pi}{\lambda} \frac{\{N[h(t)]-1\}}{\cos[\theta(t)]} v_{vert} \ , \tag{2.35}$$

which now contains the vertical rocket velocity $v_{vert} = \mathrm{d}h/\mathrm{d}t$. Knowing the rocket trajectory as a function of time, this expression can be solved for the desired $N(h)$, which contains the information about the local electron density.

This technique was used, e.g., in the DEOS rocket-campaign [9], where a radio beacon instrument could be compared with local Langmuir probe measurements. The results are shown in Fig. 2.4(b). The radio beacon confirms the general shape of the density profile and finds closer agreement with the Langmuir probe data at the maximum density in the F-layer (350 km). The F-layer extends from approximately 200–100 km altitude and is the region of highest electron density in the ionosphere [10].

Density Measurements for Dense Plasmas

Modern femtosecond lasers with intensities of $10^{19}\,\mathrm{W\,cm^{-2}}$ are capable to produce laser plasmas at metal or dielectric surfaces, which reach electron densities beyond $10^{23}\,\mathrm{cm^{-3}}$. This density corresponds to a plasma frequency of $\omega_{pe} = 1.78 \times 10^{16}\,\mathrm{s^{-1}}$ and a cut-off wavelength of $\lambda = 106\,\mathrm{nm}$. Therefore interferometry of such plasmas would require stable lasers at vacuum-UV wavelength ($\lambda \ll 100\,\mathrm{nm}$), which are not available.

Such laser-produced plasmas can be analyzed by checking the cut-off frequency of the plasma with a large number of harmonics of a powerful laser in the visible. Such a method has been developed by Sauerbrey's group [11]. High odd numbered laser harmonics are generated in a gas jet, which then penetrate the laser produced plasma under study (see Fig. 2.5). The electron density decays from $3.2 \times 10^{12}\,\mathrm{cm^{-3}}$ at 0 ps to $6.7 \times 10^{22}\,\mathrm{cm^{-3}}$ at 12 ps delay time. The natural intensity ratio of the 5th and 7th harmonic is found at $\Delta t = 12\,\mathrm{ps}$. In the early discharge, the 5th harmonic is practically in the cut-off regime and its intensity markedly suppressed whereas the 7th harmonic is still strong. Therefore the cut-off wavelength lies between the 5th and 7th harmonic and can be used as density diagnostics [12].

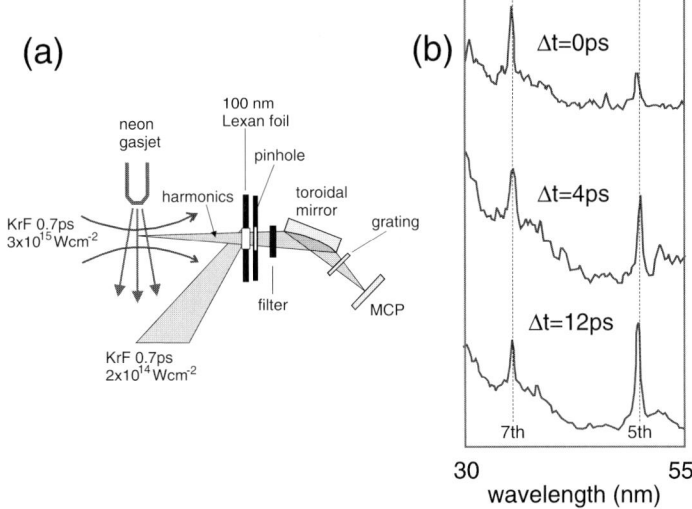

Fig. 2.5. (a) Sketch of the apparatus and results from [11]. High laser harmonics are produced in a neon gas jet and penetrate the laser produced plasma. The x-ray spectrum is recorded with a toroidal mirror, grating and micro channel plate detector. (b) Transmission spectra of the early phase of the plasma discharge, the 5th harmonic is weaker than the 7th harmonic while, at 12 ps delay time, the natural intensity ratio of the harmonics is recovered

2.3.2 Longitudinal Waves

Ion Acoustic Waves

At low frequencies electron inertia effects become negligible whereas it is no longer justified to neglect pressure forces, as we did for the high-frequency waves. A novel longitudinal wave appears in this regime, the ion-acoustic wave. Because the motion is one-dimensional the equation of motion for positive ions and electrons reads

$$m_i \frac{\partial v_i}{\partial t} = eE - \frac{1}{n_{i0}} \frac{\partial}{\partial z}(n_i k_B T_i) \tag{2.36}$$

$$0 = -eE - \frac{1}{n_{e0}} \frac{\partial}{\partial z}(n_e k_B T_e) \,, \tag{2.37}$$

in which electron inertia forces are neglected and the pressure gradient is added on the right hand side. In Fourier notation, i.e., with a harmonic field of type (2.6) this pair of equations becomes

$$-\mathrm{i}\omega m_i \hat{v}_i = e\hat{E} - \frac{\mathrm{i}k}{n_{i0}}(\gamma_i k_B T_i)\hat{n}_i \tag{2.38}$$

$$0 = -e\hat{E} - \frac{\mathrm{i}k}{n_{e0}}(k_B T_e)\hat{n}_e \,, \tag{2.39}$$

where we have assumed that the ion gas experiences an adiabatic compression with $\gamma_i = 3$ and the electrons, because of their high thermal velocity, form a heat bath for the system and react by isothermal compression.

The oscillating ion streaming leads to compression effects, which are described by the equation of continuity

$$\frac{\partial n_i}{\partial t} + \frac{\partial}{\partial z}(n_i v_i) = 0 , \tag{2.40}$$

which in Fourier notation $-i\omega \hat{n}_i + ik n_{i0} \hat{v}_i = 0$ gives

$$\hat{v}_i = \frac{\omega}{k} \frac{\hat{n}_i}{n_{i0}} . \tag{2.41}$$

Using this relation to eliminate $\hat{v}_i$ in the ion equation of motion (2.38), we obtain the density fluctuations $\hat{n}_i$ and $\hat{n}_e$ for a given wave field $\hat{E}$ as

$$\hat{n}_i = \frac{e k n_{i0}}{-i\omega^2 m_i + ik^2 \gamma_i k_B T_i} \hat{E} \tag{2.42}$$

$$\hat{n}_e = \frac{-e n_{e0}}{ik k_B T_e} \hat{E} . \tag{2.43}$$

Since we are searching for a longitudinal sound wave, the wave vector and the electric field vector are parallel. Hence, $\mathbf{k} \times \mathbf{E} = 0$ and there are no induction effects. Therefore, we can use the Poisson equation $\nabla \cdot \mathbf{E} = \rho/\epsilon_0$ in Fourier form

$$ik\hat{E} = \frac{e}{\epsilon_0}(\hat{n}_i - \hat{n}_e) \tag{2.44}$$

as condition for the self-consistency of the wave field $\hat{E}$ yielding

$$ik\hat{E} = \left(\frac{n_{i0} e^2}{\epsilon_0 m_i}\right) \frac{k}{-i\omega^2 + ik^2 \gamma_i k_B T_i/m_i} \hat{E} + \left(\frac{n_{e0} e^2}{\epsilon_0 k_B T_e}\right) \frac{1}{ik} \hat{E} . \tag{2.45}$$

Introducing the ion plasma frequency $\omega_{pi} = (n_{i0} e^2/\epsilon_0 m_i)^{1/2}$ and the electron Debye length $\lambda_{De} = (n_{e0} e^2/\epsilon k_B T_e)^{1/2}$ we find the following dielectric function

$$\epsilon(k,\omega) = 1 - \frac{\omega_{pi}^2}{\omega^2 - k^2 \gamma_i k_B T_i/m_i} + \frac{1}{k^2 \lambda_{De}^2} . \tag{2.46}$$

The dispersion relation of the electrostatic wave is again given by $\epsilon(k,\omega) = 0$ and can be solved for ω^2

$$\omega^2 = k^2 \frac{\gamma_i k_B T_i}{m_i} + \frac{k^2 C_s^2}{1 + k^2 \lambda_{De}^2} . \tag{2.47}$$

Here $C_s = (k_B T_e/m_i)^{1/2}$ is the ion sound velocity. In most gas discharge plasmas we have $T_e \gg T_i$. In that limit the ion pressure term can be dropped and we find $\omega \approx k C_s (1 + k^2 \lambda_{De}^2)^{-1/2}$. For small wavenumbers ($k^2 \lambda_{De}^2 \ll 1$)

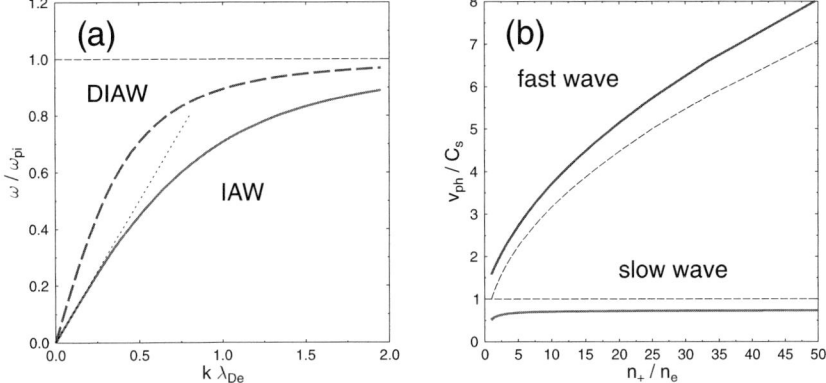

Fig. 2.6. (a) Ion-acoustic waves (IAW, *solid line*) and dust-ion-acoustic wave (DIAW, *dashed line*). The acoustic dispersion of the IAW is indicated by the *dotted line*. The DIAW has an increased phase velocity. (b) Ion acoustic waves in a negative ion plasma. The phase velocity ω/kC_s of the fast wave increases with the fraction of negative ions n_-/n_+ while the slow wave remains nearly unaffected. For comparison the dashed curve indicates the scaling of the DIAW

this wave has acoustic dispersion $\omega = kC_s$ (see Fig. 2.6). In the opposite case of large wavenumbers, the wave frequency approaches ω_{pi}.

Comparing the ion acoustic velocity $C_s = (k_B T_e/m_i)^{1/2}$ with the sound speed in a neutral gas $c_s = (\gamma p/\rho)^{1/2}$, one is used to consider an ion-acoustic wave as a kind of sound wave, where the electron temperature represents a pressure force $n_{e0}k_B T_e$ and the ions an inertia $n_{i0}m_i$, which acts as inhibitor. This picture, however, is misleading as we will see below.

There is a long tradition of experiments, where ion acoustic waves were excited with Langmuir probes or grids [13, 14]. The waves are excited by applying a pulse or a continuous wave. In a refined scheme, a swept frequency is used [15] and the beat frequency is detected. Care has to be taken that the exciter voltage is small enough to suppress "pseudo-waves" [16], which are bunches of ions accelerated in the sheath around the probe or grid. Modern techniques for detecting ion-acoustic waves are based on laser-induced fluorescence (LIF). Recently, ion-acoustic solitary waves [17] were analyzed with this technique.

Dusty Plasmas

When the plasma contains dust particles, which are very much heavier than ions and carry a large negative charge, the electron density becomes different from the ion density to maintain quasi-neutrality. It is instructive to study the ion acoustic waves under such conditions. Noting that $\lambda_{De}^2 \omega_{pi}^2 = (n_{i0}/n_{e0})C_s^2$ and assuming $T_e \gg T_i$ we see that the dust-ion-acoustic wave

$$\omega \approx \left(\frac{n_{i0}}{n_{e0}}\right)^{1/2} \frac{kC_s}{(1+k^2\lambda_{De}^2)^{1/2}} \quad (2.48)$$

has a phase velocity that increases with $(n_{i0}/n_{e0})^{1/2}$. In comparison with the ordinary sound wave discussed above this is a paradoxical situation, because the phase speed increases when there are less electrons. One would have expected that a decreasing electron pressure $n_{e0}k_BT_e$ reduces the phase velocity. Hence the above identification with ordinary sound waves is wrong. This paradox can be resolved by considering the electrons not as a gas that exerts a pressure but rather as a fluid of the opposite charge that shields the electric repulsion between the ions. From (2.42), (2.43) and (2.47) we obtain the relation

$$\frac{\hat{n}_i - \hat{n}_e}{\hat{n}_i} = \frac{k^2\lambda_{De}^2}{1+k^2\lambda_{De}^2} . \quad (2.49)$$

This ratio represents the relative space charge in units of the ion wave amplitude. For small wavenumbers $k^2\lambda_{De}^2 \ll 1$, where the dispersion is acoustic, the relative space charge in units of the ion fluctuation is proportional to λ_{De}^2, which again is proportional to T_e/n_e. Hence the relative space charge increases, when there are less electrons. The increase of the space charge with electron temperature means that more energetic electrons can escape from the ion potential well and this reduces the electron space charge as well (see Chap. 11 in Part II).

Two Ion Species

Plasmas that contain an additional component of negative ions can be described by adding the response term for the negative ions to the dielectric function

$$\epsilon = 1 - \frac{\omega_{p+}^2}{\omega^2 - k^2\gamma_i k_B T_+/m_+} - \frac{\omega_{p-}^2}{\omega^2 - k^2\gamma_i k_B T_-/m_-} + \frac{1}{k^2\lambda_{De}^2} \quad (2.50)$$

where the indices $+$ and $-$ refer to the positive and negative ions. This equation has two independent wave branches, a slow wave and a fast wave [Fig. 2.6(b)]. The fast wave is apparently sensitive to the density of free electrons whereas the slow wave is insensitive. When the mass of the negative ion becomes large ($m_- \to \infty$) the fast wave approaches the DIAW and the phase velocity of the slow wave decreases dramatically. Hence we recover the dusty plasma as a limiting case of the negative ion situation.

The sudden increase of the phase velocity in the presence of negative ions was observed by many authors [18, 19, 20]. Recently, Nakamura et al. [21] made an exhaustive study of this effect over a wide range of free electron density and found close agreement with the fluid model outlined above. Thus measuring the phase velocity of fast wave and slow wave can be used as a diagnostics for the ratio n_+/n_e.

2.3.3 Electron Beam Driven Waves

In Sect. 2.3.1 longitudinal particle motion in a cold plasma was found to form oscillations at the plasma frequency rather than waves. Their dispersion was governed by $\epsilon_{xx} = 1 - \omega_{pe}^2/\omega^2 = 0$. When a source of free energy is added to the cold plasma, e.g. an additional "beam" of electrons with a drift speed v_0, the plasma oscillations can be destabilized and grow in time.

The electrostatic dispersion relation for a beam-plasma system can be derived starting from the definition of the displacement current in 1-d

$$\frac{\partial D}{\partial t} = \epsilon \epsilon_0 \frac{\partial E}{\partial t} = \epsilon_0 \frac{\partial E}{\partial t} + j \; . \tag{2.51}$$

Because we study longitudinal waves all vector quantities are aligned in the x-direction and the vector notation is dropped. The displacement current is the sum of the vacuum displacement $\epsilon_0 \dot{E}$ and the current j from the oscillating particles. With the wave ansatz (2.6) we can gain insight in the structure of the dielectric constant ϵ

$$\epsilon = 1 - \frac{\hat{\jmath}}{i\omega\epsilon_0 \hat{E}} \; . \tag{2.52}$$

Hence, the "1" in ϵ is the normalized vacuum displacement current and $-\omega_{pe}^2/\omega^2$ represents the ratio of the particle current to the vacuum displacement current. Bearing in mind that $\omega_{pe}^2 = n_e e^2/(\epsilon_0 m_e)$ is a measure of the electron density, we can immediately write down the contribution from the beam electrons to the dielectric constant as $-\omega_{pb}^2/(\omega - kv_0)^2$. Here, $\omega_{pb}^2 = n_b e^2/(\epsilon_0 m_e)$ is calculated from the density n_b of beam electrons. The trick lies in guessing the proper frequency in the denominator, which is the Doppler-shifted frequency that the beam electrons experience in their moving frame of reference. Thus the complete dielectric function reads

$$\epsilon(\omega) = 1 - \frac{(1-\alpha)\omega_{pe}^2}{\omega^2} - \frac{\alpha \omega_{pe}^2}{(\omega - kv_0)^2} \; . \tag{2.53}$$

Here, ω_{pe} corresponds to the total electron density n_e and $\alpha = n_b/n_e$ is the fraction of beam electrons.

The dispersion relation $\epsilon = 0$ is equivalent to a polynomial of fourth order with real coefficients. Therefore there are exactly four roots $\omega(k)$, which are either real or complex conjugate pairs. For complex $\omega = \omega_r + i\omega_i$ the wave solution $E = \hat{E}\exp[i(kx - \omega_r t)]\exp(\omega_i t)$ is exponentially growing in time, when $\omega_i > 0$. The topology of these roots is compiled in Fig. 2.7.

Instability is found for wavenumbers $k < k_{crit}$. The maximum growth rate ω_i occurs at the intersection of the asymptotes $kv_0/\omega_{pe} = 1$. When $\alpha \ll 1$, we can estimate the maximum growth rate from a Taylor expansion of the dielectric function about this intersection, with $\delta\omega = \omega - \omega_{pe}$

$$0 \approx 2\frac{\delta\omega}{\omega_{pe}} - \alpha\frac{\omega_{pe}^2}{(\delta\omega)^2} \; . \tag{2.54}$$

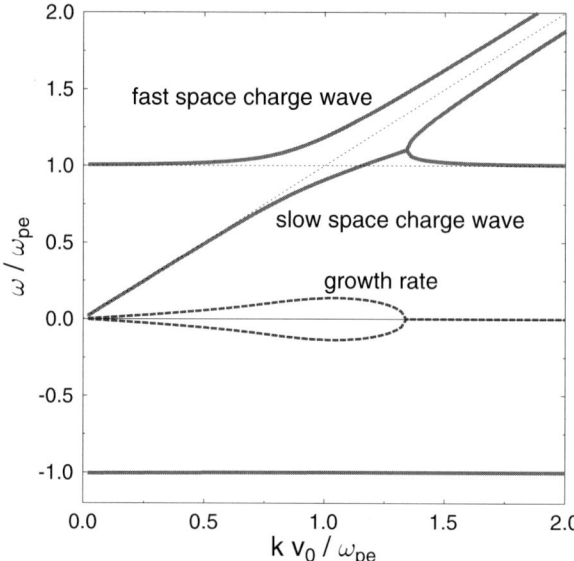

Fig. 2.7. Topology of space charge waves in the presence of an electron beam. The frequency and wavenumber are given in normalized units. For kv_0/ω_{pe} below a critical value, which is defined by the "triple point" in the real part of ω/ω_{pe} the beam space charge waves are unstable

This equation has three complex roots

$$\delta\omega = \left(\frac{\alpha}{2}\right)^{1/3} \omega_{pe} \exp\left(2\pi i \frac{n}{3}\right) \qquad n = 0, 1, 2 \ . \tag{2.55}$$

The real positive solution corresponds to the fast space-charge wave, the two conjugate complex solutions represent a damped and a growing slow space-charge wave. Frequency and growth rate of the unstable slow space-charge wave are given by

$$\omega_r = \left[1 - \frac{1}{2}\left(\frac{\alpha}{2}\right)^{1/3}\right]\omega_{pe} \tag{2.56}$$

$$\omega_i = \frac{1}{2}\sqrt{3}\left(\frac{\alpha}{2}\right)^{1/3}\omega_{pe} \ . \tag{2.57}$$

The wave frequency is therefore very close to the plasma frequency. The dependence of the growth rate on the third root of the beam fraction makes this beam-plasma interaction a violent instability. The presence of even a tiny electron beam manifests itself by excitation of waves close to the plasma frequency. (See Sect. 4.2 in Chap. 4 in Part I.)

The Plasma Oscillation Method

An electron beam can be created in the vicinity of a heated filament, which emits electrons and is held at a negative bias with respect to the ambient plasma. The excited waves at the plasma frequency are detected by a small probe at a few millimeter distance (see Fig. 2.8). This method was introduced by Sugai's group [22] as a suitable technique for measuring the electron density in reactive plasmas, where ordinary Langmuir probes (see Chap. 5 in Part I) cannot be used because of probe contamination. The heated filament used here evaporates any deposited material and deposition on the receiver probe does not affect the microwave signal, which couples to the probe by displacement currents. The technique was even capable to measure the negative ion concentration in a reactive plasma by detecting the increase in electron density after photodetachment by a powerful laser [23]. It is also in current use by other groups [24, 25, 26]. The method is limited to the very low pressure regime $p < 1\,\text{Pa}$ because of the damping of the excited plasma waves by electron-neutral collisions. For the same reason, the plasma density must exceed a pressure dependent minimum value to ensure $\omega_{pe}^2 \gg \nu_{en}^2$.

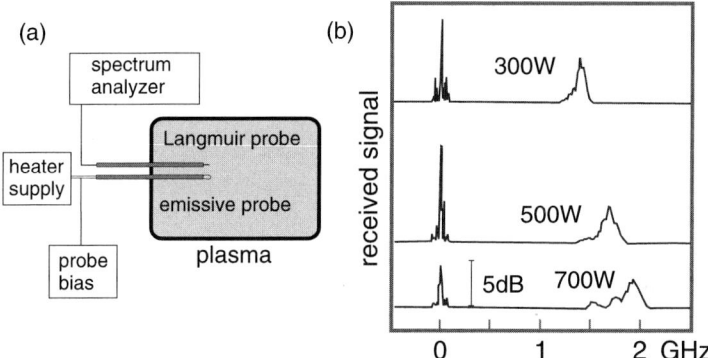

Fig. 2.8. (a) Arrangement for the plasma oscillation method. (b) A typical frequency spectrum [22]. The largest peak between 1 GHz and 2 GHz is identified as the plasma oscillation and shifts with increasing microwave power for plasma production

2.4 Waves in Magnetized Plasmas

In this Section we are only interested in the influence of a static magnetic field $\mathbf{B}_0 = (0, 0, B_0)$ on the possible plasma waves and neglect pressure forces for simplicity. The equation of motion then takes the form

$$\frac{\partial \mathbf{v}_1^{(j)}}{\partial t} = \frac{q_j}{m_j}\left(\mathbf{E}_1 + \mathbf{v}_1^{(j)} \times \mathbf{B}_0\right) \qquad j = e, i, \qquad (2.58)$$

where the index $j = e, i$ distinguishes between electrons and ions. The second order Lorentz force $\mathbf{v}_1^{(j)} \times \mathbf{B}_1$ arising from the wave magnetic field is neglected. Our task is to calculate the first-order particle velocities in the first-order wave electric field $\mathbf{E}_1$ including the deflection by the Lorentz force. From this follow the alternating currents of electrons and ions and hence the conductivity tensor. Now we drop the index 1 again.

Considering the motion in the plane perpendicular to $\mathbf{B}_0$ the resulting particle velocities in the wave field are

$$\hat{v}_x^{(j)} = i\frac{q_j}{\omega m_j}(\hat{E}_x + \hat{v}_y^{(j)} B_0)$$
$$\hat{v}_y^{(j)} = i\frac{q_j}{\omega m_j}(\hat{E}_y - \hat{v}_x^{(j)} B_0) . \qquad (2.59)$$

These two equations are usually decoupled by introducing rotating coordinates $\hat{v}^\pm = \hat{v}_x \pm i\hat{v}_y$ and $\hat{E}^\pm = \hat{E}_x \pm i\hat{E}_y$ to give

$$\hat{v}^\pm = i\frac{q_j}{\omega m_j}(\hat{E}^\pm \mp i\hat{v}^\pm B_0) . \qquad (2.60)$$

Introducing the cyclotron frequencies $\omega_{cj} = |q_j| B_0/m_j$ the solution takes the compact form

$$\hat{v}^\pm = i\frac{q_j}{m_j}\hat{E}^\pm \frac{1}{\omega \mp s_j \omega_{cj}} . \qquad (2.61)$$

Here s_j is the sign of the charge. The transform to Cartesian coordinates is accomplished by $\hat{v}_x^{(j)} = (\hat{v}^+ + \hat{v}^-)/2$ and $\hat{v}_y^{(j)} = (\hat{v}^+ - \hat{v}^-)/(2i)$. In matrix notation the relationship between the velocity amplitude and wave electric field reads

$$\begin{pmatrix} \hat{v}_x^{(j)} \\ \hat{v}_y^{(j)} \\ \hat{v}_z^{(j)} \end{pmatrix} = i\frac{q_j}{\omega m_j} \begin{pmatrix} \frac{\omega^2}{\omega^2 - \omega_{cj}^2} & i\frac{s_j \omega \omega_{cj}}{\omega^2 - \omega_{cj}^2} & 0 \\ -i\frac{s_j \omega \omega_{cj}}{\omega^2 - \omega_{cj}^2} & \frac{\omega^2}{\omega^2 - \omega_{cj}^2} & 0 \\ 0 & 0 & 1 \end{pmatrix} \begin{pmatrix} \hat{E}_x \\ \hat{E}_y \\ \hat{E}_z \end{pmatrix} . \qquad (2.62)$$

Using the definition of the total electric current $\hat{\mathbf{j}} = \sum_j n_j q_j \hat{\mathbf{v}}^{(j)}$ we obtain the conductivity tensor as:

$$\sigma = i\omega\epsilon_0 \begin{pmatrix} \sum_j \frac{\omega_{pj}^2}{\omega^2 - \omega_{cj}^2} & i\sum_j s_j \frac{\omega_{pj}^2}{\omega^2 - \omega_{cj}^2}\frac{\omega_{cj}}{\omega} & 0 \\ -i\sum_j s_j \frac{\omega_{pj}^2}{\omega^2 - \omega_{cj}^2}\frac{\omega_{cj}}{\omega} & \sum_j \frac{\omega_{pj}^2}{\omega^2 - \omega_{cj}^2} & 0 \\ 0 & 0 & \sum_j \frac{\omega_{pj}^2}{\omega^2} \end{pmatrix} . \qquad (2.63)$$

With the aid of (2.11) the dielectric tensor has the structure

$$\epsilon = \begin{pmatrix} S & -iD & 0 \\ iD & S & 0 \\ 0 & 0 & P \end{pmatrix}, \qquad (2.64)$$

which contains the so-called Stix-parameters [1]

$$S = 1 - \sum_j \frac{\omega_{pj}^2}{\omega^2 - \omega_{cj}^2} \qquad D = \sum_j s_j \frac{\omega_{pj}^2}{\omega^2 - \omega_{cj}^2} \frac{\omega_{cj}}{\omega} \qquad P = 1 - \sum_j \frac{\omega_{pj}^2}{\omega^2}. \qquad (2.65)$$

The parameter $P = \epsilon_{zz}$ is our familiar dielectric constant of the unmagnetized plasma. This coincidence is obvious because the motion of the charged particles along the magnetic field does not involve the Lorentz force. At last, using the definition of the refractive index $N = kc/\omega$ and introducing the angle ψ between wave vector and magnetic field direction the wave equation (2.13) becomes

$$\begin{pmatrix} S - N^2 \cos^2 \psi & -iD & N^2 \cos \psi \sin \psi \\ iD & S - N^2 & 0 \\ N^2 \cos \psi \sin \psi & 0 & P - N^2 \sin^2 \psi \end{pmatrix} \cdot \begin{pmatrix} \hat{E}_x \\ \hat{E}_y \\ \hat{E}_z \end{pmatrix} = 0. \qquad (2.66)$$

Here the wave vector $\mathbf{k} = (k \sin \psi, 0, k \cos \psi)$ has been used to define the x-z-plane.

2.4.1 Propagation Along the Magnetic Field

When the wave propagates along the magnetic field, we have $\psi = 0$ and the wave equation takes the form

$$\begin{pmatrix} S - N^2 & -iD & 0 \\ iD & S - N^2 & 0 \\ 0 & 0 & P \end{pmatrix} \cdot \begin{pmatrix} \hat{E}_x \\ \hat{E}_y \\ \hat{E}_z \end{pmatrix} = 0. \qquad (2.67)$$

We immediately see, that $P\hat{E}_z = 0$ describes longitudinal plasma oscillations along the magnetic field lines. The remaining set of equations for transverse polarization can be written in rotating coordinates as

$$(S - D - N^2)\hat{E}^+ + (S + D - N^2)\hat{E}^- = 0. \qquad (2.68)$$

Here $\hat{E}^+$ is a left-hand circularly polarized wave (L-wave) and requires $S - D - N^2 = 0$ while $\hat{E}^-$ is right-hand circularly polarized (R-wave) and has a different dispersion described by $S + D - N^2 = 0$. Using the Stix-parameters, we find for the refractive index of the R-wave and L-wave

$$N_R = \left[1 - \frac{\omega_{pe}^2}{\omega(\omega - \omega_{ce})} - \frac{\omega_{pi}^2}{\omega(\omega + \omega_{ci})}\right]^{1/2}$$

$$N_L = \left[1 - \frac{\omega_{pe}^2}{\omega(\omega + \omega_{ce})} - \frac{\omega_{pi}^2}{\omega(\omega - \omega_{ci})}\right]^{1/2} . \qquad (2.69)$$

2.4.2 Cut-Offs and Resonances

The refractive index for the R-wave has a resonance $N^2 \to \infty$ at the electron cyclotron frequency ω_{ce}. Here the electric field vector and the electron have the same sense of rotation angular velocity about the magnetic field. The L-wave has a corresponding resonance at the ion cyclotron frequency ω_{ci} (Fig. 2.9).

In the high density limit $\omega_{pe}^2 \gg \omega_{ce}^2$, which is typically realized in the ionospheric F-layer and in many laboratory plasmas with weak magnetic fields, there is only one mode propagating between the ion and electron cyclotron frequency. This regime is called the whistler mode.

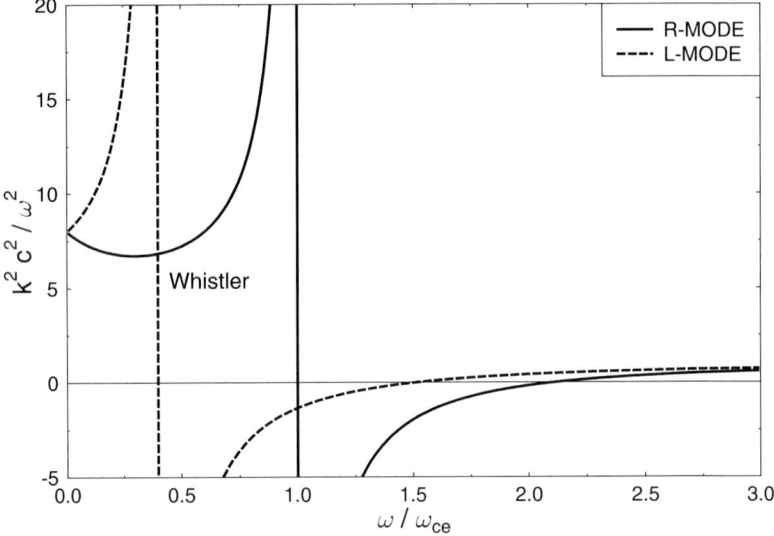

Fig. 2.9. The square of the refractive index for wave propagation along the magnetic field as a function of frequency. For clarity, an artificial mass ratio $m_e/m_i = 1/4$ is chosen. The R-mode has a resonance, $N^2 \to \infty$, at the electron cyclotron frequency whereas the L-wave shows a resonance at (*the lower*) ion cyclotron frequency. In the high density limit $\omega_{pe}^2 \gg \omega_{ce}^2$ considered here, only the R-wave is propagating between ion and electron cyclotron frequency while the L-wave is in the cut-off, $N^2 < 0$

Just above their respective cyclotron frequencies, the L-wave and the R-wave are in a cut-off band, $N^2 < 0$, until they reach a propagating band beyond the cut-off frequency. At the highest frequencies, both refractive indices approach $N = 1$. The small difference between N_R and N_L at high frequencies gives rise to the Faraday effect, namely the plane of polarization of a linearly polarized electromagnetic wave that propagates along the magnetic field line is rotated about the field.

Electron Cyclotron Emission

In a tokamak (see Chap. 7 in Part II on Magnetic Confinement) the toroidal magnetic field falls monotonically as $B \propto 1/R$ with the major radius R. Therefore, plasma radiation from gyrating electrons at the electron cyclotron frequency $\omega_{ce} = eB(r)/m_e$ can be remapped to a position in the plasma. This is the fundamental idea behind electron cyclotron emission spectroscopy (ECE). Closer inspection of the radiation from a gyrating electron [27] shows that the spectrum consists of a series of discrete harmonics

$$\omega_m = \frac{m\,\omega_{ce}}{\gamma(1 - \beta_z \cos\theta)} \,. \tag{2.70}$$

Here, $\gamma = (1-v^2/c^2)^{1/2}$ describes the relativistic mass increase, θ is the angle between the direction of observation and the magnetic field and $\beta_z = v_z/c$. The fundamental wave $m = 1$ is in most cases affected by cut-offs for the X-wave (see Fig. 2.11). Usually the second harmonic at $\omega = 2\omega_{ce}$ is monitored from the low-magnetic-field side of the torus. This wave propagates as X-mode and is in most cases optically thick in the line center. This means that the radiation at $2\omega_{ce}$ has reached the Planck curve for blackbody radiation. Hence, the radiation intensity is a measure of the local electron temperature and the ECE-spectrum can be used to recover the electron temperature profile. This diagnostic method is presently used at most magnetic fusion devices. For example, modern multichannel instruments were used on magnetic fusion devices such as Alcator C-Mod [28], W7-AS [29] and DIII-D [30]. Two dimensional resolution is achieved with an imaging system [31, 32], which is used in TEXT-U and RTP and was also recently installed in TEXTOR [33]. ECE emission spectroscopy has also been applied to the heliotron device LHD [34].

Faraday Rotation

A linearly polarized transverse wave propagating along the magnetic field can be decomposed into a pair of R-wave and L-wave. Using circular coordinates we find for the electric field components $\mathbf{E}_\pm$ [27]

$$\mathbf{E}_\pm = \frac{\hat{E}}{2} \begin{pmatrix} \exp[ik_\pm z] \\ \exp[i(k_\pm z \mp \frac{\pi}{2})] \end{pmatrix} \tag{2.71}$$

and the electric field pattern is then

$$\mathbf{E}(z) = \mathbf{E}_+(z) + \mathbf{E}_-(z) = \hat{E}\exp(i\bar{k}z)\begin{pmatrix}\cos(\delta k\,z)\\ \sin(\delta k\,z)\end{pmatrix}. \tag{2.72}$$

Here $\bar{k} = (k_+ + k_-)/2$ and $\delta k = (k_+ - k_-)/2$. The plane of polarization, which is aligned with the x-axis at $z = 0$ is obviously rotating at a rate $\alpha(z) = \delta k z$ about the magnetic field direction. This is the Faraday effect in a medium with circular bi-refringence. In the high frequency limit $\omega \gg \omega_{pe}, \omega_{ce}$ we have $N_\pm \approx 1 - \omega_{pe}^2/2\omega(\omega \pm \omega_{ce})$. This gives a rotation of the plane of polarization

$$\alpha(L) \approx \frac{\omega_{pe}^2 \omega_{ce} L}{2c\omega^2}, \tag{2.73}$$

which is proportional to the product of the plasma density ($\propto \omega_{pe}^2$), magnetic field ($\propto \omega_{ce}$) and path length L. In an inhomogeneous medium, the local product of density and magnetic field has to be integrated along the ray path.

Faraday rotation is a standard technique to study galactic magnetic fields (e.g. [35]). The magnetic structure of the solar corona was investigated with back-illumination by a satellite-borne transmitter [36] or by polarized radiation from natural radio-sources [37]. In the ionospheric plasma, modern techniques comprise Faraday rotation imaging with multiple satellites [38] or polarization analysis of coherent backscatter of radar signals [39].

In fusion devices, polarimetry with many ray paths, after its demonstration in the TEXTOR device [40], is now a well established method, which is capable of measuring the poloidal component of the magnetic field. From the sensitivity point of view, long wavelengths in the far infrared are preferred [41]. These wavelengths were also applied in a reversed field pinch [42] or in the Compact Helical System (CHS) [43]. Presently, the shorter wavelength of the CO_2 laser is used, e.g. on the Large Helical Device (LHD) [44]. The faster detector response at this wavelength even allows studying magnetic fluctuations [45]. Two-wavelength CO_2 lasers (9.27 µm and 10.6 µm) were used on the JT-60U tokamak [46] as a feasibility study for ITER. Dense plasmas, like the MAGPIE z-pinch [47] require frequency doubled Nd:YAG radiation at 532 nm. Polarimetry is also applied to laser produced plasmas, where the stimulated Brillouin backscatter signal demonstrates the self-generated magnetic fields in the plasma [48]. At high intensity, a circular polarized wave can even generate magnetic fields through the inverse Faraday effect [49].

Whistler Waves

In most parts of the ionosphere and plasmasphere the R-wave is the only propagating mode, because $\omega_{pe}^2 \gg \omega_{ce}^2$. A lightning event, e.g., on the southern hemisphere triggers a wave pulse, which is dispersed into a low frequency

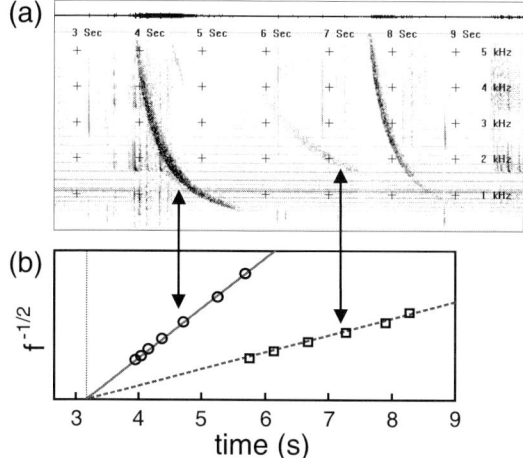

Fig. 2.10. (a) Sonogram of whistler wave events. The strong whistler starting at 4s is followed by a weak echo of much larger dispersion. (b) The evaluation shows that both signals follow a $f^{-1/2}$ law, which extrapolates to a common starting point. The example is event c09m04 from Stephen Mc Greevy's recordings [51]

($\omega^2 \ll \omega_{ce}^2$) wave train while it propagates along a magnetic field line according to the refractive index $N_R \approx \omega_{pe}/\sqrt{\omega\omega_{ce}}$. The group delay time for these low-frequency wave packets becomes $T_g \propto \omega^{-1/2}$. For an observer in the northern hemisphere this gives an electric wave field in the audible range with a slowly decaying pitch (see Fig. 2.10), which explains the name "Whistler" wave for this phenomenon [50]. The analysis of the sonogram shows that the decaying tone follows the $t \propto \omega^{-1/2}$ law and that the subsequent weaker "echo" has the same origin and, because its time scale is three times longer, represents a signal that has bounced three times between southern and northern hemisphere.

2.4.3 Propagation Across the Magnetic Field

In this case the propagation vector forms an angle $\psi = \pi/2$ with the magnetic field and we have two different choices for the polarization of the wave. The ordinary or O-mode has the electric field vector along $\mathbf{B}_0$, whereas the extraordinary or X-mode has $\mathbf{E} \perp \mathbf{B}_0$. The refractive index for the O-mode is not affected by the magnetic field, because the ion and electron motion is purely along the magnetic field, and is given by (2.26). This is why the mode is called ordinary.

For the X-mode we allow for two E-field components in the plane perpendicular to $\mathbf{B}_0$ and obtain the wave equation in the form

$$\begin{pmatrix} S & -iD \\ iD & S - N^2 \end{pmatrix} \cdot \begin{pmatrix} \hat{E}_x \\ \hat{E}_y \end{pmatrix} = 0 \,. \tag{2.74}$$

The vanishing of the determinant of this matrix yields the refractive index for the X-mode

$$N_X = \left(\frac{S^2 - D^2}{S}\right)^{1/2}. \tag{2.75}$$

This mode has resonances when $S = 0$. For high frequencies $\omega > \omega_{ce}$ we neglect the ion contributions in S and obtain $S \approx 1 - \omega_{pe}^2/(\omega^2 - \omega_{ce}^2)$. The root of this expression yields the *upper hybrid resonance* frequency

$$\omega_{uh} = (\omega_{ce}^2 + \omega_{pe}^2)^{1/2}. \tag{2.76}$$

This normal mode of the plasma invokes only electron properties, namely the coupling with the magnetic field and the electrostatic coupling with the immobile ion background.

There is a second root of S at intermediate frequencies, which we can estimate in the limiting case $\omega^2 \ll \omega_{ce}^2$ from

$$S \approx 1 + \frac{\omega_{pe}^2}{\omega_{ce}^2} - \frac{\omega_{pi}^2}{\omega^2 - \omega_{ci}^2}. \tag{2.77}$$

This lower root is then found at the *lower hybrid resonance* frequency

$$\omega_{lh} = \left(\omega_{ci}^2 + \frac{\omega_{pi}^2 \omega_{ce}^2}{\omega_{pe}^2 + \omega_{ce}^2}\right)^{1/2}. \tag{2.78}$$

In the high density limit $\omega_{pe}^2 \gg \omega_{ce}^2$ the lower hybrid frequency becomes the geometric mean of the cyclotron frequencies $\omega_{lh} \approx (\omega_{ce}\omega_{ci})^{1/2}$.

The square of the refractive index for the X-mode and O-mode as a function of frequency is shown in Fig. 2.11, which gives the resonances and cut-offs of these modes. For clarity, an artificial mass ratio $m_e/m_i = 1/4$ is chosen.

Ionosondes

The existence of a conducting layer in the high atmosphere, which is now called the ionosphere, was demonstrated by Marconi's transmission of radio signals across the Atlantic in 1901. Kennelly and Heaviside promoted the idea that the reflecting layer contained free electrons produced by the action of solar radiation. The altitude of this reflecting layer was measured by Breit and Tuve in 1925 with vertical sounding of radio waves, an early type of radar. The wave is reflected at an altitude, where the wave frequency equals the cut-off frequency. Nowadays, the electron density profile of the ionospheric plasma can be studied with ground-based (digital) ionosondes [52], which emit wave bursts at various frequencies and determine the height of the reflective layer for each particular frequency from the echo delay time. The example in Fig. 2.12 shows the echoes from the O-mode and X-mode

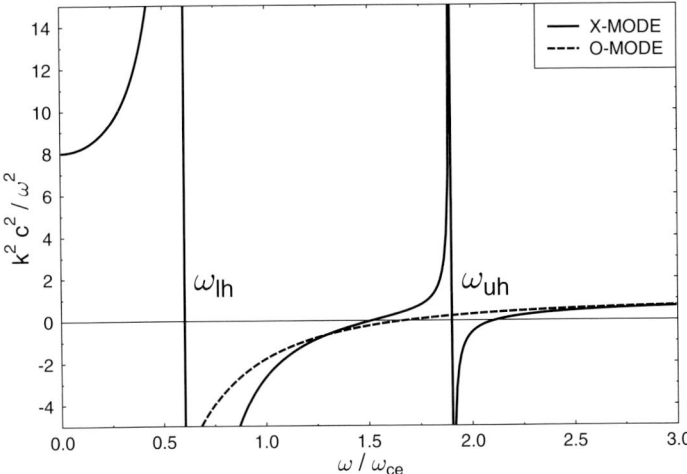

Fig. 2.11. The square of the refractive index for wave propagation perpendicular to the magnetic field as a function of frequency. An artificial mass ratio $m_e/m_i = 1/4$ is chosen. The X-mode has resonances at the lower hybrid frequency ω_{lh} and the upper hybrid frequency ω_{uh}

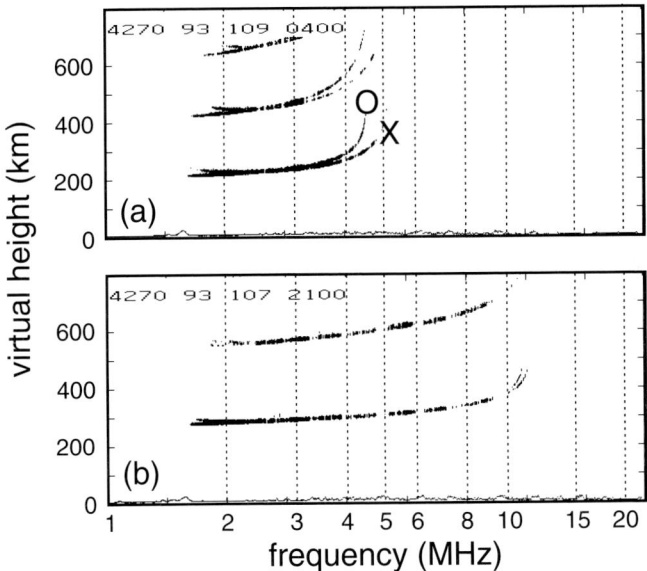

Fig. 2.12. (a) Typical night-time ionogram in the equatorial ionosphere. There appear two pass and three pass reflections between ground and ionosphere. The splitting into O-mode and X-mode is clearly visible. (b) Typical evening ionogram

as well as multiple reflections between ionosphere and ground. The examples show times, where only the F-layer is present.

The night-time ionogram (taken on day 109 of 1993 at 04:00h local time over Shriharikota, India), where the electron cyclotron frequency is $f_{ce} = 980\,\text{kHz}$, shows an O-mode cut-off at 4.5 MHz, which corresponds to an electron density of $2.5 \times 10^{11}\,\text{m}^{-3}$. The X-mode cut-off is found at 4.8 MHz while its theoretical value is 5.0 MHz. According to the higher electron density in the evening ionosphere, the O-mode and X-mode cut-offs shift to higher frequencies. The maximum electron density at 21:00h local time over the same location reaches $1.6 \times 10^{12}\,\text{m}^{-3}$.

The time delay for a particular reflection is commonly expressed in terms of the "virtual height" $h' = cT/2$, in which the refractive index of the plasma is not yet corrected. The density profile results from the reflection condition at the O-mode (X-mode) cut-off and the traversed part of the plasma up to the cut-off is used to convert the virtual height to the real height of the reflecting layer. There are analytical methods [53] and computer programs [54, 55] available for this evaluation. A survey of this technique can be found in [56]. Ionosondes have recently been applied, e.g., in the equatorial ionosphere [57] to study the "Equatorial Spread-F" phenomenon (a break-up of the F-layer at nighttime by the Rayleigh-Taylor instability).

Impedance Probes

Besides ground-based wave propagation studies of the ionosphere, in-situ measurements of plasma resonances, preferably the upper hybrid resonance, have been used on sounding rockets [58, 59, 60] and satellites [61]. For this measurement an impedance probe is used (Fig. 2.13), which consists of a sensor strip or sphere that is immersed in the ambient plasma. The sheath around the plasma has a resonance close to the upper hybrid frequency [62]. The sensor is part of a capacitance bridge and the resonance is detected in a stepped frequency sweep from 800 kHz to 12 MHz. The shift of the resonance with altitude gives the electron density profile [60].

Resonance Cones

Besides the fundamental resonances of a plasma that occur at the cyclotron frequencies for parallel wave propagation or at the hybrid frequencies for the X-mode, there are additional resonance effects for oblique wave propagation that are suitable for plasma diagnostics. For wave frequencies between the lower hybrid frequency and the electron cyclotron frequency, a lower oblique resonance is observed, which occurs at a propagation angle θ_c with respect to the magnetic field direction [63]. When the radiation field is excited with a very small antenna the resonance occurs on the surface of a cone whose apex is at the transmitting antenna. The wave field is conveniently scanned by moving a receiver antenna on a circle about the transmitter.

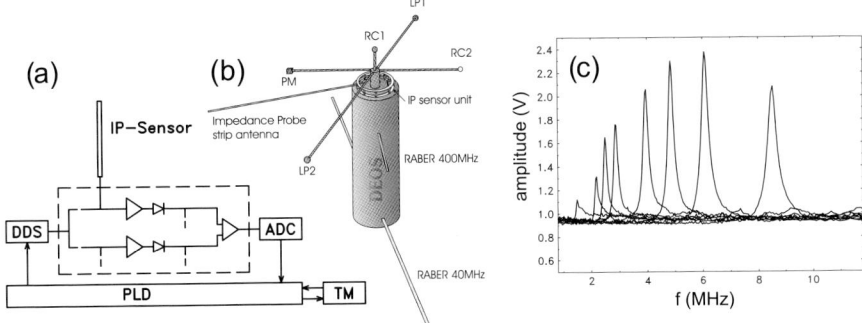

Fig. 2.13. (a) Schematic of the rocket-borne impedance probe instrument in the DEOS project [60]. The strip sensor of 1 m length is part of a capacitance bridge. The radio frequency signal is produced by a Direct Digital Synthesis (DDS) oscillator, which is controlled by a Programmable Logic Device (PLD) that also handles the Analog Digital Conversion (ADC) and telemetry (TM) interface. (b) Mechanical arrangement of the impedance probe (IP), Langmuir probes (LP), resonance cone antennas (RC) and radio beacon (RABER) on the DEOS payloads. (c) The samples show, how the resonance at the upper hybrid frequency gives a signal peak that shifts during the flight according to the ambient plasma density

The resonance angle is in a cold plasma given by [63]

$$\sin^2 \theta_c = \frac{\omega^2(\omega_{pe}^2 + \omega_{ce}^2 - \omega^2)}{\omega_{pe}^2 \omega_{ce}^2} \:. \tag{2.79}$$

An example of resonance cones in the ionospheric plasma from the COREX-campaign [64] is shown in Fig. 2.14. The cone is excited with an antenna at the tip of the rocket and the receiver antenna is on a radial boom (see Fig. 2.13). The wave field is scanned with the spin of the rocket. There are two pronounced maxima at ±15 degrees. Thermal effects are responsible for interference peaks (θ_{int}) inside the resonance cone. A simultaneously recorded set of resonance cones with rapidly interchanged role of transmitter and receiver shows a shift of the entire cone pattern caused by plasma drift.

Resonance cone diagnostics has been used in laboratory plasmas to measure electron density [63], density decay [65] or drift effects [66, 67]. Interference structures on resonance cones [68] can be used for estimating electron temperature or non-thermal effects [67]. On sounding rockets, resonance cones were used for plasma diagnostics or for detecting plasma drifts (see [69] and references therein). The method was also used aboard a satellite [70].

2.5 Concluding Remarks

This tutorial has shown that plasma waves are proven tools for the diagnostics of laboratory, ionospheric and space plasmas. The author hopes that this

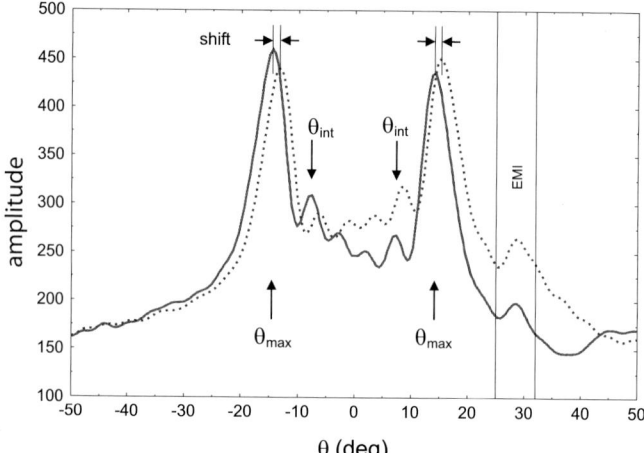

Fig. 2.14. Resonance cones from the COREX-I rocket experiment [64]. The *solid* and *dashed curve* are interlaced measurements of the resonance cone with reversing the role of transmitter and receiver antenna. The main resonance angle θ_{max} is sensitive to the electron density. The first-order interference maximum θ_{int} can be used for electron temperature estimates. The shift between the resonance cones is caused by plasma drift effects. The peak on the right hand side is an electromagnetic interference from another instrument

approach gives the reader some familiarity with the various wave types as well as access to recent applications and forms a platform for further studies of plasma instabilities, kinetic effects, nonlinear wave effects or wave heating of plasmas.

References

1. T. H. Stix: *The theory of plasma waves* (Mc Graw-Hill, New York 1962)
2. T. H. Stix: *Waves in plasmas* (AIP, New York 1992)
3. D. G. Swanson: *Plasma waves* (Academic Press, Boston 1989)
4. H. G. Booker: *Cold plasma waves* (Martinus Nijhoff, Dordrecht 1984)
5. M. Brambilla: *Kinetic theory of plasma waves* (Clarendon Press, Oxford 1998)
6. F. A. Hopf, M. Cervantes: Appl. Opt. **21**, 668 (1982)
7. N. Bretz, F. Jobes, J. Irby: Rev. Sci. Instrum. **68**, 713 (1997)
8. F. C. Jobes, N. L. Bretz: Rev. Sci. Instrum. **68**, 709 (1997)
9. M. Hirt, A. Piel, C. T. Steigies et al: Comparison of in-situ and radio beacon measurements during DEOS flight F07. In: *Proceedings of the 15th ESA Symposium on European Rocket and Balloon Programmes and Related Research*, vol ESA-SP 471, ed by B. Schürmann (European Space Agency Publications Division, Nordwijk 2001) pp 389–392
10. M. C. Kelly: *The Earth's ionosphere: Plasma physics and electrodynamics*, International Geophysics Series vol 43, (Academic Press, San Diego 1989)

11. W. Theobald, R. Häßner, C. Wülker et al: Phys. Rev. Lett. **77**, 298 (1996)
12. U. Teubner, K. Eidmann, U. Wagner et al: Phys. Rev. Lett. **92**, 185001 (2004)
13. A. Y. Wong, R. W. Motley, N. D'Angelo: Phys. Rev. **133**, A436 (1964)
14. G. Joyce, K. Lonngren, I. Alexeff et al: Phys. Fluids **12**, 2592 (1969)
15. N. E. Abt, G. Guethlein, J. W. Stearns: Phys. Fluids **25**, 2359 (1982)
16. I. Alexeff, W. D. Jones, K. E. Lonngren: Phys. Rev. Lett. **21**, 878 (1968)
17. N. Claire, G. Bachet, F. Skiff: Phys. Plasmas **9**, 4887 (2002)
18. A. Y. Wong, D. L. Mamas, D. Arnush: Phys. Fluids **18**, 1489 (1975)
19. J. L. Cooney, M. T. Gavin, K. E. Lonngren: Phys. Fluids B **3**, 2758 (1991)
20. B. Song, N. D'Angelo, R. L. Merlino: Phys. Fluids B **3**, 284 (1991)
21. Y. Nakamura, T. Odagiri, I. Tsukabayashi: Plasma Phys. Control Fusion **39** 105 (1997)
22. T. Shirakawa, H. Sugai: Jpn. J. Appl. Phys. **32**, 5129 (1993)
23. T. H. Ahn, K. Nakamura, H. Sugai: Jpn. J. Appl. Phys. **34**, L1405 (1995)
24. A. Schwabedissen, E. C. Benck, J. R. Roberts: Plasma Sources Sci. Technol. **7**, 119 (1998)
25. A. Schwabedissen, C. Soll, A. Brockhaus et al: Plasma Sources Sci. Technol. **8**, 440 (1999)
26. M. Kanoh, J. Tonotani, K. Aoki et al: Jpn. J. Appl. Phys. **41**, 3963 (2002)
27. I. H. Hutchinson: *Principles of plasma diagnostics* (Cambridge University Press, Cambridge 1987)
28. J. W. Heard, C. Watts, R. F. Gandy et al: Rev. Sci. Instrum. **70**, 1011 (1999)
29. C. Fuchs, H. J. Hartfuss: Rev. Sci. Instrum. **72**, 383 (2001)
30. M. E. Austin, J. Lohr: Rev. Sci. Instrum. **74**, 1457 (2003)
31. B. H. Deng, C. W. Domier, N. C. Luhmann jr. et al: Phys. Plasmas **8**, 2163 (2001)
32. B. H. Deng, C. W. Domier, N. C. Luhmann jr. et al: Rev. Sci. Instrum. **72**, 301 (2001)
33. B. H. Deng, C. W. Domier, N. C. Luhmann jr. et al: Rev. Sci. Instrum. **72**, 368 (2001)
34. P. C. de Vries, K. Kawahata, Y. Hagayama et al: Phys. Plasmas **7**, 3707 (2000)
35. J. A. Eilek, F. N. Owen: Astrophys. J. **567**, 202 (2002)
36. J. V. Hollweg, M. K. Bird, H. Volland et al: J. Geophys. Res. **87**, 1 (1982)
37. S. Mancuso, S. R. Spangler: Astrophys. J. **539**, 480 (2000)
38. S. Ganguly, G. H. Van Bavel, A. Brown: J. Geophys. Res. **105**, 16063 (2000)
39. D. L. Hysell, J. L. Chau: J. Geophys. Res. **106/A12**, 30371 (2001)
40. H. Soltwisch: Rev. Sci. Instrum. **57**, 1939 (1986)
41. A. J. H. Donné, T. Edlington, E. Joffrin et al: Rev. Sci. Instrum. **70**, 726 (1999)
42. D. L. Brower, Y. Jiang, W. X. Ding et al: Rev. Sci. Instrum. **72**, 1077 (2001)
43. K. Tanaka, K. Kawahata, A. Ejiri et al: Rev. Sci. Instrum. **70**, 730 (1999)
44. T. Akiyama, S. Tsuji-Iio, R. Shumada et al: Rev. Sci. Instrum. **74**, 2695 (2003)
45. W. X. Ding, D. L. Brower, S. D. Terry et al: Phys. Rev. Lett. **90**, 035002 (2003)
46. Y. Kawano, S. Chiba, H. Shirai et al: Rev. Sci. Instrum. **70**, 1430 (1999)
47. M. Tatarakis, R. Aliaga-Rossel, A. E. Dangor et al: Phys. Plasmas **5**, 682 (1998)
48. M. Khan, C. Das, B. Chakraborty et al: Phys. Rev. E **58**, 925 (1998)
49. Y. Horovitz, S. Eliezer, A. Ludmirsky et al: Phys. Rev. Lett. **78**, 1707 (1997)
50. L. R. O. Storey: Philos. Trans. R. Soc. Lond. Ser. A–Math. Phys. Eng. Sci. **246**, 113 (1953)
51. S. McGreevy: http://www-pw.physics.uiowa.edu/mcgreevy/

52. H. Rishbeth, O. K. Garriot: *Introduction to Ionospheric Physics* (Academic Press, New York 1969)
53. E. R. Schmerling: J. Atmospheric Terr. Phys. **12**, 8 (1958)
54. J. E. Titheridge: Radio Sci. **23**, 831 (1988)
55. J. E. Titheridge: Ionogram Analysis with the Generalised Program POLAN. Report UAG-93, World Data Center A for Solar-Terrestrial Physics, U.S. Dept. of Commerce, Boulder CO (1985)
56. B. W. Reinisch: *Modern ionosondes* (EGS, Katlenburg-Lindau, 1996) pp 440
57. P. E. Argo, M. C. Kelley: J. Geophys. Res. **91**, 5539 (1986)
58. H. Oya, T. Obayashi: Space Research **20**, 481 (1966)
59. E. Neske, R. Kist: Journal of Geophysics **40**, 593 (1974)
60. C. T. Steigies, D. Block, M. Hirt et al: J. Geophys. Res. **106**, 12/765 (2001)
61. T. Takahashi, H. Oya, S. Watanabe et al: J. Geomag. Geoelectr. **37**, 389 (1985)
62. S. Ullrich: *Plasmaphysikalische Grundlagen und Methoden des Einsatzes von Plasmasonden in der Raumforschung* (Akademie Verlag, Berlin 1976)
63. R. K. Fisher, R. W. Gould: Phys. Rev. Lett. **22**, 1093 (1969)
64. V. Rohde, A. Piel, H. Thiemann et al: J. Geophys. Res.**98**, 19163 (1993)
65. G. Derra, A. Piel: Plasma Phys. Control Fusion **25**, 435 (1983)
66. K. Lucks, M. Krämer: Plasma Phys. **22**, 879 (1980)
67. G. Oelerich-Hill, A. Piel: Phys. Fluids B **1**, 275 (1989)
68. R. K. Fisher, R. W. Gould: Phys. Fluids **14**, 857 (1971)
69. E. Michel, C. Béghin, A. Gonfalone et al: Ann. Géophys. **31**, 463 (1975)
70. H. C. Koons, D. C. Pridmore-Brown, D. A. McPherson: Radio Sci. **9**, 541 (1974)
71. K. H. Burrell: Phys. Fluids **18**, 897 (1975)
72. H. J. Doucet: Phys. Lett. A **33**, 283 (1970)
73. N. D'Angelo, S. von Goeler, T. Ohe: Phys. Fluids **9**, 1605 (1966)
74. M. O'Gorman, E. Zilli, L. Giudicotti et al: Rev. Sci. Instrum. **72**, 1063 (2001)
75. I. Alexeff, W. D. Jones, K. Lonngren: Phys. Fluids **12**, 345 (1969)
76. H. Oya, T. Takahashi, S. Watanabe: J. Geomag. Geoelectr. **38**, 111 (1986)
77. Y. Kawano, S. Chiba, A. Inoue: Rev. Sci. Instrum. **72**, 1068 (2001)
78. G. Breit, M. A. Tuve: Phys. Rev. **28**, 554 (1926)
79. D. L. Brower, W. X. Ding, S. D. Terry et al: Rev. Sci. Instrum. **74**, 1534 (2003)
80. C. T. Steigies, D. Block, M. Hirt et al: J. Phys. D–Appl. Phys. **33**, 405 (2000)
81. D. B. Muldrew, A. Gonfalone: Radio Sci. **9**, 873 (1974)
82. D. A. Poletti-Liuzzi, K. C. Yeh, C. H. Liu: J. Geophys. Res. **82**, 1106 (1977)
83. N. Balan, K. N. Iyer: J. Geophys. Res. **88**, 259 (1983)
84. N. Sato, T. Mieno, T. Hirata et al: Phys. Plasmas **1**, 3480 (1994)
85. K. Davies, R. B. Fritz, T. B. Gray: J. Geophys. Res. **81**, 2825 (1976)
86. A. Airoldi, G. Ramponi: Rev. Sci. Instrum. **68**, 509 (1997)
87. H. Jakowski: *Modern ionospheric science* (EGS, Katlenburg-Lindau 1996) pp. 371
88. P. Gibbon: Phys. Rev. Lett. **76**, 50 (1996)
89. P. Gibbon, D. Altenbernd, U. Teubner et al: Phys. Rev. E **55**, R6352 (1997)
90. R. Häßner, W. Theobald, S. Niedermeier et al: Opt. Lett. **22**, 1491 (1997)

3 An Introduction to Magnetohydrodynamics (MHD), or Magnetic Fluid Dynamics

B.D. Scott

Abstract. The physics of hot plasmas is based on understanding of the interdependency of magnetic and hydrodynamics properties of plasmas.

This Section gives theoretical background of MHD. It is basic to the following Chap. 4 in Part I and turbulence and transport phenomena as dealt with in Part II.

3.1 What MHD Is

Magneto-hydro-dynamics (MHD) means magnetic fluid dynamics. It is a model system designed to treat macroscopic dynamics of an electrically neutral fluid which is nevertheless made up of moving charged particles, and hence reacts to magnetic fields. Because the magnetic field is in turn produced by electric currents – here, a relative drift between the two fluids of opposite charge density which permeate each other – the resulting dynamical system is rich in nonlinear character.

This is a heuristic introduction to the ideas of fluid dynamics and MHD. Each fluid – they may be termed electrons and ions – is separately treated as a perfect fluid which reacts in a dissipation-free way to the electric and magnetic fields. By perfect fluid, the concept of thermodynamic equilibrium is implied: the effects of dissipation in all forms are taken to be negligible. These include resistivity, thermal conductivity, and viscosity. The two fluids interact only through the electric and magnetic fields they induce. These fields in turn react to changes in the distribution of sources, which are the charge densities and currents represented by the two fluids. Externally imposed fields may also be present. An infinitesimal element of each fluid is assumed to contain an arbitrarily large number of charged particles of the corresponding species, but to be small compared to the spatial scale over which macroscopic thermodynamic or field quantity varies. The picture is a meaningful one if this scale is large compared to the mean-free path between particle collisions, and the radius of gyration of each particle about magnetic field lines is negligibly small. This is the *ideal two-fluid model* of plasma dynamics.

Then, because the mass of an individual electron is so much smaller than that of an individual ion, the contribution of the electrons to fluid inertia may be neglected. With a few additional assumptions which in effect define a parameter regime, the system is treated as a single fluid which responds to a magnetic force of the Lorentz type, except that the plasma current enters instead of the fluid velocity. The magnetic field, responding to the electric field as usual, is actually advected by the velocity field, since the latter is related in a constitutive way to the electric field. This is the *ideal MHD model* of plasma dynamics, a subset of the two-fluid model.

3.2 The Ideas of Fluid Dynamics

A fluid is in essence a continuous medium. Rather than an ensemble of particles, each individually treated, *populations* of particles are treated. The system is described in terms of a velocity flow field, a density field, and a few thermodynamic state variables. In the simplest picture, the fluid is an ideal gas, in which all state variables are simple relations of the density and temperature.

3.2.1 The Density in a Changing Flow Field – Conservation of Particles

We are first interested in how the density of the fluid evolves, when the flow field is known. This involves the deformation of volumes by the flow. Consider first a volume, V, which is fixed in space. The conservation of particles follows simply from counting: the change of the number of particles, N, in V is given by the number of particles entering V, less those that leave V. Consider that V is bounded by a surface, S, an element of which is ΔS with unit normal $\hat{\mathbf{e}}$. For each area element, the number of particles which cross it in a time interval, Δt, is given by

$$\delta N = n\,\delta\mathbf{x}\,\Delta S\,, \qquad \delta\mathbf{x} = v\cos\theta\,\Delta t = \mathbf{v}\cdot\hat{\mathbf{e}}\,\Delta t\,, \qquad (3.1)$$

where $n = N/V$ is the density of particles per unit volume, θ is the angle between the velocity vector and the unit normal, and $\delta\mathbf{x}$ gives the thickness of the region containing particles which cross ΔS during Δt (a note on signs: the unit normal points outward, so a positive δN, indicating particles leaving V, contributes negatively to the change in the total number of particles, δN).

Since S is a closed surface, we may add up all the contributions from the ΔS's by performing a surface integral:

$$\delta N = -\oint_S \Delta S\, n\mathbf{v}\cdot\hat{\mathbf{e}}\,\Delta t\,. \qquad (3.2)$$

Since V is fixed, a change in N affects the density:

$$\delta N = \oint_V dV\,\delta n\,, \qquad (3.3)$$

so upon setting these two expressions equal, we have

$$\oint_V dV\, \delta n = -\oint_S d\mathbf{S}\cdot(n\mathbf{v})\,\Delta t\,, \qquad (3.4)$$

where ΔS and $\hat{\mathbf{e}}$ have been combined into $d\mathbf{S}$. Note now that since S is the surface which encloses V, we may apply the Gaussian divergence theorem:

For any volume V enclosed by surface S, with directional element $d\mathbf{S}$, any continuously differentiable vector $\mathbf{F}$ satisfies

$$\oint_V dV\, \nabla\cdot\mathbf{F} = \oint_S d\mathbf{S}\cdot\mathbf{F}\,. \qquad (3.5)$$

Therefore, we may write:

$$\oint_V dV\, \delta n = -\Delta t\oint_V dV\, \nabla\cdot n\mathbf{v}\,. \qquad (3.6)$$

This is valid for arbitrary V; specifically, it is valid for an infinitesimal volume about any point. This means the integrands must themselves be equal, since in general they do not vanish. Further taking $\Delta t \to 0$, we obtain

$$\frac{\partial n}{\partial t} = -\nabla\cdot n\mathbf{v} \qquad (3.7)$$

as a statement of conservation of particles of number density n, advected by velocity $\mathbf{v}$. It holds for a *fixed* reference frame. The partial time derivative refers to the fact that V is fixed in space.

So what is meant by divergence? Suppose now that we take the same flow and particle distribution, but now let the boundary surface elements of the volume move with the fluid. Note now that no particles enter or leave V:

$$\delta N = 0\,. \qquad (3.8)$$

The volume does, however, change with time: since the density is given by $n = N/V$, its change is inverse to that in V:

$$\frac{\delta n}{n} = -\frac{\delta V}{V}\,. \qquad (3.9)$$

At each point, the boundary element moves with the velocity:

$$\Delta\mathbf{x} = \mathbf{v}\Delta t\,. \qquad (3.10)$$

The separation of points changes according to spatial variations in the velocity:

$$\Delta\mathbf{x}_2 - \Delta\mathbf{x}_1 = (\mathbf{v}_2 - \mathbf{v}_1)\Delta t\,. \qquad (3.11)$$

A volume may be expanded in a combination of two ways: spreading or stretching. If the direction of a surface element $\mathbf{dS}$ is taken as the z-direction at some point, spreading is due to the expansion perpendicular to the element: $\partial v_x/\partial x + \partial v_y/\partial y$. Stretching is due to longitudinal variation: $\partial v_z/\partial z$. The sum of these is called the *divergence*, and the divergence of a velocity means the same thing as the expansion of the advected volume:

$$\delta V = V \nabla \cdot (\Delta \mathbf{x}), \tag{3.12}$$

or, including Δt, which commutes with spatial derivatives,

$$\frac{\delta V}{\Delta t} = V \nabla \cdot \frac{\Delta \mathbf{x}}{\Delta t} = V \nabla \cdot \mathbf{v}. \tag{3.13}$$

With this for the volume change, the density change is given by

$$\frac{\delta n}{\Delta t} = -\frac{n}{V}\frac{\delta V}{\Delta t} = -n \nabla \cdot \mathbf{v}, \tag{3.14}$$

or

$$\frac{dn}{dt} = -n \nabla \cdot \mathbf{v}, \tag{3.15}$$

as a statement of conservation of particles of number density n, advected by velocity $\mathbf{v}$. It holds for the *co-moving* reference frame. Note that the total time derivative, d/dt, gives changes in the frame moving with the fluid at each point, hence the name co-moving.

One final note: a common name for the equation for the conservation of particles is often called the *continuity equation*, since the fact that n and $\mathbf{v}$ are continuously differentiable field quantities has been implicitly used to get it.

3.2.2 The Advective Derivative and the Co-moving Reference Frame

We now have two statements of the general conservation of particles in terms of the density, n, and the velocity with which they are advected, $\mathbf{v}$. One is in the fixed reference frame:

$$\frac{\partial n}{\partial t} = -\nabla \cdot n\mathbf{v}. \tag{3.16}$$

The other is in the co-moving reference frame:

$$\frac{dn}{dt} = -n \nabla \cdot \mathbf{v}. \tag{3.17}$$

These are actually the same statement in two different coordinate systems. Let the fixed system be denoted as (t, x, y, z) and the co-moving frame as

(t', x', y', z'). The transformation between them, which is a *local* transformation, may be written:

$$dx' = dx - v_x dt, \qquad dy' = dy - v_y dt, \qquad dz' = dz - v_z dt. \qquad (3.18)$$

The time derivative transforms as

$$\begin{aligned}\frac{d}{dt} &= \frac{\partial}{\partial t} + \frac{dx}{dt}\frac{\partial}{\partial x} + \frac{dy}{dt}\frac{\partial}{\partial y} + \frac{dz}{dt}\frac{\partial}{\partial z} \\ &= \frac{\partial}{\partial t} + v_x \frac{\partial}{\partial x} + v_y \frac{\partial}{\partial y} + v_z \frac{\partial}{\partial z}\end{aligned} \qquad (3.19)$$

where the velocity components are not differentiated because the transformation is local. In general form,

$$\frac{d}{dt} = \frac{\partial}{\partial t} + \mathbf{v} \cdot \nabla. \qquad (3.20)$$

This is the time derivative in the co-moving frame, written in terms of the time derivative in the fixed frame and the local, instantaneous velocity. It is also known as the *advective derivative*, or in some texts the substantiative derivative. Its usefulness is that the laws governing the motion or constitutive change of fluid elements are often most easily formulated in the co-moving frame (or the fluid element's rest frame) and then transformed into the fixed frame. One can readily see that the two equations for the fluid density are equivalent.

3.2.3 Forces on the Fluid – How the Velocity Changes

We now allow the velocity to change, and see how it is caused to change. We examine the forces on the fluid element in the co-moving frame, and then transform to the fixed frame. We consider only fluid pressure and electromagnetic (Lorentz) forces, neglecting such others as gravitation or the effects of moving boundaries.

Consider first the fluid pressure, p. Each of the fluid particles is moving with a velocity, $\mathbf{w}$, in the co-moving frame, and the average of $\mathbf{w}$ over all the particles in a fluid element is zero. Nevertheless, the fluid exerts pressure on any surface, according to the rate at which momentum is transferred by the random particle motion. If p is prescribed, we may compute the net force, $\mathbf{F}$, on any fluid element by adding up the forces, $\Delta \mathbf{F}$, on each surface element, $\mathbf{dS}$:

$$\Delta \mathbf{F} = p\, \mathbf{dS}. \qquad (3.21)$$

Summing all the force elements over a closed surface gives an area integral:

$$\mathbf{F} = \oint_S \mathbf{dS}\, p. \qquad (3.22)$$

Since the force is exerted on the enclosed volume, V, we may apply another vector theorem:

For any volume V enclosed by surface S, with directional element $\mathbf{dS}$, any continuously differentiable scalar ζ satisfies

$$\oint_V dV\, \nabla \zeta = \oint_S \mathbf{dS}\, \zeta\,. \tag{3.23}$$

It is important to get the correct sign: the force is exerted against the surface from the surrounding fluid; hence the signs in the above equation should all be negative (except for the theorem itself). Considering a cubical fluid element, oriented such that the pressure on the right is higher than that on the left: the forces on the top and bottom are equal and opposite, the forces on the front and back are equal and opposite, but the force from the right towards the left is greater than the force from the left toward the right. There is a net force from right to left, given by the pressure difference, $\Delta p = p_{\text{right}} - p_{\text{left}}$, times the surface area, ΔS. If the coordinates are set up such that z increases towards the right, we can see that the force from right to left is *opposite* to the directional gradient in the pressure. Since the size of the element is Δz, then the force is given by

$$\mathbf{F} = -\Delta p\, \Delta S\, \hat{\mathbf{z}} = -\Delta z\, \Delta S\, (\Delta p/\Delta z)\hat{\mathbf{z}} = -\delta V\, \nabla p \rightarrow -\oint_V dV\, \nabla p\,. \tag{3.24}$$

This force acts to accelerate the fluid element. The mass in the element is given by the mass of each particle, m, times the number of particles, N, in the volume V. The element accelerates according to

$$Nm \frac{d\mathbf{v}}{dt} = \mathbf{F}\,, \tag{3.25}$$

or taking the volume to zero about a given point, the velocity change can be taken inside the integral:

$$\oint_V dV\, nm \frac{d\mathbf{v}}{dt} = -\oint_V dV\, \nabla p\,. \tag{3.26}$$

Dispensing with the volumes as before, we obtain

$$nm \frac{d\mathbf{v}}{dt} = -\nabla p\,, \tag{3.27}$$

in the co-moving frame, and

$$nm \left(\frac{\partial}{\partial t} + \mathbf{v} \cdot \nabla \right) \mathbf{v} = -\nabla p\,, \tag{3.28}$$

in the fixed frame.

This is the force law which determines how the fluid velocity field changes in reaction to gradients in its pressure. Standing alone, it is the force law for a neutral fluid.

Adding the Lorentz force is straightforward, since the force is a body force, one which acts on the element as a whole and not through the surface. The force on N particles of charge q in the fixed frame is

$$\mathbf{F} = Nq\left(\mathbf{E} + \frac{\mathbf{v}}{c} \times \mathbf{B}\right) = \oint_V dV \, nq\left(\mathbf{E} + \frac{\mathbf{v}}{c} \times \mathbf{B}\right), \quad (3.29)$$

so that the integrand is added to the neutral fluid force law to obtain the force law for a charged fluid:

$$nm\left(\frac{\partial}{\partial t} + \mathbf{v} \cdot \nabla\right)\mathbf{v} = -\nabla p + nq\left(\mathbf{E} + \frac{\mathbf{v}}{c} \times \mathbf{B}\right). \quad (3.30)$$

Note that one could just as easily derive the Lorentz force term in the co-moving frame as well, with $\mathbf{v}$ zero, and then transform to the fixed frame via the Lorentz transformation (the low-velocity form of which is the transformation used in this chapter).

We now have equations describing the evolution of the density and velocity of a fluid in general, so we could build up the two-fluid system consisting of one set of these for each species, and then close the system with Maxwell's equations for the electric and magnetic fields. This would only be complete for an isothermal system, though, so we do need to discuss how the pressure changes.

3.2.4 Thermodynamics of an Ideal Fluid – How the Temperature Changes

The simplest system is the ideal fluid: one which evolves quasi-statically through successive states of local thermodynamic equilibrium. Changes are expected to be slow enough such that equilibrium is maintained, but fast enough so that fluid elements do not exchange entropy.

In thermodynamic equilibrium with only isentropic changes considered, the first law reads:

$$\Delta E + p\,\delta V = \Delta Q \to 0. \quad (3.31)$$

E is the internal energy of a given fluid element of N particles. For the ideal gas law,

$$E = \frac{3}{2}NkT. \quad (3.32)$$

V is the volume occupied by the N particles. Over infinitesimal time intervals,

$$\frac{dE}{dt} + p\frac{dV}{dt} = 0, \quad (3.33)$$

in the co-moving frame.

We already know how the volume changes:

$$\frac{dV}{dt} = V\nabla \cdot \mathbf{v} . \tag{3.34}$$

Given that the number of particles is kept fixed,

$$\frac{3}{2}Nk\frac{dT}{dt} + pV\nabla \cdot \mathbf{v} = 0 , \tag{3.35}$$

and since the density is $n = N/V$,

$$\frac{3}{2}nk\frac{dT}{dt} + p\nabla \cdot \mathbf{v} = 0 . \tag{3.36}$$

Now, transforming to the fixed frame, we have the relation governing the change of the fluid's temperature given the flow field and all the thermodynamic state variables:

$$\frac{3}{2}nk\left(\frac{\partial T}{\partial t} + \mathbf{v} \cdot \nabla T\right) + p\nabla \cdot \mathbf{v} = 0 . \tag{3.37}$$

3.2.5 The Composite Fluid Plasma System

Under the preceding conditions (perfect fluid with an ideal gas law, no reactions that create or destroy particles of any species, maintenance of local thermodynamic equilibrium...) the system of several fluids, each made up of charged particles of species α, evolves according to

$$\frac{\partial n_\alpha}{\partial t} + \nabla \cdot n_\alpha \mathbf{v}_\alpha = 0 , \tag{3.38}$$

$$n_\alpha m_\alpha \left(\frac{\partial}{\partial t} + \mathbf{v}_\alpha \cdot \nabla\right) \mathbf{v}_\alpha = -\nabla p_\alpha + n_\alpha q_\alpha \left(\mathbf{E} + \frac{\mathbf{v}_\alpha}{c} \times \mathbf{B}\right) , \tag{3.39}$$

$$\frac{3}{2}n_\alpha k\left(\frac{\partial T_\alpha}{\partial t} + \mathbf{v}_\alpha \cdot \nabla T_\alpha\right) + p_\alpha \nabla \cdot \mathbf{v}_\alpha = 0 , \tag{3.40}$$

while the electric and magnetic fields evolve according to Maxwell's equations:

$$\nabla \cdot \mathbf{E} = 4\pi \sum_\alpha n_\alpha q_\alpha , \tag{3.41}$$

$$\frac{1}{c}\frac{\partial \mathbf{E}}{\partial t} = \nabla \times \mathbf{B} - \frac{4\pi}{c}\sum_\alpha n_\alpha q_\alpha \mathbf{v}_\alpha , \tag{3.42}$$

$$\nabla \cdot \mathbf{B} = 0 , \tag{3.43}$$

$$\frac{1}{c}\frac{\partial \mathbf{B}}{\partial t} = -\nabla \times \mathbf{E} . \tag{3.44}$$

Note that throughout this chapter cgs units are used and the charge density and the current have been specified in terms of the fluid variables:

$$\rho_{ch} = \sum_\alpha n_\alpha q_\alpha , \qquad \mathbf{J} = \sum_\alpha n_\alpha q_\alpha \mathbf{v}_\alpha , \qquad (3.45)$$

The above constitutes a closed system governing the evolution of several charged ideal fluids and the electric and magnetic fields they induce. Note that external fields may also be imposed.

3.3 From Many to One – the MHD System

In many cases, the regime of parameters in which the plasma finds itself allows considerable simplification of the multi-fluid system. The fact that the electron mass is so much smaller than that of any ion allows one to neglect the electron inertia in comparison to that of the ions, with the result that the electron force equation becomes a relation for the electric field in terms of the velocity and magnetic field.

For purposes of illustration it is useful to consider that there is only one ion species.

The following assumptions are made:

1. The *displacement current*, which is the term involving the time derivative of $\mathbf{E}$, is negligible because the time for a light wave to cross the system is much shorter than any relevant dynamical scale.
2. The electron mass is sufficiently small that parallel force balance on the electrons is maintained at all times. This allows the neglect of electron inertia. For the ideal MHD system we assume that the parallel electron dynamics are generally negligible; this means the components of both ∇p_e and $\mathbf{E}$ parallel to the magnetic field.
3. The drift velocity of the electrons relative to the ions due to the current is small compared to the ion velocity. This allows one to assume that all species move with the same velocity, when specifying velocities.
4. Pressure forces are negligible compared to Lorentz forces on all the fluids.
5. The system is approximately neutral, such that the total charge density is negligible compared to that of any constituent. Note that this does not mean that $\mathbf{E} \to 0$, but that the spatial scale of any variation is large compared to the Debye length of the plasma, and that the plasma frequency is faster than any rate of change. This means, however, that we can use the relation of zero charge density as a good approximation for the electron density in terms of the ions. For one ion species with charge Ze, this means

$$n_e \approx Z n_i .$$

These assumptions will be checked once we explore the dynamical scales of the MHD system.

3.3.1 The MHD Force Equation

The electron and ion force equations appear as

$$n_i M_i \left(\frac{\partial}{\partial t} + \mathbf{v}_i \cdot \nabla \right) \mathbf{v}_i = -\nabla p_i + n_i Z e \left(\mathbf{E} + \frac{\mathbf{v}_i}{c} \times \mathbf{B} \right), \quad (3.46)$$

$$0 = -\nabla p_e - n_e e \left(\mathbf{E} + \frac{\mathbf{v}_e}{c} \times \mathbf{B} \right), \quad (3.47)$$

Adding these and using the charge neutrality relation, we obtain

$$\rho \left(\frac{\partial}{\partial t} + \mathbf{v} \cdot \nabla \right) \mathbf{v} = -\nabla p + \frac{1}{c} J \times \mathbf{B}. \quad (3.48)$$

This is the MHD force equation. To get it, we have written p for the total pressure, $p = p_e + p_i$, and $\rho = n_i M_i$ for the mass density. We have further dropped the subscript on the ion velocity, using this as the bulk fluid velocity. The electron velocity is obtainable from $\mathbf{v}$ and $\mathbf{J}$. Note that the current is

$$\mathbf{J} = n_i Z e v_i - n_e e v_e. \quad (3.49)$$

The electrons have two roles:

1. They provide pressure, and can be dominant in doing so (but only if the temperatures are allowed to be unequal).
2. They keep the system quasi-neutral through their ability to move arbitrarily fast along the magnetic field lines.

3.3.2 Treating Several Ion Species

One can easily generalize the MHD force equation to a system of several ion species by defining the total velocity as the velocity of the *center of mass*:

$$\rho \mathbf{v} = \sum_i n_i M_i \mathbf{v}_i, \quad (3.50)$$

where the sum is over all the ion species. The mass density evolves according to the continuity equations for all ion species:

$$\frac{\partial \rho}{\partial t} + \nabla \cdot \rho \mathbf{v} = 0. \quad (3.51)$$

Provided all the temperatures are equal, this can be done for the energy equation as well:

$$C_v \left(\frac{\partial}{\partial t} + \mathbf{v} \cdot \nabla \right) T + p \nabla \cdot \mathbf{v} = 0, \quad (3.52)$$

where the specific heat at constant volume is

$$C_v = \frac{3}{2} k \sum_\alpha n_\alpha , \qquad (3.53)$$

and the sum is over all species, including the electrons. Alternatively, this may be combined with the continuity equation to reflect the fact that p/ρ^γ, with $\gamma = 5/3$, is conserved in the dynamics, even for several species (see below).

One very important note: the system can be treated as ideal either if the heat exchange among the particle populations is (1) negligible or (2) so fast that the temperatures are all kept equal. In laboratory plasmas the first of these limits is usually well-satisfied for dynamics, and the second is usually valid for quasi-static equilibria.

3.3.3 The MHD Kinematic Equation

We need now only determine how the magnetic field evolves, and since the current is given by

$$\mathbf{J} = \frac{c}{4\pi} \nabla \times \mathbf{B} , \qquad (3.54)$$

that will be enough to close the system.

The magnetic field evolves according to

$$\frac{1}{c}\frac{\partial \mathbf{B}}{\partial t} = -\nabla \times \mathbf{E} , \qquad (3.55)$$

and from the electron force equation we have

$$\mathbf{E} + \frac{\mathbf{v}_e}{c} \times \mathbf{B} = -\frac{1}{n_e e} \nabla p_e . \qquad (3.56)$$

For the dynamical scales of interest, it will become clear that the pressure gradient is negligibly small in this equation (recall we are neglecting parallel electron dynamics). This is assumption (assumption 4.) above. We have already assumed that $\mathbf{J}$ is small enough that the electron and ion velocities are equal to high accuracy (assumption 3.). This implies that the component of the fluid velocity across the magnetic field is given by the *E×B velocity*:

$$\mathbf{v} = \frac{c}{B^2} \mathbf{E} \times \mathbf{B} , \qquad (3.57)$$

which means that all the particles E × B drift together across the magnetic field lines, preventing any significant charge build-up. Substituting for $\mathbf{E}$ in the equation for $\mathbf{B}$, we obtain

$$\frac{\partial \mathbf{B}}{\partial t} = \nabla \times (\mathbf{v} \times \mathbf{B}) . \qquad (3.58)$$

This is the MHD kinematic equation. Whether or not the magnetic field has important effect on the dynamics, it describes how the magnetic field is advected by the flow velocity. Much of the character of MHD springs from this equation, especially its most important consequence: flux conservation. We will see what this is and what it means in a moment.

3.3.4 MHD at a Glance

The complete ideal MHD system is collected here for clarity.
The continuity equation:

$$\frac{\partial \rho}{\partial t} + \nabla \cdot \rho \mathbf{v} = 0 \ . \tag{3.59}$$

The MHD force equation:

$$\rho \left(\frac{\partial}{\partial t} + \mathbf{v} \cdot \nabla \right) \mathbf{v} = -\nabla p + \frac{1}{c} \mathbf{J} \times \mathbf{B} \ . \tag{3.60}$$

The adiabatic pressure equation:

$$\left(\frac{\partial}{\partial t} + \mathbf{v} \cdot \nabla \right) p + \frac{5}{3} p \nabla \cdot \mathbf{v} = 0 \ . \tag{3.61}$$

The MHD kinematic equation:

$$\frac{\partial \mathbf{B}}{\partial t} = \nabla \times (\mathbf{v} \times \mathbf{B}) \ . \tag{3.62}$$

Ampere's law:

$$\mathbf{J} = \frac{c}{4\pi} \nabla \times \mathbf{B} \ . \tag{3.63}$$

3.4 The Flux Conservation Theorem of Ideal MHD

There is an important result that arises immediately from the MHD kinematic equation. This is that the magnetic flux through any surface element advected by the fluid remains constant no matter what the flow field. Closely related is that the flux through the surface defined by any closed curve within the fluid is also conserved. The result gives rise to the concept of the magnetic flux tube.

3.4.1 Proving Flux Conservation

Consider an arbitrary, infinitesimal surface defined by a triangle of infinitesimal sides. Three points, $\mathbf{x}_0$, $\mathbf{x}_1$, $\mathbf{x}_2$, are given, and the surface element is

$$d\mathbf{S} = \frac{1}{2} (\mathbf{x}_1 - \mathbf{x}_0) \times (\mathbf{x}_2 - \mathbf{x}_1) \ . \tag{3.64}$$

The magnetic flux through the surface is given by $\mathbf{B} \cdot d\mathbf{S}$, and it changes according to

$$\frac{d}{dt} \mathbf{B} \cdot d\mathbf{S} = \frac{d\mathbf{B}}{dt} \cdot d\mathbf{S} + \mathbf{B} \cdot \frac{d}{dt} d\mathbf{S} \ . \tag{3.65}$$

Remembering that the surface and magnetic field are advected by the fluid,
$$\frac{d\mathbf{B}}{dt} = \frac{\partial \mathbf{B}}{\partial t} + \mathbf{v} \cdot \nabla \mathbf{B}. \tag{3.66}$$

Now find how the surface element changes:
$$\frac{d}{dt}\mathbf{dS} = (\mathbf{x}_1 - \mathbf{x}_0) \times \frac{1}{2}(\mathbf{v}_2 - \mathbf{v}_1) + \frac{1}{2}(\mathbf{v}_1 - \mathbf{v}_0) \times (\mathbf{x}_2 - \mathbf{x}_1), \tag{3.67}$$

noting that $d\mathbf{x}/dt = \mathbf{v}$. Reform this in terms of contributions to $\Delta \mathbf{x} \times \mathbf{v}$:

$$\begin{aligned}\frac{d}{dt}\mathbf{dS} &= -(\mathbf{x}_1 - \mathbf{x}_0) \times \frac{1}{2}(\mathbf{v}_1 + \mathbf{v}_0) - (\mathbf{x}_2 - \mathbf{x}_1) \times \frac{1}{2}(\mathbf{v}_2 + \mathbf{v}_1) \\ &\quad - (\mathbf{x}_0 - \mathbf{x}_2) \times \frac{1}{2}(\mathbf{v}_0 + \mathbf{v}_2) \end{aligned} \tag{3.68}$$

$$= -\sum_j \Delta \mathbf{x}_j \times \mathbf{v}_j, \tag{3.69}$$

where j tracks the midpoint of each line segment. Employ a vector identity to replace this expression by

$$\frac{d}{dt}\mathbf{dS} = -(\mathbf{dS} \times \nabla) \times \mathbf{v}. \tag{3.70}$$

We require to further re-form this; using the component-index representation of the right side, we have with the the totally antisymmetric Levi-Civita tensor ϵ_{ijk}

$$\begin{aligned}\left[(\mathbf{dS} \times \nabla) \times \mathbf{v}\right]_i &= \epsilon_{ipq}(\epsilon_{plm}\, \Delta S_l\, \partial_m)\, v_q \\ &= \epsilon_{pqi}(\epsilon_{plm}\, \Delta S_l\, \partial_m)\, v_q \\ &= \Delta S_q\, \partial_i\, v_q - \Delta S_i\, \partial_q\, v_q, \end{aligned} \tag{3.71}$$

which may be written in vector form as

$$\frac{d}{dt}\mathbf{dS} = \mathbf{dS}(\nabla \cdot \mathbf{v}) - (\nabla \mathbf{v}) \cdot \mathbf{dS}. \tag{3.72}$$

Inserting this back into the original expression for the flux evolution, we obtain

$$\begin{aligned}\frac{d}{dt}\mathbf{B} \cdot \mathbf{dS} &= \left(\frac{\partial \mathbf{B}}{\partial t} + \mathbf{v} \cdot \nabla \mathbf{B}\right) \cdot \mathbf{dS} + \mathbf{B} \cdot \left[\mathbf{dS}(\nabla \cdot \mathbf{v}) - (\nabla \mathbf{v}) \cdot \mathbf{dS}\right] \\ &= \mathbf{dS} \cdot \left(\frac{\partial \mathbf{B}}{\partial t} + \mathbf{v} \cdot \nabla \mathbf{B} - \mathbf{B} \cdot \nabla \mathbf{v} + \mathbf{B}\nabla \cdot \mathbf{v}\right) \\ &= \mathbf{dS} \cdot \left(\frac{\partial \mathbf{B}}{\partial t} - \nabla \times \mathbf{v} \times \mathbf{B}\right) = 0, \end{aligned} \tag{3.73}$$

since the expression in parentheses vanishes according to the MHD kinematic equation.

This proves the flux conservation theorem for an infinitesimal surface element advected by the fluid. It follows that the flux through any *surface* advected by the fluid is also conserved; simply add up all the surface elements.

3.4.2 Magnetic Flux Tubes

As a result of the fact that the magnetic flux through any surface advected by the fluid is conserved, we may find a group of field lines which serve as the boundary for a definite volume. Define a closed curve which is the boundary for a small but finite surface, the magnetic flux through which is not zero. Follow each field line an arbitrary distance away from the original curve, and define another curve which intersects the same field lines. This is a *magnetic flux tube*. Note that the magnetic flux through the sides of the tube is zero, and because the flux is conserved it stays zero.

Under advection of a flow which deforms the flux tube, the identity of the flux tube is maintained, even though the flux tube may be very greatly twisted and tangled with several other flux tubes. The consequence of this is that the field line topology is not allowed to change.

Consider two flux tubes which may be defined in a sheared magnetic field. Shear in the magnetic field may be thought of as follows: Consider a horizontal plane in which parallel lines are drawn. Now consider a plane immediately above or below the first one, in which parallel lines are also drawn. If the orientation of each set of parallel lines changes from plane to plane, then the field represented by the drawn lines is said to be *sheared*, and the shear can be quantified by giving the rate of change of this angle of orientation with perpendicular distance. Label two very narrow flux tubes, one lying in one such plane and another lying initially in a plane immediately below the first one, "a" and "b", respectively. Due to the shear, when the flux tubes are brought together by a flow field, they cross. Propose that they might pass through each other as they are forced together. Before the interaction, the magnetic flux through the sides of both tubes is zero. If they are allowed to pass through each other, the field lines due to tube "a" would intersect the sides of tube "b", and vice versa. The magnetic flux through the sides of the tubes would no longer be zero. This is in obvious contradiction to the flux conservation theorem, so the conclusion is that the flux tubes are never allowed to cross. Note that this conclusion holds as well for flux tubes of infinitesimal cross-section, and hence for individual field lines.

In ideal MHD, magnetic flux tubes and field lines cannot be advected through each other, because of the magnetic flux conservation theorem.

Below, we will explore how this constraint is relaxed by a finite plasma resistivity.

3.5 Dynamics, or the Wires-in-Molasses Picture of MHD

Note that the magnetic force term in the MHD force equation can be split into two pieces using an elementary vector identity:

$$\frac{1}{c}\mathbf{J} \times \mathbf{B} = \frac{1}{c}\left(\frac{c}{4\pi}\nabla \times \mathbf{B}\right) \times \mathbf{B} = -\nabla\frac{B^2}{8\pi} + \frac{\mathbf{B}\cdot\nabla\mathbf{B}}{4\pi} \ . \qquad (3.74)$$

These two contributions denote *magnetic pressure* and *magnetic tension*, respectively. Magnetic pressure may be combined with gas pressure:

$$\rho\left(\frac{\partial}{\partial t} + \mathbf{v}\cdot\nabla\right)\mathbf{v} = -\nabla\left(p + \frac{B^2}{8\pi}\right) + \frac{\mathbf{B}\cdot\nabla\mathbf{B}}{4\pi} \ , \qquad (3.75)$$

where one notes that it behaves like an energy density with two degrees of freedom (for which the pressure and energy density are equal).

The other contribution is magnetic tension. It is part of what gives MHD its unique character (flux conservation, or the advection of magnetic field lines is the other part). To see why it is called tension, consider the magnetic field due to a wire carrying current, surrounded by a vacuum. The field lines are described by loops centered upon the wire, and the forces are obviously zero because it is a vacuum. A cylindrical coordinate system may be defined, with z in the direction of the wire and r perpendicular to it. The only nonzero component of $\mathbf{B}\cdot\nabla\mathbf{B}$ is

$$\mathbf{B}\cdot\nabla\mathbf{B} = -\frac{B^2}{r}\hat{\mathbf{e}}_r \ . \qquad (3.76)$$

This shows that magnetic tension is a force which acts in the direction of the curvature vector (towards the loop's center), a general result for a curved field line. Since the current outside the wire vanishes, the tension and pressure forces must be in balance: since $B \sim I/r$,

$$-\nabla\frac{B^2}{8\pi} = 2\frac{B^2}{8\pi}\frac{1}{r}\hat{\mathbf{e}}_r = \frac{B^2}{4\pi r}\hat{\mathbf{e}}_r = -\mathbf{B}\cdot\nabla\mathbf{B} \ . \qquad (3.77)$$

In such a situation, which can also exist in a plasma where the current is not zero, magnetic pressure and tension are in balance, and the configuration is termed *force-free*.

3.5.1 Magnetic Pressure Waves

Consider a compression in a magnetized plasma in equilibrium, perpendicular to the magnetic field. The velocity depends only on the direction perpendicular to the field, and it is itself directed perpendicular to the field. In this situation,

$$\mathbf{B}\cdot\nabla\mathbf{B} = 0 \ , \qquad (3.78)$$

since $\mathbf{B}$ is compressed perpendicular to its direction. The field lines are compressed together with the fluid, according to the MHD kinematic equation:

$$\frac{\partial\mathbf{B}}{\partial t} = -\mathbf{v}\cdot\nabla\mathbf{B} - \mathbf{B}\nabla\cdot\mathbf{v} \ , \qquad (3.79)$$

where the third piece, $\mathbf{B} \cdot \nabla v$, vanishes due to the geometry. This equation states that $\mathbf{B}$ reacts exactly as would a density:

$$\frac{\partial \mathbf{B}}{\partial t} + \nabla \cdot \mathbf{v} \mathbf{B} = 0 \,. \tag{3.80}$$

With the vanishing magnetic tension, the force equation reads

$$\rho \left(\frac{\partial}{\partial t} + \mathbf{v} \cdot \nabla \right) \mathbf{v} = -\nabla \left(p + \frac{B^2}{8\pi} \right) \,. \tag{3.81}$$

The result is exactly analogous to sound waves, since both the pressure and magnetic field are perturbed in the same way. Consider small perturbations of short wavelength, on which scale the equilibrium pressure and magnetic field is homogeneous. We *linearize* the equations by retaining terms only to first order in the perturbations, which are denoted by a tilde symbol. For example, B^2 becomes $B^2 + 2\mathbf{B} \cdot \widetilde{\mathbf{B}}$, with the term quadratic in $\widetilde{\mathbf{B}}$ neglected. The velocity requires no symbol, since it is understood to belong to the perturbations. Both pressure and magnetic perturbations are induced by compression in the velocity:

$$\frac{1}{\gamma p} \frac{\partial \widetilde{p}}{\partial t} = \frac{1}{B} \frac{\partial \widetilde{B}}{\partial t} = -\nabla \cdot \mathbf{v} \,, \tag{3.82}$$

with $\gamma = 5/3$. The perturbed force equation reads

$$\rho \frac{\partial \mathbf{v}}{\partial t} = -\nabla \left(\widetilde{p} + \frac{B \widetilde{B}}{4\pi} \right) \,, \tag{3.83}$$

and note the factor of two arising from perturbing B^2. These may be combined into a wave equation:

$$\frac{\partial^2 \widetilde{B}}{\partial t^2} - \left(\frac{B^2}{4\pi \rho} + \frac{\gamma p}{\rho} \right) \nabla_\perp^2 \widetilde{B} = 0 \,, \tag{3.84}$$

where it is noted that the derivatives are all perpendicular to $\mathbf{B}$.

The second term in the parentheses will be recognized as the square of the adiabatic sound velocity v_s. The first term introduces the characteristic velocity of MHD in general, and the velocity of propagation of small magnetic disturbances in particular. It is the square of the *Alfvén velocity*:

$$v_A^2 = \frac{B^2}{4\pi \rho} \,, \tag{3.85}$$

after Hannes Alfvén, who is recognised as the founder of the MHD description of plasma fluid dynamics. Magnetic pressure waves, like sound waves, are longitudinal waves, but unlike sound waves they are in their pure form only when the disturbance propagates perpendicular to the magnetic field. In general, when both gas pressure and magnetic pressure are present, they both contribute to the actual wave speed: $v^2 = v_s^2 + v_A^2$. More on the ratio of gas to plasma pressure shortly.

3.5.2 Alfvén Waves: Magnetic Tension Waves

Because the magnetic field also exhibits tension, transverse waves similar to waves on a taut string also occur in MHD. Assume now that there is a divergence-free perturbation of a magnetized plasma in equilibrium, still perpendicular to the magnetic field. The velocity now depends only on the direction parallel to the field, although it is itself directed perpendicular to the field. In this situation,

$$\nabla \cdot \mathbf{v} = \widetilde{p} = \mathbf{B} \cdot \widetilde{\mathbf{B}} = 0 \;, \tag{3.86}$$

since the disturbance is a transverse-shear perturbation. The field lines are not compressed, but are bent, according to the MHD kinematic equation:

$$\frac{\partial \mathbf{B}}{\partial t} = -\mathbf{v} \cdot \nabla \mathbf{B} + \mathbf{B} \cdot \nabla \mathbf{v} \;, \tag{3.87}$$

where the divergence piece, $\mathbf{B}\nabla \cdot \mathbf{v}$, vanishes due to the geometry. Field line bending becomes clearer when this expression is re-cast in the co-moving frame:

$$\frac{\mathrm{d}\mathbf{B}}{\mathrm{d}t} = \mathbf{B} \cdot \nabla \mathbf{v} \;. \tag{3.88}$$

The single contribution arises due to the fact that the field line is moved in alternate directions according to position along it.

This is a new situation, in which the pressure is unperturbed, and the magnetic field is perturbed only through the component perpendicular to the equilibrium field. For homogeneous perturbations:

$$\frac{\partial \widetilde{\mathbf{B}}_\perp}{\partial t} = \mathbf{B} \cdot \nabla \widetilde{\mathbf{v}}_\perp \;, \tag{3.89}$$

$$\rho \frac{\partial \widetilde{\mathbf{v}}_\perp}{\partial t} = \frac{\mathbf{B} \cdot \nabla \widetilde{\mathbf{B}}_\perp}{4\pi} \;, \tag{3.90}$$

where there is no factor of two since the perturbed field component is perpendicular. These may be combined into a wave equation:

$$\frac{\partial^2 \widetilde{\mathbf{B}}_\perp}{\partial t^2} - \frac{B^2}{4\pi\rho} \frac{(\mathbf{B} \cdot \nabla)^2}{B^2} \widetilde{\mathbf{B}}_\perp = 0 \;, \tag{3.91}$$

where note now that the derivatives are all parallel to $\mathbf{B}$.

This type of disturbance propagates parallel to $\mathbf{B}$, and it is a transverse wave. It is called an *Alfvén wave*, since it is the type of propagating wave which exists in MHD but not in neutral fluid dynamics. Its propagation speed is purely the Alfvén velocity. The detection of Alfvén waves in the solar wind by spacecraft in the 1960s gave evidence that MHD phenomena do occur in nature and are not a theoretical artifice.

The concept of field line tension is now clear, since the behavior of the field line in MHD is the same as that of a taut string: "pluck it, and transverse waves run down the line". The concept of *field line bending* is closely related: curvature of a field line gives rise to the magnetic tension force. However, field line bending should not be considered as a force in itself, but a cause of one.

A general description of MHD can be that of "molasses threaded by wires". The wires are magnetic field lines, which exert force on the fluid and which are advected by the fluid as it moves. Additions to this are that the wires exert pressure as well as tension, tending to repel each other, and that in the ideal limit the fluid is not viscous.

One final note: the attractive force between two wires is due to magnetic tension overcoming magnetic pressure.

3.6 The Validity of MHD

The Alfvén velocity as just introduced determines the natural time scale of any MHD phenomena of a confined system. With this in hand, we are in a position to judge the validity of MHD by providing *a posteriori* checks on its fundamental assumptions. The Alfvén times are discussed first, and then the checks are made.

3.6.1 Characteristic Time Scales of MHD

Although they were derived for small disturbances on a homogeneous background, the wave velocities for propagation perpendicular and parallel to the magnetic field indicate the characteristic time scales for adjustment to equilibrium for general perturbations of any confined MHD system. This is much the same as the way the time it takes for a sound wave to cross a neutral fluid in hydrostatic equilibrium under gravity gives the characteristic adjustment time for that fluid. The reason is that a global free oscillation is nothing more than the longest-wavelength limit of the appropriate wave. Examples of the neutral fluid case would be an ocean layer, the Earth's atmosphere, or the Sun. In that case, for a scale length a, the sound-wave transit time is

$$\tau_s = a/c_s , \tag{3.92}$$

which is also the inverse of the frequency of the fundamental global mode of oscillation (cf. the five-minute oscillation observed on the Sun).

For a confined plasma in equilibrium the sound speed is replaced by the Alfvén velocity, v_A. It must be noted, though, that the geometry of the confined system is important, since different types of waves propagate parallel to and perpendicular to the magnetic field. With characteristic scale lengths

$L_\perp$ perpendicular to the field and $L_\parallel$ parallel to the field, two time scales are of interest. Due to the propagation of magnetic pressure waves, we have the

$$\text{\textit{fast Alfvén}, or compressional Alfvén time:} \qquad \tau_A = L_\perp/v_A \ . \qquad (3.93)$$

From the propagation of magnetic tension waves (Alfvén waves) we have the

$$\text{\textit{slow Alfvén}, or shear Alfvén time:} \qquad \tau_A = L_\parallel/v_A \ . \qquad (3.94)$$

Both of these are usually written as τ_A in the literature, and one has to extract the meaning from the context. The name *shear Alfvén time* originates from the fact that parallel length scales for many laboratory plasma instabilities arise from the existence of magnetic shear in the equilibrium configuration.

These time scales and characteristic velocities can be deduced directly from the MHD force and kinematic equations by *scaling* them. Assuming that the spatial scale is a, the kinematic equation yields:

$$\frac{\partial \mathbf{B}}{\partial t} = \nabla \times (\mathbf{v} \times \mathbf{B}) \qquad \rightarrow \qquad \frac{B}{\tau} \sim \frac{Bv}{a} \ , \qquad (3.95)$$

and the force equation yields:

$$\rho \frac{d\mathbf{v}}{dt} = -\nabla p + \frac{1}{c} \mathbf{J} \times \mathbf{B} \qquad \rightarrow \qquad \rho \frac{v}{\tau} \sim \frac{B^2}{4\pi a} \ , \qquad (3.96)$$

if the gas pressure is negligible. These similarity relations may be solved for v and τ, given B and a, and the result is $v \sim v_A$ with $\tau \sim a/v_A$.

3.6.2 Checking the Assumptions

With a as a representative spatial scale and τ_A as the corresponding time scale, we may now examine the MHD assumptions listed in Sect. 3.3, a little out of order.

1. The displacement current is neglected because $v_A \ll c$. This is easily satisfied for laboratory and space plasmas.
5. Quasi-neutrality depends on both $\tau_A \omega_{pe} \gg 1$ and $\rho_i \gg \lambda_D$. The latter three parameters in this list are the electron plasma frequency ($\omega_{pe}^2 = 4\pi n e^2/m_e$), the ion Larmor gyroradius ($\rho_i^2 = c^2 M_i T_i/e^2 B^2$), and the Debye screening length ($\lambda_D^2 = T/4\pi n e^2$). The former inequality is needed to neglect the parallel electric field, as the time scale for the electrons to equilibrate charge is ω_{pe}. The latter is needed in order to use the Lorentz force in the fluid description, since the Lorentz force has the gyroradius as its implicit length scale and the particles are not supposed to individually interact.

3. Fluid elements of all species move with velocity **v**, with the relative drift implied by **J** negligible. This requires $J \ll nev_A$, or using Ampere's law to express J in terms of B (as magnitudes), we find

$$\frac{\rho_i^2}{a^2} \equiv (\text{drift parameter})^2 \ll \text{plasma beta} = \beta \equiv \frac{8\pi p}{B^2} .$$

(The plasma beta is discussed below.) Neglecting ion inertia as a correction to the E-cross-B velocity for ions depends on the same limit, since $\tau_A \Omega_i \gg 1$ is an equivalent statement ($\Omega_i = eB/M_i c$ is the ion gyrofrequency). The statement that the drift parameter be smaller than some limit is a requirement on how strongly the plasma is magnetized, and that it must be smaller than $\beta^{1/2}$ is usually well-satisfied in any fusion or space plasma application.

2. Electron inertia is negligible because $m_e \ll M_i$ in general. It is negligible compared to the pressure gradient if $nm_e v_A/\tau \ll p/a$, which implies $\beta \gg m_e/M_i$. This has no impact on the MHD kinematic equation, since ∇p is negligible to that anyway. But this serves as a check that electron inertia is negligible even when ion inertia is very strong compared to pressure corrections, which is why a limit on β makes sense (see below).

4. That ∇p is negligible compared to the Lorentz force term is implied as well by (3), since $\mathbf{J} \times \mathbf{B}/c$ is comparable to ∇p and $J \ll nev_A$. If $\beta \ll 1$, then this is even more so.

3.6.3 A Comment on the Plasma Beta

The *plasma beta*, defined by

$$\frac{\text{gas pressure}}{\text{plasma pressure}} = \frac{8\pi p}{B^2} \equiv \beta , \qquad (3.97)$$

is very important in plasma fluid dynamics. It gives the relative importance of the gas pressure to the magnetic field as the restoring force to any disturbance. If $\beta \gg 1$, then the magnetic force has a negligible effect on the dynamics, but the magnetic field is still advected by the flow. This is called *MHD kinematics*. It is important in studies of the generation of a magnetic field by a conducting fluid undergoing motion forced by other means, such as convection in a gravitational field. The generation of a large-scale magnetic field by convective turbulence is called the *dynamo effect*, and it is thought to be the most likely scenario for the origin of planetary and stellar magnetic fields. (A note: some treatments define β with 4π instead of 8π.)

In the opposite limit, $\beta \ll 1$, the gas pressure drops out of the MHD force equation to lowest order in β, but it remains as a slight correction to the geometry of any equilibrium. For such a *low-beta* plasma, the gas pressure is still important because it can break the tendency of magnetic pressure and tension to cancel. A consequence is that a finite gas pressure prevents the establishment of a force-free equilibrium, to which the plasma tends to relax in many important configurations, even in a situation with a nonzero current density. In addition, the gas pressure can cause instability in an equilibrium which would otherwise be stable to MHD perturbations (*MHD-stable*). In the tokamak configuration for a fusion plasma, for example, this fact is responsible for limiting the plasma beta to quite low values.

The tendency of magnetic pressure and tension to cancel is ultimately the reason that assumptions (3) and (5) have to be separately checked.

3.7 Parallel Dynamics and Resistivity, or Relaxing the Ideal Assumption

One limit which was not examined above is the role of parallel dynamics: the general case of flows parallel to the magnetic field. If we compare the drift velocity implied by **J** not to v_A but to v_s, we get a different limit: $\beta \gg \rho_i/a$ (the reader is invited to check this). The drop of one power of the drift parameter places a rather strict limit on how low the beta can be allowed to go. In solar plasmas the drift parameter is really very small, so this point can be ignored. In fusion plasmas, especially tokamak and stellarator plasmas in which β is quite limited by effects arising from the gas pressure, the limit can be violated, and it is nearly always violated in the boundary regions of the plasma. It is important to realize that a perturbation compressing the gas purely parallel to the magnetic field involves no disturbance of the field. A small disturbance of this type merely leads to longitudinal sound waves propagating parallel to the field lines. Such parallel effects may be neglected for any disturbance of the general equilibrium or violent instability, since these evolve on Alfvén time scales. However, transport phenomena may give rise to force imbalances along the field lines, and these would relax on the parallel sound transit time scale. If phenomena on this scale are of global importance, then the kinematic MHD equation will be affected since although these effects are parallel, the forces they cause may have a nonzero perpendicular curl. When this is the case, studies of the consequences must treat the electron and ion fluids separately, on an equal footing. Further inquiry along these lines is beyond the scope of this introduction to MHD.

A simpler effect which breaks the ideal MHD constraints but still allows treatment of the system as MHD and as a single fluid is electrical resistivity. Resistivity means an exchange of momentum between electrons and ions as their respective fluids drift past each other. The details of the electron-ion collision process are complex, due to the fact that the angle through which an electron is scattered upon close approach to an ion depends strongly on the relative velocity. Nevertheless, by the inclusion of a simple momentum exchange term loosely based on a collision frequency, a qualitative picture of the most important consequence can be given: magnetic field line diffusion, or how those flux tubes tangled together at arbitrarily small scale eventually relax. MHD with electrical resistivity is called *resistive MHD*.

Consider a plasma with a single ion species of charge e, and allow the ions and electrons to exchange momentum. The continuity and energy equations are unaffected, but the electrons lose momentum to the ions on a time scale given by a collision frequency, ν_{ei}:

$$0 = -\nabla p_e - n_e e \left(\mathbf{E} + \frac{\mathbf{v}_e}{c} \times \mathbf{B} \right) - n_e m_e \nu_{ei} (\mathbf{v}_e - \mathbf{v}_i) , \qquad (3.98)$$

in which inertia is still neglected. To conserve momentum, the same term appears with opposite sign in the ion momentum equation:

$$n_i M_i \frac{d\mathbf{v}_i}{dt} = -\nabla p_i + n_i e \left(\mathbf{E} + \frac{\mathbf{v}_i}{c} \times \mathbf{B} \right) + n_e m_e \nu_{ei} (\mathbf{v}_e - \mathbf{v}_i) . \qquad (3.99)$$

We now add these two to obtain the resistive MHD force equation:

$$\rho \left(\frac{\partial}{\partial t} + \mathbf{v} \cdot \nabla \right) \mathbf{v} = -\nabla p + \frac{1}{c} J \times \mathbf{B} . \qquad (3.100)$$

Note that since the MHD force equation is one for *total* momentum, the addition of resistivity does not alter its form.

Now consider the MHD kinematics. Solving the electron momentum equation for $\mathbf{E}$, obtain

$$\mathbf{E} + \frac{\mathbf{v}_e}{c} \times \mathbf{B} = -\frac{\nabla p_e}{n_e e} - \frac{m_e \nu_{ei}}{n_e e^2} n_e e (\mathbf{v}_e - \mathbf{v}_i) . \qquad (3.101)$$

As before, we neglect the pressure force and assume that $\mathbf{v}_e$ is $\mathbf{v}$. Note as well that $n_e e (\mathbf{v}_e - \mathbf{v}_i) = \mathbf{J}$. (The astute reader will note the slight complications that arise when the charge state of the ions differs from $+1$; let this be left as an exercise.)

The electron force balance, modified by resistivity, now reads

$$\mathbf{E} + \frac{\mathbf{v}}{c} \times \mathbf{B} = \eta \mathbf{J} , \qquad (3.102)$$

where $\eta = m_e \nu_{ei} / n_e e^2$ is the resistivity. Inserting this relation into the MHD kinematic equation, we find

$$\frac{\partial \mathbf{B}}{\partial t} = \nabla \times (\mathbf{v} \times \mathbf{B}) - \nabla \times \frac{\eta c^2}{4\pi} \nabla \times \mathbf{B} . \qquad (3.103)$$

Assuming for the moment that η is homogeneous, the double-curl operation may be reduced, re-casting the kinematic equation as

$$\frac{\partial \mathbf{B}}{\partial t} = \nabla \times (\mathbf{v} \times \mathbf{B}) + \frac{\eta c^2}{4\pi} \nabla^2 \mathbf{B} . \qquad (3.104)$$

This has the form of a diffusion, which is the role that resistivity plays. The eventual fate of tangled magnetic flux tubes is now apparent. Supposing that

$$\frac{v_A}{a} \gg \frac{\eta c^2}{4\pi} \frac{1}{a^2}, \tag{3.105}$$

or in terms of the *Lundquist number*, S, and *resistive decay time*, τ_R,

$$S \equiv \frac{\tau_A}{\tau_R} \gg 1, \qquad \tau_R = \frac{a^2}{\eta c^2 / 4\pi}, \tag{3.106}$$

the system evolves according to ideal MHD on large scales. When flux tubes are tangled on ever-smaller scales, however, some scale, λ, is reached at which $S(a \to \lambda) \sim 1$. At that scale the magnetic field lines lose their identity through diffusion and *re-connection*, and the tangles are smoothed out.

3.8 Towards Multi-Fluid MHD

Further relaxing the assumptions of MHD brings one eventually to the necessity of treating all the constituent fluids on an equal footing, especially when further *collisional* phenomena become important. This is in any case beyond the goal of this study, which is to introduce the ideas of MHD. Interested readers will no doubt find it stimulating both to explore MHD phenomena further, and to consider in more depth the different effects one finds in two- or more-fluid dynamics. For this purpose a set of references is provided.

3.9 Further Reading

A good MHD text, both for an introduction to MHD and for further study of the basic phenomena (it is the very best book containing reconnection and MHD turbulence) is the monograph by Biskamp [1]. It cites all of the references below.

A perhaps more introductory text which is less like a review is the one by Freidberg [2].

The standard text for astrophysical MHD applications is *Cosmical Magnetic Fields* by Parker [3]. Convection of magnetic flux tubes and dynamo theory form the centerpiece of this book.

A tokamak-oriented text which includes the basic MHD problems as well as the more complicated phenomena one encounters is the one by White [4].

The best source on two-fluid dynamics is still the review article by S.I. Braginskii [5]. It contains a systematic derivation of the equations from kinetic theory as well as a clear introduction to the effects represented by the individual terms (and why they appear). One obtains a good understanding of the dependence of both the two-fluid and MHD models on the kinetic theory of plasmas, and their range of validity.

References

1. D. Biskamp: *Nonlinear Magnetohydrodynamics* (Cambridge University Press, Cambridge 1993)
2. J.P. Freidberg: *Ideal Magnetohydrodynamics* (Plenum Press, New York 1987)
3. E.N. Parker: *Cosmical Magnetic Fields* (Clarendon Press, Oxford 1979)
4. R.B. White: *Theory of Tokamak Plasmas* (NorthHolland, Amsterdam 1989)
5. S.I. Braginskii: Transport Processes in a Plasma. In:*Reviews of Plasma Physics*, vol 1, ed by M.A. Leontovich (Consultants Bureau, New York 1965) pp 205–311

4 Physics of "Hot" Plasmas

H. Zohm

Abstract. The physics of hot plasmas, i.e. nearly collisionless plasmas with finite pressure as they occur in astrophysical or fusion applications, is discussed. Description in terms of a kinetic equation, but also using velocity averaged fluid equations is introduced and examples for results obtained using this description are given. Hot plasmas are shown to be fascinating objects in which complex many-body interaction occur.

4.1 What is a Hot Plasma?

The term "hot plasma" is not unambiguously defined. However, it is commonly used for plasmas which are fully ionized, and in which the particles have a long mean free path. Also, a non-negligible kinetic pressure is often a signature of a hot plasma. More precisely, we may express the above mentioned criteria with respect to typical energy, length or time scales of the object under consideration. Two typical examples, namely the Sun and a fusion plasma in the ASDEX Upgrade tokamak experiment in Garching, Germany, are shown in Fig. 4.1.

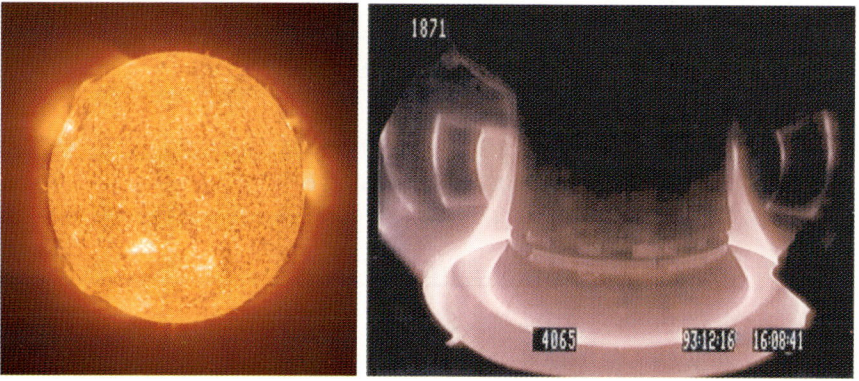

Fig. 4.1. Two examples of "hot" plasmas: the Sun ($T = 1\,\text{keV}$) and the fusion plasma in the ASDEX Upgrade tokamak ($T = 10\,\text{keV}$)

The condition for full ionization usually means that the thermal energy of the particles exceeds the ionization energy of the atoms from which the plasma is formed

$$E_{therm} = \frac{3}{2}k_B T > E_{ion} \;. \tag{4.1}$$

In the case of a pure hydrogen plasma, $E_{ion} = 13.6\,\text{eV}$, but in practical applications, we often have a small amount of light elements such as C or O in a hydrogen plasma. This implies that the temperature should really be in excess of several $100\,\text{eV}$ to fully ionize these as well. Below these temperatures, excessive power is needed to sustain such a plasma because line radiation from partially ionized C or O atoms represents a major loss channel (see also Chap. 7 in Part II and Chap. 17 in Part III).

Temperatures of the order of $1\,\text{keV}$ or above usually imply a low collisionality, i.e. the mean free path λ_{mfp} of the particles exceeds the typical system length L:

$$\lambda_{mfp} = v_{therm} \tau_{coll} \sim \sqrt{T}\, \frac{T^{3/2}}{n} = \frac{T^2}{n} > L \;. \tag{4.2}$$

For example, in a tokamak plasma used for laboratory fusion research, at $T = 10\,\text{keV}$ and $n_e = 10^{20}\,\text{m}^{-3}$, the mean free path is $\lambda_{mfp} \sim 20\,\text{km}$ and a particle passes 2000 times around the torus of typical circumference $L = 10\,\text{m}$ before it undergoes a collision.

This low collisionality also implies a high electrical conductivity, since the collisions are responsible for the resistivity of a plasma. The consequence is that the current distribution in a hot plasma can only vary slowly. This can be seen by rewriting Ampère's law

$$\nabla \times (\nabla \times \mathbf{B}) = \nabla(\nabla \cdot \mathbf{B}) - \Delta \mathbf{B} = \mu_0 \nabla \times \mathbf{j} = -\mu_0 \sigma \frac{\partial \mathbf{B}}{\partial t}\;, \tag{4.3}$$

where we have used $j = \sigma E$ and assumed that σ has no spatial variation. This equation can be read as

$$\Delta \mathbf{B} = \mu_0 \sigma \frac{\partial \mathbf{B}}{\partial t} \tag{4.4}$$

which is a diffusion equation for the magnetic field with diffusion coefficient $D = 1/(\mu_0 \sigma)$. We can estimate the typical timescale τ_R for resistive diffusion of the magnetic field in a plasma of typical dimension L as

$$\tau_R \approx \mu_0 \sigma L^2 \sim T^{3/2} \;. \tag{4.5}$$

In the ASDEX Upgrade tokamak plasma, τ_R is of the order of several seconds, which is comparable to the discharge duration. In the Sun, a value of 10^7 years is found, implying that global changes in the magnetic field are not possible on the timescale on which we observe the Sun. However, local changes take place in solar flares or other phenomena. We will return to this in Sect. 4.4 of this chapter.

Finally, we discuss the issue of finite kinetic pressure in a hot plasma. It will be shown in Sect. 4.3 that a magnetic field can balance the kinetic pressure of a plasma, thus confining it. This gives rise to the definition of the magnetic pressure $B^2/(2\mu_0)$. The significance of the kinetic pressure in a plasma can thus be quantified by the dimensionless number β, which is the ratio of kinetic and magnetic pressure

$$\beta = \frac{p}{B^2/(2\mu_0)} \,. \qquad (4.6)$$

In a magnetically confined fusion plasma, β can reach several per cent before the kinetic pressure can drive instabilities that deteriorate the confinement (see Sect. 4.4). In astrophysical plasmas, situations with $\beta \approx 1$ may also occur.

4.2 Kinetic Description of Plasmas

4.2.1 The Kinetic Equation

The question arises how a plasma can be adequately described. Clearly, there is interaction between the individual particles implying that a single-particle treatment is not enough. On the other hand, solving coupled equations of motions for all particles including their interaction is impossible. Thus, a reduction of the number of equations is needed. This is achieved by a statistical approach where we look at all alike particles of species α as an ensemble in six dimensional phase space $d^3x\, d^3v$ that is described by its distribution function $\hat{f}_\alpha(\mathbf{x}, \mathbf{v}, t)$ with the definition (see also Chap. 1 in Part I):

$$\hat{f}_\alpha(\mathbf{x}, \mathbf{v}, t)\, d^3x\, d^3v = \text{Number of particles in } d^3x\, d^3v \text{ at time } t\,. \qquad (4.7)$$

If the number of particles is conserved, this means that along the trajectories of the system the 6-dimensional equation of continuity must be fulfilled

$$\frac{d\hat{f}_\alpha}{dt} = \frac{\partial \hat{f}_\alpha}{\partial t} + \mathbf{v} \cdot \nabla \hat{f}_\alpha + \frac{q}{m}\left(\hat{\mathbf{E}} + \mathbf{v} \times \hat{\mathbf{B}}\right) \cdot \nabla_v \hat{f}_\alpha = 0\,, \qquad (4.8)$$

where ∇_v is the gradient in velocity space and we have used the equations of motion $\dot{\mathbf{x}} = \mathbf{v}$ and $\dot{\mathbf{v}} = \mathbf{F}/m$. The electromagnetic field $\hat{\mathbf{E}}, \hat{\mathbf{B}}$ is the exact field at the position of the particle and must be calculated self-consistently from Maxwell's equations using the particles as sources for current and charge distribution. To resolve this problem, we assume that the field has a macroscopic and a microscopic part, where the microscopic part dominates the interaction of the particles and the macroscopic fields are "mean fields" averaged over, e.g. a Debye sphere. We then arrive at a "smoothed" equation of continuity

$$\frac{\partial f_\alpha}{\partial t} + \mathbf{v} \cdot \nabla f_\alpha + \frac{q}{m}\left(\mathbf{E} + \mathbf{v} \times \mathbf{B}\right) \cdot \nabla_v f_\alpha = \left(\frac{\partial f_\alpha}{\partial t}\right)_{coll}\,, \qquad (4.9)$$

where the left hand side contains only averaged quantities and the so-called collision term on the right hand side contains all microscopic interactions. Thus, the problem of correctly describing the particle interaction has been transferred to the problem of formulating an adequate collision term. Equation (4.9) is also called *kinetic equation*. Depending on the collision term, it has different names. There is also a more rigorous formalism to derive the kinetic equation, the so-called BBGKY-hierarchy, but this is not treated here.

For low particle density or on timescales short compared to the collision time, as can be the case for high-frequency waves, we may set $(\partial f_\alpha/\partial t)_{coll} = 0$ and obtain the so-called *Vlasov equation*. A collision term based on the collision of hard spheres yields the so-called *Boltzmann equation*, whereas inserting Coulomb collisions in a plasma leads to the *Fokker–Planck equation*.

4.2.2 Landau Damping

As an example for the use of the kinetic equation, we consider the simple electrostatic plasma oscillation introduced before. We use the Vlasov equation in one dimension, considering the electrons only. With a perturbation ansatz $f = f_0 + f_1$ and $E = 0 + E_1$ (no electric field in equilibrium) we obtain to first order

$$\frac{\partial f_1}{\partial t} + v_x \frac{\partial f_1}{\partial x} - \frac{e}{m_e} E_1 \frac{\partial f_0}{\partial v_x} = 0 \, . \tag{4.10}$$

Next, we assume the perturbed quantities to be plane waves (Fourier decomposition), i.e.

$$E_1 = \tilde{E} \exp\left[\mathrm{i}(kx - \omega t)\right] \, , \qquad f_1 = \tilde{f} \exp\left[\mathrm{i}(kx - \omega t)\right] \tag{4.11}$$

to obtain

$$\tilde{f} = \frac{\mathrm{i}e}{m_e} \frac{1}{\omega - k v_x} \frac{\partial f_0}{\partial v_x} \tilde{E} \, . \tag{4.12}$$

The perturbed electric field follows from the Poisson equation

$$\frac{\mathrm{d}\tilde{E}}{\mathrm{d}x} = \mathrm{i}k\tilde{E} = \frac{\tilde{\rho}}{\epsilon_0} = -\frac{e}{\epsilon_0}\tilde{n} = -\frac{e}{\epsilon_0} \int \tilde{f} \, \mathrm{d}v_x \, , \tag{4.13}$$

where we have expressed the perturbed density by integrating the distribution function over the spatial coordinates. Inserting (4.12) into (4.13), we obtain

$$1 = -\frac{\omega_p^2}{n_0 k^2} \int \frac{\partial f_0/\partial v_x}{\omega/k - v_x} \, \mathrm{d}v_x \tag{4.14}$$

with $\omega_p^2 = ne^2/(m_e \epsilon_0)$. This equation relates ω and k; it is thus a dispersion relation for the electrostatic waves. The integral in (4.14) has a pole on the v_x axis and must be solved by integrating around the pole in the complex plane. Under the assumption that f_0 is a Maxwell distribution, we obtain

$$\omega = \omega_p(1 + 3k^2\lambda_D^2)^{1/2} - i\sqrt{\frac{\pi}{8}}\frac{\omega_p}{(k\lambda_D)^3}\exp\left[-\frac{1}{2}\left(\frac{1}{k^2\lambda_D^2} + 3\right)\right] . \quad (4.15)$$

In contrast to the simple relation $\omega = \omega_p$ derived before, we now have a finite dispersion and also an imaginary part. This means that the wave is also damped, which is somewhat of a paradox because we have neglected collisions, i.e. dissipation. Such a dissipationless damping is called Landau damping and is only obtained when the full kinetic description is applied.

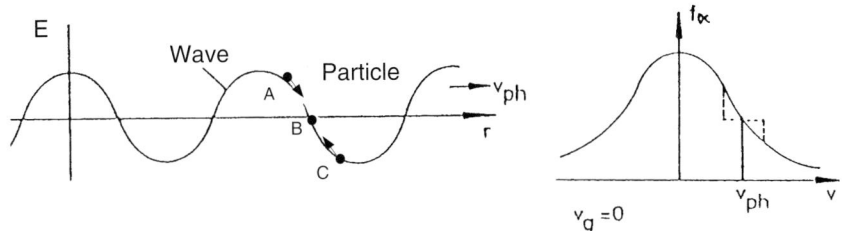

Fig. 4.2. Landau damping in a single particle picture. *Left*: Acceleration of particles with $v < v_{ph}$ (**A**) or deceleration of particles with $v > v_{ph}$ (**C**) leads to a flattening of the distribution function around $v = v_{ph}$ (see distribution function on the right). Particles with $v = v_{ph}$ do not experience any acceleration (**B**). *Right*: distribution function

A simple picture for Landau-damping is that of a surfer on a wave as shown schematically in Fig. 4.2: particles moving with the wave phase velocity have a fixed phase relation to its E-field and just travel with the wave. Particles moving somewhat slower start to lag behind the wave and see an accelerating electric field, whereas somewhat faster particles see a retarding field. Thus, the wave transfers energy to the slower particles, whereas faster particles transfer energy to the wave. The net effect is a flattening of the distribution function around v_{ph}. Damping thus occurs when there are more slower particles, i.e. $df_0/dv < 0$ as is the case in thermodynamic equilibrium (Maxwell distribution). On the other hand, if $df_0/dv > 0$, i.e. a non-thermal population of fast particles exists, waves may be excited by fast particles (see e.g. Chap. 2 in Part II). This so-called inverse Landau-damping is observed under various experimental conditions. One example is the excitation of Alfvén eigenmodes by a beam of fast particles with velocity close to the Alfvén velocity injected into a fusion plasma.

4.3 Fluid Description of Plasmas

4.3.1 The MHD Equations

In the previous Section, the kinetic equation was used to describe a plasma. A kinetic description is mandatory whenever non-thermal effects are dominant as, e.g. in the presence of a non-thermal energy source such as beam or wave heating, but also when thermalization is not effective, as, e.g. at very low density or high temperature, when the mean free path λ_{mfp} is longer than the system length. Differently to the previous chapter on MHD (Chap. 3 in Part I), where a fluid picture was employed to arrive at the MHD equations, we derive the MHD equations from moments of the distribution function.

Contrary to this, in thermodynamic equilibrium, the distribution function is close to a Maxwellian and we can characterize the plasma species α by the concept of a temperature, defined by the kinetic energy of the thermal motion

$$T_\alpha = \frac{1}{n_\alpha} \frac{m_\alpha}{3} \int \mathbf{w}_\alpha^2 f_\alpha(\mathbf{v}) \, d^3v , \qquad (4.16)$$

where the thermal motion has been defined as $\mathbf{w}_\alpha = \mathbf{v} - \mathbf{u}_\alpha$ with $\mathbf{u}_\alpha$ the center of mass velocity (or fluid velocity)

$$\mathbf{u}_\alpha = \frac{1}{n_\alpha} \int \mathbf{v} f_\alpha(\mathbf{v}) \, d^3v \qquad (4.17)$$

and the particle density

$$n_\alpha = \int f_\alpha(\mathbf{v}) \, d^3v . \qquad (4.18)$$

More generally, we can define the k-th moment of the distribution function by

$$\int \mathbf{v}^k f_\alpha(\mathbf{v}) \, d^3v \qquad \text{with} \qquad k = 0, 1, 2, 3, \ldots . \qquad (4.19)$$

These moments are functions in configuration space, but no longer contain the explicit velocity dependence. Note that by defining the temperature according to (4.16), we have assumed that the distribution function is close to a Maxwellian.

A formal way of deriving equations that describe the plasma in configurational space is thus to multiply the kinetic equation by $\mathbf{v}^k$ and integrate over velocity space. By doing this, one encounters the moments of the distribution function as introduced above.

Integration of the kinetic equation itself ($k = 0$) yields the equation of continuity

$$\frac{\partial n_\alpha}{\partial t} + \nabla(n_\alpha \mathbf{u}_\alpha) = 0 . \qquad (4.20)$$

Integration of the kinetic equation, multiplied by $m_\alpha \mathbf{v}$ (i.e. $k = 1$) yields the force balance

$$m_\alpha n_\alpha \left[\frac{\partial \mathbf{u}_\alpha}{\partial t} + (\mathbf{u}_\alpha \cdot \nabla) \mathbf{u}_\alpha \right] - q_\alpha n_\alpha (\mathbf{E} + \mathbf{u}_\alpha \times \mathbf{B}) + \nabla p_\alpha = \mathbf{R}_{\alpha\beta} , \qquad (4.21)$$

where we have introduced the scalar pressure $p_\alpha = n_\alpha T_\alpha$ (note that more generally, the pressure is a tensor). $\mathbf{R}_{\alpha\beta}$ is the friction force exerted from species β on species α. This term arises from the integration of the collision term, multiplied by $\mathbf{v}$ (collisions of alike particles do not exert a net force since thermal motion has zero center of mass velocity). Equation (4.21) just states that the acceleration of a volume element is given by the sum of all forces. In usual hydrodynamics, it is known as Euler equation. The temporal derivative is expressed in a fixed lab frame, thus changes can occur due to explicit temporal variation ($\partial/\partial t$), or by convection ($\mathbf{u}_\alpha \cdot \nabla$).

A closer inspection of the equations derived reveals that in the k-th equation, a $k+1^{st}$ moment occurs, which is only described by the $k+1^{st}$ equation. Thus, the system of equations must be closed by an additional assumption. If one wants to use the 0th and 1st equation only, the assumption has to be made about the pressure p (2nd moment), i.e. an equation of state for the plasma. This can, e.g. be adiabatic or isothermal, depending on the situation one wants to describe.

In a two-component plasma, the system of equations can be further simplified by defining one-fluid variables, i.e. the mass density ρ, the center of mass velocity $\mathbf{v}$ and the electric current $\mathbf{j}$ according to

$$\rho = m_i n_i + m_e n_e \approx m_i n \,, \tag{4.22}$$

$$\mathbf{v} = \frac{1}{\rho}(m_i n_i \mathbf{u}_i + m_e n_e \mathbf{u}_e) \approx \mathbf{u}_i \,, \tag{4.23}$$

$$\mathbf{j} = en(\mathbf{u}_i - \mathbf{u}_e) \,, \tag{4.24}$$

where we have made use of quasi-neutrality ($n_i = n_e = n$) and $m_e \ll m_i$. For a hydrogen plasma, adding and re-expressing the equations in the one-fluid variables gives the following system of equations

$$\frac{\partial \rho}{\partial t} + \nabla(\rho \mathbf{v}) = 0 \,, \tag{4.25}$$

$$\rho \left[\frac{\partial \mathbf{v}}{\partial t} + (\mathbf{v} \cdot \nabla) \mathbf{v}\right] = -\nabla p + \mathbf{j} \times \mathbf{B} \,, \tag{4.26}$$

$$\mathbf{E} + \mathbf{v} \times \mathbf{B} = \frac{1}{\sigma}\mathbf{j} \,, \tag{4.27}$$

$$\nabla \times \mathbf{B} = \mu_0 \mathbf{j} \,, \tag{4.28}$$

$$\nabla \cdot \mathbf{B} = 0 \,, \tag{4.29}$$

$$\nabla \times \mathbf{E} = -\frac{\partial \mathbf{B}}{\partial t} \,, \tag{4.30}$$

$$\frac{d}{dt}\left(\frac{p}{\rho^\gamma}\right) = 0 \,, \tag{4.31}$$

where we have neglected some terms in Ohm's law and added the low-frequency Maxwell equations as well as the adiabatic equation of state (with γ being the adiabaticity coefficient). This system of equations is often referred to as *MHD* (Magneto Hydro Dynamics). If the conductivity σ is assumed to be infinite, it is called ideal MHD. These fluid equations are valid when the distribution is close to Maxwellian, i.e. the spatial scales are large compared to the mean free path length and the temporal scales are long with respect to typical collision times. Also, for a fluid description it is required that there are many particles within a volume, i.e. the number of particles in a fluid element is much larger than one and the Larmor radius of the particles is smaller than a typical spatial scale. It turns out that in hot plasmas, the condition of a short mean free path is not simple to fulfill. In the case of magnetic confinement, it is usually not fulfilled along the field lines, whereas perpendicular to **B**, the mean free path is effectively restricted to the Larmor radius and MHD is valid.

4.3.2 Consequences of the MHD Equations

MHD Equilibrium

An application of the MHD equations is the calculation of MHD equilibrium, i.e. configurations, in which the sum of all forces is zero. For cases without flow ($\mathbf{v} = 0$), the force balance reduces to

$$\nabla p = \mathbf{j} \times \mathbf{B} \tag{4.32}$$

which states that in equilibrium, a pressure gradient is balanced by currents across the magnetic field. Using Ampère's law, this can be rearranged to yield

$$\nabla p = \frac{1}{\mu_0} (\nabla \times \mathbf{B}) \times \mathbf{B} = \frac{1}{\mu_0} \left[\mathbf{B} \nabla B - \nabla \left(\frac{B^2}{2} \right) \right]$$

$$\rightarrow \nabla_\perp \left(p + \frac{B^2}{2\mu_0} \right) + \frac{B^2}{\mu_0 R_c} \mathbf{e}_{R_c} = 0 . \tag{4.33}$$

This can be interpreted as two contributions to the force balance: the magnetic field can exert

1. a *magnetic pressure* $B^2/(2\mu_0)$ perpendicular to the field lines and,
2. if the field lines are curved, a so-called *magnetic field line tension* which provides a restoring force if a field line is bent.

R_c is the radius of curvature of the field line.

A simple example is the equilibrium in a linear configuration, as e.g. a so-called z-pinch. Here, a current flows along the z-axis, creating a poloidal field that in turn acts to confine the plasma (see Fig. 4.3).

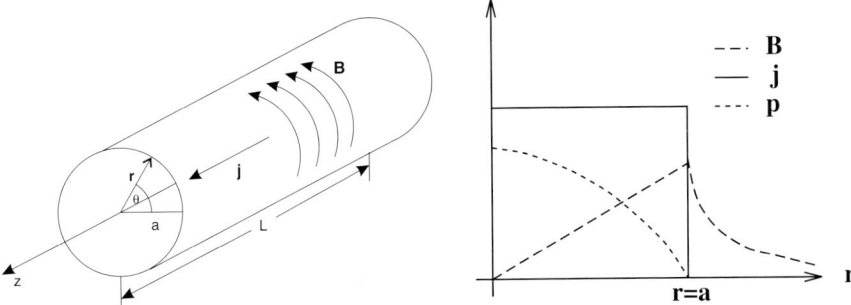

Fig. 4.3. Schematic configuration of a z-Pinch (*left*) and typical profiles (*right*)

From Ampère's law, we get:

$$\frac{1}{\mu_0 r}\frac{\mathrm{d}}{\mathrm{d}r}(rB_\theta) = j_z \tag{4.34}$$

and this can be used to rewrite the force balance

$$\frac{\mathrm{d}p}{\mathrm{d}r} = -j_z B_\theta = -\frac{1}{\mu_0 r}B_\theta \frac{\mathrm{d}}{\mathrm{d}r}(rB_\theta) = -\frac{B_\theta^2}{\mu_0 r} - \frac{\mathrm{d}}{\mathrm{d}r}\frac{B_\theta^2}{2\mu_0}. \tag{4.35}$$

We see that here, both magnetic pressure and field line tension contribute to the equilibrium (in this example, the radius of curvature is just the minor radius r). It is easy to calculate the equilibrium in the z-pinch for a constant current density $j_z = I_p/(\pi a^2)$ (where a is the radius of the plasma cylinder and I_p is the total plasma current). We obtain the profiles depicted in Fig. 4.3:

$$B_\theta(r) = \frac{\mu_0 I_p}{2\pi a^2}r \quad \text{if } r \leq a$$

$$B_\theta(r) = \frac{\mu_0 I_p}{2\pi r} \quad \text{if } r > a$$

$$p = \frac{\mu_0 I_p^2}{4\pi^2 a^2}\left[1 - \left(\frac{r}{a}\right)^2\right]. \tag{4.36}$$

Thus, knowledge of the current distribution completely determines the kinetic profile. For more sophisticated geometries, such as toroidal devices, the equilibrium must often be calculated numerically (see Chap. 7 in Part II).

Alfvén Waves

The concept of magnetic pressure and field line tension can also be used to understand special types of plasma waves that are covered within the framework of ideal MHD, the so-called *Alfvén waves*. As discussed in the

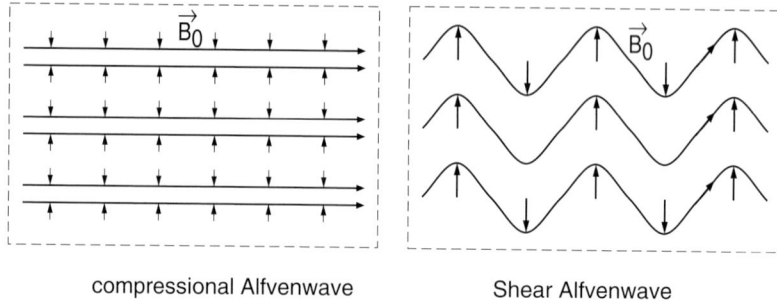

Fig. 4.4. Compressional Alfvén waves (*left*) are longitudinal waves, whereas shear Alfvén waves (*right*) are transversal

previous chapter (Chap. 3, Part I), two branches exist, namely compressional and shear Alfvén waves as shown in Fig. 4.4.

Compressional Alfvén waves are the MHD analogue to sound waves, where the restoring force is given by the magnetic rather than the kinetic pressure. Analogous to the mechanical case, the wave velocity is given by

$$c_A = \left(\gamma \frac{p}{\rho}\right)^{1/2} = \left(\frac{B^2}{\mu_0 \rho}\right)^{1/2}, \quad (4.37)$$

where we have used the magnetic pressure $B^2/(2\mu_0)$ and the adiabaticity coefficient $\gamma = 2$ for a two dimensional motion. The shear Alfvén wave shown in the right part of Fig. 4.4 is the MHD analogue to the mechanical vibration of a string; it travels along the field line with c_A due to the restoring force of field line tension.

c_A is also referred to as the *Alfvén velocity*. It defines the typical timescale of ideal MHD, the Alfvén timescale

$$\tau_A = L/c_A. \quad (4.38)$$

It is determined by the inertia of the plasma attached to the field line. For a hot plasma, it is usually much shorter than the resistive time scale (see (4.5) introduced above).

The "Frozen in" Magnetic Flux

Using the ideal Ohm's law $\mathbf{E} + \mathbf{v} \times \mathbf{B} = 0$, we can prove an important rule for the motion of plasma fluid elements. Consider a closed loop C which moves through the plasma with velocity $\mathbf{u}$. Then, the change of magnetic flux $\Psi = \int \mathbf{B}\, d\mathbf{A}$ is

$$\frac{d\Psi}{dt} = \int \partial_t \mathbf{B} d\mathbf{A} - \int \mathbf{u} \times \mathbf{B} d\ell. \quad (4.39)$$

Using the ideal Ohm's law and Stokes' theorem, we obtain

$$\frac{d\Psi}{dt} = \int (\mathbf{v} - \mathbf{u}) \times \mathbf{B} d\ell \, . \tag{4.40}$$

This means that magnetic flux is conserved if the contour moves with the fluid velocity **u** or, turning the argument around, the magnetic flux moves with the fluid velocity. This leads to the concept of a flux tube, which is a cylinder mantled by field lines as shown in the left part of Fig. 4.5. Since in ideal MHD, the flux within an individual tube cannot change, these tubes cannot intersect or overlap during their motion.

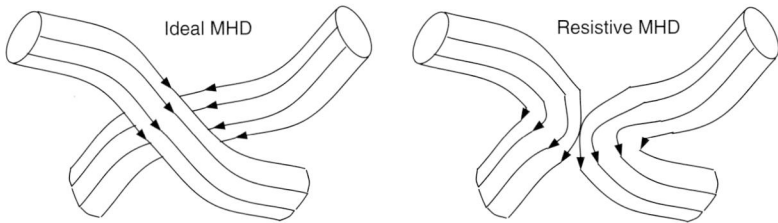

Fig. 4.5. Concept of a flux tube (*left*): in ideal MHD, flux tubes cannot intersect. With finite conductivity, *field-lines* of opposite sign can reconnect to give a new topology (*right*)

In particular, if a perfectly conducting sphere with a dipole magnetic field contracts, the field lines cannot escape and are compressed so that the field is increased. This happens when a star explodes in a supernova and contracts to a neutron star. The flux contained in the initial sphere is approximately $B_1 \pi R_1^2$ and after contraction it is $B_2 \pi R_2^2$. Thus, contraction increases the field by a factor of $(R_1/R_2)^2$. For $R_1 = 10^9$ m (star radius) and $R_2 = 10^4$ m (radius of a neutron star), the field is therefore enhanced by a factor of 10^{10}. Thus, an initial field of 10^{-5} T (comparable to the Earth's magnetic field) will increase to 10^5 T. Magnetic fields of this magnitude are actually deduced for neutron stars from the analysis of the emitted radiation, proving the effectiveness of magnetic field compression for ideally conducting materials. Following the first section of this chapter, this means that the resistive timescale τ_R is much longer than the typical time scale of the process, a condition particularly well fulfilled for the supernova explosion, lasting about 15 minutes, which has to be compared with $\tau_R \approx 10^7$ years.

If the timescale under consideration is long compared to τ_R, finite electrical conductivity becomes important and field lines may rearrange to form a new topology. This process leads to *reconnection* of field lines of opposite sign and may be the source of an instability, if the new configuration has lower free energy than the initial one. An example is shown on the right hand side of Fig. 4.5. Since for macroscopic length scales, τ_R is often very large in high temperature plasmas, this implies that reconnection happens only in a thin current sheet where the field lines of opposite sign touch and merge. It should be mentioned that nevertheless, reconnection is often observed to take place on a faster time scale than that predicted by τ_R, a fact that makes reconnection a very active research topic in plasma physics.

4.4 MHD Instabilities

An important property of a hot plasma is that it contains free energy that can potentially be released or lowered when transiting from one equilibrium state to another. In other words, hot plasmas are often subject to instabilities. By solving the force balance, we find a solution which is in equilibrium, but we cannot predict if this solution is unstable to a small perturbation. This is the goal of MHD stability analysis. Linear stability analysis assumes that the perturbation is small, linearizes the equations and assumes a temporal variation like $\exp(\gamma t)$. This typically results in an eigenvalue problem for γ^2 which determines the stability properties: negative eigenvalues correspond to oscillating, i.e. stable, solutions, whereas positive values of γ^2 result in exponentially growing, i.e. unstable, solutions.

While linear stability analysis indicates if a particular perturbation is initially unstable or not, the effect of the instability on the plasma can usually only be inferred from the nonlinear analysis, since the back effect of the perturbation on the equilibrium decides about the final stage. Usually, the perturbation acts to decrease the gradient that has driven it unstable, so that a new equilibrium state with an instability saturated at finite amplitude may exist. In the case of small scale instabilities, this state may be fully turbulent as is the case for heat and energy transport in magnetically confined fusion plasmas (see Chaps. 7, 8 and 9 in Part II). However, the effect on the plasma may be so large that it can no longer be sustained, thus leading to a termination of the discharge. Here, we will not proceed with the formal analysis, but give a discussion of the different types of large scale instabilities and then show some examples. Small scale turbulence, although a very important subject in high temperature plasma physics, will not be treated here due the limited scope of this article.

4.4.1 Classification of MHD Instabilities

Free Energies

MHD instabilities mainly come from two different sources of free energy, namely

1. the kinetic pressure of the plasma and
2. the currents flowing in the plasma.

A third mechanism that can sometimes drive MHD instabilities is the presence of a population of fast (non-thermal) particles that can excite instabilities via inverse Landau damping as outlined above.

Perhaps the most general pressure driven instability is the interchange instability. Here, one has to differ between different regions of curvature of the field lines, as shown in Fig. 4.6.

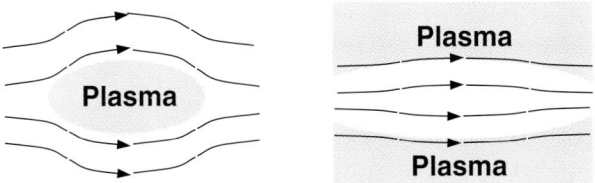

Fig. 4.6. Interchange instability in a magnetically confined plasma. *Left*: stable situation. *Right*: unstable situation

In the configuration shown in the left part of Fig. 4.6, an interchange of plasma and field lines leads to an expansion of plasma and a shortening of the field lines. Thus, the energy of both ingredients is lowered and the situation is unstable with respect to the *interchange instability*. This type of curvature is called "unfavorable" curvature. On the other hand, for the situation in the right part of Fig. 4.6, an expansion of the plasma into the central volume would lead to a lengthening of field lines and thus increase the energy in the field. This is a potentially stabilizing contribution and thus the curvature in the right hand side is called "favorable". The interchange instability limits the achievable pressure in magnetic mirror devices, where the simplest geometry is similar to that on the left hand side of Fig. 4.6. Since field line bending is a stabilizing contribution, MHD instabilities are often found to have very small wavenumbers along the magnetic field. In toroidal systems, MHD modes in fact often have constant phase along the equilibrium magnetic field, i.e. do not at all bend the equilibrium field lines. Furthermore, a tilting of neighboring field lines, also called magnetic shear, tends to suppress the interchange instability.

Concerning current driven instabilities, a gradient in current density is potentially unstable towards a redistribution (flattening) of the current. The prototype of a current gradient driven mode is the *kink instability*, which can also occur in a current carrying wire. It is illustrated in Fig. 4.7.

The kink distortion weakens B on the outside of the bend and increases it on the inside, creating a higher magnetic pressure there which tends to further bend the cylinder. Thus, a small perturbation produces a force that acts in the direction of the initial perturbation – an unstable situation. A kink can be stabilized by an axial field of sufficient strength. A mechanical

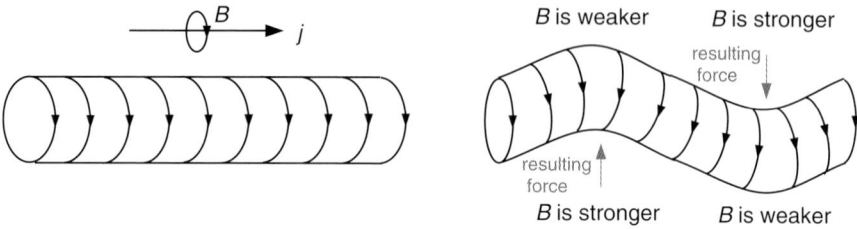

Fig. 4.7. Kink instability of a current carrying plasma cylinder (see text)

analogue to this instability is the kinking that occurs when wringing out a towel.

Topological Changes

Another classification of MHD instabilities is related to the topology change introduced by the instability. As outlined above, ideal MHD motion cannot change the topology of the flux tubes, whereas finite resistivity plasmas can change their topology by reconnection. Ideal and resistive MHD processes occur on very different time scales: for ideal MHD instabilities, the growth rate is only limited by the inertia, hence they occur on the fast Alfvén time scale τ_A, whereas for resistive instabilities, the change of magnetic flux is the limiting element and therefore they happen on the resistive time scale τ_R which, in a hot plasma, usually is much longer than τ_A. For example, in a fusion plasma, the ratio τ_R/τ_A, also referred to as *Lundquist number*, can be as large as 10^8.

Thus, there can be two variants of the same type of instability, an ideal and a resistive one. Figure 4.8 shows an example of a kink instability in a torus: In the ideal case, all flux surfaces are distorted in a similar manner, conserving the topology, whereas in the resistive case, the perturbation flips phase across the flux surface on which the helicity is equal to the perturbation's helicity and hence field lines can reconnect, forming so-called magnetic islands. The importance of these processes will be discussed in the next section.

4.4.2 Examples of MHD Instabilities

Coronal Loops and Solar Flares

A prominent example of an MHD phenomenon is the activity at the Sun's surface. On pictures taken from the Sun's surface, one can sometimes see loops of plasma standing above the surface (see Fig. 4.9). This so-called *coronal loop* is interpreted as a flux tube which has risen from a deeper region of the Sun due to the so-called magnetic buoyancy instability: a flux tube in thermal equilibrium is not in mechanical equilibrium if its magnetic field is stronger

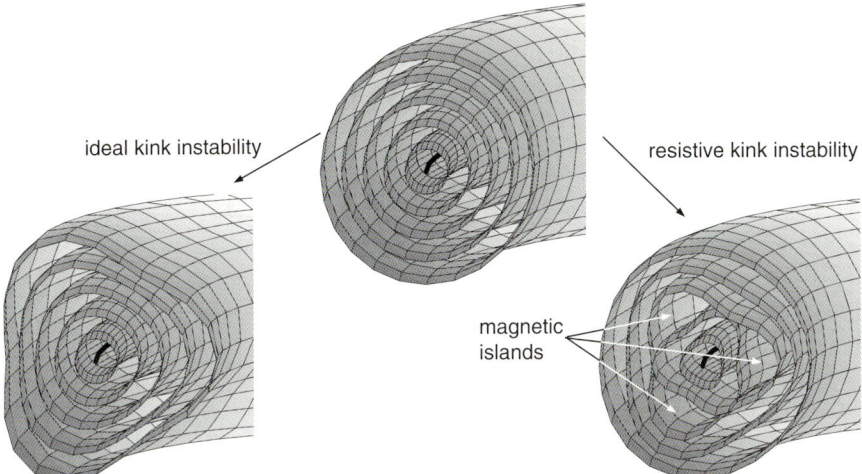

Fig. 4.8. Kink instability of a toroidal plasma: in the ideal case (*left*), the topology is conserved, whereas with finite resistivity (*right*), reconnection leads to new topological objects, so-called magnetic islands

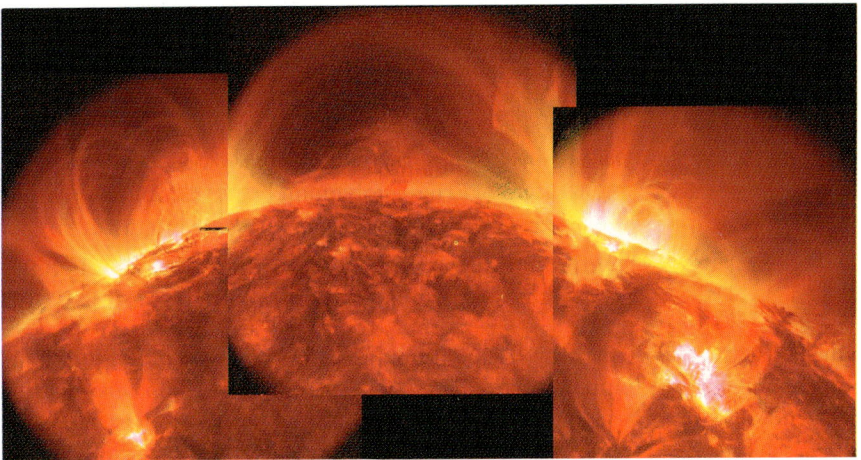

Fig. 4.9. Coronal loops on the Sun's surface – magnetic flux tubes filled with plasma have risen from deeper regions and extend above the photosphere

than that outside of the tube. The total pressure inside the tube is given by $\rho_{in}T/m + B_{in}^2/(2\mu_0)$ whereas outside, it is given by $\rho_{out}T/m + B_{out}^2/(2\mu_0)$ and thus $B_{in} > B_{out}$ requires $\rho_{in} < \rho_{out}$ for pressure equilibrium. This in turn leads to a buoyancy force $g(\rho_{in} - \rho_{out})$ on the tube. If the tube rises locally, the resulting field line bending produces a restoring force and a new equilibrium is possible. However, it can be shown that if the magnetic field strength decreases with height, the situation is unstable for a tube longer than the scale length of the magnetic field.

The buoyancy effect described before may be responsible for the formation of coronal loops, but the observation of solar flares, i.e. rapid variations in the brightness of these filaments is attributed to another effect, namely the kink instability of the flux tube when the twist exceeds a critical value and can no longer be stabilized by the line-tying boundary condition in the photosphere. In eruption events of even bigger size [coronal mass ejection (CME)], large masses of plasmas are ejected from the Sun. These events lead to a considerable variation of the intensity of the solar wind that arrives at the Earth ("space weather").

Magnetic Islands in Tokamaks

An important MHD phenomenon in magnetically confined fusion plasmas are the magnetic islands introduced in Fig. 4.8. In the tokamak configuration, a toroidal confinement device with a strong toroidal current (Chap. 7 in Part II), the radial gradient of the current density may lead to the so-called *tearing mode* instability that in turn creates magnetic islands. They are of special concern since they connect different radial regions of the plasma with field lines, thus effectively shortening out the radial gradients in these regions. This leads to flat spots in the temperature profile and reduced energy confinement. An example is shown in Fig. 4.10, where isothermals measured by Electron Cyclotron Emission (ECE) spectroscopy (Chap. 2 in Part I) are shown as function of time in a tokamak discharge. Since the plasma rotates past the observer, this effectively means a map of the toroidal co-ordinate along time. Note that above 50–100 eV, the heat conduction along the magnetic field is much higher than the perpendicular one, so that the isothermals are effectively also a visualization of the flux surfaces.

It can clearly be seen that two chains of magnetic islands have formed, leading to pronounced flattening of the temperature profile. The two chains of magnetic islands initially co-exist without any interaction and slowly grow until at $t = 1.719$ s, suddenly a heat wave is observed to travel across the inner island chain and, 1 ms later, also across the outer island chain. The completely flattened gradients indicate that the plasma has lost its thermal insulation property in this region. This can be explained by the onset of stochastization due to the interaction of the two island chains. This means that the field lines no longer map nested surfaces, but cover ergodically a region between the two islands.

A sudden thermal loss as shown in Fig. 4.10 is of great importance for the control of the plasma discharge in a tokamak, since it can lead to a rapid cooling and hence a rapid increase in resistivity. If a major loss of thermal energy occurs, the whole plasma can collapse from the 1–5 keV range to temperatures in the 10–20 eV range, meaning an increase of resistivity by more than a factor of 1000. This means that the current can no longer be sustained

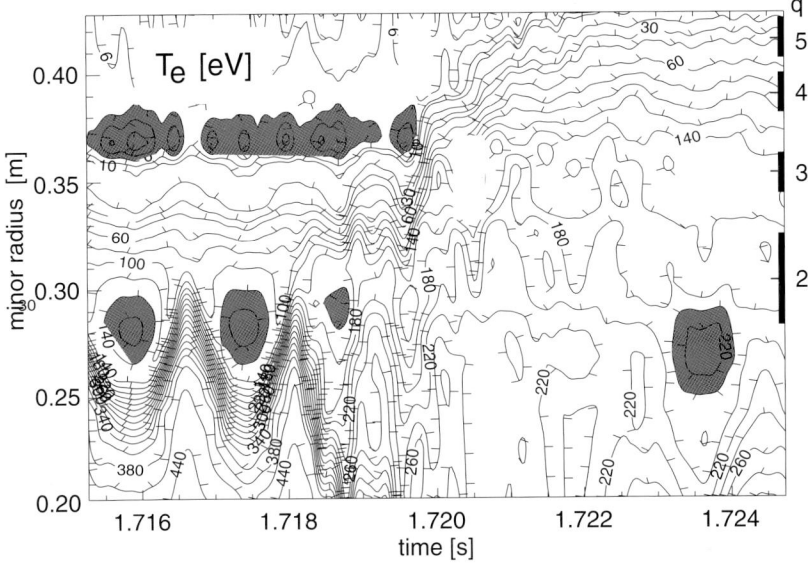

Fig. 4.10. Magnetic islands in a tokamak discharge: the flat spots in the temperature profile indicate the formation of two island chains with different helicity. At $t = 1.72$ s, the interaction of the island chains leads to stochastization, inducing a loss of global confinement as seen by the outward travelling heat wave [1]

and the plasma disrupts on the time scale of several ms. These *disruptions* can lead to large thermal and mechanical loads on the tokamak components and are therefore highly undesirable. Active research is undertaken to find out how disruptions can be avoided or ameliorated.

Nonlinear Cycles

In the previous examples, MHD instabilities have led to disruptive phenomena that destroy the original configuration and eventually terminate the plasma discharge under consideration. This is not necessarily the case; MHD instabilities may also occur in non-linear cycles that lead to a new quasi-stationary equilibrium. An example is the Edge Localised Mode (ELM) instability in fusion plasmas which rapidly expels plasma from the edge region of the plasma discharge, thereby removing the pressure and current gradients that have driven it unstable.

However, the plasma survives this loss and the gradients build up again on a transport timescale until the instability criterion is violated again. Figure 4.11 shows an example from which we can see that this nonlinear cycle can be very stable and lead to quasi steady state conditions. There are numerous other examples of these nonlinear cycles and the characterization of their dynamics is a research topic of ongoing interest.

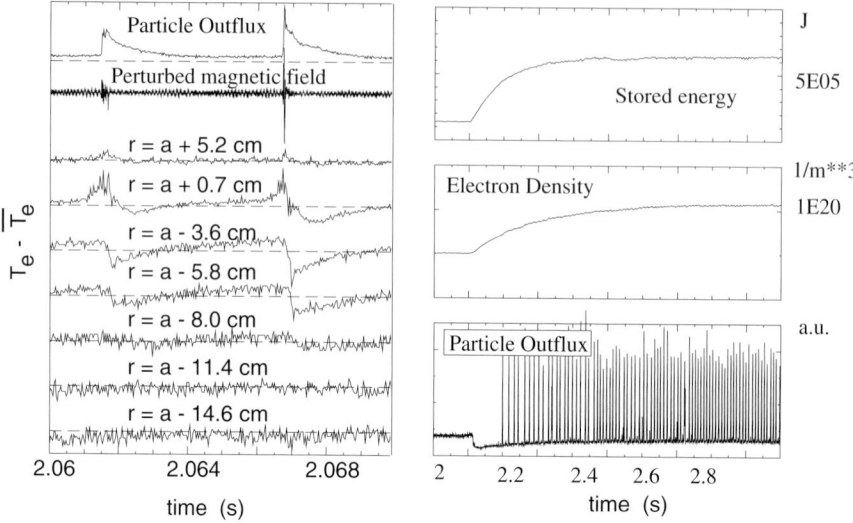

Fig. 4.11. Edge localized modes (ELMs) in a fusion plasma: this MHD instability is repetitively triggered by edge pressure and current gradients which it removes on a fast timescale (1 ms, see *left figure* where the edge temperature relative to the plasma edge at $a = 50$ cm is shown). The recovery of the gradients happens on a local transport timescale (10–100 ms). The global plasma parameters (*right part*) are hardly affected; the plasma is in quasi steady-state [2]

4.5 Summary

Hot plasmas are plasmas that are essentially fully ionized, have few collisions and non-negligible kinetic pressure. Examples are plasmas in an astrophysical context, such as in stars, or in magnetically confined fusion experiments on Earth. Due to the low collision frequency and hence the long mean free path of particles, the distribution function may deviate significantly from Maxwellian and the adequate description often requires a kinetic treatment. If these effects can be neglected, a fluid description (MHD) is adequate and many gross equilibrium and stability properties can be covered with this description.

Due to the free energy associated with the gradients of kinetic pressure and current density distribution, instabilities are a common element in hot plasmas, often leading to non-linear reorganization that makes the quantitative description of plasma properties a challenging task, often only accessible by numerical simulation. Thus, the physics of hot plasmas is a very interesting and active research field with many new discoveries yet to be made.

References

1. W. Suttrop, K. Buchl, J.C. Fuchs et al: Nucl. Fusion **37**, 119 (1997)
2. H. Zohm: Plasma Phys. Control. Fusion **38**, 105 (1996)
3. F.F. Chen: *Introduction to Plasma Physics* (Plenum Press, New York 1988)
4. A.R. Choudhuri: *The Physics of Fluids and Plasmas* (Cambridge University Press, Cambridge 1998)
5. R. Kippenhahn, C. Möllenhoff: *Elementare Plasmaphysik*, in German (BI Wissenschaftsverlag, Zürich 1975)
6. N.A. Krall, A.W. Trivelpiece: *Principles of Plasma Physics* (San Francisco Press, San Francisco 1986)
7. W.M. Stacey: *Fusion Plasma Analysis* (Wiley and Sons, New York 1981)
8. U. Schumacher: *Fusionsforschung – Eine Einführung* (Wissenschaftliche Buchgesellschaft, Darmstadt 1993)
9. R.J. Goldston, P.H. Rutherford: *Introduction to Plasma Physics* (Intitute of Physics Publishing, Bristol and Philadelphia 1995), in German (Vieweg Verlag, Braunschweig 1998)
10. M. Kaufmann: *Plasmaphysik und Fusionsforschung* (B.G. Teubner, Stuttgart 2003)
11. B.W. Caroll, D.A. Ostlie: *Modern Astrophysics* (Addison-Wesley, Reading 1996)
12. M.S. Longair: *High Energy Astrophysics* (Cambridge University Press, Cambridge 1981)
13. A.L. Perrat: *Physics of the Plasma Universe* (Springer, Berlin Heidelberg New York 1992)
14. G. Bateman: *MHD Instabilities* (The MIT Press, Cambridge MA 1978)
15. D. Biskamp: *Nonlinear MHD* (Cambridge University Press, Cambridge 1993)
16. J. Freidberg: *Ideal Magnetohydrodynamics* (Plenum Press, New York London 1987)
17. E.R. Priest: *Solar Magnetohydrodynamics* (D. Reidel, Berlin 1982)

5 Low Temperature Plasmas

J. Meichsner

Abstract. Characteristic properties and generation mechanisms of low temperature plasmas in different gas discharges are presented. A special part is focussed to the plasma surface transition including electric probes for plasma diagnostics. Reactive plasma surface interaction is exemplarily shown in polymer surface modification and deposition of thin organic films.

5.1 Introduction

In contrast to the high temperature plasmas (see Chap. 4 in Part I, e.g., fusion plasma) the classic low temperature plasmas are divided into thermal and non-thermal plasmas. The thermal plasma (e.g. arc discharge at atmospheric pressure) is characterized by the same temperature of the plasma particles, $T_e \approx T_{ion} \approx T_{gas} \approx 10^4$ K (Local Thermodynamic Equilibrium, LTE). The non-thermal plasma of electrical gas discharge (e.g. low-pressure glow discharge) is weakly ionised (degree of ionisation 10^{-6}–10^{-4}) and is characterized by a significant non-equilibrium state. The electron temperature is much higher than the ion and neutral gas temperature, $T_e \approx 10^4$ K $\gg T_{ion} \approx T_{gas} \approx 300$ K (Partial Thermodynamic Equilibrium, PTE). The reasons are the small kinetic energy transfer in elastic collisions between electrons and heavy particles and the heavy particle confinement time. Because of the low collision frequency of electrons with heavy particles at low pressure, the heavy particle temperatures remain in the order of room temperature and we observe a so-called cold plasma. In such non equilibrium plasma the electrons represent the energetic particles which have a temperature in the order of 10^4 K or mean kinetic energy of a few eV.

The physical and chemical processes in the plasma bulk and at surrounding surfaces have to be described by means of many elementary processes. For example, following processes can contribute to generation and loss of charged particles:

Volume:

Generation
(1) ionisation by electrons
(2) ionisation by ions
(3) ionisation by metastable excited neutrals (Penning-effect)
(4) associative ionisation
(5) charge transfer collision
(6) photo ionisation
(7) electron attachment

These elementary processes can be combined with electronic excitation, and in the case of molecular gases with dissociation, vibrational and rotational excitation.

Loss
(8) electron–ion recombination (two body collision)
(9) ion–ion recombination
(10) electron–ion–neutral recombination (three body collision)

Surface:

Generation
(11) secondary electron emission by impinging electrons, ions, metastable excited neutrals and fast neutrals
(12) photoemission
(13) thermal electron emission
(14) field electron emission

Loss
(15) direct absorption of electrons at metallic surface
(16) Auger-neutralisation of positive ions at metallic surface
(17) charging of insulated surfaces (leads to floating potential)
(18) recombination (surface acts as third body)

The listed elementary processes have to be extended for gas discharges in reactive, molecular gases. Many chemical reactions between neutral and charged radicals produce new chemical compounds in the gas phase, and at surfaces.

Therefore, fundamental information is necessary about collision cross sections and energy distribution functions of plasma particles for calculation of rate coefficients of the different elementary processes. The knowledge of the electron energy distribution function f (EEDF) is the fundamental task in low temperature plasma physics. The most important kinetic equation for calculation of EEDF is the Boltzmann equation (see also (1.45) in Chap. 1, Part I and [1]):

$$\frac{\mathrm{d}f}{\mathrm{d}t} = \frac{\partial f}{\partial t} + \mathbf{v} \cdot \frac{\partial f}{\partial \mathbf{r}} + \frac{\mathbf{F}}{m_e} \cdot \frac{\partial f}{\partial \mathbf{v}} = C^{el} + \sum_n C_n^{inel}, \tag{5.1}$$

where $\mathbf{F}$ denotes a general force and the collision rate C is separated into elastic and inelastic collisions.

The ionisation of atoms and molecules due to inelastic collisions with electrons is the main generation mechanism in the gas discharge volume. For a given electron energy distribution function $f(\epsilon)$ and ionisation cross section $\sigma_i(\epsilon)$ the rate of ionisation of neutrals (N) in the ground state is

$$\left[\frac{\mathrm{d}n_e}{\mathrm{d}t}\right]_i = n_e \nu_i = N n_e k_i = N n_e \langle v\sigma_i \rangle = N n_e \frac{\int f(\epsilon) v \sigma_i(\epsilon) \mathrm{d}\epsilon}{\int f(\epsilon) \mathrm{d}\epsilon} \;, \qquad (5.2)$$

where n_e is the electron density, ν_i is the ionisation frequency and k_i is the rate coefficient for ionisation.

Gas discharge plasmas are generated and maintained by the input of electrical power. The used various electric power sources includes devices running at dc, standard line frequency ($\sim$50 Hz), low frequency ($\sim$50 kHz), radio frequency (rf) ($\sim$13.56 MHz) and microwave ($\sim$2.45 GHz). In connection with the wide range of gas pressure ($10^{-1} \ldots 10^6$ Pa), nature of gas, and boundary conditions at discharge electrodes and walls the manifold phenomena of electrical gas discharge plasmas can be found. In particular, the characteristic time constants for production and loss of charged particles, as well as charge carrier transport (field drift, diffusion) determine the evolution and relaxation of the electron energy distribution function. The well known discharge types can be observed like glow, arc, spark, corona, dielectric barrier, radio frequency or microwave discharges. An overview of discharges is given in Fig. 5.1.

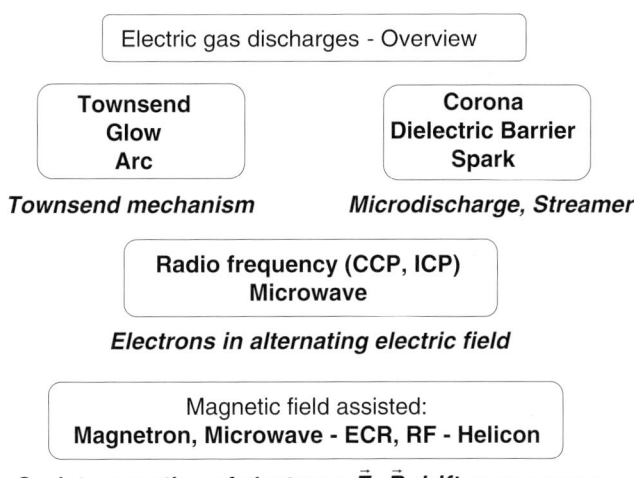

Fig. 5.1. Overview of different gas discharge types and main sustaining mechanisms

5.2 Gas Discharges and Low Temperature Plasmas: Basic Mechanisms and Characteristics

5.2.1 Classical Townsend Mechanism and Electric Breakdown in Gases

The Townsend mechanism describes the electrical breakdown in gases between a discharge gap consisting of two parallel metal plate electrodes with separation d_E and uniform electric field strength (E). Starting point is the generation of a free electron at the cathode due to an external source. These electrons drift towards the anode in the uniform electric field. In a gas at the pressure p the first Townsend coefficient α describes the number of electron–ion pairs per length unit which are produced by one electron moving to the anode (the first Townsend coefficient is related to the ionisation frequency by $\nu_i = \alpha v_d$, where v_d is the drift velocity of the electrons). The number of electrons rises on exponentially with increasing path length z of the electron.

$$N_e = 1 \exp(\alpha z) \:. \tag{5.3}$$

The new electrons are accelerated in the electric field and they produce new electron–ion pairs again and so on (charge carrier multiplication).

The total production of electron ion pairs depends on electrode separation and the total gas pressure. In this way the expression α/p represents the pressure reduced number of ionising collisions of an electron per length unit. The mean kinetic energy gain of the electron within the mean free path length of the electron can be expressed by the reduced electric field strength E/p.

$$W_{kin} = eE\lambda_e \sim \frac{E}{N} \sim \frac{E}{p}, \tag{5.4}$$

where λ_e is the mean free path length of electrons.

If the kinetic energy exceeds the threshold for ionisation, new electrons can be produced. Townsend suggested an empirical relation for α/p, see also Fig. 5.2 (a):

$$\frac{\alpha}{p} = C_1 \exp\left(-C_2 \frac{p}{E}\right) \:. \tag{5.5}$$

C_1 and C_2 are constant for a special kind of gas and limited range of reduced electric field strength [2].

On the other hand, the generated positive ions are accelerated in opposite direction towards the cathode. The ionization of neutral gas atoms by ion–neutral collisions (second Townsend coefficient, β) is in most cases negligible because of low ionisation cross section in comparison with that of the electrons in the considered range of kinetic energy. Positive ions bombard the cathode surface and generate secondary electrons. This mechanism is characterized by the third Townsend coefficient γ which describes the number of secondary electrons per impinging ion. Therefore the terms

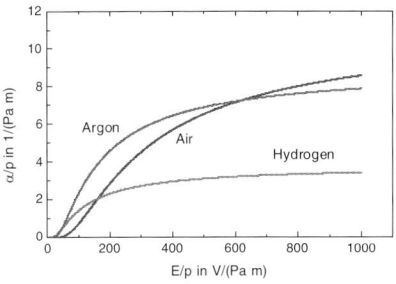
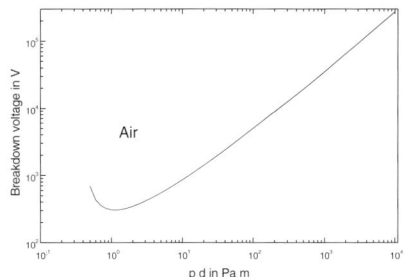

Fig. 5.2. *Left*: Pressure reduced Townsend coefficient a/p in dependence on E/p. *Right*: breakdown voltage (air) in dependence on $p\,d_E$

$$\exp(\alpha\,d_E) - 1 \tag{5.6}$$

and

$$\gamma\,[\exp(\alpha\,d_E) - 1] \tag{5.7}$$

represent the total number of the generated positive ions from one electron arrived the anode (5.6), and the number of secondary electrons by ion bombardment (5.7), respectively. The coefficient γ depends on the kind of gas, kinetic energy of ions and cathode material. Typical values for γ are in the order of magnitude of $0.01\ldots0.1$. These secondary electrons, starting at the cathode, produce new electron–ion pairs due to collisions with neutral gas atoms discussed above.

In glow discharges, the coefficient γ implies rather an effective secondary electron emission coefficient including secondary electron emission caused by photons and metastable species.

In steady state, the electron current density at the anode represents the total discharge current density according to:

$$j_{eA} = j_{e0}\frac{\exp(\alpha d_E)}{1 - \gamma\,[\exp(\alpha d_E) - 1]}\,, \tag{5.8}$$

where j_{e0} is the electron current density at the cathode due to an external source.

If the denominator approaches zero the current density tends to infinity. This increase of the current density leads to the electrical breakdown in the gas.

$$\gamma\,[\exp(\alpha d_E)] = 1\,. \tag{5.9}$$

The condition for electrical breakdown (5.9) reflects the reproduction of one electron by gas ionisation and secondary electron emission at the cathode. Furthermore, this relation defines the transition to the self sustaining gas discharge.

The breakdown voltage (U_{Br}) can be derived by inserting α/p (5.5) in the condition for breakdown (5.9):

$$U_{Br} = E_{Br}\, d_E = \frac{C_2(p\,d_E)}{\ln(p\,d_E) + \ln[C_1/\ln(1+1/\gamma)]} \ . \qquad (5.10)$$

The breakdown voltage in terms of the product $p\,d_E$ represents the well known Paschen law. The characteristic Paschen minimum of the breakdown voltage [cf. Fig. 5.2 right] represents optimal relation between α/p and E/p:

$$U_{Br}^{min} = \frac{2.72\,C_2}{C_1}\ln(1+1/\gamma) \ , \qquad (5.11)$$

$$(p\,d_E)_{min} = \frac{2.72}{C_1}\ln(1/\gamma + 1) \ , \qquad (5.12)$$

$$\left(\frac{E}{p}\right)_{min} = C_2 \ . \qquad (5.13)$$

Gas discharges at various geometric dimensions and total pressures can be compared by use of the similarity laws. For example, the discharges are similar if the following expressions have the same value:

$$p\,d_E = \text{const}\ ; \quad E/p = \text{const}\ ; \quad \alpha/p = \text{const}\ ; \quad j/p^2 = \text{const}\ . \qquad (5.14)$$

For example, the cosmic radiation and natural radioactivity leads approximately to $10^3\,\text{cm}^{-3}$ charged particles in air at atmospheric pressure. The breakdown voltage amounts to about $30\,\text{kV}$ per cm [Fig. 5.2 (right)].

5.2.2 Townsend and Glow Discharge

At very low discharge current the space charge effect is negligible and the electric field is approximately uniform. Taking into calculation the charge carrier transport by electric field drift, the estimated axial density distribution between the discharge gap represents no plasma (!) because of the significantly higher concentration of positive ions nearly in the complete discharge volume. This stationary discharge regime represents the dark discharge or Townsend discharge.

With increasing discharge current the positive space charge dominates more and more and determines the electric field strength in the discharge gap. The electric field strength increases in the vicinity of the cathode and decreases in the vicinity of the anode. The transition to the glow discharge regime is defined for the case that the electric field tends to the value zero at the anode [2].

The glow discharge at low pressure (1–10^4 Pa) is characterised by low current (10^{-4}–10^{-1} A) and voltages up to few 10^3 V. The subnormal glow discharge with negative U/I-characteristic represents a transition regime to the normal glow discharge. The normal glow discharge shows a strong increasing discharge current at little raising voltage, but the discharge current density remains constant due to increasing active cathode surface area. The well known discharge regions like cathode layer, negative glow, Faraday dark space, positive column and anode layer can be observed. The applied discharge voltage drops nearly complete over the space charge region in front of the cathode. The cathode layer is affected by the type of gas and cathode material. The positive column and approximately the negative glow represent quasi-neutral regions that means a weakly ionised low temperature plasma [2].

The anomalous glow discharge regime (complete cathode surface involved, high discharge voltage) shows a positive U/I-characteristic. The electron multiplication in the cathode layer is low (only between one and two). The beam-like electrons (run away electrons) enter into the negative glow. In the negative glow with approximately zero electric field strength these electrons lose their kinetic energy in elastic and inelastic collisions, and by ionisation of gas atoms. The negative glow is characterised by an axial inhomogeneous plasma density with a dominant maximum ($\sim 10^{10}$–10^{11} cm^{-3}) nearby the cathode layer boundary. The mean energy of plasma electrons is in the order of about ~ 0.1 eV. The charge carrier transport is mainly determined by ambipolar diffusion. Positive ions may enter into the cathode layer and drift in the inhomogeneous electric field towards the cathode. At the cathode, the kinetic energy of the ions is high enough for secondary electron emission (γ-process). The ion energy distribution is influenced by charge transfer and elastic collisions which depends on total pressure and kind of gas. In collision free case, the ion energy corresponds directly to the voltage drop over the cathode layer.

At sufficient distance between cathode and anode, the positive column is observed. This discharge region is characterized by a small axial potential gradient (low axial electric field strength). The loss of charged particles due to field drift, radial ambipolar diffusion to the wall, and recombination is replaced by electron impact ionisation, arising from electrons of the high energetic tail of the electron energy distribution function. In cylindrical discharge tubes (radius R) a simple modelling describes the positive column by an axially uniform plasma density and the radial change of plasma density according to a Bessel function of zero order

$$n_e(r) = n_{e0} J_0 (2.405 \, r/R) \ . \tag{5.15}$$

The electron temperature is in the order of a few eV and the typical plasma density (at the discharge axis) amounts to about 10^9–10^{11} cm^{-3}.

A special type of (glow) discharge is the hollow cathode discharge. The hollow cathode consists of a cylindrical cathode cup with the diameter d_H, or two parallel metal plates operating as double cathode with separation d_H. Simplified, the beam-like electrons entered into the negative glow will be reflected from the opposite cathode layer and vice versa. For optimal parameter $p \, d_H$ the negative glows overlap, and the oscillating electrons have many inelastic collisions. In result the effective excitation and ionisation is performed. The hollow cathode discharge is characterised by multiplication of the discharge current in comparison with usual (glow) discharge arrangements at the same discharge voltage, and generates an intensive low temperature plasma.

5.2.3 Arc Discharge

At high discharge current the heating of the cathode surface is high enough that the secondary electron emission by ion bombardment is replaced by the thermal electron emission. Besides the thermal secondary electron emission the field electron emission can essentially contribute due to the high electric field in front of the cathode. Consequently, low discharge voltage of about 10–50 V and high current density ($\sim 10^2$–10^4 Acm^{-2}) are typical discharge parameters.

Because of the negative U/I-characteristics, a sufficiently large series resistor is required in order to provide a stable operation point. Arc discharges can be maintained over large pressure range, from vacuum arc to high pressure arc.

The vacuum arc is a gas discharge in the vapour of cathode material and it is maintained by evaporation of cathode material by the arc itself. The arc discharge can form small cathode spots (current density $\sim 10^4$–10^7 Acm^{-2}) which locally heat the cathode surface. At pressures below 10^3 Pa or electron densities lower than 10^{12} cm^{-3}, the plasma of the positive column shows different temperatures ($T_e \gg T_i \approx T_N$), whereas at higher pressure ($>10^4$ Pa) or electron density ($>10^{15}$ cm^{-3}) a thermal plasma is observed (local thermodynamic equilibrium, $T_e = T_i = T_N$). The axis temperature of the positive column amounts to about 6000 K. In steady state the power balance is characterised by the Joule's heating as source term, as well as radiation, thermal conductivity and convection for loss processes. At very high pressure ($>10^6$ Pa) about 80–90% of the Joule's heat is converted into radiation. This has found application for high pressure lamps (e.g. HID lamps, see Chap. 15 in Part III).

5.2.4 Streamer Mechanism
and Micro-Discharges, Dielectric Barrier and Corona Discharge

Generally, these types of discharges operate non-stationary, they need high voltages (high electric field strength) and they appear typically at higher (atmospheric) pressure. The Townsend breakdown mechanism by charge carrier multiplication and secondary electron emission at the cathode at low pressure condition, approximately at $p\,d_E < 10^4$–10^5 Pa cm, must be replaced by the concept of streamer and micro discharge. It is based on the growth of thin ionised channels between the discharge electrodes. The streamer can grow in one or both directions towards the electrodes (cathode directed, anode directed streamer). Similar to the electrical breakdown at low pressure the starting point is a free electron. Because of the high electric field and pressure it will be produced a local plasma, which emits energetic photons. The reabsorption of the photons produces new electrons and ions by photo ionisation and leads to the growing of the streamer or creates secondary streamer. Recent modelling of the streamer evaluation indicates non-linear processes for streamer branching [3]. The generation rate of electrons is proportional to the electron density, but also depends on the local field strength in a non-linear way. It becomes substantial if both the electron density is non-vanishing and the local electric field is sufficiently strong. Since the motion of the free electrons and ions is damped by collisions with the gas atoms, the current is composed of a drift and a diffusive term. The proposed simple but realistic model can exhibit already spontaneous spark branching, in contrast to previous expectations, discussed above. The high-voltage breakdown between discharge electrodes may be performed within about 10 ns. Following, a spark discharge (high current arc, 10^4–10^5 A) is observed which is limited by the power source capacity.

In dielectric barrier discharge (DBD) the micro discharges are created between a small discharge gap (separation ≈1 mm) in which one or two of the electrodes are covered with an insulating material (dielectric) [4]. Typically, it can be observed a large number of short-lived discharge filaments. Such micro-discharges have duration of 1–10 ns, diameter ∼0.1 mm, and current density ∼10^2–10^3 Acm^{-2}. Discharge voltage of between 5–100 kV and excitation frequency in the range 50 Hz–1 MHz are applied. The plasma of such micro-discharge is non-thermal, i.e. it is characterised by $n_e \approx 10^{14}$–10^{15} cm^{-3} and $T_e \approx 1$–10 eV. The dielectric limits the charge carrier supply and produces a local electric field by charging the dielectric. Consequently, the effective electric field is reduced and the micro-discharge will be quenched. Recent investigations have been done in the field of a diffuse barrier discharge or atmospheric pressure glow discharge (APGL). In this case the metastable excited species of the used gases (N_2, He) should play an important role.

The corona discharge (positive or negative corona) can be observed in regions with large electric field strength nearby a thin wire or other strong curved surfaces. In this region the condition for breakdown is given. The negative corona (wire on negative potential) may also be assisted by a cathode mechanism similar with that of the glow discharge. Away from the wire the electric field goes down rapidly. In this drift region the low energetic electrons may be attached by neutrals (dust particles) and form negative ions. The positive corona starts by streamer mechanism nearby the wire but no cathode mechanism assists the discharge. In the low field region away from the wire mainly positive ions are present.

Corona and barrier discharge have found increasing interest again for plasma chemical gas cleaning and surface treatment because of their operation at atmospheric pressure (no vacuum equipment is necessary) and their good up-scaling from laboratory to industrial devices. Furthermore, the dielectric barrier discharges are applied for efficient UV and VUV radiation sources, and plasma display panels.

5.2.5 Glow Discharge at Alternating Electric Field, RF and Microwave Discharge

If an alternating electric field (angular frequency $\omega = 2\pi f$) is applied at the discharge gap, the charged particles start to oscillate (polarisation drift). The breakdown voltage depends on $p\, d_E$ and $\omega\, d_E$ [5]. In the low frequency case the processes can be considered to be similar to the static electric field (quasi-stationary conditions). For example, at the standard line frequency of 50 Hz the discharge represents an alternating dc glow discharge. For each half period we have an electric breakdown (ignition) and the development of the discharge regions (e.g. cathode layer, negative glow). If the electric field goes down (zero point of ac cycle) no new charged particles are produced. The charged particles disappear mainly by ambipolar diffusion to the electrode surface or walls. At higher frequencies ($\omega\, d_E > 1$ MHz cm) the loss by ambipolar diffusion and recombination per time unit is low in comparison with the electric field frequency. No new ignition is necessary because of the remaining charged particles in the discharge centre between the electrodes. The cathode layer and negative glow alternate between the two electrodes. At higher frequencies ($\omega\, d_E > 10$ MHz cm) the ions are not able to follow instantaneously the change in electric field (transition to rf discharges). The number of ions impinging the electrode decreases and secondary electron emission goes down. In the high frequency case ($\omega\, d_E > 100$ MHz cm) electrons are trapped (rf, microwave discharges). These electrons cause ionisation of neutrals in the bulk.

The input of electrical power to plasma electrons in the bulk can be described by the equation of motion:

$$m_e \ddot{x} + \nu_{coll} m_e \dot{x} = eE_o \exp(i\omega t) , \qquad (5.16)$$

$$P = jE = en_e \dot{x} E . \qquad (5.17)$$

The power which is absorbed per volume unit is expressed by $P_{abs} = \text{Re}[j] \times \text{Re}[E]$:

$$P_{abs} = \frac{n_e e^2}{m_e \nu_{coll}} \frac{(\nu_{coll}/\omega)^2}{1 + (\nu_{coll}/\omega)^2} \frac{E_0^2}{2} . \qquad (5.18)$$

Without collisions ($\nu_{coll} = 0$) no power input is possible (the phase shift of the electron motion related to the electric filed can be interpreted by a non-magnetic inductivity, imaginary part only). If the excitation frequency $\omega = 0$ we have the classical Drude-model for free electrons, $P_{abs} = P_{dc}$. The power input will be maximal for the case $\omega = \nu_{coll}$ [see Fig. 5.3(a)]:

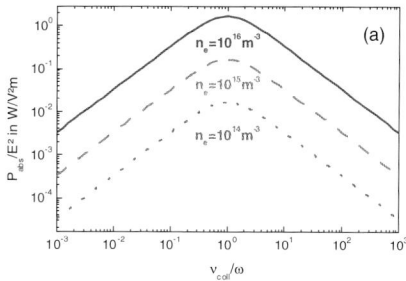 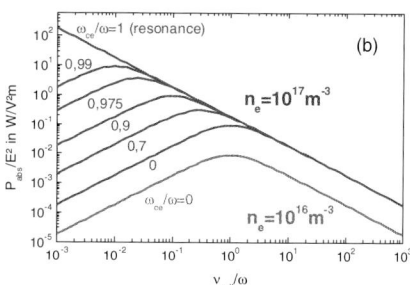

Fig. 5.3. (a) Absorbed electric power P_{abs} per m^3 and E^2 for 13.56 MHz-discharge. (b) 2.45 GHz-discharge in dependence on the reduced collisions frequency. (b) shows the influence of the magnetic field (ω_{ce}) on the power input in addition

$$P_{abs}^{max} = \frac{1}{2} \frac{n_e e^2}{m_e \nu_{coll}} \frac{E_0^2}{2}. \tag{5.19}$$

Rf discharges at frequency of 13.56 MHz or 27.12 MHz are widely used in plasma technologies for surface treatment and thin film deposition. The rf discharges may be operate by means of internal or external electrodes and they are classified into two types according to the method of coupling the high frequency power, capacitively or inductively. The coupling is mostly capacitively by the electric field between two internal parallel plate discharge electrodes (CCP). In the case of the inductive coupling the high frequent current passes trough a solenoid coil and leads to an oscillating magnetic field which induces the electric field. The induced electric field generates dense low temperature plasma (ICP).

The generation of dense low temperature plasma can be also reached by microwave discharges, for example at 2.45 GHz. The microwaves produce much larger electric fields in the plasma bulk. On the other hand microwave discharges are of special interest in combination with magnetic fields. If the electron cyclotron motion (ω_{ce}) is in phase with the rotating electric field vector of the R-wave and the electron collision frequency is not too high the electrons absorb effective energy [6]. Then, the power input per volume is given by [cf. Fig. 5.3(b)]

$$P_{abs} = \frac{n_e e^2}{m_e \nu_{coll}} \frac{(\nu_{coll}/\omega)^2}{[1-(\omega_{ce}/\omega)]^2 + (\nu_{coll}/\omega)^2} \frac{E_0^2}{2}. \tag{5.20}$$

The power input reaches a maximum for the case $\omega = \omega_{ce}$, that means the electron cyclotron resonance (ECR). At a frequency of 2.45 GHz the corresponding magnetic field amounts $B_{ECR} = 87.5$ mT.

Besides the power absorption due to electron collisions (Ohmic heating), a stochastic heating can be found in capacitively coupled rf discharges. Electrons are heated without collisions due to interaction of electrons moving towards the rf electrode during the expanding space charge sheath [7].

5.3 Plasma Surface Transition

5.3.1 Plasma Boundary Sheath, Bohm Criterion

We consider a stationary, uniform low temperature plasma consisting of single charged positive ions and electrons with Maxwell energy distributions ($T_e \gg T_i$). The discharge electrodes, surrounding surfaces or immersed substrates/probes will be negatively charged in respect to the plasma potential (V_{Pl}) because of higher mobility of electrons in respect to the ions ($\propto [m_i T_e/(m_e T_i)]^{1/2}$). The negative charged surface is shielded by a positive space charge sheath in front of the surface. This plasma boundary sheath, the potential of which is shown in Fig. 5.4, determines the charge carrier transport to the surface and may influence the discharge mechanism by secondary particle emission from the surface.

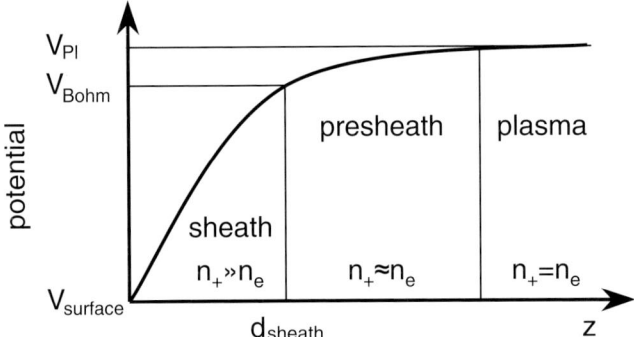

Fig. 5.4. Plasma boundary sheath. Typical potential change in dependence on the distance from the surface

More in detail, the plasma surface transition is characterised by a pre-sheath and the space charge sheath. Coming from the quasi-neutral plasma at the potential V_{Pl}, a small potential drop over the quasi-neutral pre-sheath accelerates positive ions to the Bohm velocity.

$$\text{Potential drop}: \quad \Delta V_{\text{Bohm}} = \frac{1}{2}\frac{k_B T_e}{e}, \quad (5.21)$$

$$\text{Bohm} - \text{velocity}: \quad v_{\text{Bohm}} = \left(\frac{k_B T_e}{m_i}\right)^{1/2}, \quad (5.22)$$

$$\text{Bohm} - \text{criterion}: \quad \frac{m_i}{2} v_{\text{Bohm}}^2 \geq \frac{1}{2} k_B T_e. \quad (5.23)$$

In the space charge region the ions will be accelerated and electrons retarded according their kinetic energy. For an insulating surface the net current in steady state will be zero:

$$j_e + j_i = 0 \,, \tag{5.24}$$

$$j_e = e\, n_{e0} \exp(-1/2) \left(\frac{k_B T_e}{2\pi m_e}\right)^{1/2} \exp\left(-\frac{e\, \Delta V_{sh}}{k_B T_e}\right), \tag{5.25}$$

$$j_i = e\, n_{i0} \exp(-1/2) \left(\frac{k_B T_e}{m_i}\right)^{1/2} \,; \quad n_{i0} = n_{e0} \,. \tag{5.26}$$

Then, the potential drop across the space charge sheath is

$$\Delta V_{sh} = \frac{1}{2} \frac{k_B T_e}{e} \ln\left(\frac{m_i}{2\pi\, m_e}\right) \approx 3 \cdots 4 \, \frac{k_B T_e}{e} \,, \tag{5.27}$$

$$V_{fl} = V_{Pl} - \Delta V_{\text{Bohm}} - \Delta V_{sh} \,. \tag{5.28}$$

Typical electron temperature of a few 10^4 K (few eV) leads to a floating potential V_{fl} of about 10–20 V negative to the plasma potential. Using Poisson's equation and a plane geometry the positive ion current density can be calculated in the collision free and collision dominated case:

$$\lambda_i \gg d_{sh} \quad \text{collision free}: \qquad j_i = \frac{4}{9} \epsilon_0 \left(\frac{2e}{m_i}\right)^{1/2} \frac{V_{sh}^{3/2}}{d_{sh}^2} \,,$$
$$\text{(Child – Langmuir)} \tag{5.29}$$

$$\lambda_i \ll d_{sh} \quad \text{collision dominated}: \qquad j_i = \frac{9}{8} \epsilon_0 \mu_i \frac{V_{sh}^2}{d_{sh}^3} \,. \tag{5.30}$$

Furthermore, taking into calculation the ion current density j_i according to the Bohm criterion at the sheath boundary, the space charge sheath thickness can be expressed in terms of the Debye length λ_D [see (1.7) in Chap. 1] and the sheath voltage (5.34).

In a similar way the plasma sheaths are formed in front of discharge electrodes (dc cathode layer, rf electrodes), immersed samples for plasma diagnostics or samples for material surface treatment, and walls.

In low pressure dc discharges, the ion current at the cathode represents nearly the total discharge current. Davis and Vanderslice [8] have presented firstly an analytical expression for the ion energy distribution function at the cathode. Rickards [9] has modified this model.

Linear increasing electric field strength towards the cathode surface, no ionisation in the cathode layer, and charge transfer collisions with constant cross section are included in the model.

$$F(\epsilon^*) = \frac{dN_i}{d\epsilon^*} = \frac{N_{i0}}{2} \frac{L}{\lambda_{ex}} (1-\epsilon^*)^{-1/2} \exp\left[-\frac{L}{\lambda_{ex}} + \frac{L}{\lambda_{ex}}(1-\epsilon^*)^{1/2}\right], \tag{5.31}$$

where λ_{ex} is the mean free path for charge transfer, L thickness of the cathode layer, $\epsilon^* = \epsilon/(eV_c)$ the ion energy relative to the maximal ion energy for collision free transport in the cathode layer ($\epsilon^* = 0 \ldots 1$).

5.3.2 RF Plasma Sheath

More complicated is the situation at the electrodes in rf discharges, characterised by excitation frequencies between ion and electron plasma frequency. In a capacitively coupled, unconfined and for that reason asymmetrical rf discharge, a negative self bias voltage is created at the powered electrode which is little less than half the peak to peak rf voltage. Assuming capacitive sheath model the simplified potential change with time is represented in Fig. 5.5.

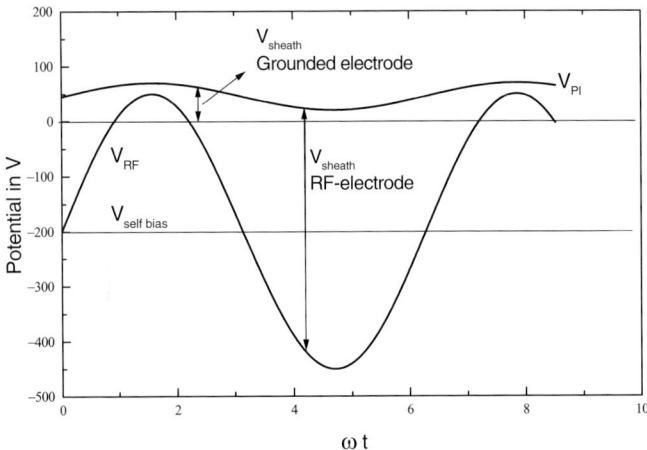

Fig. 5.5. Simplified potential change with time and resulting sheath voltage at the rf and grounded electrode. The difference between the plasma potential V_{Pl} and V_{rf} respectively ground potential ($V = 0$) represents the sheath voltage at the powered and grounded electrode

$$V_{SB} \leq \frac{1}{2}V_{PP} = V_{rf0} ; \quad V_{rf} = V_{SB} + V_{rf0}\sin(\omega t) ; \tag{5.32}$$

$$V_{Pl} = V_{fl} + \frac{1}{2}(V_{SB} + V_{rf0})[1 + \sin(\omega t)] . \tag{5.33}$$

The negative self bias voltage results from the fact that no net dc current can be flow over one rf cycle at capacitive coupling. The electrons can follow the alternating electric field and leads to a moving electron front which modulates the positive space charge region, respectively the voltage and the thickness of the sheath. Increasing rf voltage amplitude mainly affect the self bias voltage and have only little influence on the plasma potential in strong asymmetric rf discharges. As the result the self bias voltage at the powered electrode increases with rf voltage, whereas at the grounded electrode no influence can be seen. The sheath properties at the grounded electrode are comparable with a dc plasma sheath.

The transport of positive ions is influenced by the oscillating plasma sheath voltage, charge transfer collisions and elastic collisions.

The sheath thickness is determined by the Debye length λ_D and the voltage drop across the sheath V_{sh}. For collisionless regime the sheath thickness is given by (5.34). The factor C_e represents a correction term due to the time modulated electron density [10].

$$d_{sh} = \frac{2^{5/4}}{3} C_e \lambda_D \left(\frac{e V_{sh}}{k_B T_e} \right)^{3/4} \quad \text{with} \quad 1 \leq C_e \leq \left(\frac{50}{27} \right)^{1/2}. \quad (5.34)$$

The value $C_e = 1$ is given for the dc sheath.

The energy of impinging positive ions at the rf electrodes depends on the ratio between the ion transition time in the rf sheath and the rf-cycle. Therefore, the ion mass, rf frequency, and collisions in the sheath regions have significant influence on the shape of the ion energy distribution function. Direct ion extraction at the discharge electrode and energy selective mass spectrometry can be used for experimental determination of ion energy distributions. Figure 5.6 (a) and Fig. 5.6 (b) show the time averaged Ar$^+$ energy distributions at the driven and grounded electrode of an rf discharge in argon. As expected from the potential structure, the ion energy at the powered electrode is strongly coupled to the rf voltage due to the self bias voltage. The corresponding maximal ion energy is of the order of several hundred eV, whereas the energy of positive ions at the grounded electrode is lower then 20 eV.

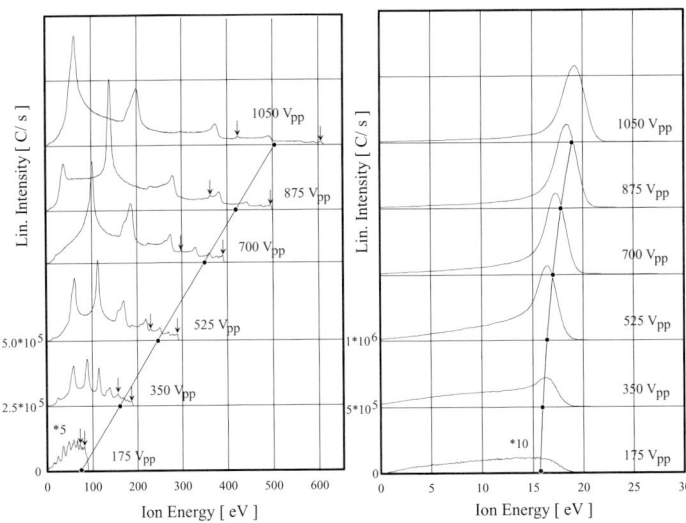

Fig. 5.6. Ion energy distribution function (Ar$^+$) at the rf electrode **(a)** and grounded electrode **(b)** (13.56 MHz-discharge in argon at 5 Pa, parameter: peak-to-peak voltage [11])

Significant differences in the shape of the ion energy distribution at the two electrodes are observed. Ions coming from the bulk plasma need several rf cycles for transition to the rf electrode. In Fig. 5.6(a) the arrows mark the situation for entering the ions from the bulk plasma into the sheath at the low and high sheath voltage, respectively. In result a saddle shaped structure in the ion energy distribution is found. The observed multiple peak structure in the low energetic part comes from charge transfer collisions in the sheath region. In the time averaged ion energy distribution the saddle shaped structures overlap from ions directly from bulk plasma and ions from charge transfer collisions. At the grounded electrode the single peak at the high energy end is seen, only.

With increasing pressure the elastic collisions will have more influence. This is connected with increasing ion intensity at the low energy part and disappearing (multiple) peak structure.

In microwave discharges and inductively coupled rf discharges (ICP) the lower sheath voltages represents conditions similar to the conditions in front of surfaces at floating potential. Using magnetized plasmas the cyclotron motion must be taken into the consideration.

5.3.3 Electric Probes

Electric probes are widely applied to determine experimentally the electron energy distribution function $f(\epsilon)$, or T_e in the case of Maxwellian distribution, electron and ion density and potentials. The I/U-characteristic is taken from a circuit consisting of a small metallic electrode (probe) of plane, cylindrical or spherical shape immersed in the plasma, a reference electrode, and variable external dc power supply. In single probe measurement (Langmuir probe) the reference electrode is one of the discharge electrodes, and in double probe measurements a second immersed probe is applied, respectively. Generally, the immersed electric probe represents a disturbance of the plasma around the probe position due to space charge sheath in front of the probe surface and the extraction of charged particles. Despite the relative simple technique, specific conditions have to be fulfilled for accurate analysis of the plasma parameters derived from the I/U-characteristic [12].

Single (Langmuir) Probe

We consider an uniform low temperature plasma consisting of single charged positive ions and electrons at Maxwellian energy distribution with $T_e \gg T_i$. A typical I/U-characteristic of a cylindrical Langmuir probe is shown in Fig. 5.7. Principally, the characteristic can be divided into three parts:

I: Ion current saturation region $I_i \gg I_e$ ($V \ll V_{fl}$)
II: Transition region
III: Electron current saturation region $I_i \ll I_e$ ($V > V_{pl}$)

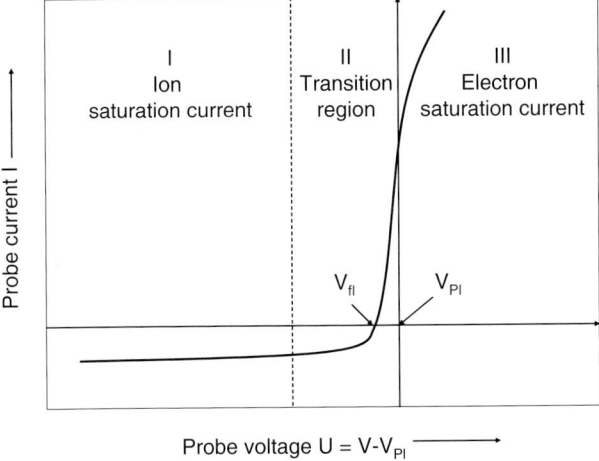

Fig. 5.7. Different regions of a single probe characteristic

The working regime of the electric probe is determined by the parameters probe radius r_P, the mean free path length of charge carrier $\lambda_{i/e}$ for ions and electrons respectively, and the Debye length λ_D. For the classic collisionless single probe theory the two working regimes are defined by

$\lambda_{i/e} \gg r_P \gg \lambda_D$ (thin sheath)
$\lambda_{i/e} \gg \lambda_D \gg r_P$ (thick sheath, orbital motion limit – OML)

Ion Current Saturation Region (I)

A simple solution is given for the OML case, and energy and angular momentum conservation law. Not all ions entering the sheath are collected by the probe. For cylindrical probes the ion saturation currents is given by:

$$I_i = n_i\, e 2\pi\, r_p\, l_p \left(\frac{k_B T_e}{m_i}\right)^{1/2} \left(1 - \frac{eU}{k_B T_e}\right)^{1/2} \qquad (5.35)$$

with $U = V - V_{pl}$. Furthermore, Sonin [13] has introduced the so-called Sonin-plot for calculation the ion density. In this plot the term $\xi_P^2\, i_i$ is plotted against the dimensionless current i_i for a pre-selected value of $\eta_{fl} - 10$ [14]. Taking into calculation the ion current:

$$I_i = n_i\, e\, A_p \left(\frac{k_B T_e}{m_i}\right)^{1/2} i_i\, . \qquad (5.36)$$

The multiplication of (5.36) with $\xi_P^2 = (r_p/\lambda_D)^2$ and conversion results in

$$\xi_P^2\, i_i = \left(\frac{r_p^2}{e}\right) \left(\frac{2m_i}{e}\right)^{1/2} \left(\frac{e}{k_B T_e}\right)^{3/2} \left[\frac{I_i\,(\eta_{fl} - 10)}{A_p}\right] \qquad (5.37)$$

with $\eta_{fl} = (eV_{fl} - eV_{pl})/(k_B T_e)$. The right side of (5.37) includes experimental data, only. From the determined value $\xi_p^2 i_i$ the dimensionless current i_i can be found in the Sonin-plot, and from (5.36) the ion density.

Transition Region (II)

The electron current increases due to the reduced retarding negative probe potential in respect to the plasma potential. Therefore, the information about the EEDF is included in this part of the probe characteristic. Assuming Maxwellian distribution of the electron energy, the electron current is

$$I_e = n_e \, e \, A_p \left(\frac{k_B T_e}{2\pi m_e} \right)^{1/2} \exp\left(-\frac{eU}{k_B T_e}\right) . \tag{5.38}$$

The electron temperature can be calculated from the slop of the semi-logarithmic plot according to

$$T_e = -\frac{e}{k_B} \frac{d}{dU} \ln\left(\frac{I_e}{I_{e0}}\right) \tag{5.39}$$

and the electron density from the electron current at plasma potential:

$$n_e = \frac{I_e(U=0)}{A_p \, e} \left(\frac{2\pi m_e}{k_B T_e} \right)^{1/2} . \tag{5.40}$$

If the electron energy distribution function is non-Maxwellian, the Druyvesteyn method can be used by second derivative of the electron current:

$$\frac{d^2 I_e}{dU^2} = \left(\frac{2e}{m_e} \right)^{1/2} \frac{e \, A_p}{4} n_e \, (-U)^{-1/2} \, f(\epsilon) V . \tag{5.41}$$

with $U = V - V_{pl}$.

Electron Current Saturation Region (III)

The electron saturation current in the OML case can be described similar to the ion saturation current. The electron current is expressed by

$$I_e = n_e \, e \, A_p \, \frac{2}{\sqrt{\pi}} \left(\frac{k_B T_e}{2\pi m_e} \right)^{1/2} \left(1 + \frac{eU}{k_B T_e}\right)^{1/2} . \tag{5.42}$$

Symmetrical Double Probe

The I/U-characteristic is taken from two small symmetric probes immersed in the plasma. This characteristic resembles the ion current part of the single probe characteristic in both direction of the applied voltage, see Fig. 5.8.

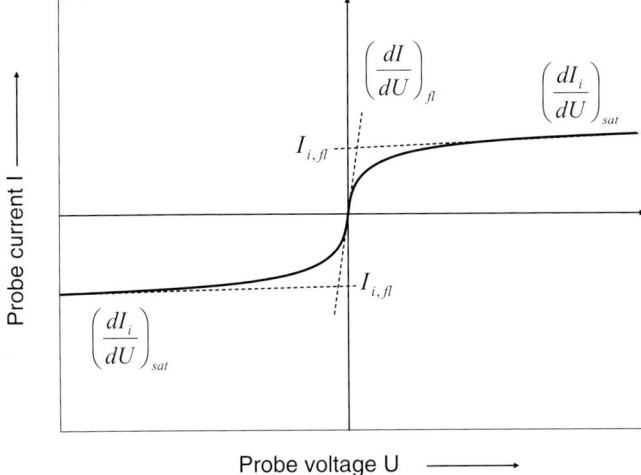

Fig. 5.8. I/U-Characteristic of a symmetric double probe

The double probe theory was firstly given by Johnson and Malter [15]. The advantage is that the complete probe system is on floating potential, and it can be applied in electrode-less generated plasmas without a big reference electrode. The disadvantage is that no information can be obtained about the complete electron energy distribution function. Taking into calculation a Maxwellian EEDF the electron temperature is given by:

$$\frac{k_B T_e}{e} = I_{i,fl} \left[2 \left(\frac{\mathrm{d}I}{\mathrm{d}U} \right)_{fl} - \left(\frac{\mathrm{d}I_i}{\mathrm{d}U} \right)_{sat} \right]^{-1} . \qquad (5.43)$$

The ion density is obtained similar to the procedure at single probe measurement.

Special Cases, Problems in Probe Diagnostics

The analysis of I/U-characteristics and calculation of plasma parameters is more complicated because of the influence by:

1. collisions in the sheath,
2. different kinds of positive ions,
3. presence of negative ions,
4. magnetized plasmas (anisotropic transport),
5. rf plasmas (modulated plasma potential needs compensation methods and damping of the rf amplitude in the probe circuit) and
6. deposition plasmas.

5.4 Reactive Plasmas and Plasma Surface Interaction

Low temperature plasmas are widely applied for plasma etching, surface modification, and thin film deposition. The processes at the surface depend on the kind of gas discharge and plasma boundary sheath which control the transport of charged particle to the surface, as well as the used processing gases.

The plasma etching of silicon material for wafer patterning in the microelectronics is a well known and successfully applied plasma technology. In this case low pressure plasmas in fluorocarbons e.g. CF_4, C_2F_6 or CHF_3 are applied for plasma etching of silicon or silicon dioxide.

On the other hand many conventional synthetic polymers (e.g. polyethylene, polystyrene) and natural polymers (e.g. cellulose, wool) have to be equipped with specific surface properties like hydrophilic, hydrophobic or oleophobic surfaces. Using low pressure or cold plasmas a thin layer (nm scale) of the polymer surface can be modified due to incorporation of new atoms and formation of functional groups, which determine the surface properties.

Generally, the interaction of the low temperature plasma with the material surface includes the surface modification and plasma etching simultaneously. Figure 5.9 represents the plasma surface modification of polyethylene in low

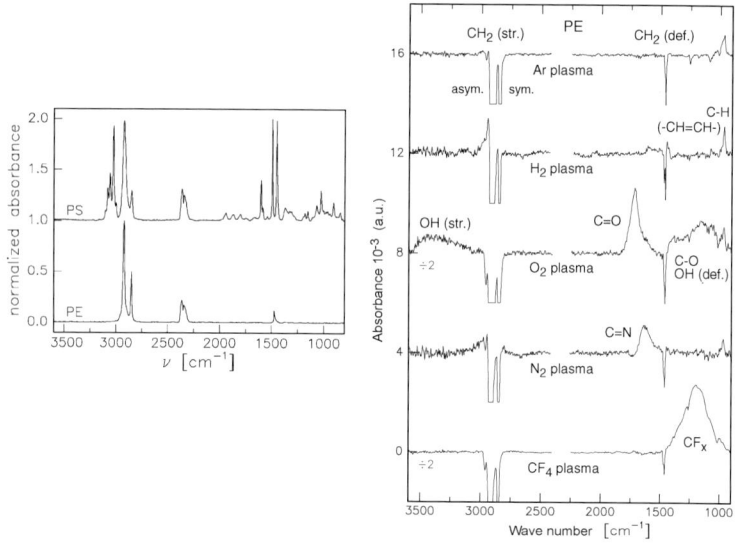

Fig. 5.9. *Left*: Infrared absorption spectrum of polystyrene (ps) and polyethylene (pe). *Right*: Difference of the infrared absorption spectra after/before rf plasma treatment of polyethylene in different process gases. Degradation and etching of the polymer structure (CH_2-groups) and formation of new functional molecular groups (e.g. O_2-plasma: OH, C-O, C=O)

pressure discharge at different process gases. Shown are the differences in infrared absorption spectra of polyethylene. In this case the negative bands in relation to the base line represents the degradation or etching of the polymer structure and the positive bands show new molecular structure due to incorporation of plasma particles and formation of new molecular groups.

Using molecular organic gases (e.g. hydrocarbons, fluorocarbons, organo silicon compounds) as processing gas in low pressure discharges the dissociation of the precursor molecules results in a lot of reactive plasma species

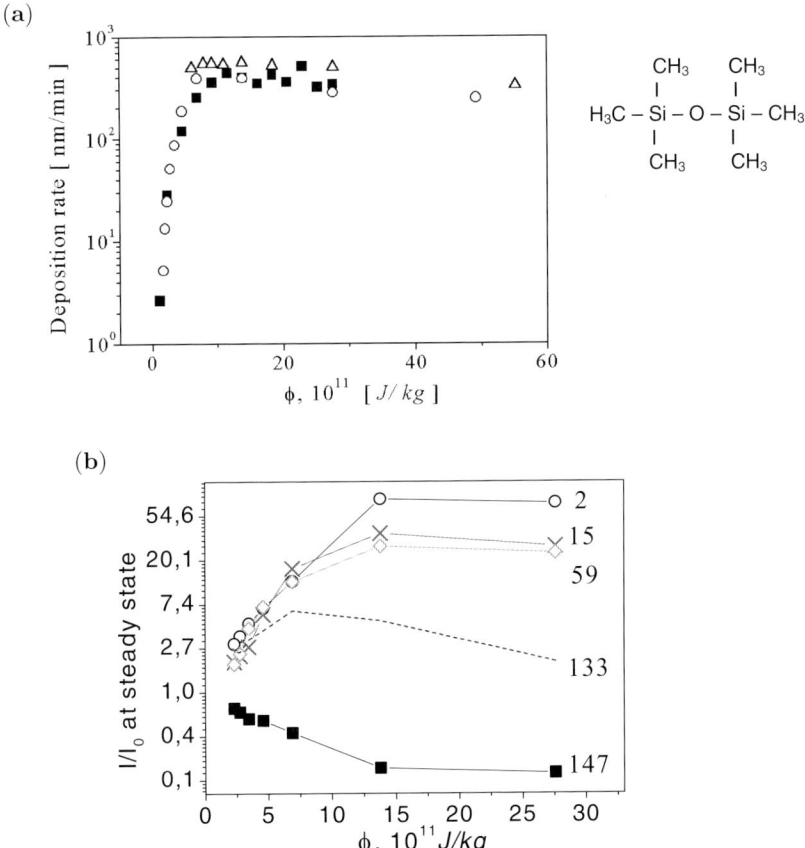

Fig. 5.10. (a) Deposition rate of organo silicon films from hexamethyldisiloxane (HMDSO) rf plasma (13.56 MHz) against the energy parameter ($\Phi \sim W/(p f)$; W: rf power, p: pressure, f: gas flow rate). (b) Mass spectroscopic analysis of the plasma chemical gas conversion. Shown are relative intensities of selected fragment ions (147 amu: precursor molecule HMDSO, 2 amu Hydrogen, 15 amu, 59 amu and 133 amu represent intermediate reaction products) against the energy parameter

(neutral and charged atoms and molecule fragments) which produce new stable gas compounds (plasma chemical gas conversion) and leads to the deposition of thin organic films (plasma polymerization) on discharge electrodes, walls and immersed substrates. By controlling the plasma processes and chemical reactions the macroscopic film properties can be varied.

Figure 5.10 shows the change in the deposition rate of organo silicon films from hexamethyldisiloxane (HMDSO) rf plasma in dependence on the energy input per precursor molecule. At lower input energy the strong increasing deposition rate results from relatively high content of the precursor in the discharge, which contribute to the film growth. In the other case at higher energy input the strong precursor fragmentation and consumption limit the deposition rate. This can be seen in the mass spectrometric analysis of the plasma process gas. The two deposition regimes result in different molecular structure and properties of the films.

References

1. R. Winkler: Electron Kinetics in Weakly Ionized Plasmas. In: *Low Temperature Plasma Physics – Fundamental Aspects and Applications*, ed by R. Hippler, S. Pfau, M. Schmidt et al (Wiley-VCH, Berlin 2001) pp 29–54
2. Y.P. Raizer: *Gas Discharge physics* (Springer, Heidelberg Berlin New York 1991)
3. B. Meulenbroek, A. Rocco, U. Ebert: Phys. Rev. E **69**, 067402 (2004)
4. U. Kogelschatz: Plasma Chem Plasma P **23**, 1 (2003)
5. M. Konuma: *Film Deposition by Plasma Techniques* (Springer, Heidelberg Berlin New York 1992)
6. G. Janzen: *Plasmatechnik* (Hüthig Buchverlag, Heidelberg 1992)
7. M.A. Lieberman and A.J. Lichtenberg: *Principles of Plasma Discharges and Materials Processing* (John Wiley and Sons, New York 1994)
8. W.D. Davis, T.A. Vanderslice: Phys. Rev. **131**, 219 (1964)
9. J. Rickards: Vacuum **34**, 559 (1984)
10. M.A. Liebermann: IEEE Trans. Plasma Sci. **16**, 638 (1988)
11. M. Zeuner, H. Neumann, J. Meichsner: J. Appl. Phys. **81**, 2985 (1997)
12. S. Pfau, M. Tichy: Langmuir probe diagnostics of low temperature plasmas. In: *Low Temperature Plasma Physics – Fundamental Aspects and Applications*, ed by R. Hippler, S. Pfau, M. Schmidt et al (Wiley-VCH, Berlin 2001) pp 131–172
13. A.A. Sonin: AIAA J. **4**, 1588 (1966)
14. S. Klagge, M. Tichy: Czech. J. Phys.B **55**, 988 (1985)
15. E.O. Johnson, L. Malter: Phys. Rev. **80**, 58 (1950)

6 Strongly Coupled Plasmas

R. Redmer

Abstract. We give a brief introduction into the basic many-particle effectes in strongly coupled plasmas and their theoretical treatment. Exemplary results are shown for the composition and equation of state as well as for electrical conductivity.

6.1 Introduction

The physical properties of strongly coupled plasmas have gained much interest over the last decades [1]. We show the density–temperature plane for the non-relativistic domain in Fig. 6.1 [2]. As it was discussed in Chap. 1 in Part I, the ion coupling parameter Γ_c and the electron degeneracy parameter Θ are taken as plasma parameters [3],

$$\Gamma_c = \frac{(Ze)^2}{4\pi\varepsilon_0 k_B T a}, \quad \Theta = \frac{k_B T}{E_F}, \tag{6.1}$$

where $a = (4\pi n_i/3)^{-1/3}$ is the mean distance between ions (Wigner–Seitz radius), and $E_F = \hbar^2 (3\pi^2 n_e)^{2/3}/(2m_e)$ is the Fermi energy of electrons.

We distinguish between ideal plasmas ($\Gamma_c \ll 1$) and non-ideal or strongly coupled plasmas ($\Gamma_c \geq 1$). In the latter case, the correlations between the particles are important for the physical properties so that consistent methods of many-particle physics have to be applied. The electron subsystem can be divided into the classical region ($\Theta \gg 1$) and the degenerate domain ($\Theta \ll 1$) where quantum effects are important.

Dilute astrophysical plasmas are, e.g., the interstellar gas (cold), the Solar corona and accretion discs of active galactic nuclei (hot). Tokamaks and stellarators as facilities for magnetic confinement fusion experiments are examples for hot ideal plasmas, see Chap. 7 in Part II for details. Matter inside giant planets such as Jupiter and brown dwarfs is strongly coupled and degenerate (see Chap. 12 in Part II), while white dwarfs and the core of the Sun are examples for weakly coupled plasmas.

Due to the enormous progress in shock wave experimental technique, the region of strong coupling with $\Gamma_c \geq 1$ has been accessed in the recent years. For instance, the electrical conductivity of various metal plasmas has been determined by using the rapid wire evaporation technique [4, 5, 6]. Gas guns

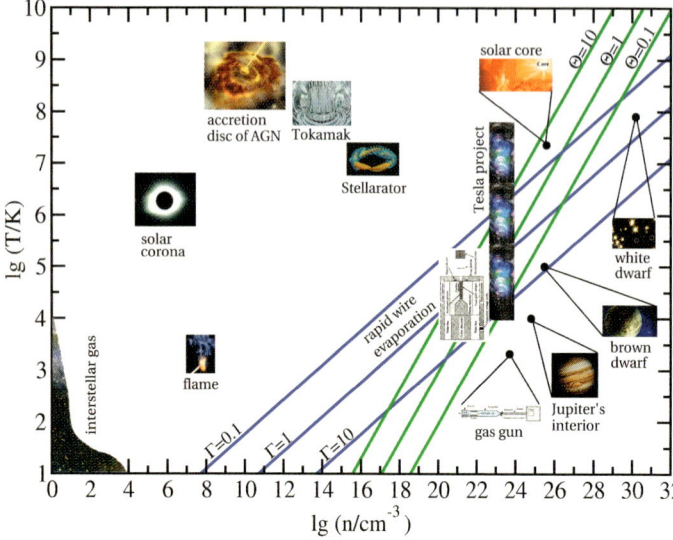

Fig. 6.1. Density–temperature plane with typical values for the plasma parameters Γ_c and Θ. Examples for astrophysical and laboratory plasmas are given (see [2])

have proven their capacity to compress materials up to megabar pressures and, most spectacular, metallic-like hydrogen has been observed for the first time [7]. Other shock wave experiments to generate strongly coupled plasmas are driven by lasers [8], magnetically-launched flyers [9], or chemical explosions [10]. A new technique to generate strongly coupled plasmas will be available soon at the free electron laser (FEL) facility at DESY Hamburg (Tesla project) [11]. There, the interaction of matter with intense and ultra-short laser pulses at wavelengths from 100 nm to the X-ray regime of 0.1 nm will be studied. Especially, time-resolved Thomson spectroscopy (see Chap. 13 in Part III) will allow the determination of plasma parameters such as density and temperature [12].

We give here an overview of theoretical methods for the treatment of strongly coupled plasmas and review their thermophysical properties. Besides the composition and equation of state (EOS), the electrical conductivity is discussed. Examples are given for dense metal and noble gas plasmas. We finish with a summary and give an outlook.

6.2 Many-Particle Effects and Plasma Properties

Plasmas as many-particle systems of charged particles can be treated by consistent methods of statistical physics. For instance, a quotidian EOS based on the Thomas–Fermi model has been developed for applications in hydrodynamic simulations [13]. It gives smooth and usable predictions for the EOS data and the ionization state. A density-response formalism has been elaborated for arbitrary electron degeneracy and strong Coulomb-coupling effects [14] that was successfully applied to calculate the correlation functions, the EOS, the stopping power, and the conductivities of dense, high-temperature plasmas.

Numerical simulations give more and more reliable results for the behavior of many-particle systems due to the enormous increase in computer capacity. Although the particle number is always finite in actual simulations, they approach the limit of exact results, and are thus very helpful. For instance, quantum molecular dynamics simulations can be performed by combining the classical MD technique with a quantum treatment of the electrons as wave packets [15] or ab-initio electronic structure calculations within density functional theory (DFT) [16, 17, 18, 19, 20]. Another attempt is the path-integral Monte Carlo (PIMC) technique [21, 22] which gives the exact high-temperature limit but requires approximative treatments for lower temperatures.

Another possibility to treat systems of charged particles is given by the Green's function technique [23]. The respective correlation functions can be represented by Feynman diagrams which allow a systematic evaluation of relevant contributions to the physical properties. In this way, many-particle effects such as self-energy or dynamic screening as well as the formation and decay of bound states can be treated in a systematic way [24]. We will give here a brief summary of the relations between physical properties and Green's functions which have successfully been applied to strongly coupled plasmas [25].

6.2.1 Green's Function Technique: Spectral Function

The relation between the thermodynamic Green's function $G_c(k,\omega)$ and the particle number density $n_c(\beta_T, \mu_c)$ is used to derive thermodynamic functions of the plasma [23],

$$n_c(\beta_T, \mu_c) = \frac{1}{\Omega_0} \sum_k \int_{-\infty}^{\infty} \frac{d(\hbar\omega)}{\pi} \, \text{Im}\, G_c(k, \omega - i0^+) f_c(\hbar\omega) , \qquad (6.2)$$

where $f_c(E) = \{\exp[\beta_T(E - \mu_c) + 1]\}^{-1}$ is the Fermi distribution function, Ω_0 the system volume, μ_c the chemical potential, and $\beta_T = 1/k_B T$. The one-particle Green's function is defined as the thermodynamic average of time-ordered products of creation and annihilation operators and determined via the Dyson equation

$$G_c(k,\omega)^{-1} = \hbar\omega - E_c(k) - \Sigma_c(k,\omega) . \qquad (6.3)$$

The self-energy $\Sigma_c(k,\omega)$ is introduced to decouple the hierarchy of the equations of motions for the Green's functions at least formally. The main problem is then the determination of the dynamical self-energy $\Sigma_c(k,\omega)$ which contains the interactions with all other particles so that the energy spectrum of free particles $E_c(k)$ is modified. For that purpose, the efficient technique of Feynman diagrams (see [23]) can be applied. The propagator of free particles

$$G_c^0(k,\omega)^{-1} = \hbar\omega - E_c(k) \qquad (6.4)$$

is denoted by a single arrow. Vertex points are given by dots and represent the interaction between particles in space–time coordinates so that energy and momentum conservation holds at the vertex in Fourier space. The diagrams for the infinite number of interaction contributions to the total Green's function can be constructed by means of these elements. For practical applications, a perturbation expansion can be performed. Especially, partial summations of classes of diagrams are a very efficient tool to include relevant physical effects. Examples are the sum of ladder-type diagrams describing the formation of bound states such as atoms, and the sum of bubble-type diagrams which represent the screening of the long-range Coulomb interaction. These approximations will be presented in the next section.

The "dressed" one-particle Green's function (double arrow) according to the Dyson equation (6.3) is given via an integral equation, see Fig. 6.2.

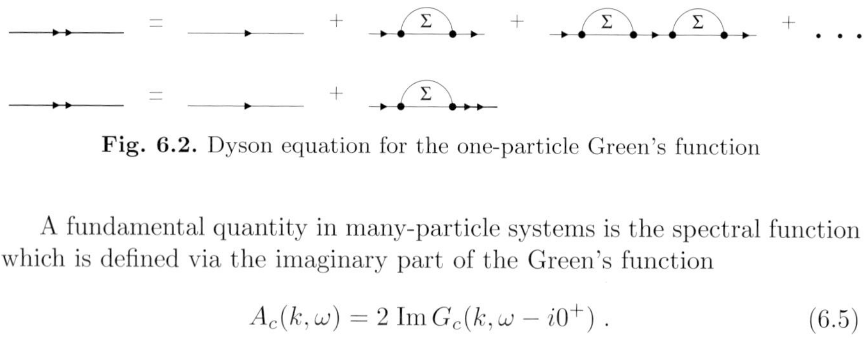

Fig. 6.2. Dyson equation for the one-particle Green's function

A fundamental quantity in many-particle systems is the spectral function which is defined via the imaginary part of the Green's function

$$A_c(k,\omega) = 2\,\mathrm{Im}\,G_c(k,\omega - i0^+)\,. \qquad (6.5)$$

This quantity is normalized according to

$$\int_{-\infty}^{\infty} \frac{\mathrm{d}(\hbar\omega)}{2\pi}\, A_c(k,\omega) = 1 \qquad (6.6)$$

and describes the contribution of frequencies $\hbar\omega$ to the occupation of states with wave vector k. In this way, the spectral function $A_c(k,\omega)$ contains all relevant information on the many-particle system.

For an ideal system, the self-energy vanishes, $\Sigma_c^0(k,\omega) = 0$, and we have just a delta-function, i.e. $A_c^0(k,\omega) = 2\pi\delta[\hbar\omega - E_c(k)]$. Therefore, only free states with energy $\hbar\omega = E_c(k)$ contribute to the density. In Hartree–Fock approximation, the self-energy yields a constant energy shift $\Sigma_c^{\mathrm{HF}}(k,\omega) = \Delta_c^{\mathrm{HF}}$ and we get $A_c^{\mathrm{HF}}(k,\omega) = 2\pi\delta[\hbar\omega - E_c(k) - \Delta_c^{\mathrm{HF}}]$, i.e. shifted but undamped one-particle states. In the general case, the self-energy is a complex dynamical quantity, $\Sigma_c(k,\omega) = \Delta_c(k,\omega) + i\widetilde{\Gamma}_c(k,\omega)$, and the spectral function describes shifted and damped one-particle states:

$$A_c(k,\omega) = \frac{2\widetilde{\Gamma}_c(k,\omega)}{[\hbar\omega - E_c(k) - \Delta_c(k,\omega)]^2 + [\widetilde{\Gamma}_c(k,\omega)]^2} \, . \tag{6.7}$$

Summation over the frequency yields the mean occupation number for the state with wave number k,

$$\langle n_c(k) \rangle = \int_{-\infty}^{\infty} \frac{d(\hbar\omega)}{2\pi} \, f_c(\hbar\omega) \, A_c(k,\omega) \, , \tag{6.8}$$

and, alternatively, the density of states (DOS) follows from the summation over all wave numbers,

$$D(\omega) = \frac{1}{\Omega_0} \sum_k A(k,\omega) \, . \tag{6.9}$$

The DOS for hydrogen plasma is shown schematically in Fig. 6.3. A shift and broadening of the discrete energy levels occurs for a given density, and the continuum edge is lowered with increasing density. This leads to effective

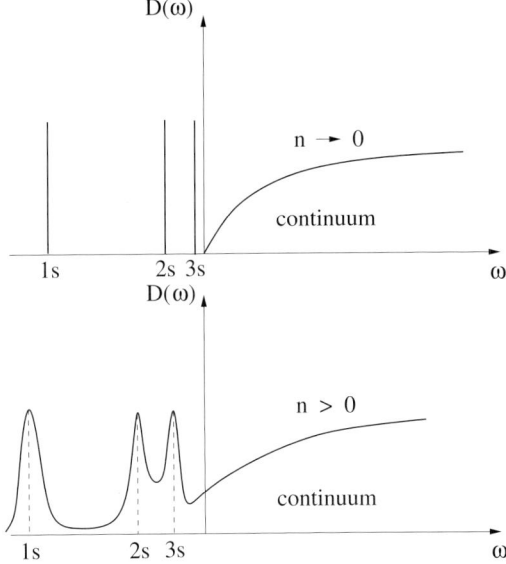

Fig. 6.3. Density of states for hydrogen (schematically). Shown are the lowest discrete s-levels and the continuum of scattering states. *Upper figure*: isolated H atom, *lower figure*: H atom embedded in a dense plasma

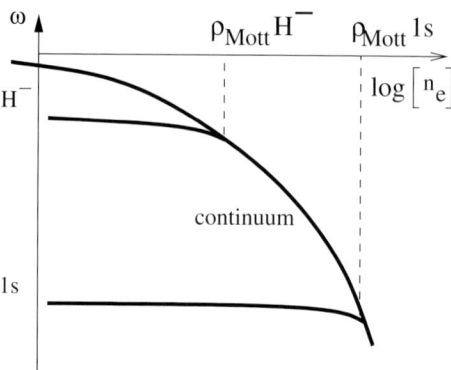

Fig. 6.4. Mott effect in hydrogen plasma (schematically). Shown are the energies ($\propto \omega$) of the 1s ground state of the atom and that of the hydrogen ion H$^-$ as function of the electron density n_e. The continuum edge is lowered, while the binding energies are shifted and broadened (not shown). At a certain density, the bound states merge into the continuum of scattering states

ionization energies so that the bound states break up at a certain density – the Mott density. Due to the broadening of the energy levels and the thermal distribution of free carriers, this transition is not sharp. We show this behavior schematically again for hydrogen plasma in Fig. 6.4.

The Mott effect leads to a transition from a weakly ionized to a fully ionized plasma and affects the physical properties strongly, especially the electrical conductivity and the reflectivity. The related nonmetal-to-metal transition may be accompanied by a plasma phase transition (PPT) so that the phase diagram at ultra-high pressures can show some peculiarities, see Chap. 12 in Part II for a more detailed discussion and further references.

6.2.2 Cluster Decomposition of the Self-Energy

The self-energy is the main ingredient for the determination of the spectral function and, thus, for the physical properties of the system. For strongly coupled plasmas, a perturbation expansion of the self-energy, i.e. considering only the lowest orders of the Born approximation, fails to describe the system adequately. We have to focus on important effects in a strongly coupled plasma, i.e. screening, strong collisions, as well as the formation and decay of bound states. This can be described within a so-called cluster decomposition of the self-energy which has originally been developed for nuclear matter [26] and then was also applied to dense plasmas [25]. The corresponding set of integral equations is shown in Fig. 6.5.

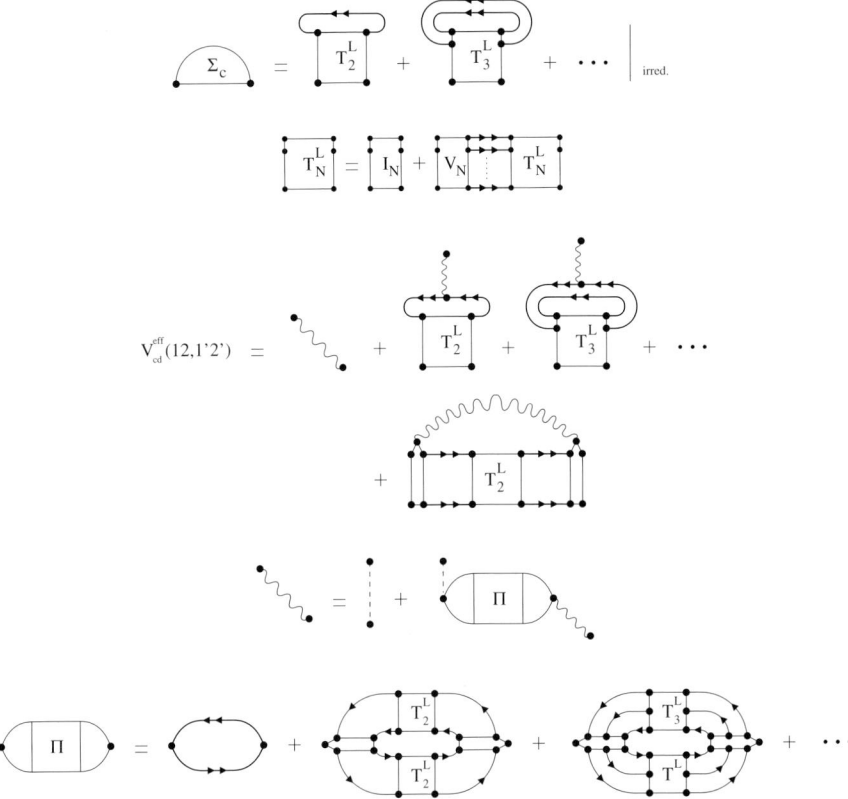

Fig. 6.5. Cluster decomposition of the self-energy $\Sigma_c(k,\omega)$ and the polarization function $\Pi(k,\omega)$ [25, 26]. A ladder approximation T_N^L is applied for each N-particle contribution with respect to the effective two-particle potential V_{cd}^{eff}. *Wavy line*: dynamically screened potential, *broken line*: (naked) Coulomb potential

The basic idea is that in a partially ionized plasma we have to consider the contribution of every bound state formed out of N particles ("N-particle cluster") to the self-energy which is, in principle, given by the sum of all irreducible diagrams (first row of diagrams in Fig. 6.5). The N-particle cluster itself is described by a N-particle ladder-T-matrix T_N^L (second row of diagrams) with respect to an effective, two-particle interaction potential V_{cd}^{eff}. This potential includes the dynamically screened potential given by a wavy line and contributions of every N-particle cluster, again via the respective N-particle ladder-T-matrix (third row of diagrams). The dynamically screened potential is determined by the polarization function $\Pi(k,\omega)$ (fourth row of diagrams) for which a similar cluster decomposition has to be applied [27] (fifth row of diagrams). The lowest-order contribution, a bubble-type diagram, gives the standard random phase approximation (RPA).

A self-consistent solution of the set of integral equations shown in Fig. 6.5 is in any case limited to clusters of lowest order, i.e. consisting of $N = 2, 3, \ldots$ particles, but has not been achieved even for this level yet [28]. However, one can focus on dominant contributions of one- and two-particle states so that a dense, partially ionized plasma is tractable within the so-called chemical picture. Pioneering work has been done on the level of the two-particle spectral function [29, 30] which was then applied to dense plasmas, see [25] for a review.

As a result, the total density of the system is given by a sum over the partial densities of N-particle clusters,

$$n_c(\beta_T, \mu_c) = n_c^{(1)} + n_c^{(2)} + 2n_c^{(3)} + \ldots , \quad (6.10)$$

which are given by respective partition functions,

$$n_c^{(1)} = \frac{1}{\Omega_0} \sum_k f_c[\varepsilon_c(k)] , \quad (6.11)$$

$$n_c^{(m)} = \frac{1}{\Omega_0} \sum_{\alpha, P} g_m[\varepsilon_{\alpha,P}^{(m)}] . \quad (6.12)$$

The Fermi distribution function of one-particle states is $f_c(E)$ and

$$g_m(E) = \{\exp[\beta_T(E - \mu_1 - \mu_2 - \cdots - \mu_m)] - (-1)^m\}^{-1} \quad (6.13)$$

is the distribution function of m-particle clusters (m odd/even: Fermi/Bose statistics). The argument of the Fermi distribution function is the quasi-particle energy of one-particle states $\varepsilon_c(k)$ determined by the Dyson equation (6.3). The energy spectrum $\varepsilon_{\alpha,P}^{(m)}$ of m-particle clusters with the total momentum $P = p_1 + p_2 + \cdots + p_m$ and the complete set of internal quantum numbers α follows from a respective m-particle Bethe–Salpeter equation (see [24]), and we have the following relations:

$$\varepsilon_c(k) = E_c(k) + \Delta_c(k) , \quad \Delta_c(k) = \operatorname{Re} \Sigma_c(k, z - i0^+)|_{z=\varepsilon_c(k)} , \quad (6.14)$$

$$\varepsilon_{\alpha,P}^{(m)} = E_{\alpha,P}^{(m)} + \Delta E_{\alpha,P}^{(m)} , \quad \Delta E_{\alpha,P}^{(m)} = \sum_{i<j} \langle \Phi_{\alpha,P}^{(m,0)} | V_{ij}^{\text{eff}} - V_{ij} | \Phi_{\alpha,P}^{(m,0)} \rangle . \quad (6.15)$$

$E_{\alpha,P}^{(m)}$ and $\Phi_{\alpha,P}^{(m,0)}$ are the eigenvalues and eigenfunctions of isolated m-particle clusters, whereas $\Delta E_{\alpha,P}^{(m)}$ contains the corrections due to interactions with the surrounding particles. The quasi-particle energies $\varepsilon_c(k)$ take into account the interaction corrections via quasi-particle shifts $\Delta_c(k)$ which yield a lowering of the ionization energy and, thus, the Mott transition as illustrated in Fig. 6.4. A solution of the self-consistent system of equations described above is rather complex. However, an approximate solution can be given within the chemical picture.

6.3 Composition of Strongly Coupled Plasmas

In order to derive the composition and the EOS of strongly coupled plasmas, the partial densities of electrons and ions as well as of bound states ("clusters") such as atoms, molecules, and molecular ions have to be calculated according to (6.10)–(6.15). One has to focus on the main physical effects in order to get tractable expressions. One usually considers the screening of the Coulomb interaction and the self-energy of the free particles (electrons and ions) as well as the reactions in the plasma. Pressure dissociation occurs in molecular systems such as hydrogen and has been presented in another contribution (see Chap. 12 in Part II) so that we concentrate here on ionization processes in noble gas and metal plasmas. We treat ionization up to higher stages, i.e.

$$M^0 \rightleftharpoons M^{1+} + e \ ,$$
$$M^{1+} \rightleftharpoons M^{2+} + e \ , \qquad (6.16)$$
$$M^{2+} \rightleftharpoons M^{3+} + e \ \ldots \ .$$

Every reaction is described by a chemical equilibrium according to

$$\mu_0 + E_{\text{ion}}^{(1)} = \mu_{1+} + \mu_e \ ,$$
$$\mu_{1+} + E_{\text{ion}}^{(2)} = \mu_{2+} + \mu_e \ , \qquad (6.17)$$
$$\mu_{2+} + E_{\text{ion}}^{(3)} = \mu_{3+} + \mu_e \ \ldots \ .$$

where $E_{\text{ion}}^{(i)}$ is the ionization energy of the ith ionization stage. The chemical potentials are split into an ideal and an interaction part $\mu_\alpha = \mu_\alpha^{\text{id}} + \mu_\alpha^{\text{cor}}$ with $\{\alpha\} = \{e, 0, 1+, 2+, 3+, 4+, 5+\}$ so that the following mass action laws are derived for the partial densities [31, 32]:

$$n_0 = 2n_{1+} \exp\left[\beta_T \left(\mu_e^{id} + E_{ion}^{(1)} + \Delta\mu_1\right)\right] \ ,$$
$$n_{1+} = \frac{1}{2} n_{2+} \exp\left[\beta_T \left(\mu_e^{id} + E_{ion}^{(2)} + \Delta\mu_2\right)\right] \ , \qquad (6.18)$$
$$n_{2+} = 2n_{3+} \exp\left[\beta_T \left(\mu_e^{id} + E_{ion}^{(3)} + \Delta\mu_3\right)\right] \ \ldots \ .$$

The main ingredients are the correlation parts of the chemical potentials,

$$\Delta\mu_k = \mu_e^{\text{cor}} + \mu_k^{\text{cor}} - \mu_{k-1}^{\text{cor}} \ , \qquad (6.19)$$

which are given by the quasi-particle shifts Δ_c of (6.14). We have used efficient Padé approximations [33] for the correlation parts of electrons and the various ion species. A hard sphere model [34] can be applied for the extended particles. A more sensitive approach to these contributions is given by fluid variational theory which is discussed in (see Chap. 12 in Part II) for hydrogen and helium plasmas. Taking beryllium plasma as an example, heavy particles with the same number of occupied shells are considered to have the same radii. We have chosen $r_1 \approx r_0 = 0.106$ nm and $r_2 \approx r_3 = 0.03$ nm. In the low density limit, the quasi-particle shifts Δ_c can be neglected and (6.18) gives the usual Saha equation, see also Chap. 1 in Part I.

The respective composition of beryllium plasma as function of the temperature is shown in Fig. 6.6 for a constant density of 0.1 g/cm^3. The occurrence of different ionization stages Be^{Z+} ($Z = 0 \ldots 4$) is characterized by the partial fraction $\alpha_Z = n_Z/n_{\text{heavy}}$ of ions with charge Z on the total number density of heavy particles $n_{\text{heavy}} = \sum_Z n_Z$. The average ionization degree $\alpha_e = n_e/n_{\text{heavy}}$ is given by the number of free electrons per heavy particle. The ionization stages are excited one after another with increasing temperature. At about 10^6 K, the maximum charge state Be^{4+} is reached for nearly all heavy particles. The average ionization degree tends to $\alpha_e = 4$. The large gap between the occurrence of Be^{2+} and Be^{3+} and the following plateau for α_e is a typical shell effect. The twofold charged ion Be^{2+} has a closed 1s shell for which a much higher ionization energy is needed to create Be^{3+} compared with the first two ionization stages.

In Fig. 6.7, our theoretical result is compared with data derived from recent spectrally resolved X-ray Thomson scattering experiments [12]. A pump pulse generates Be plasma at $\rho_{\text{solid}} = 1.84 \text{ g/cm}^3$. The 4.75 keV He$_\alpha$ line from Ti is used as X-ray source for the probe pulse. Separation of the intensities of Rayleigh and Compton scattering allows one to determine the ionization degree intensities. The overall agreement between experimental data and the

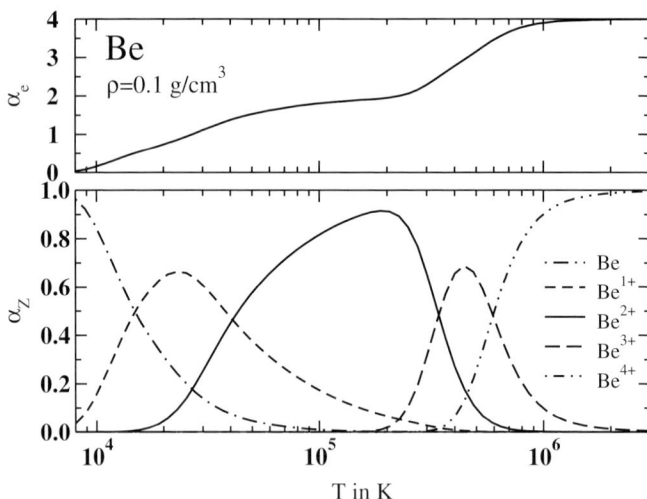

Fig. 6.6. Composition of beryllium plasma at a fixed density of 0.1 g/cm^3 as function of the temperature [2, 35]

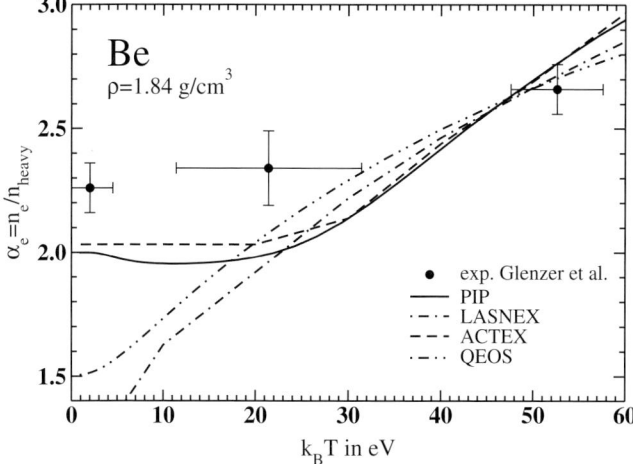

Fig. 6.7. Average ionization degree as a function of temperature as derived from the PIP model [2, 35] compared with pump-probe experiments [12] and other theoretical results [13, 36, 37]

result within the PIP model is good. It coincides with results of Rogers [36], where the interaction contributions to the equation of state were determined within an activity expansion (ACTEX). Hydrodynamic-based calculations (LASNEX) [37] and a modified Thomas–Fermi model (QEOS) [13] do not reproduce the experimental data at low temperatures.

Extensive calculations were performed for a variety of metal plasmas [2] and for noble gases [38]. For instance, the EOS calculated for helium can be compared with double-shock experiments [39]. The respective Hugoniot curves which connect all possible final states of a shock wave experiment starting from a given initial state, are shown in Fig. 6.8. The consideration of ionization is important only for compressions higher than could be achieved in these experiments, and the second ionization stage has almost no influence in that domain as can be seen from the inset. However, the consideration of ionization processes yields a maximum compression at about 400 GPa that is higher than the ideal gas limit $4\varrho_1$ for pure neutral helium indicated by a grey line; ϱ_1 is the density after the first shock in point B.

6.4 Electrical Conductivity

Besides the EOS, the electrical conductivity is an important diagnostic quantity, for which numerous experiments were performed up to the region of strong coupling $\Gamma_c \geq 1$. The transport coefficients of a partially ionized plasma can be determined within linear response theory in the formulation

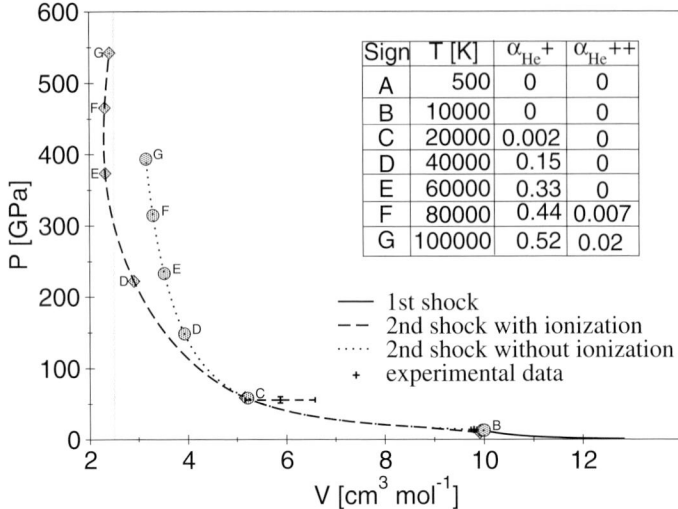

Fig. 6.8. Hugoniot curve for helium [38] compared with double-shock experiments [39]. After the first shock from (from A to B), helium is still neutral. For the second shock starting at B, several points along the Hugeniot curve are characterized by the information given in the inset: temperature, fraction of He$^+$ and He^{2+} ions. The consideration of ionization processes is important for strong shocks. The maximum compression in the ideal gas limit starting from point B is indicated by a *grey line*. The compression is described in terms of the molar volume V

of Zubarev [40] which is valid for arbitrary degeneracy of the system; see [31, 32, 41] for details. A generalized Boltzmann equation is derived, and the drift and collision term are expressed in terms of correlation functions. The correlation functions are related to thermodynamic Green's functions and similar techniques as used for the EOS in Sect. 6.2.2 can be used to evaluate these quantities in a systematic way, see also [42, 43].

The transport coefficients are then given in a compact determinant representation known from standard kinetic theory and the electrical conductivity reads

$$\sigma = -\frac{\beta_T e^2}{m_e^2 \Omega_0} \frac{1}{|(D_{nm})|} \begin{vmatrix} 0 & (N_{0m}) \\ (N_{n0}) & (D_{nm}) \end{vmatrix}. \tag{6.20}$$

The elements of the matrices, the correlation functions $N_{nm} = (\mathbf{P}_n, \mathbf{P}_m)$ and $D_{nm} = \langle \dot{\mathbf{P}}_n(\varepsilon), \dot{\mathbf{P}}_m \rangle + (\dot{\mathbf{P}}_n, \mathbf{P}_m)$, are defined in [41]. The moments of the distribution function $\mathbf{P}_n = \sum_k \hbar \mathbf{k} (\beta_T E_k)^n a_k^\dagger a_k$ with $E_k = \hbar^2 k^2/(2m_e)$ characterize, as a set of relevant observables, the non-equilibrium state. Creation and annihilation operators are given by $a_k^\dagger$ and a_k, and the time derivative of operators $\mathbf{P}$ is denoted by $\dot{\mathbf{P}}$.

The correlation functions N_{nm} and D_{nm} can be evaluated for arbitrary degeneracy. The N_{nm} are given by multiples of the electron particle number

N_e. The force–force correlation functions D_{nm} contain contributions due to electron–ion, electron–electron, and electron scattering at neutrals (atoms, molecules), i.e. $D_{nm} = D_{nm}^{\text{ei}} + D_{nm}^{\text{ee}} + D_{nm}^{\text{en}}$. They can be evaluated on different levels also known from kinetic theory: the Boltzmann (T matrix), Lenard–Balescu (dynamic screening), and Landau collision term (static Born approximation).

The scattering processes between charged particles are treated on T matrix level performing a phase-shift analysis with respect to the Debye potential. In this way, also strong collisions are incorporated. Elastic electron–neutral scattering is treated in Born approximation with respect to a statically screened polarization potential. Three moments $n, m = \{0, 1, 2\}$ are taken for practical evaluations which is sufficient to reach convergence of the method within few percent, see [25]. Similar expressions for other transport coefficients such as thermal conductivity and thermopower can be given [41].

We compare as illustrative examples the calculated isotherms for the electrical conductivity of aluminum and xenon plasma with data from shock wave experiments [45] in Figs. 6.9 and 6.10, respectively. As one can see, the present linear response formalism is applicable for a large domain of densities and temperatures. It gives the correct low-density (Spitzer results [52]) and high-density behavior (Ziman–Faber formula [53]). The region of strong coupling in between is characterized by a nonmetal-to-metal transition, a

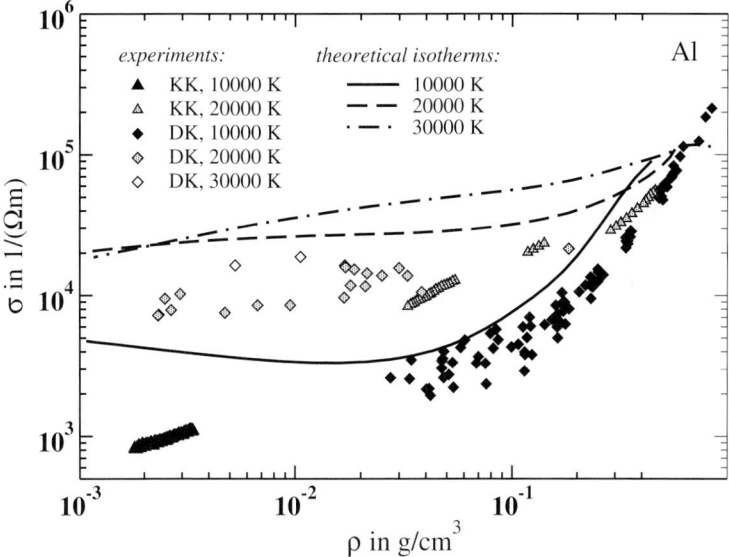

Fig. 6.9. Isotherms for the electrical conductivity of partially ionized aluminum plasma as function of the density [2] compared with rapid wire evaporation experiments. KK refers to Krisch and Kunze [5], DK to De Silva and Kunze [6]

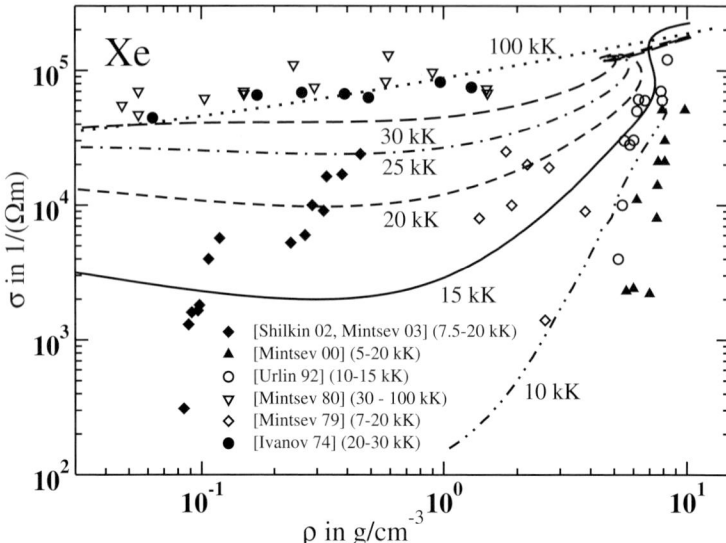

Fig. 6.10. Calculated electrical conductivity in xenon plasma [2, 44] compared with shock wave experiments, see [45] for an overview. The experimental data are taken from [46, 47, 48, 49, 50, 51]

steep increase of the electrical (and thermal) conductivity, which is caused by pressure ionization in our chemical model. This transition is very pronounced at low temperatures. The question whether or not it is accompanied by a thermodynamic instability, the PPT, is discussed in Chap. 12 in Part II.

Other quantities such as the reflectivity are also used for plasma diagnostics. The dc conductivity (6.20) can be related via a Drude formula to the reflectivity which has been studied in detail for xenon plasma [54, 55]. Further applications of the linear response theory and the Green's function technique presented here are related to the dielectric function, especially the local field factors [56], and the inverse bremsstrahlung [57] in strongly coupled plasmas.

6.5 Conclusion

We have given here a brief overview of theoretical approaches to the thermophysical properties of strongly coupled plasmas. Especially, the Green's function technique allows a systematic evaluation of relevant contributions to the self-energy, the key quantity in this context. However, state-of-the art evaluations are limited to certain classes of diagrams and, thus, always approximate solutions of the problem.

The enormous progress in simulation technique allows already now a rather complete treatment of many-particle systems. Especially, the combination of DFT and MD simulations is very promising for a genuine quantum molecular dynamics (QMD) technique which has already been applied to strongly coupled plasmas and warm dense matter [16, 17, 18, 19]. Alternatively, the accurate high-temperature results of the PIMC method can be extended to lower and lower temperatures [21, 22].

Pioneering experiments to the physics of strongly coupled plasmas are expected from new facilities as, e.g., the National Ignition Facility (NIF) in Livermore [58] and the FEL at DESY Hamburg [11]. Furthermore, high-power lasers or intense heavy ion beams are used at present to generate and diagnose strongly coupled plasmas [59]. Physical problems in this field are closely related to astrophysics where a growing number of high-resolution observatories on Earth and in space will give new insight into the behavior of matter under extreme conditions [60].

I thank Mike Desjarlais, Siegfried Glenzer, Hauke Juranek, André Kietzmann, Sandra Kuhlbrodt, Christian Neißner, Heidi Reinholz, Gerd Röpke, Volker Schwarz, and Thomas Tschentscher for stimulating discussions, their help in preparing the lecture and this manuscript. I thank the organizers of the Summer School for the invitation and the WE–Heraeus Foundation for financial support.

References

1. *Papers presented at the International Conference on Strongly Coupled Coulomb Systems*, In: J. Phys. A–Math. Gen. **36**, 5827–6280 (2003)
2. S. Kuhlbrodt: Transport properties of dense plasmas. PhD Thesis, University of Rostock, Rostock (2003)
3. S. Ichimaru: *Plasma Physics* (Addison–Wesley, Redwood City 1988)
4. A.W. DeSilva, H.-J. Kunze: Phys. Rev. E **49**, 4448 (1994)
5. I. Krisch, H.-J. Kunze: Phys. Rev. E **58**, 6557 (1998)
6. A.W. DeSilva, J.D. Katsouros: Phys. Rev. E **57**, 5945 (1998); J. Phys. IV **10**, Pr5-209 (1999)
7. S.T. Weir, A.C. Mitchell, W.J. Nellis: Phys. Rev. Lett. **76**, 1860 (1996)
8. L.B. Da Silva, P. Celliers, G.W. Collins et al: Phys. Rev. Lett. **78**, 483 (1997)
9. M.D. Knudson, D.L. Hanson, J.E. Bailey et al: Phys. Rev. Lett. **87**, 225501 (2001); M.D. Knudson, D.L. Hanson, J.E. Bailey et al: Phys. Rev. Lett. **90**, 035505 (2003)
10. S.I. Belov, G.V. Boriskov, A.I. Bykov et al: JETP Lett. **76**, 433 (2002)
11. The X-Ray Free Electron Laser. In: *TESLA Technical Design Report, part V* ed by G. Materlik, Th. Tschentscher (DESY, Hamburg 2001); First Stage of the X-Ray Laser Laboratory. In: *TESLA XFEL Technical Design Report (supplement)*, ed by R. Brinkmann et al (DESY, Hamburg 2002)
12. S.H. Glenzer, G. Gregori, R.W. Lee et al: Phys. Rev. Lett. **90**, 175002 (2003); S.H. Glenzer, G. Gregori, F.J. Rogers et al: Phys. Plasmas **10**, 2433 (2003)

13. R.M. More, K.H. Warren, D.A. Young et al: Phys. Fluids **31**, 3059 (1988); A. Kemp, J. Meyer-ter-Vehn: Das Zustandsgleichungsmodell QEOS für heiße Materie. MPQ-Report 229, Max-Planck-Institut für Quantenoptik, Garching, Germany (1998)
14. S. Ichimaru, S. Mitake, S. Tanaka et al: Phys. Rev. A **32**, 1768 (1985); S. Mitake, S. Tanaka, X. Yan et al: ibid 1775; S. Tanaka, S. Mitake, X. Yan et al: ibid 1779; X. Yan, S. Tanaka, S. Mitake et al: ibid 1785; S. Ichimaru, S. Tanaka: ibid 1790
15. M. Knaup, P.-G. Reinhard, C. Toepffer: Contrib. Plasma Phys. **41**, 159 (2001); M. Knaup, P.-G. Reinhard, C. Toepffer et al: J. Phys. A–Math. Gen. **36**, 6165 (2003)
16. M.P. Desjarlais, J.D. Kress, L.A. Collins: Phys. Rev. E **66**, 025401(2002)
17. M.P. Desjarlais: Phys. Rev. B **68**, 064204 (2003)
18. L. Collins, I. Kwon, J. Kress et al: Phys. Rev. E **52**, 6202 (1995)
19. T.J. Lenosky, S.R. Bickham, J.D. Kress et al: Phys. Rev. B **61**, 1 (2000)
20. G. Galli, R.Q. Hood, A.U. Hazi et al: Phys. Rev. B **61**, 909 (2000)
21. B. Militzer, D. Ceperley: Phys. Rev. Lett. **85**, 1890 (2000)
22. B. Militzer, D. Ceperley, J.D. Johnson et al: Phys. Rev. Lett. **87**, 275502 (2001)
23. L.P. Kadanoff, G. Baym: *Quantum Statistical Mechanics* (WA Benjamin Inc., New York 1962)
24. W.D. Kraeft, D. Kremp, W. Ebeling et al: *Quantum Statistics of Charged Particle Systems* (Akademie-Verlag, Berlin 1986; Plenum, New York 1986)
25. R. Redmer: Phys. Rep.-Rev. Sec. Phys. Lett. **282**, 35 (1997)
26. G. Röpke, L. Münchow, H. Schulz: Nucl. Phys. A **379**, 526 (1982); G. Röpke, M. Schmidt, L. Münchow et al: Nucl. Phys. A **399**, 587 (1983); G. Röpke, M. Schmidt, H. Schulz: Nucl. Phys. A **424**, 594 (1984)
27. G. Röpke, R. Der: Phys. Status Solidi B-Basic Res. **92**, 501 (1979)
28. A. Wierling: Vielteilchentheoretische Beschreibung des Plasmas im Sonneninneren und die Elektroneneinfangreaktion an Be VII. PhD Thesis, University of Rostock, Rostock (1997)
29. H. Stolz, R. Zimmermann: Phys. Status Solidi B-Basic Res. **94**, 135 (1979)
30. D. Kremp, W.D. Kraeft, A.J.D. Lambert: Physica A **127**, 72 (1984) D. Kremp, M.K. Kilimann, W.D. Kraeft et al: Physica A **127**, 646 (1984)
31. R. Redmer: Phys. Rev. E **59**, 1073 (1999)
32. S. Kuhlbrodt, R. Redmer: Phys. Rev. E **62**, 7191 (2000)
33. A. Förster, T. Kahlbaum, W. Ebeling: Laser Part. Beams **10**, 253 (1992)
34. G.A. Mansoori, N.F. Carnahan, K.E. Starling et al: J. Chem. Phys. **54**, 1523 (1971)
35. S. Kuhlbrodt, H. Juranek, V. Schwarz et al: Contrib. Plasma Phys. **43**, 342 (2003)
36. F.J. Rogers: Phys. Rev. A **24**, 1531 (1981); Astrophys. J. **310**, 723 (1986); Phys. Plasmas **7**, 51 (2000)
37. J.A. Harte, W.E. Alley, D.S. Bailey et al: LASNEX–A 2-d Physics Code for Modelling ICF. Report UCRL-LR-105821-96-4, Lawrence Livermore National Laboratory, Livermore (1996) pp 150–164
38. V. Schwarz: Zustandsgleichung von Wasserstoff-Helium-Gemischen. Diploma thesis, University of Rostock, Rostock (2003)
39. W.J. Nellis, N.C. Holmes, A.C. Mitchell et al: Phys. Rev. Lett. **53**, 1248 (1984)
40. D.N. Zubarev, V. Morozov, G. Röpke: *Statistical Mechanics of Nonequilibrium Processes*, vols 1 and 2 (Akademie–Verlag, Berlin 1996 and 1997)

41. H. Reinholz, R. Redmer, S. Nagel: Phys. Rev. E **52**, 5368 (1995)
42. F.E. Höhne, R. Redmer, G. Röpke et al: Physica A **128**, 643 (1984)
43. G. Röpke: Phys. Rev. A **38**, 3001 (1988)
44. R. Redmer, H. Juranek, S. Kuhlbrodt et al: Z. Phys. Chemie-Int. J. Res. Phys. Chem. Chem. Phys. **217**, 783 (2003)
45. V.E. Fortov, V.Ya. Ternovoi, M.V. Zhernokletov et al: J. Exp. Theor. Phys. **97**, 259 (2003)
46. N.S. Shilkin, S.V. Dudin, V.K. Gryaznov et al: J. Exp. Theor. Phys. **97**, 922 (2003)
47. V.B. Mintsev, V.Ya. Ternovoi, V.K. Gryaznov et al: Electrical Conductivity of Shock Compressed Xenon. In: *Shock Compression of Condensed Matter, AIP Conference Proceedings 505* ed by M.D. Furnish, L.C. Chhabildas, R.S. Hixson (Woodbury, New York 2000) pp 987–990
48. V.D. Urlin, M.A. Mochalov, O.L. Mikhailova: High Pressure Res. **8**, 595 (1992)
49. V.B. Mintsev, V.E. Fortov, V.K. Gryaznov: Sov. Phys. JETP **52**, 59 (1980)
50. V.B. Mintsev, V.E. Fortov: Jetp Lett. **30**, 375 (1979)
51. Y.V. Ivanov, V.E. Fortov, V.B. Mintsev et al: Sov. Phys. JETP **44**, 112 (1976)
52. L. Spitzer, R. Härm: Phys. Rev. **89**, 977 (1953)
53. J.M. Ziman: Philos. Mag. **6**, 1013 (1961)
54. V.B. Mintsev, Y.B. Zaporoghets: Contrib. Plasma Phys. **29**, 493 (1989)
55. H. Reinholz, Yu. Zaporoghets, V. Mintsev et al: Phys. Rev. E **68**, 036403 (2003)
56. G. Röpke, R. Redmer, H. Reinholz et al: Phys. Plasmas **7**, 39 (2000); H. Reinholz, R. Redmer, G. Röpke et al: Phys. Rev. E **62**, 5648 (2000)
57. A. Wierling, Th. Millat, G. Röpke et al: Phys. Plasmas **8**, 3810 (2001)
58. See http://www.llnl.gov/nif
59. D.H.H. Hoffmann, V.E. Fortov, I.V. Lomonosov et al: Phys. Plasmas **9**, 3651 (2002)
60. See http://www.eso.org for the European Southern Observatory in Chile and http://www.esa.int/export/esaCP/index.html for the European Space Agency

Part II

Confinement, Transport and Collective Effects

7 Magnetic Confinement

F. Wagner and H. Wobig

Abstract. This chapter describes the basic principles of magnetic confinement. The main representatives of magnetic confinement are the mirror machine, the tokamak device and the stellarator. A short description of the technical layout and the theory of plasma confinement in these concepts will be given. The chapter ends with the problem of plasma losses in toroidal devices and discusses the various transport processes.

7.1 Conditions for Fusion

The great example for energy production by the fusion processes is the sun. And in order to gain energy by fusion in a similar way as the sun specific conditions must be fulfilled. For example, in the case of a deuterium–tritium plasma the triple product $nT\tau_E$, which is the product of density, temperature and confinement time, must be larger than 3.3×10^{21} m^{-3}keVs in the ideal case of a clean plasma. As a side condition the temperature must be about 15 keV, where the maximum of the fusion output is. The goal of magnetic confinement is to reach a Lawson parameter of $n\tau_E > 2.2 \times 10^{20}$ m^{-3}s at a density of about 2×10^{20} m^{-3} and a confinement time of several seconds.

The energy confinement time τ_E is a figure of merit for the thermal insulation of the high-temperature plasma. It results from the energy balance of the plasma, where plasma heating by alpha-particles is in equilibrium with the plasma losses which consist of radiation and transport losses by heat conduction and convection. The confinement time of the plasma must be large enough so that the energy released by the fusion products to the plasma is larger or equal to the energy lost by thermal conduction and radiation.

The situation is comparable to the cooling-down time in a house if one turns off the heating system at night. In the simplest case the temperature decays exponentially in a time τ_E. If the house is well isolated, this decay time – or energy confinement time of the house – is long. If the house is badly isolated, one needs a more powerful oven to maintain the same room temperature. Fusion scientists study and optimize the confinement properties of the plasma aiming at a large confinement time of the energy and a small confinement time of the "ash", which is helium.

A more detailed discussion is to be found in Chap. 17 in Part III.

7.2 The Need for Magnetic Confinement

Some simple considerations demonstrate the capability of a magnetic field to confine charged particles. Let us compare the transport properties of a plasma under reactor conditions with magnetic field and without magnetic field. The physics of hot plasmas is introduced in Chap. 4 in Part I. A key element is the mean free path of the particles which is about $\lambda_{DD} = 10^4$ m for deuterons and electrons at a temperature of 15 keV and a density of 2.5×10^{20} m^{-3}. These numbers are computed on the basis of Coulomb collisions between charged particles and, since heat conduction is also of interest, like-particle collisions between deuterons (or tritons) and electrons. In analogy to Brownian motion we can make an estimate of the energy diffusion coefficient combining the mean free path λ and the collision time τ. The result is $D = \lambda^2/\tau$. Collision time and mean free path are correlated by $\lambda = v\tau$; v is the thermal velocity. In a reactor plasma the thermal velocity is $v = 1.2 \times 10^6$ ms^{-1} for deuterons and 7.26×10^7 ms^{-1} for electrons. Typical numbers of the energy diffusion coefficient are $D_D = \lambda v = 1.2 \times 10^{10}$ m^{2}s^{-1} and $D_e = 7.26 \times 10^{11}$ m^{2}s^{-1}. The confinement of a plasma without magnetic field is characterized by large diffusion coefficients. The estimates of the confinement time $\tau_E = a^2/D$ based on these diffusion coefficients yield hopelessly small values (a is the plasma diameter).

Now let us consider the issue the other way around: how large must a D-T plasma be in order to reach a confinement time of about $\tau_E = 1$ s? The answer is 700 km, which proves that a fusion reactor on this basis is an illusion. The way out of this dilemma is magnetic confinement. Charged particles move freely along magnetic field lines, however, perpendicular to the magnetic field they encircle the field lines on a small circular orbit; they gyrate around the field lines. Perpendicular to the magnetic field perfect confinement exists and, as long as the particles do not hit a material target, they are well confined. The gyro radius plays the role of the mean free path. The gyro or Larmor radius is given by $\rho_L = [(2mT)/(eB)]^{1/2}$, which, in a fusion plasma with a magnetic field of 5 T and a temperature of 15 keV, is $\rho_D = 5 \times 10^{-3}$ m for deuterons and $\rho_e = 8 \times 10^{-5}$ m for electrons.

Let us now replace the mean free path by the gyro radius and use the Coulomb collision time again as above. The energy confinement time of a plasma with 1 m diameter is now 1×10^3 s for ions and 3×10^4 s for electrons. The conditions are reversed compared to the field-free plasma: The heavy ions represent the loss channel, their confinement time is shorter than the confinement time of the electrons. As a consequence of good confinement small dimensions are sufficient to reach the goal of a self-sustained fusion plasma. However, these estimates are only useful to demonstrate the importance of the magnetic field. In reality the confinement time of a magnetically confined plasma is not determined by the gyro radius and Coulomb collisions, but by turbulent processes and, under specific conditions, by direct particle losses (see Chap. 9 in Part II).

Comment 1. Let us consider a cylindrical plasma confined in a tube and imbedded in an axial magnetic field. On the axis we have fusion conditions with high density and on the boundary – strictly speaking in a small sheath of the order gyro radius – we have zero density. There is a density and a temperature gradient which implies that a fusion plasma isolated from the wall is not in thermodynamic equilibrium.

Comment 2. In a confined plasma there are particle and energy fluxes opposite to the gradients of density and temperature and without sources the plasma would decay on the time scale of confinement times. In order to sustain a stationary state we need particle and energy sources. In a fusion plasma highly energetic alpha-particles provide the energy source in the plasma centre. An external source is required for refuelling the particles.

Comment 3. A magnetically confined plasma is highly anisotropic. Particles can freely move along magnetic field lines; in perpendicular direction they are tightened to field lines. Parallel to the magnetic field the mean free path is the characteristic length, the gyro radius is the characteristic length in perpendicular direction. Along field lines the plasma is nearly homogeneous since any deviation from constancy is instantly levelled off; only in perpendicular direction strong gradients are possible.

7.3 Particle Motion in Electro-Magnetic Fields

A plasma consists of charged particles which strongly interact with each other. However, in a high-temperature fusion plasma the mean free path for binary collisions is very large and the orbits in between collisions are subject only to the forces of the averaged magnetic and electric fields. Therefore, understanding plasma confinement in magnetic fields means, at first, to understand the unperturbed orbits of charged particles. Good plasma confinement requires good confinement of particle orbits. In the following section we will describe the motion of single particles in a given electric and magnetic field. The equation of motion of a particle with charge q is

$$\mathbf{F} = \frac{d\mathbf{p}}{dt} = q(\mathbf{E} + \mathbf{v} \times \mathbf{B}) + \mathbf{K}_{ext} \qquad (7.1)$$

$\mathbf{p}$ is the momentum, $\mathbf{E}$ is the electric field, $\mathbf{v} \times \mathbf{B}$ is the Lorentz-force and $\mathbf{K}_{ext}$ is an external force, e.g. a friction force. In homogeneous magnetic fields the orbit of the particles is a circle with radius ρ_L and frequency ω_c.

Case 1: Motion in a Homogeneous Magnetic Field

We consider the equation of motion $m\dot{\mathbf{v}} = q\mathbf{v} \times \mathbf{B}$ which reads in Cartesian coordinates

$$\ddot{x} = \frac{qB}{m}\dot{y}; \quad \ddot{y} = -\omega_c \dot{x}; \quad \ddot{z} = 0 \tag{7.2}$$

$\omega_c = qB/m$ is the cyclotron or Larmor frequency; it does not depend on the energy of the particles. The solution of the equation of motion is

$$x(t) = x_0 + \frac{mv_\perp}{qB}\sin(\omega_c t); \quad y(t) = y_0 + \rho_L \cos(\omega_c t); \quad z = z_0 + v_\| t \tag{7.3}$$

$v_\|$, $v_\perp$ are the velocity components parallel and perpendicular to the magnetic field (see Fig. 7.1), which together with the spatial coordinates x_0, y_0 and z_0 are the initial values of the solution. ρ_L is the Larmor-Radius: $\rho_L = v_\perp/\omega_c = mv_\perp/qB$, $h = v_\|/\omega_c$ is the pitch of the helix. The velocity $\mathbf{v}$ is the sum of the perpendicular and parallel components

$$\mathbf{v} = \mathbf{v}_\perp + \mathbf{v}_\|; \quad \mathbf{v}_\| = v_\| \mathbf{b}; \quad \mathbf{v}_\perp = \mathbf{b} \times (\mathbf{v} \times \mathbf{b}) \tag{7.4}$$

$\mathbf{b}$ is the unit vector in the direction of the magnetic field. The cyclotron frequency of electrons and protons is

$$\omega_{ce} = 1.76 \times 10^{11} B \,(\mathrm{s}^{-1}); \quad \omega_{cp} = 9.58 \times 10^7 B \,(\mathrm{s}^{-1}) \quad \text{with } B \text{ in (T)} \tag{7.5}$$

x_0, y_0 and $z_0 + v_\| t$ is the centre of the circular motion, it is called the *guiding centre*. In most cases one is more interested in the motion of this guiding centre than in the actual orbit. In particular in fusion experiments, where the Larmor radius is small in comparison to the size of the plasma, it is sufficient to compute the orbit of the guiding centre. The orbit of the guiding centre may be considered as the average over the gyration; extensive theoretical studies have shown that this interpretation is justified.

Comment: Magnetic Moment of Gyrating Particles

The circular motion of electrons and ions is equivalent to electric currents, which generate a magnetic field themselves. This dipole field is opposite to the external field and weakens the external magnetic field. This diamagnetic effect is characterized by the *magnetic moment* of the guiding centre

$$\mu = -\frac{m\mathbf{v}_\perp^2}{2B}\mathbf{b} = -\frac{W_\perp}{B}\mathbf{b}. \tag{7.6}$$

The magnetic moment μ grows linearly with the perpendicular energy $W_\perp$; it is independent of the charge. At equal temperatures the magnetic moments of electrons and ions are equal. The magnetic field inside the gyrating orbit has been reduced.

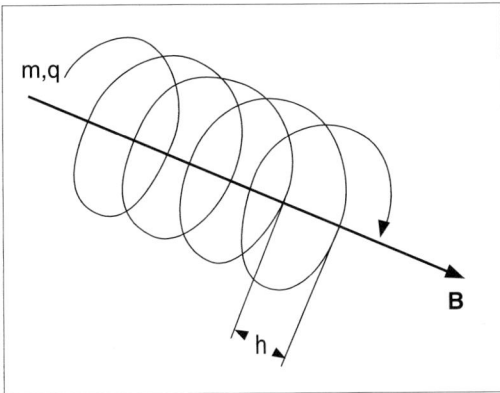

Fig. 7.1. Gyration of a charged particle in a homogeneous magnetic field. The particle orbit in the $x-y$-plane is a circle with radius ρ_L. The velocity in z-direction is constant and after one gyro period the particle has travelled over the distance h which is the pitch of the helical orbit

Case 2: Force on Charged Particles in a Magnetic Field

Next we consider the general case of a particle in a magnetic field and neglect the acceleration in the equation of motion (7.1) and thus we neglect the gyration of the particle

$$0 = \mathbf{K}_\perp + q\left(\mathbf{v}_D \times \mathbf{B}\right)_\perp . \tag{7.7}$$

$\mathbf{K}$ is an external force resulting from an electric field or a gravitational field. $\mathbf{K}_\perp$ is the component perpendicular to $\mathbf{B}$. The equation describes the drift velocity $\mathbf{v}_D$ which is perpendicular to $\mathbf{v}$ and $\mathbf{K}$

$$\mathbf{v}_D = \frac{\mathbf{K} \times \mathbf{B}}{qB^2} . \tag{7.8}$$

In general the drift velocity depends on the charge and its sign. The drift of particles with opposite charge leads to a polarisation of the plasma.

Case 3: Constant Electric Field and Constant Magnetic Field

The magnetic field may point into the z-direction and the electric field into the x-direction. The electric field accelerates the particle in the x-direction and the resulting motion is a superposition of a drift and the gyration. The equation of motion is

$$\ddot{x} = \omega_c \dot{y} + \frac{E}{m} ; \quad \ddot{y} = -\omega_c \dot{x} \tag{7.9}$$

which has the solution

$$x(t) = \rho_L \sin(\omega_c t); \quad y(t) = \rho_L \cos(\omega_c t) - v_E t \quad \text{with } v_E = \frac{E}{B}. \quad (7.10)$$

The new element in this solution is the drift into the $\mathbf{E} \times \mathbf{B}$-direction. This drift velocity $\mathbf{v}_D = \mathbf{E} \times \mathbf{B}/B^2$ does not depend on the charge or mass of the particles, which implies that the total plasma, ions and electrons, drift with the same velocity into the y-direction.

Case 4: Drift in an Inhomogeneous Magnetic Field

The orbit of the gyrating particle has a constant curvature if the magnetic field is homogeneous resulting in an orbit which is a circle. However, if the magnetic field is inhomogeneous the curvature is no longer constant. It is larger in regions of strong magnetic field. The orbit is not closed any more and in average the particle drift into a direction which is perpendicular to $\mathbf{B}$ and ∇B (see Fig. 7.2). Replacing the gyrating particle by a dipole with magnetic moment μ leads to a force $\mathbf{K}_{\nabla B} = -\mu \nabla B$ and the drift is the resulting $\mathbf{K}_{\nabla B} \times \mathbf{B}$-motion:

$$\mathbf{v}_{\nabla B} = \frac{\mathbf{K} \times \mathbf{B}}{qB^2} = -\mu \frac{\nabla B \times \mathbf{B}}{qB^2} \quad (7.11)$$

or

$$\mathbf{v}_{\nabla B} = \frac{1}{2} v_\perp \rho_L \frac{\mathbf{B} \times \nabla B}{qB^2}. \quad (7.12)$$

This ∇B-drift depends on mass and charge of the particles and since particles with opposite charges drift in opposite directions there is a tendency for charge separation. However, a build-up of an electric field $\mathbf{E}$ prohibits a charge separation and the associated $\mathbf{K} \times \mathbf{B}$-drift shifts the plasma in $\mathbf{E} \times \mathbf{B}$-direction.

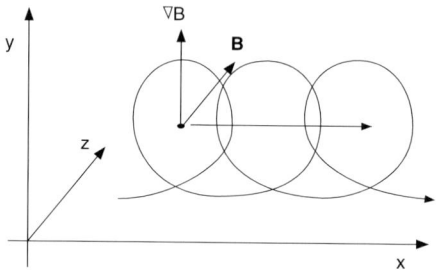

Fig. 7.2. Drift in inhomogeneous magnetic field. The magnetic field in this figure is $\mathbf{B} = [0, 0, B_z(y)]$. The curvature of the particle orbit grows with field strength $B_z(y)$ in the y-direction; for this reason the orbits in the $x-y$-plane are no longer closed as in the case of a homogeneous magnetic field

Case 5: Curvature Drift in Inhomogeneous Fields

If the magnetic field is inhomogeneous field lines are also curved, in general. Particles, which move along field lines, experience a centrifugal force $\mathbf{K}_c = mv_\parallel^2 \mathbf{n}/R_c$. R_c is the local radius of curvature and $\mathbf{n}$ is the normal vector of the field line. The result of this force is another drift perpendicular to $\mathbf{B}$ and $\mathbf{K}_c$

$$\mathbf{v}_c = \frac{\mathbf{K}_c \times \mathbf{B}}{qB^2} = \frac{mv_\parallel^2}{qB^2} \frac{\mathbf{n} \times \mathbf{B}}{R_c}. \tag{7.13}$$

There exists a relation between the radius of curvature and the inhomogeneity of the field. Let $\mathbf{n}$ be the normal vector of the field line. The following relation follows from Frenet's formula

$$\frac{\mathbf{n}}{R_c} = \left(\frac{\mathbf{B}}{B} \cdot \nabla\right) \frac{\mathbf{B}}{B}. \tag{7.14}$$

In a vacuum magnetic field the right hand side is

$$\left(\frac{\mathbf{B}}{B} \cdot \nabla\right) \frac{\mathbf{B}}{B} = \frac{\nabla B}{B} - \mathbf{b}\left(\mathbf{b} \cdot \frac{\nabla B}{B}\right) = \frac{\nabla_\perp B}{B}; \quad \mathbf{b} = \frac{\mathbf{B}}{B}. \tag{7.15}$$

Superposition of the $\nabla_\perp B$-drift and the curvature drift yields

$$\mathbf{v}_{\nabla B} = \frac{m}{q}\left(v_\parallel^2 + \frac{v_\perp^2}{2}\right) \frac{\mathbf{B} \times \nabla_\perp B}{B^2}. \tag{7.16}$$

Due to these drifts the particles do not stay on field lines; there is the danger that they leave the confinement region and drift towards the wall. Therefore, a major task of confining a plasma is to optimize the magnetic field in view of reducing the drift to a tolerable level.

Case 6: Parallel Motion in Inhomogeneous Fields

Next, we consider the effect of a gradient of B in parallel direction. Again, we replace the gyrating particle by its magnetic moment and find a parallel force

$$K_\parallel = -\mu \nabla_\parallel B \tag{7.17}$$

A particle which moves into a region of stronger magnetic field will be slowed down by this force, since the increase of B along field lines acts like a mirror. This magnetic mirror effect is used to confine the plasma in mirror machines.

7.4 Constants of Motion

If a magnetic field has continuous symmetry like axisymmetry, particle orbits are much more easily to compute than in non-symmetric fields. In such a case one or several spatial coordinates are ignorable, which means that the field does not depend on these coordinates. According to Noether's theorem in such a case one or several integrals of motion exist. In case of an axisymmetric magnetic field the toroidal canonical momentum is a constant of motion. The earth's angular momentum is such an example since the gravitational field of the sun does not depend on the angular coordinate.

7.4.1 Exact Invariants

Case 1: Exact Integrals of Motion: Kinetic Energy

In steady state magnetic fields the kinetic energy $W = 1/2\,mv^2$ is a conserved quantity. This can easily be derived by multiplying the force balance with the velocity $\mathbf{v}$

$$m\mathbf{v} \cdot \frac{d\mathbf{v}}{dt} = \frac{d}{dt}\frac{m}{2}v^2 = q\mathbf{v} \cdot (\mathbf{v} \times \mathbf{B}) = 0 \;. \tag{7.18}$$

The Lorentz-force is always perpendicular to the velocity and for this reason the work done by the Lorentz-force is zero. It should be noted that in time-dependent fields the energy is not conserved; particles can gain or loose energy.

Case 2: Conservation of Canonical Momentum

An axisymmetric magnetic field in cylinder coordinates r, θ, z is described by $\mathbf{B} = [B_r(r,z), 0, B_z(r,z)] = \nabla \times \mathbf{A}$. The vector potential has only one component $\mathbf{A} = [0, A_\theta(r,z), 0]$. The components of the canonical momentum are

$$p_r = m\dot{r}\,; \quad p_\theta = mr^2\dot{\theta} + qrA_\theta\,; \quad p_z = m\dot{z}\;. \tag{7.19}$$

Because of the toroidal invariance the canonical momentum p_θ is conserved. Note that not the mechanical momentum $mr^2\dot{\theta}$ is conserved but the canonical momentum. This is a special case of the general theorem in mechanics, which says that, if the Hamiltonian does not depend on a coordinate q_k, the conjugate canonical momentum p_k is conserved.

7.4.2 Adiabatic Invariants

In addition to the constants of motion the adiabatic invariants are of great interest. The notion of an adiabatic invariant is meaningful if a system with fast oscillatory motion has a slowly varying parameter. Slow means slow compared with the oscillation. A well-known mechanical example is the pendulum with slowly varying length $l(t)$. If the amplitude is small the pendulum behaves like a harmonic oscillator with slowly varying frequency $\omega(t)$. Small means here

$$\frac{1}{\omega}\frac{d\omega}{dt} = o(\epsilon) \ll 1 . \tag{7.20}$$

It already has been found by Einstein 1911 that the ratio H/ω is an adiabatic invariant, i.e. it is constant of the order ϵ. H is the total energy of the pendulum.

In the general theory of adiabatic invariants one starts from a Hamiltonian $H(p, q, \epsilon t)$ which describes a fast oscillation in the p, q-plane. In lowest order in ϵ the orbits are closed. The theory states that the action $J = \oint p\, dq$ is constant of the order ϵ. Adiabatic invariants are useful tools to understand the confinement of charged particles, they help to compute the long-term behaviour of particles and the guiding centres. Details of the gyro-motion are averaged out. In the strong magnetic field of fusion experiments there are three adiabatic invariants which will be described shortly in the following.

Case 1: The Magnetic Moment μ

As has been shown in (7.2) the motion of the gyrating particle is equivalent to a harmonic oscillator. Now let us assume that the magnetic field is slowly varying, either in time or in space. As quoted above the ratio $W_\perp/\omega_c$ is an adiabatic invariant. Explicitly this reads

$$J_\perp = \frac{mv_\perp^2}{2\omega_c} \propto \frac{mv_\perp^2}{2B} = \mu . \tag{7.21}$$

The magnetic moment defined in (7.6) is an adiabatic invariant. Since the gyro radius is small in fusion experiments and the magnetic field variation over the gyro radius is very small the magnetic moment may be considered as a constant in nearly all cases of interest (however, not in collisions).

Case 2: The Longitudinal Invariant

This invariant is of great importance in the context of particle confinement in magnetic mirrors. As has been derived above the two conserved quantities of a gyrating particle are the energy W and the magnetic moment μ. Writing the energy in the form $W = mv_\parallel^2/2 + \mu B$ shows that the parallel motion comes to a halt when the point $\mathbf{x}_0$ given by the condition $W = \mu B(\mathbf{x}_0)$ has been reached. At this point the parallel motion changes sign and the particle is reflected. If there are two such mirror points the particle oscillates between these points. However, because of the inhomogeneity of the field the particle drifts away from the initial field line. An example is the oscillation of particles in the earth's dipole field between the northern and southern hemisphere.

Let be $l, p_l = mv_\|$ the spatial coordinate and the associated canonical momentum along the magnetic field line and x, y the remaining spatial coordinates. The motion of the guiding centre is governed by a fast oscillation along l between turning points and a slow drift in x- and y-direction. The longitudinal invariant is defined by

$$J_\| = \oint mv_\| \, dl \, , \tag{7.22}$$

where the integration is taken between turning points. Using this invariant the description of the motion can be reduced to two equations for the variables x and y. These equations are useful to describe the long-term behaviour of trapped particles – trapped between turning points – and to check whether they stay confined or intersect the wall.

Figure 7.3 (left) shows the orbit of a charged particle coming from the solar wind in the magnetic field of the earth. All options discussed above occur: gyration, oscillations between mirror points because of $\nabla_\| B$ and the drift originating from $\nabla_\perp B$. Figure 7.3 (right) shows the azimuthal drift around the earth.

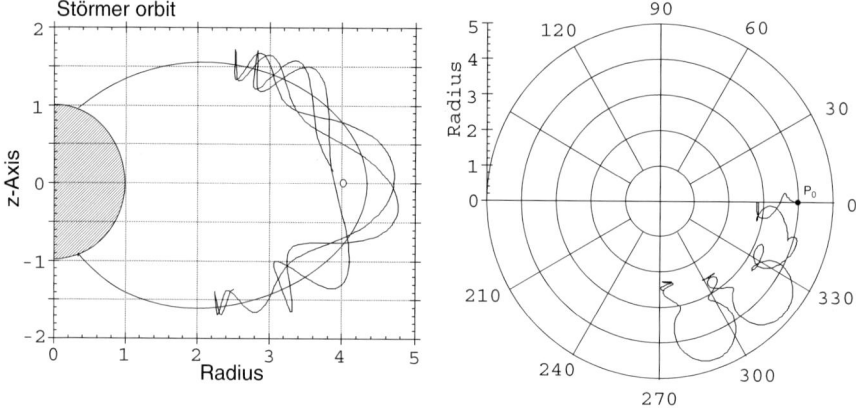

Fig. 7.3. Orbit of a 184 MeV proton in the dipole field of the earth. *Left*: orbit in r, z-coordinates, *right*: Orbit in r, φ-coordinates

Case 3: The Third Adiabatic Invariant

This invariant occurs if one of the remaining variables of the previous Section, x or y, undergo a fast periodic motion, for instance the azimuthal precession of the particle in the earth's magnetic field. As above, the associated action integral, the longitudinal magnetic flux, is an adiabatic invariant.

7.5 Concepts of Magnetic Confinement

7.5.1 Introduction

In the following we will discuss several concepts how to confine a plasma by magnetic fields. Roughly speaking one can distinguish between two concepts: one type are the open configurations where magnetic field lines leave the plasma region and the other ones are the toroidal configurations where field lines stay in the plasma region and never intersect the wall. The mirror machine is the representative of the first category, its efficiency depends on the mirror effect described in the previous sections. A characteristic feature of toroidal confinement is the existence of magnetic surfaces which are densely covered by magnetic field lines. In first approximation charged particles follow the field lines and if these stay on toroidally closed surfaces the particles are confined absolutely. Particles trapped in local magnetic mirrors, however, do not stay on these surfaces and it is a challenging task to optimize the magnetic field as to minimize the radial drift of these trapped particles. The result of this optimisation, the advanced stellarator, will be described in Sect. 7.5.9.

The best-known representative of toroidal confinement is the tokamak configuration which utilizes an axisymmetric magnetic field. According to the theory sketched in the previous section the toroidal canonical momentum and the energy of the particles are conserved and the particles stay on toroidally closed surfaces. Even particles trapped in local magnetic mirrors precess around the torus and are absolutely confined.

7.5.2 The Mirror Machine

The basis of the mirror concept is the conservation of the magnetic moment of gyrating particles. As described above the guiding centre is reflected at the turning points where the parallel velocity is zero. The most simple configuration to achieve confinement is a set-up of two circular coils – a Helmholtz system – as shown in Fig. 7.4. In cylindrical coordinates r, θ, z the magnetic field is independent of the poloidal coordinate θ.

The magnetic field has a maximum in the plane of the coils and the gyrating particles oscillate between the turning points. However, curvature and ∇B-drift move the particle away from the initial field lines. Due to the axial symmetry of the configuration this drift does not destroy the confinement, the orbits of the drifting particles are poloidally closed. A drawback of the mirror concept is the existence of a loss cone. In order to understand this effect let us assume that a particle starts in the middle of the device where the magnetic field has a minimum B_{min}. The choice of the initial velocity $v_\parallel$ and $v_\perp$ fixes the energy and magnetic moment μ. If the parallel velocity is too large the particles cannot be reflected and they escape along the field lines. The condition for escape is

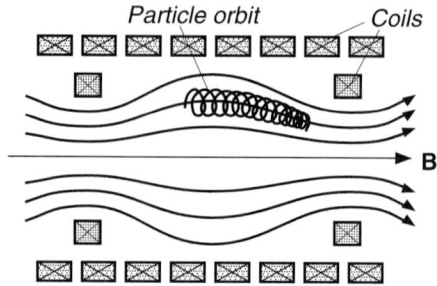

Fig. 7.4. Scheme of a magnetic mirror. The configuration has cylinder symmetry

$$v_\parallel^2 > v_\perp^2 \frac{B_{max} - B_{min}}{B_{min}}. \qquad (7.23)$$

B_{max} is the maximum field on the field line. The pitch angle at the initial point is defined by $\tan\phi = v_\perp/v_\parallel$ which explains the notion of a loss cone: If the pitch angle is too small the particles are lost. The ratio $R_M = B_{max}/B_{min}$ is called the mirror ratio and making this ratio large improves the confinement of particles, but there are technical limitations to this ratio. So far Coulomb collisions have been neglected, however, in addition to direct orbit losses they pose another limitation to the confinement. Coulomb collisions change the magnetic moment and the pitch angle and for this reason there is a permanent diffusion flux in velocity space into the loss cone. This mechanism determines the confinement time of a plasma in a magnetic mirror and in order to meet the condition of fusion the requirements on mirror ratio and plasma conditions are rather severe. Another obstacle to sufficient confinement in simple mirror machines is the curvature of field lines which leads to magneto-hydrodynamic instabilities deteriorating the confinement even faster than Coulomb collisions. These discouraging results have made the mirror concept a less promising candidate for fusion plasmas than toroidal confinement concepts.

7.5.3 Toroidal Confinement

As shown in the previous Section the mirror concept suffers from the loss of particles parallel to the field lines. Toroidally closed field lines present a natural path out off this deadlock, however, it was soon discovered that a simple toroidal field of a solenoid is not suited to confine a plasma. Let us consider the field of a solenoid which is generated by coils placed in a circle. All field lines are closed and circular. The field lines are curved and the gradient of B points radially towards the vertical axis of the configuration. This inhomogeneity of the field creates a vertical drift of the particles as described above, and this drift moves the particles with opposite charge in opposite directions. A vertical electric field is the consequence and the resulting $\mathbf{E}\times\mathbf{B}$-drift shifts

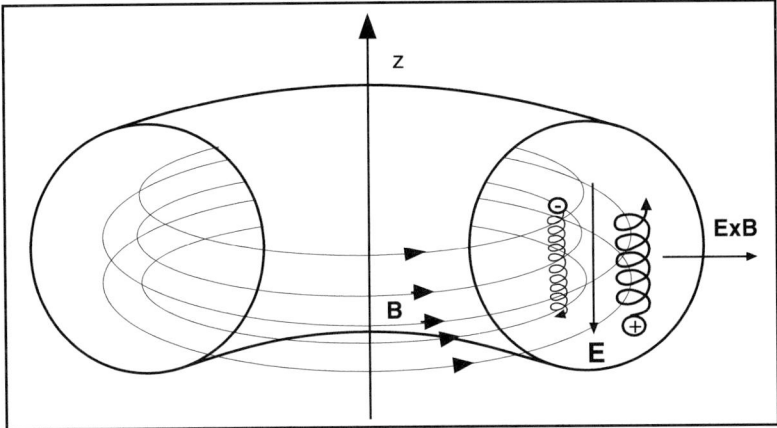

Fig. 7.5. Drift of charged particles in the field of a solenoid

the plasma radially outward. In this configuration plasma equilibrium is not possible (see Fig. 7.5).

This example demonstrates that in toroidal confinement one has to shape the magnetic field in such a way that single particle orbits taking into account curvature and ∇B-drift are confined, they either stay in a bounded domain or on toroidally closed surfaces, the so-called *drift-surfaces*. This requirement has led to the idea of *rotational transform*, where, in some sense, the vertical drift is outwitted and all particle orbits are well confined.

7.5.4 Magnetic Surfaces and Toroidal Equilibrium

Up to now we discussed magnetic confinement solely on the basis of particle orbits, however, a plasma is more than just an ensemble of charged particles. Due to the large number of particles of opposite charge per volume element ($\approx 10^{20}\,\text{m}^{-3}$) any attempt of charge separation would introduce large electric fields which prohibit large spatial separation. Only on a small scale characterized by the Debye length charge separation is possible and on larger scales the plasma is quasi-neutral: In volume elements larger than the Debye length there is no net electric charge. Static electric fields applied from outside are screened off and do not penetrate into the plasma. Therefore, we may regard the plasma as a neutral fluid, which is capable to carry large electric currents and is subject to magnetic forces only. The forces on the plasma are the gradient of the plasma pressure ∇p and the $\mathbf{j} \times \mathbf{B}$-force. Here $\mathbf{B}$ consists of the external field produced by coils and the field produced by currents flowing in the plasma. Plasma equilibrium in its simplest approximation means that the pressure gradient is balanced by the magnetic force

$$\nabla p = \mathbf{j} \times \mathbf{B}\,; \quad \nabla \times \mathbf{B} = \mu_0 \mathbf{j}\,; \quad \nabla \cdot \mathbf{B} = 0\,. \tag{7.24}$$

Since we are dealing with a toroidal plasma without loss cones we may assume that the pressure is isotropic. Toroidal plasma confinement in this fluid model of ideal magneto-hydrodynamics means to shape the magnetic field as to provide toroidally nested pressure surfaces with a maximum pressure in the centre and small or zero pressure in the boundary region. An immediate consequence of the equilibrium condition (7.24) is $\mathbf{B} \cdot \nabla p = 0$ and $\mathbf{j} \cdot \nabla p = 0$. This implies that the field lines and the current lines cover the pressure surface $p = $ const densely. Either all field lines on the pressure surface are closed upon themselves or one field line covers the pressure surface densely (or ergodically), because of $B \neq 0$ field lines cannot cross each other. In general, a toroidally closed surface $\psi = C$ defined by $\mathbf{B} \cdot \nabla \psi = 0$ is called a *magnetic surface*, a notion which is also applicable to vacuum magnetic fields. These magnetic surfaces are akin to the integral surfaces of mechanical systems with a continuous symmetry, and indeed, the theory of magnetic surfaces can be formulated in terms of Hamiltonian mechanics showing all the properties of Hamiltonian systems: existence of surfaces in case of continuous symmetry, destruction by island formation and ergodicity, existence of KAM-surfaces [1]. Magnetic surfaces are also flux surfaces, the toroidal magnetic flux is independent of the position of the poloidal cut. The flux enclosed in a magnetic surface can be used as a label of the surfaces, another option is to use the volume inside the surface. The innermost surface has zero volume, it is called the *magnetic axis*.

The *rotational transform* is a figure of merit which describes an integral property of the field line on a magnetic surface. The rotational transform ι is the ratio between the poloidal and toroidal turns of the field line before it closes, $\iota = n_p/n_t$. In such a case the rotational transform is a rational number. If the field lines are not closed the rotational transform is defined as the limit $\iota = n_p/n_t$; $n_t \longrightarrow \infty$. In this case ι is an irrational number. In our example above of the simple solenoid with closed field lines the rotational transform is zero. The example above shows that magnetic surfaces with a finite rotational transform are required to confine a toroidal plasma.

Comment 1: Drift Surfaces and Magnetic Surfaces

Above, the drift-surface has been introduced as a toroidally closed surface covered by guiding center orbits. When particles follow field lines they tend to stay on magnetic surfaces, due to curvature and ∇B-drifts, however, drift surfaces deviate from magnetic surfaces. This deviation is small and of the order ρ_L/ι, but this is true only for circulating particles with $v_\parallel \neq 0$; trapped particles can deviate from magnetic surfaces up to larger distances. These distances represent the stepsize in a random walk view of collisional transport (see Sect. 7.6.2).

Comment 2: The Virial Theorem

One is tempted to speculate whether a toroidal equilibrium might exist with magnetic fields generated by the plasma currents alone thus avoiding the technically complicated external coil system. However, this is ruled out by the virial theorem. This theorem can be easily derived by multiplying the equilibrium condition $\nabla p = \mathbf{j} \times \mathbf{B}$ with the spatial vector $\mathbf{x}$ and integrating over the whole space. If there are no external coils with currents generating an external field the result is

$$\iiint_V \left(3p + \frac{B^2}{2\mu_0}\right) d^3\mathbf{x} = 0; \quad V \longrightarrow \infty. \tag{7.25}$$

The integration extends over the whole space. Therefore, without external coils and currents only the trivial solution $p = 0$; $\mathbf{B} = 0$ exists.

7.5.5 Confinement in Tokamaks

A tokamak is an axisymmetric configuration, where all plasma parameters and the magnetic field do not depend on the toroidal coordinate φ. This feature offers many advantages to engineers and physicists. The coil system is axisymmetric and consists mainly of coils of the same shape, however the toroidal and poloidal coils are interlinked. The equations of plasma equilibrium (7.24) can be reduced to one scalar equation, which can be easily treated numerically, and the particles are absolutely confined because of the toroidal invariance. In principle, plasma diagnostics can be restricted to one poloidal r, z-plane. All these properties have contributed to the world-wide success of tokamak experiments.

Figure 7.6 shows the scheme of magnetic surfaces in a tokamak. The magnetic surfaces are described by a function $\chi(r, z) = \text{const}$. The cross section of the surfaces can be either circular, elliptic, triangular, bean-shaped, and a combination of all.

An unavoidable consequence of the axisymmetry is the existence of a net toroidal current in tokamak equilibria, a current which must be maintained by an externally induced toroidal voltage, since plasma resistivity will lead to a decay of any toroidal current. For this reason a tokamak experiment necessarily runs in pulsed operation, however, experimental efforts are being made to maintain the toroidal current by non-inductive current drive.

Recalling the axisymmetry, it is easy to understand why a tokamak equilibrium without toroidal current does not exist. As said above, toroidal confinement requires a finite rotational transform. We decompose the magnetic field on a magnetic surface in a toroidal and a poloidal component B_φ and B_θ (θ is the poloidal angular coordinate). The poloidal component is non-zero and does not change sign on a surface. If it where zero somewhere, it would be zero on a line closed toroidally; this is a field line with $\iota = 0$ – a

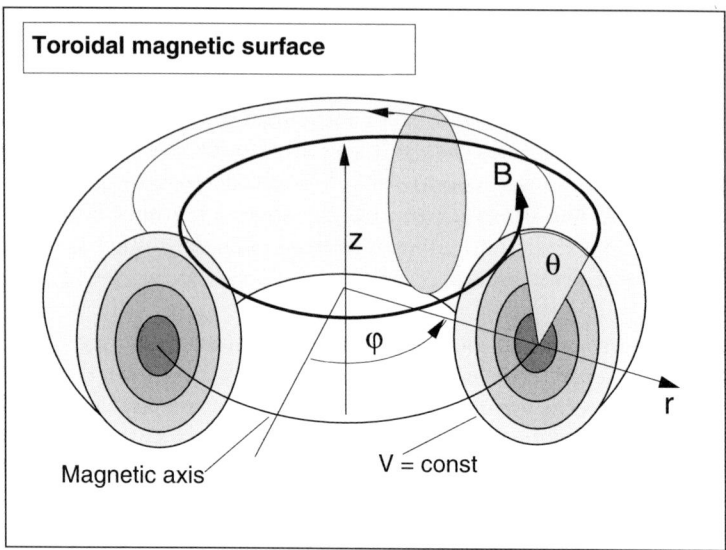

Fig. 7.6. Axisymmetric magnetic surface in cylinder coordinates. φ is the toroidal and θ the poloidal coordinate

clear contradiction to the assumption of finite rotational transform. Because of $B_\theta \neq 0$ the line integral $\oint B_\theta \, d\theta$ is non-zero which is just the enclosed toroidal current.

7.5.6 Coil System of Tokamaks

Basically the magnetic field of a tokamak consists of two components: the toroidal field and the poloidal field generated by the toroidal plasma current. Furthermore, a vertical field is needed in order to control the radial position of the plasma column.

The coil system of the toroidal field (TF) is the largest technical component and the simplest version is a set of 16 or 20 circular coils. The number of coils is a compromise between the goal of axisymmetry and the necessity to provide access for diagnostic equipments and plasma heating. For the reason of minimizing mechanical stresses coils in modern tokamaks are D-shaped. Furthermore, some tokamaks (Tore Supra) are equipped with superconducting windings; in future tokamaks and fusion reactors this is the only option.

The vertical field coils are large circular coils, they are also used to shape the plasma column into an elliptical or triangular cross section. The geometry of the plasma in tokamaks is characterized by the major radius R_0 and (in the case of a circular cross section) the minor radius a; the *aspect ratio* $A = R_0/a$ is the ratio between major and minor radius. The aspect ratio in present tokamaks is of the order three to six; in so-called spherical tokamaks the aspect ratio approaches one. Non-circular tokamaks are described by the elongation b/a (major and minor axis of the ellipse) and a triangularity parameter δ. A non-circular cross section is helpful in optimizing equilibrium and stability of the plasma column.

The toroidal current in tokamaks is of the order of several MAmps; in the JET experiment it reaches 7 MA. External transformer coils are installed in order to induce the plasma current. A transformer with an iron yoke as shown in Fig. 7.7 is often used to enhance the magnetic flux, however, for the sake of axisymmetry of the magnetic field an air core transformer is more favourable. Modern tokamaks like ASDEX Upgrade are equipped with an air core transformer. Various modern tokamaks are shown in Fig. 7.8.

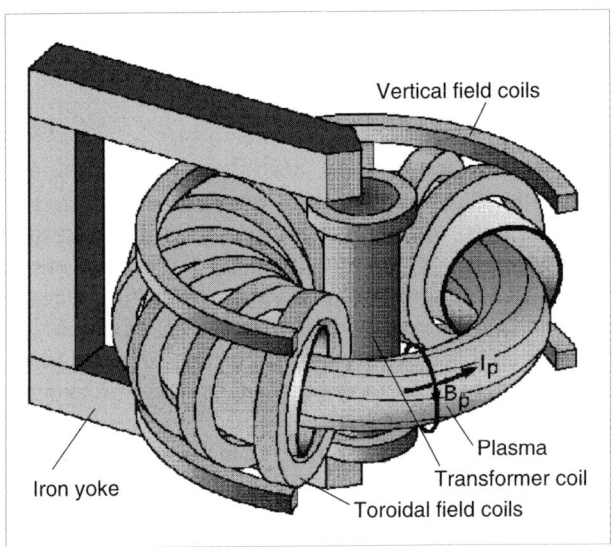

Fig. 7.7. Scheme of a tokamak device

7.5.7 Theory of Tokamak Equilibria

In a cylindrical coordinate system r, z, φ the magnetic field of an axisymmetric equilibrium is described by

$$\mathbf{B} = \mu_0 J(\chi) \nabla\varphi + \nabla\chi \times \nabla\varphi ; \quad \chi = \chi(r, z) \tag{7.26}$$

where contour lines $\chi(r, z) = C$ display the cross section of the magnetic surfaces. $J(\chi)$ is an arbitrary function of its argument; taking $J = J_0$ yields the vacuum field of the TF-coils $\mathbf{B}_0 = \mu_0 J_0 \nabla\varphi$. The poloidal magnetic field is $\mathbf{B}_p = \nabla\chi \times \nabla\varphi$. The plasma pressure is also a free function of the flux function χ. Free means that in the frame of ideal MHD-equilibrium this function is

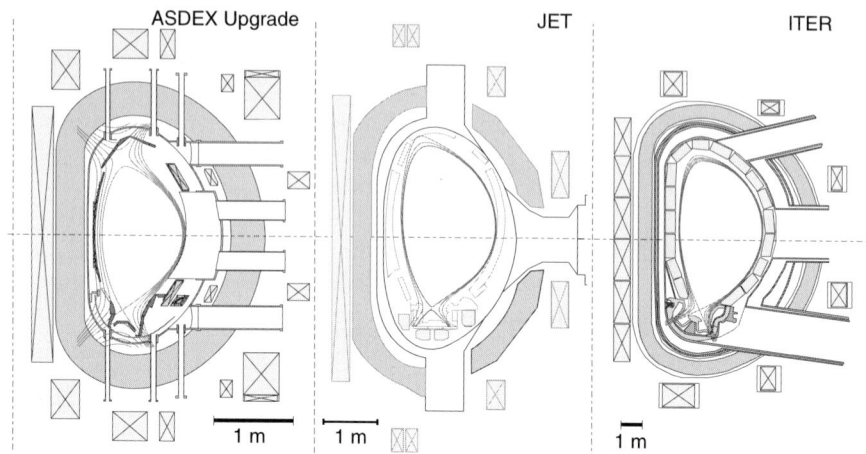

Fig. 7.8. Poloidal cut of some tokamaks with divertor. *Left*: ASDEX Upgrade, *middle*: JET, *right*: ITER. Note that the scale in the three figures is not the same

not uniquely determined and can be chosen arbitrarily. In a model taking into account transport processes the pressure profile and also the function $J(\chi)$ are uniquely determined.

The task to compute a tokamak equilibrium is reduced to solve a differential equation for χ. A straightforward analysis of the equilibrium condition (7.24) together with (7.26) leads to the "Lüst-Schlüter-Grad-Shafranov-equation" [2, 3]:

$$-\left(\frac{\partial^2}{\partial z^2} + \frac{\partial^2}{\partial r^2} - \frac{1}{r}\frac{\partial}{\partial r}\right)\chi = \mu_0 r^2 p'(\chi) + \mu_0^2 JJ'(\chi) \ . \qquad (7.27)$$

Shafranov has given a simple solution of this equation starting from the linear ansatz $p(\chi) \propto \chi$ and $J = J_0$

$$\chi(r,z) = \frac{r^2}{R^4}\left(2R^2 - r^2 - 4a^2 z^2\right) \qquad (7.28)$$

where R and a are constants. The contour lines $\chi = C$ are shown in Fig. 7.9. This figure shows a characteristic feature of toroidal equilibria: The nested magnetic surfaces are not concentric, the surfaces around the centre are horizontally displaced relative to the outer surfaces in the direction of the major radius. This Shafranov shift may become large in high-pressure plasmas and in order to reduce this Shafranov shift modern tokamaks exhibit elongated and D-shaped cross sections.

7.5.8 Cylindrical Approximation

In describing tokamak equilibria often the approximation of a cylindrical plasma as shown in Fig. 7.10 is made which can be justified if the aspect ratio

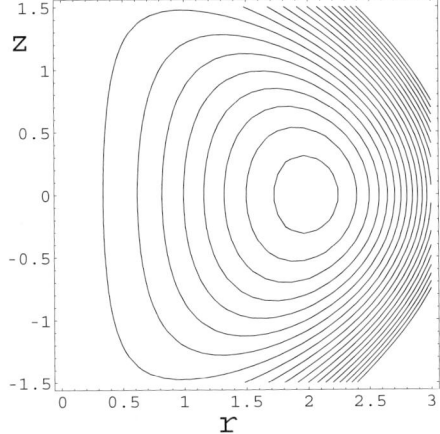

Fig. 7.9. Contour lines of $\chi(r,z)$, $R=2$, $a=0.8$

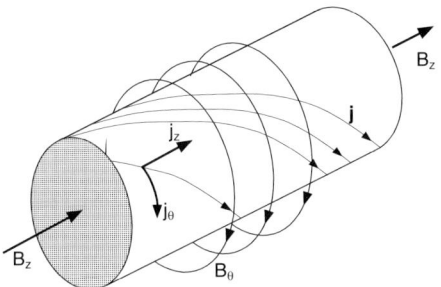

Fig. 7.10. Plasma cylinder

is large. In a cylindrical plasma all quantities depend on a radial coordinate r alone. The components of the magnetic field are $[0, B_\theta(r), B_z(r)]$ and the current density is given by $\mathbf{j} = [0, j_\theta(r), j_z(r)]$.

The equilibrium condition (7.24) has the simple form

$$\frac{d}{dr}\left(p + \frac{B^2}{2\mu_0}\right) = \frac{B_\theta^2}{r\mu_0}; \quad B^2 = B_\theta^2 + B_z^2. \tag{7.29}$$

This equation can be integrated providing a relation between the averaged plasma pressure $\langle p \rangle$ and the magnetic field. The volume average of a quantity g is defined by

$$\langle g \rangle = \frac{2}{a^2}\int_0^a g\, r\, dr \tag{7.30}$$

where a is the plasma radius. Applying this definition to (7.29) yields after partial integration

$$\langle p \rangle = \frac{B_\theta^2(a)}{2\mu_0} + \left(\frac{B_z^2(a)}{2\mu_0} - \left\langle \frac{B_z^2}{2\mu_0}\right\rangle\right). \tag{7.31}$$

The poloidal field $B_\theta(a)$ is proportional to the axial plasma current I_z and this relation can be written as

$$\langle p \rangle = \frac{\mu_0}{8\pi^2 a^2} I_z^2 + \frac{B_z^2(a)}{2\mu_0} - \left\langle \frac{B_z^2}{2\mu_0} \right\rangle . \tag{7.32}$$

The magnetic field lines in this cylindrical plasma are helices with constant pitch. The differential equation is

$$\frac{r d\theta}{dz} = \frac{B_\theta}{B_z} \quad \Rightarrow \quad \theta = \frac{B_\theta(r)}{r B_z(r)} z . \tag{7.33}$$

Let $L = 2\pi R$ be the period length of the plasma column. The poloidal angular displacement or *rotational transform* of the field line is

$$\iota =: \frac{\theta_L}{2\pi} = \frac{R B_\theta(r)}{r B_z(r)} . \tag{7.34}$$

In the theory of tokamaks the inverse of the rotational transform $q(r) = 1/\iota$ is called the *safety factor*.

7.5.9 Confinement in Stellarators

A stellarator is a magnetic confinement concept without induced toroidal current. In view of a future fusion reactors, which run in steady state, the exploration of toroidal configurations without induced toroidal current has begun already in the early fifties of the last century [4]. Strictly speaking the stellarator is defined by zero loop voltage since toroidal plasma currents, the *bootstrap currents* driven by the plasma pressure, do also appear in stellarators. The rotational transform in stellarator configurations is generated by external coils, in contrast to tokamaks, where the toroidal current provides the transform. The helicity of the field lines requires an equivalent helicity of the external coil system and therefore stellarator configurations are not axisymmetric.

The main technical component of stellarators is the coil system, which, in classical stellarators, consists of a toroidal field system and a set of helical windings. In modern stellarators this concept has been abandoned in favour of a modular coil system, where toroidal field coils and helical windings are combined in one system.

The Straight Stellarator

In order to understand the main principles of stellarator fields we consider a straight system with helical invariance. Here, analytical formulas of the vacuum field are available and magnetic surfaces can be presented in analytical form.

Let $r = a$ be a straight infinite cylinder along the z-direction and r, z, φ a cylindrical coordinate system. r is the radial coordinate and φ the poloidal coordinate. A magnetic field which is helically invariant has the form $\mathbf{B} = \mathbf{B}(r, \varphi - kz)$ where $\varphi - kz = $ constant defines a helical field line with the pitch k along which all quantities are invariant. The magnetic surfaces of such a field, which are defined by $\mathbf{B} \cdot \nabla \psi = 0$ can be easily found using the vector potential $\mathbf{A}$ by

$$\psi = A_z(r, \varphi - kz) + kr A_\varphi(r, \varphi - kz) = \text{const.} \quad (7.35)$$

The existence of the magnetic surfaces – although infinitely long – are the consequence of the helical invariance. The orbit of the magnetic field lines on these surfaces is not so easy to compute but nevertheless straightforward. For details the reader is referred to the excellent review paper by Morozov and Solov'ev [5]. A straight helical vacuum field is described by a scalar potential Φ satisfying the Laplace equation $\Delta \Phi = 0$. In cylinder geometry the solutions are modified Bessel functions $I_n(nkr)$ and Fourier harmonics in φ and z:

$$\Phi = B_0 z + \frac{1}{k} \sum_{n=1}^{\infty} b_n I_n(nkr) \sin[n(\varphi - kz)] \quad (7.36)$$

and the magnetic surfaces are given by

$$\psi = B_0 \frac{kr^2}{2} - r \sum_{n=1}^{\infty} b_n I'_n(nkr) \cos[n(\varphi - kz)] = \text{const.} \quad (7.37)$$

Using this expansion in Bessel functions a current distribution on the cylinder with radius a can be computed which produces the given field inside. This current sheath has helical symmetry and by discretizing this sheath we obtain the helical windings. This is one of the few non-trivial examples where magnetic field, magnetic surfaces and the coil system can be given in terms of analytic functions.

The Classical Stellarator

All these surfaces of a straight helical system including the separatrix are invariant surfaces ($\mathbf{B} \cdot \nabla \psi = 0$). Destruction of the surfaces and stochasticity only occur if the coil system deviates from helical symmetry. One unavoidable perturbation, however, occurs if one bends the straight helical stellarator into a torus. The torus curvature destroys the helical symmetry and magnetic surfaces with islands and stochastic regions occur (see later in Fig. 7.15). The axial field $B_0 \mathbf{e}_z$ is generated by circular coils, for the sake of experimental flexibility these are separated from the helical windings. The notation of the helical windings is made according to the dominating helical harmonic in (7.36). One distinguishes between $\ell = 1, 2$ and $\ell = 3$-systems. In case of

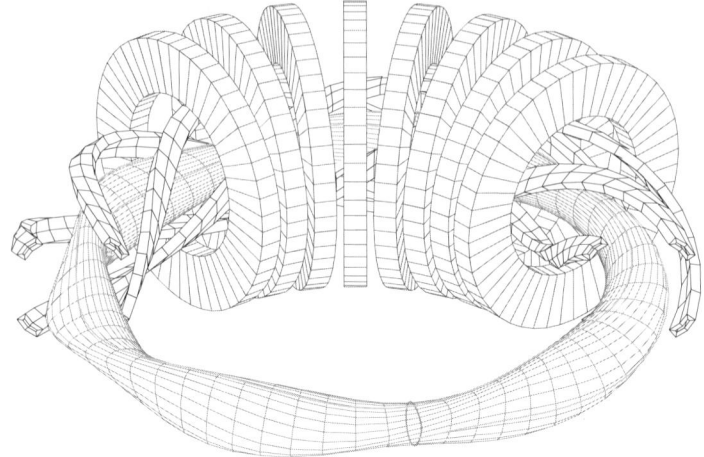

Fig. 7.11. The stellarator Wendelstein VII-A in Garching. Major radius 2 m, average plasma radius 0.1 m, number of field periods 5, rotational transform 0.23, magnetic field ≤ 3.5 T

a $\ell = 1$-system the magnetic surfaces roughly are shifted circles, in $\ell = 2$-systems the cross section of the surfaces is an ellipse and in $\ell = 3$-systems the shape is nearly triangular. The currents in the helical windings can be varied independently of the current in the main field coils and therefore the shape of the magnetic surfaces and the rotational transform can be varied in the experiment. Figure 7.11 shows an example of a $\ell = 2$ stellarator. It consists of two pairs of helical windings with currents in both directions, there is no net toroidal current.

The Torsatron

The *torsatron* [6] is a slightly different system where the currents in the helical windings are unidirectional (Fig. 7.12). Some of the difficulties of the classical stellarator are alleviated in the torsatron configuration. Here, a field with ℓ-fold poloidal symmetry is generated by ℓ helical windings, all carrying currents in the same direction. The torsatron thus generates both toroidal and poloidal field components, and, in principle, no further coils are needed. Furthermore, the complication with two sets of interwoven windings is eliminated. Problems associated with disassembly and maintenance still exist due to the toroidal continuity of the helical coils but are less severe. Since in torsatrons one set of ℓ-windings is used (rather than the 2ℓ windings of a stellarator) the access can be improved.

The torsatron coils usually generate an average vertical field, which opens the vacuum flux surfaces. Thus, unless a specific winding law is selected (the "ultimate" torsatron law), an additional compensating vertical field coil set

Fig. 7.12. Torsatron configuration

is needed. Furthermore, the basic torsatron configuration lacks experimental flexibility (i.e., variation of rotational transform, well depth, etc.), because of the use of a single set of windings. This flexibility can be restored with the use of an additional small vertical field, an additional small toroidal field or by allowing variation in the helical harmonic content.

Among the advantages of the torsatron configuration is the possibility of significant reduction of the forces on the helical windings. The forces tend to be directed radially outward, so that the support structure is no longer a severe problem. Indeed, it is possible to transfer the average outward forces onto external Helmholtz-type coils far from the plasma. For a torsatron with a certain winding law of the helical coils the radial force averaged over a field period even may be reduced to zero. For this case large forces appear on the compensation coils, but these can be located where there is adequate space for support structure.

The Modular Stellarator

The concept of helical windings is useful for experiments since it provides a large amount of flexibility, however, in large experiments or in a stellarator reactor it presents severe technical difficulties because of the interaction between these windings and the main field coils. The concept of *modular coils* – in some sense already realized in the early figure-8 stellarators – bypasses these difficulties and offers the chance to build larger devices or a reactor with coils of feasible size. The idea of modular coils [7] results from the fact that any toroidal vacuum field in a domain Ω and tangential to the surface $\partial\Omega$ can be produced by a current sheath on this surface. Discretizing this current sheath to current lines yields a set of modular and poloidally closed coils which reproduce the magnetic field in the domain Ω to high accuracy.

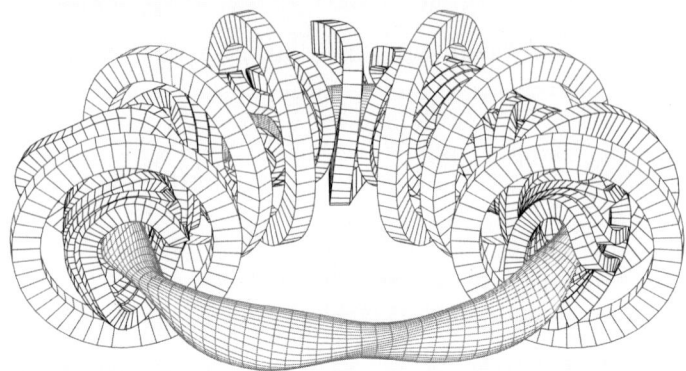

Fig. 7.13. The first modular stellarator Wendelstein 7-AS in Garching. Major radius 2 m, average plasma radius 0.2 m, number of field periods 5, rotational transform 0.4, magnetic field ≤ 3.0 T. The plane circular coils allow one to vary the rotational transform

In a next step analytic winding laws were used to model these twisted coils and to generate the desired magnetic field and the rotational transform by a proper arrangement of these coils.

In the Wendelstein 7-AS stellarator (Garching) shown in Fig. 7.13 and the IMS device (University of Wisconsin) this concept has been realized. A step forward was provided by the idea of P. Merkel [8] to calculate the system of surface currents on a second torus $\partial \Omega_c$ surrounding $\partial \Omega$ such that the vacuum field produced by these surface currents is tangential to the given surface $\partial \Omega$ ($B_n = 0$ on $\partial \Omega$).

The method to calculate the coil system after the magnetic field has been specified offers the chance to optimize the magnetic field first according to criteria of optimum plasma performance and then to compute the coil system after this procedure has come to a satisfying result. Along this line the *Advanced Stellarator* has been developed [9, 10].

MHD-Equilibrium in Stellarators

In the frame of ideal MHD plasma equilibrium in stellarators is described by the force balance equation (7.24), however, in contrast to tokamak equilibria, the problem cannot be reduced to a scalar equation of the Grad-Shafranov type. As shown by Kruskal and Kulsrud [11] the system (7.24) can be derived from a variation principle which makes the positive functional

$$U = \int \left(\frac{B^2}{2\mu_0} + \frac{p}{(\gamma - 1)} \right) dV \;; \quad \gamma = \text{adiabatic constant} \qquad (7.38)$$

stationary. The constraints of this procedure are determined by the conservation of fluxes during the motion of an ideal plasma. The procedure starts from a magnetic field with a family of nested magnetic surfaces which are not necessarily pressure surfaces. Let be ψ the toroidal flux of these surfaces and $\chi(\psi)$ the poloidal flux. These quantities are invariant under the variation of the magnetic field. A further constraint is the invariance of the mass $M = \int p^{1/\gamma} \, dV$ within every magnetic surface. Computing an extremum of the functional U is the method how to compute 3-D equilibria numerically [12].

Another scheme is the iterative method proposed by Spitzer [13]. In this scheme a sequence $\mathbf{B}_n$ of magnetic fields is constructed, which are assumed to converge towards a self-consistent equilibrium. This sequence is computed by

$$0 = -\nabla p + \mathbf{j}_{n+1} \times \mathbf{B}_n, \quad \mathbf{j}_{n+1} = \nabla \times \mathbf{B}_{n+1} \quad , \quad \nabla \cdot \mathbf{B}_{n+1} = 0 \, . \qquad (7.39)$$

Figures 7.14 and 7.15 show poloidal cross-sections of magnetic flux surfaces of Wendelstein 7-AS. However, as pointed out by A. Boozer [14], magnetic surfaces can be destroyed, even if the process begins with a set of nested surfaces.

Optimisation of Stellarators

Optimisation of plasma equilibria is an important objective of the stellarator concept. The key role is played by the Pfirsch–Schlüter (P.S.) currents which are closely connected to the Shafranov shift and to the radial drift of

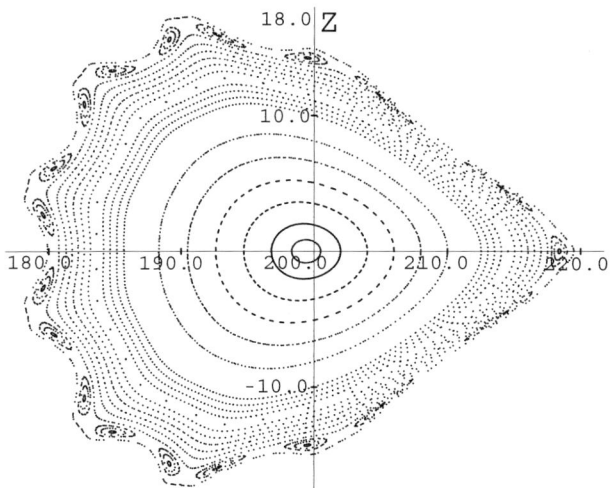

Fig. 7.14. Poincaré plot of magnetic surfaces in W7-AS

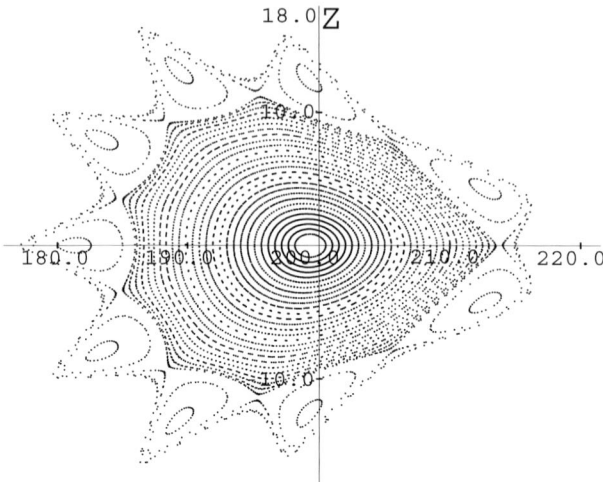

Fig. 7.15. Poincaré plot of magnetic surfaces with islands at $\iota = 5/9$

particles. In a toroidal equilibrium the plasma current $\mathbf{j}$, which is needed to satisfy the equilibrium condition $\mathbf{j} \times \mathbf{B} = \nabla p$, usually is decomposed as a diamagnetic current $\mathbf{j}_\perp = \mathbf{B} \times \nabla p / B^2$ and a parallel current $\mathbf{j}_\parallel$. Both are linked by $\nabla \cdot \mathbf{j} = 0$. The first ones to compute these currents were D. Pfirsch and A. Schlüter [15], who found the approximation $j_\parallel \approx (2 j_\perp / \iota) \cos\theta$ in axisymmetric magnetic fields $\mathbf{B}$. This approximation has been made for stellarators without net toroidal current and with large aspect ratio. The diamagnetic current is proportional to the plasma pressure gradient and indispensable for plasma confinement. The parallel current density does not contribute to the force balance, however, the magnetic field associated with these parallel currents is mainly responsible for the Shafranov shift. In order to avoid large Shafranov shifts the magnetic field must be shaped aiming at small parallel currents. The efforts of optimising stellarator equilibria along this line have led to the Helias concept [9, 10]. Figure 7.16 shows Wendelstein 7-X, an advanced stellarator under construction in Greifswald. The full list of requirements to the optimised stellarator with respect to its role as a fusion reactor is

- Good magnetic surfaces of the vacuum field without major resonances, islands and stochastic regions in the bulk plasma. Islands or stochasticity in the boundary region may be utilized for divertor action.
- Low Shafranov shift and a stiff equilibrium configuration up to $\langle \beta \rangle \approx 5\%$. The rotational transform should depend only weakly on the plasma pressure.
- MHD-stability up to $\langle \beta \rangle \approx 5\%$.
- Sufficiently small neoclassical plasma losses so that the ignition condition is satisfied.

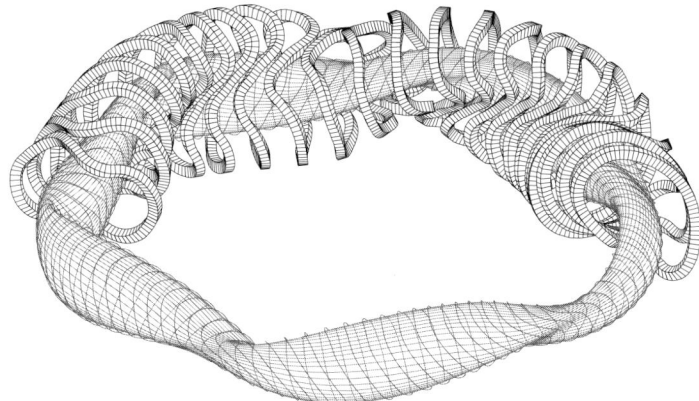

Fig. 7.16. Advanced stellarator Wendelstein 7-X. This figure shows one magnetic surface together with the magnetic field lines and the current lines. These are poloidally closed

- Sufficiently small bootstrap currents so that the magnetic field is not modified by this effect.
- Good confinement of alpha-particles.
- Good technical feasibility of the coil system and sufficiently large space between coils and plasma for blanket and shield.

The advanced stellarator combines both principles: modularity of the coil system and optimisation of plasma equilibrium. It turns out that the reduction of the parallel currents and the Shafranov shift also have some beneficial effects on the particle orbits. The parallel currents $j_\parallel$ are linked to the gradient of B by $\nabla \cdot \mathbf{j} = 0$ and the diamagnetic current. The guiding centre drift is also linked to this gradient (7.16), and a reduction of $\nabla_\perp B$ reduces the parallel currents and the radial drift of particles.

7.6 Transport in Plasmas

Magnetic confinement is not perfect. There are particle and energy losses along orbits which are not confined. This applies to particles in a mirror which escape the mirror confinement due to a large ratio in $v_\parallel / v_\perp$. Similar losses appear in the classical stellarator where particles trapped within the local mirrors between two helical coils drift away in vertical direction because of the toroidal field inhomogeneity. As the canonical angular momentum is conserved in truly symmetric systems, losses of this kind do not appear in the idealised tokamak. Also quasi-symmetric stellarators do not have orbit losses.

Another leak rate for particles and energy is established by binary collisions. Particles, which gyrate are scattered and, in the average, move radially outward. Such, a collision induced loss flux of particles and energy emerges which renders τ_E (and the equivalent particle confinement time τ_p) finite. The displacement from the field line (or the flux surface) is the relevant step size for the radial transport, which is the particle Larmor radius in the case of a simply magnetized plasma.

We will see that toroidal geometry allows particle obits with larger deviations and therefore larger step sizes, which will govern the radial fluxes and the confinement times. Under practical plasma conditions, however, the collisional losses play a role but they are dominated by losses induced by plasma turbulence. The turbulent movement of magnetic field lines induces magnetic flutter and particles are moved radially outward by flowing along the turbulent field lines. In addition, electrostatic turbulence arises owing to the correlated fluctuations of plasma density and plasma potential. Plasma turbulence and turbulence induced plasma losses are covered in Sect. 7.6.4. There is another branch of plasma losses which is caused by plasma instabilities. These losses appear transiently when the plasma equilibrium is violated. Phases like these are not suitable for fusion reactor operation and must be avoided by selecting stable operational scenarios. Such cases are not treated in this short summary and can be found in Chap. 9 (Part II).

7.6.1 Collisional Losses

Now we study the limits of plasma confinement, which yield a finite τ_E in somewhat more detail. We start with a simple magnetized plasma where plasma collisions give rise to a finite plasma resistivity η. The Spitzer resistivity η is linked to the plasma parameters by

$$\eta = \frac{m_e \nu_{ei}}{n e^2} , \tag{7.40}$$

ν_{ei} is the electron–ion collision frequency of binary collisions. We assume a quiescent plasma and an established plasma equilibrium where the equilibrium condition (7.24) is fulfilled. Ohm's law in a plasma is: $\mathbf{E} + \mathbf{v} \times \mathbf{B} = \eta \mathbf{j}$. We replace $\mathbf{j}$ in Ohm's law by the equilibrium relation and compute the perpendicular component:

$$\mathbf{v}_\perp = \frac{\mathbf{E} \times \mathbf{B}}{B^2} - \eta \frac{\nabla p}{B^2} . \tag{7.41}$$

In cylindrical geometry $\mathbf{E}$ and ∇p are vectors with radial components only. The equilibrium condition renders the flux surfaces as nested equipotential and isobaric toroids. $\mathbf{v}_\perp$ has two components: The $\mathbf{E} \times \mathbf{B}$-flow occurs within a flux surface perpendicular to the field lines. It causes the plasma to flow without impact on confinement. The $\eta \nabla p / B^2$ velocity is in radial direction and its related flux gives rise to a finite confinement. An ideal plasma without collisions, with $\eta = 0$, would have infinite confinement. The resulting collision induced flux is

$$\Gamma = nv_r = -\eta n \frac{\nabla p}{B^2} \,. \tag{7.42}$$

Under isothermal conditions the pressure gradient is proportional to the density gradient $\nabla p = k_B(T_e + T_i)\nabla n = 2k_B T \nabla n$ and the particle flux has the form of a Fick's law

$$\Gamma = nv_r = -\frac{\eta 2nk_B T}{B^2} \nabla n \tag{7.43}$$

with a diffusion coefficient

$$D = \eta \frac{2nk_B T}{B^2} \,. \tag{7.44}$$

This simple result is only valid in cylindrical geometry. In a torus the diffusion flux is larger since the $\mathbf{E} \times \mathbf{B}$-term in (7.41) gives rise to an additional flux. In toroidal geometry plasma currents also flow along magnetic field lines. Because of these parallel currents – or Pfirsch–Schlüter currents – there also exists a parallel electric field $E_\parallel = \eta j_\parallel$ and a poloidal electric field. The resulting poloidal electric field leads to a radial particle flux which changes sign on the magnetic surface, however, the surface-averaged flux is non-zero and even larger than the classical flux given in (7.42). In simplest approximation to toroidal geometry (axisymmetry and circular cross section) the net particle flux is [15]

$$\Gamma = n v_r = -\eta \frac{2nk_B T}{B^2} \left(1 + \frac{2}{\iota^2}\right) \nabla n \,, \tag{7.45}$$

ι is the rotational transform of the magnetic surface. The reason for the growth with decreasing rotational transform is the growth of parallel equilibrium currents at small rotational transform. Reducing the parallel Pfirsch–Schlüter currents is not only helpful in reducing the Shafranov shift, as shown in the previous section, but also in reducing the collisional losses in toroidal equilibria.

7.6.2 Particle Picture of Classical Diffusion

We have obtained the diffusion coefficient from the MHD equations. There is an alternative way to look at the problem of gyrating particles subject to Coulomb collisions by constructing the diffusion coefficient from a random-walk process of independent steps. Let be Δ the averaged step width of a test particle during the random walk process (Brownian motion) and τ a characteristic time step between successive collisions. Then, $D = \Delta^2/\tau$ is the diffusion coefficient and $x = (Dt)^{1/2}$ the distance which the particle has passed during the time t. In a collisional plasma the step width is the gyro radius and the Coulomb collision time $\tau_C = 1/\nu$ is the relevant time step. This identification allows us to define the diffusion coefficient by

$$D = \rho_L^2 \nu \,. \tag{7.46}$$

Since electrons and ions have different Larmor radii and collision frequencies the issue arises which one is the relevant diffusion coefficient of the entire plasma. Since a dense plasma stays quasineutral the diffusion fluxes of the particle species are equal and the component with the smallest diffusion coefficient determines the resulting flux. Thus the electron diffusion is the decisive process. The electron diffusion coefficient is

$$D_e = \rho_e^2 \nu_{ei}; \quad \rho_e = \frac{m_e v_\perp}{eB}; \quad v_\perp = \left(\frac{2k_B T}{m_e}\right)^{1/2}. \qquad (7.47)$$

The random walk perpendicular to the magnetic field is a two dimensional process which is the reason for the special ansatz of the thermal velocity. Replacing the collision frequency by the resistivity [see (7.40)] leads to

$$D_e = \rho_e^2 \nu_{ei} = \frac{\eta 2 n k_B T}{B^2} \qquad (7.48)$$

which coincides with (7.44). The transport coefficient, obtained from the random-walk picture, agrees with that from the MHD equations. However, one should be aware that this coincidence depends on the assumption of quasi-neutrality and equal fluxes which is a result of the fluid model. The random walk model fails if one neglects basic properties of the fluid model.

7.6.3 Neoclassical Transport

The transport which we have studied up to now has been termed *classical transport*. *Neo-classical transport* denotes the collisional transport in toroidal geometry for confinement and realistic particle orbits. As shown in previous Sections the field curvature causes particle drifts and additional mechanisms give rise to radial particle- and heat flux contributions which by far surpass the classical ones in homogenous fields. One has to consider now the actual particle orbits around the torus and the collisionality conditions in more detail. Since MHD equations do not describe particle orbits a correct description of neoclassical losses must start from a kinetic equation including collisions and particle orbits. The Fokker–Planck equation describes the distribution function $f(\mathbf{x}, \mathbf{v})$ which contains all information about particle and energy fluxes. Instead of computing the exact particle orbits including the gyration it is sufficient to start from the guiding center orbits and to compute the distribution function of the guiding centres. This is done by solving the drift-kinetic equation which is the result of an averaging procedure applied to the Fokker–Planck equation. Instead of going into the details of the drift-kinetic equation [16] we will use the particle picture and the random walk model in order to demonstrate the essential feature of neoclassical diffusion.

The radial drift of guiding centre orbits in normal direction to magnetic surfaces is the key in understanding the mechanism of neoclassical diffusion. In general, guiding centres move along field lines and drift in perpendicular

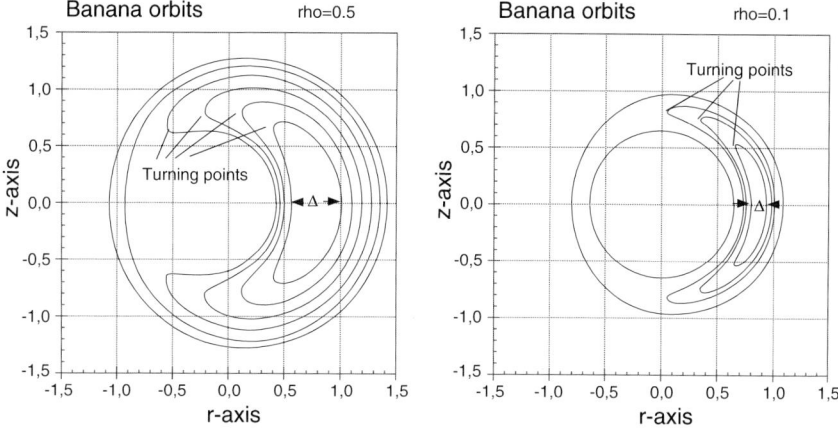

Fig. 7.17. Banana orbits in a tokamak; projection onto a poloidal plane. The gyro radius is proportional to ϱ. In the right figure the gyro radius is five times smaller than in the left figure. In addition also orbits of circulating particles are shown

direction. In particular, particles which are trapped between two magnetic mirrors oscillate back and forth between the turning points, however, due to the curvature drift the forward orbit and the backward orbit do not coincide. Let us designate the radial distance between these branches with Δ, then Δ plays the role of the step width in a random walk process and it replaces the gyro radius in the previous model above. Although these trapped particles are a minority in the ensemble of all particles they contribute the largest fraction to the neoclassical losses.

In axisymmetric tokamak geometry these guiding centre orbits are particularly easy to understand. There are only two classes of orbits: the circulating particles with $v_\parallel \neq 0$ and the trapped particles. If the parallel velocity is small enough these particles will be reflected at some turning points which are determined by the inhomogeneity of the toroidal magnetic field. Figure 7.17 shows these *banana orbits* starting from various radial positions. As these particles do not experience the full action of rotational transform, their drift gives rise to large displacements Δ. The collisionality of the plasma may cause the particles in the average to scatter before they have moved from the plasma outside to the inside. Under these circumstances, the plasma behaves rather fluid like. This is, however, not the condition of a fusion reactor. Under fusion conditions, even the trapped particles can move through their *banana orbit* before being scattered. For this case, again the diffusion coefficient can be roughly determined from a random walk ansatz $D = f_t \nu_t \Delta^2$. f_t is the fraction of trapped particles which in simple tokamak geometry is $f_t \approx \epsilon$; $\epsilon = a/R$. ϵ is the *inverse aspect ratio* $A^{-1} = a/R$; $R=$ major radius, $a=$ averaged plasma radius. While the distance of circulating particles from magnetic surfaces is of the order ρ_L/ι the deviation of the trapped particles in tokamaks is larger: $\Delta \approx \epsilon^{-0.5} \rho_L/\iota$. The characteristic time $\tau(\epsilon)$ of the random walk process is the time which is needed to scatter the particles out of the loss cone into the region of circulating particles. This time is $\epsilon \tau_{90}$, τ_{90} is the 90°-deflection time. Hence $\nu_\tau = 1/\epsilon \tau_{90} = \nu_{ei}/\epsilon$ is the relevant collision frequency. Putting all pieces together yields a diffusion coefficient

$$D_{neo} = \epsilon^{-1.5} \nu_{ei} \frac{\rho_e^2}{\iota^2} . \tag{7.49}$$

Again, the slower component, the electrons, determine the diffusion process.

7.6.4 Turbulent Transport

Classical and neoclassical diffusion theories predict rather small transport coefficients for a fusion plasma. Classical transport coefficients are of the order $0.01 \, \text{m}^2\text{s}^{-1}$. The neoclassical values are about $0.5 \, \text{m}^2\text{s}^{-1}$ for ions and overlap with measured transport coefficients; the electron values are lower by the square-root of mass ratio and are far off the experimental data. In general, experiments in toroidal geometry exhibit stronger losses than classical theory predicts. The plasma is not quiescent but turbulent. As in fluid dynamics, where the turbulence enhances the radial momentum transport and gives rise to turbulent heat transport, plasma turbulence leads to all kinds of enhanced transport processes: enhanced particle transport, enhanced energy and momentum transport. Since a plasma can become unstable for far more reasons than a simple fluid turbulence theory in plasma is a rather complex field. In practice, empirical scaling laws serve as a guideline in predicting turbulent transport and confinement times in future experiments and in the fusion plasma.

In fully developed turbulence characteristic time scales and length scales are the relevant parameters which determine the anomalous transport. Let τ_0 and λ_0 be a time scale and a length scale, then a rough estimate following random walk arguments yields $D_{an} = \lambda_0^2/\tau_0$. However, without going into the origin of turbulence and the underlying instabilities little can be said about these estimates. Plasma confinement necessarily requires gradients of density and temperature, and instabilities driven by these gradients are nearly unavoidable in fusion plasmas. In particular, ion temperature gradients are considered as a favourite candidate for causing plasma turbulence. Characteristic frequencies of these instabilities are the *drift frequency*

$$\omega_e^* = k_\perp \frac{k_B T_e}{eB} \frac{n_0'(r)}{n_0} \approx k_\perp \frac{k_B T_e}{eB \, a} . \tag{7.50}$$

$k_\perp$ is the perpendicular wave vector and a is the plasma radius. The drift frequency is proportional to the density gradient $n_0'(r) \approx 1/a$. The ansatz for the transport coefficient yields

$$D_{an} \approx \frac{1}{k_\perp a} \frac{k_B T_e}{eB} . \tag{7.51}$$

The length scale $k_\perp^{-1}$ of the drift instabilities can vary between the plasma radius as the largest wave length and the gyro radius as the smallest wave length. In the worst case of $k_\perp \approx 1/a$ the transport coefficient is the Bohm diffusion coefficient $D_{an} \propto k_B T_e/eB$. The other extreme case $k_\perp \approx 1/\rho_i$ – the characteristic wave length is of the order ion gyro radius – leads to the gyro–Bohm scaling

$$D_{an} \approx \frac{\rho_i}{a} \frac{k_B T_e}{eB} . \tag{7.52}$$

For a discussion of numerical simulation of plasma turbulence see Chap. 8 in Part II and [17].

An interesting feature of turbulent plasmas is that self-organisation can happen. In this case, the turbulence is quenched and locally steep gradients appear. The original observations were made on the ASDEX tokamak with the H-mode [18]. At a spontaneous transition, the plasma edge suddenly became quiescent and an edge transport barrier developed. The present understanding is that out of the turbulent flow field a coherent component develops which causes the plasma to strongly rotate with a sheared velocity field which feeds back to the turbulence and decorrelates the turbulent eddies into smaller and less transport relevant scales. Meanwhile, also transport barriers deeper inside the plasma can be developed when, in addition, the magnetic shear is low or reversed. Low or reversed shear can be developed in tokamaks with strong bootstrap current; in stellarators, this situation is naturally provided.

7.6.5 Empirical Scaling Laws

The lack of a generally accepted theory of turbulent plasma transport has led to a large spectrum of empirical scaling laws which describe current experiments and serve as a guide-line in designing new experiments. Since plasma confinement not only depends on plasma parameters but also on plasma wall interactions and the geometry of the material walls a general scaling law valid for all experiments cannot be expected. Furthermore, radiation of impurity ions affect the plasma energy balance and since the fraction of impurity ions discontinuously varies among the experiments a universal scaling law is not in sight. A further argument against universal scaling laws is the non-linear behaviour of plasma equations which allows for all kinds of discontinuous transitions from quiescent solutions to turbulent solutions. A typical example is the transition from L-mode confinement to H-mode confinement. A further issue in establishing scaling laws for the energy confinement time is the choice of the independent parameters. Besides the geometric parameters major radius R and plasma radius a, magnetic field B, toroidal current I and heating power P are the parameters which can be controlled externally. In tokamaks ellipticity κ and triangularity δ of the magnetic surface are additional geometrical parameters while in stellarators the toroidal current does not exist and the rotational transform is introduced instead. In tokamaks the confinement time also depends on the hydrogen isotope introducing the atomic mass A as an independent parameter. Thus, the general ansatz of an empirical scaling law is

$$\tau_E = C_0 R^{a1} a^{a2} \kappa^{a3} \delta^{a4} B^{a5} I^{a6} P^{a7} n^{a8} A^{a9} , \tag{7.53}$$

κ is the ellipticity of the magnetic surface and n is the averaged plasma density. The units are: R, a in m, B in T, n in 10^{20} m^{-3} and P in MW. The exponents $a1 \ldots a9$ and the constant C_0 are fitted to the experiment. Several arguments have been brought up against this ansatz: firstly the power law may not be the most general ansatz, secondly the number of independent parameters may not be sufficient and finally the choice of these parameters does not take into account the underlying physics. The scaling law should reflect the equations which govern the plasma confinement. As introduced by Connor and Taylor [19] plasma beta, normalized mean free path (or collisionality) and the normalized gyro radius are the dimensionless parameters characterising the plasma confinement and efforts have been made to formulate the confinement time in terms of these dimensionless parameters. However, for practical applications the formulation above is more favourable. Tokamak confinement without H-mode improvement can be well described by the so-called ITER L-mode scaling [20].

$$\tau_E = 0.048 \, R^{1.2} a^{0.3} B^{0.2} I^{0.85} \kappa^{0.5} n^{0.1} A^{0.5} P^{-0.5} \tag{7.54}$$

while in H-mode the following formula holds

$$\tau_E = 0.144 \, R^{1.39} a^{0.58} B^{0.15} I^{0.93} \kappa^{0.78} n^{0.41} A^{0.19} P^{-0.69} \tag{7.55}$$

This ELMy-H-mode scaling is presently the favourite tool to extrapolate the confinement towards future tokamak experiments. The variety of stellarator experiments in geometry and magnetic field makes the search for a general scaling law difficult. Nevertheless, a least square fit of all existing stellarator data led 1995 to the so-called ISS95-scaling [21]

$$\tau_E = 0.256 \, R^{0.65} a^{2.21} B^{0.83} \iota^{0.4} n^{0.51} P^{-0.59} . \tag{7.56}$$

Recent experiments in Wendelstein 7-AS and LHD, however, showed better confinement than predicted by this scaling law. In W7-AS the Lackner–Gottardi scaling [22] is a good fit to the experimental data.

$$\tau_E = 0.21 \, Ra^2 B^{0.8} \iota^{0.4} n^{0.6} P^{-0.6} . \tag{7.57}$$

This equation also fits to the dimensional constraints of Connor and Taylor. However, it fails to describe H-mode confinement in the high-density regime of Wendelstein 7-AS, there the confinement is better than this scaling predicts. This example demonstrates that empirical scaling laws have a limited regime of validity, and, unless a better understanding of the underlying physics is available, extrapolation towards future experiments or towards a fusion reactor should be made with caution. Plasma instabilities leading to turbulence and turbulent transport of particles, energy and momentum are still a wide field for theoretical and experimental research and the understanding of magnetic confinement remains incomplete until experimental findings, theoretical efforts and numerical simulation have come to a consensus on the physics behind.

References

1. J.R. Cary, R.G. Littlejohn: Ann. Phys. **151**, 1 (1983)
2. V.D. Shafranov: Sov. Phys. JETP **6**, 545 (1958)
3. R. Lüst, A. Schlüter: Z. Naturforsch. **12a**, 850 (1957)
4. L. Spitzer jr: A proposed stellarator. AEC Research and Development Report NYO–993, United States Atomic Energy Commision, Washington D.C. (1951)
5. A.I. Morozov, L.S. Solovev: Motion of charged particles in electro–magnetic fields. In: *Reviews of Plasma Physics*, vol 2 ed by M.A. Leontovich (Consultants Bureau, New York 1966) pp 201–297
6. C. Gourdon, D. Marty, E.K. Maschke et al: Configurations du type stellarator avec puit moyen et cisaillement des lignes magnétiques. *Proc. of the 3rd IAEA Conference on Plasma Physics and Controlled Nuclear Fusion Research*, vol 1 (IAEA, Vienna 1969) pp 847–861
7. S. Rehker, H. Wobig: A stellarator coil system without helical windings. In: *Proc. of the 7th Symposium on Fusion Technology (SOFT–Grenoble)* (Commission of the European Communities General Directorate for Dissimination of Knowledge for Information and Documentation, Luxembourg 1972) pp 345–357
8. P. Merkel: Nucl. Fusion **27**, 867 (1987)
9. J. Nührenberg, R. Zille: Phys. Lett. A **114**, 129 (1986)
10. C.D. Beidler, G. Grieger, F. Herrnegger et al: Fusion Technol. **17**, 148 (1990)
11. M.D. Kruskal, R.M. Kulsrud: Phys. Fluids **1**, 265 (1958)
12. W.I. van Rij, S.P. Hirshman: Phys. Fluids B **1**, 563 (1989)
13. L. Spitzer jr.: Phys. Fluids **1**, 253 (1958)
14. A.H. Boozer: Phys. Fluids **27**, 2110 (1984)
15. D. Pfirsch, A. Schlüter: Der Einfluß der elektrischen Leitfähigkeit auf das Gleichgewichtsverhalten von Plasmen niedrigen Drucks in Stellaratoren. Bericht MPI/PA/7/62, Max-Planck-Insitut für Physik und Astrophysik, München (1962) (in German)
16. A.A. Galeev, R.Z. Sagdeev: Sov. Phys. JETP-USSR **26**, 233 (1968); A.A. Galeev, R.Z. Sagdeev: Theory of Neoclassical Diffusion. In: *Reviews of Plasma Physics*, vol 7 ed by M.A. Leontovich (Consultants Bureau, New York 1979) pp 257–343
17. X. Garbet: Plasma Phys. Control. Fusion **39**, B91 (1997)
18. F. Wagner, G. Becker, K. Behringer et al: Phys. Rev. Lett. **49**, 1408 (1982)

19. J.W. Connor, J.B. Taylor: Nucl. Fusion **17**, 1047 (1977)
20. ITER Physics Expert Group on Disruptions, Plasma Control and MHD, ITER Physics Basis Editors: Nucl. Fusion **39**, 2251 (1999)
21. U. Stroth, M. Murakami, R.A. Dory et al: Nucl. Fusion **36**, 1063 (1996)
22. K. Lackner, N.A.O. Gottardi: Nucl. Fusion **30**, 767 (1990)

8 Introduction to Turbulence in Magnetized Plasmas

B.D. Scott

Abstract. Turbulence is a complex manifestation of collective plasma behaviour. An introduction to turbulence with particular emphasis on magnetized plasmas is given.

This chapter discusses the theoretical background to a leading edge topic in modern plasma physics – it is understood to accompany the section on experimental transport studies (Chap. 9 in Part II).

8.1 Part A – Statistical Nonlinearity and Cascade Dynamics

In general situations the dynamics of a magnetized plasma is not comprised solely of a single wave or instability. This is due to the general nonlinearity of the system when two or more eigenfunctions are excited; the various waves are coupled together and interact with each other as well as with the background. Turbulence due to a fluid flow like the basic plasma $\mathbf{E} \times \mathbf{B}$ velocity has as its dominant coupling mechanism a quadratic nonlinearity: advection of each dependent variable by the velocity, itself a dependent variable. Quadratic nonlinearities exhibit a phenomenon known as three wave coupling: in a Fourier decomposition in terms of wave-numbers, each wave $\mathbf{k}$ interacts with the other waves $\mathbf{k}'$ through beat waves $\mathbf{k}''$ satisfying the constraint $\mathbf{k} + \mathbf{k}' + \mathbf{k}'' = 0$. One can consider this as a process of modes $\mathbf{k}'$ and $\mathbf{k}''$ beating together to drive mode $\mathbf{k}$ (in the shorthand in which the word *mode* is used to signify an individual eigenmode, wave, or Fourier component). We are thinking of a relatively homogeneous system with many degrees of freedom; even if the background has gradients, the turbulence can be thought of as homogeneous if the typical scale of motion is smaller than any gradient scale length of the background.

We can distinguish between quasi-linear and nonlinear effects. Quasi-linear dynamics refers to a single wave interacting with the background; in other words, mode zero (zero wavenumber, nonzero frequency) and mode $-\mathbf{k}$ driving mode $\mathbf{k}$, as in a linear system, but now together with modes $\mathbf{k}$ and $-\mathbf{k}$ driving mode zero. The waves excite changes in the background which affect the subsequent evolution of the waves. Here, $\mathbf{k}$ and $-\mathbf{k}$ are actually the same mode, since for a real scalar field $\rho(\mathbf{x})$ we have $\rho_{-\mathbf{k}} = \rho_{\mathbf{k}}^*$ in the (complex)

Fourier components, where the superscript asterisk denotes the complex conjugate. The quasi-linear system therefore involves the same set of modes the linear system does; the only exception is that mode zero is allowed to evolve. When many modes are excited, the dynamics could still be quasi-linear if this particular wave/background interaction were still dominant. Such dynamics tends to produce relaxation oscillations in conservative systems: the action of the wave on the background is stabilizing but overshoots the point at which the growth rate goes to zero, and then the free energy in the wave all goes back to the background, reestablishing the original unstable state, and so on. The interaction typically has a coherent character, with long term memory of initial conditions.

Turbulence *per se* is very different from this. The same three wave coupling mechanism works as well between all possible triplets in the full set of waves. Each three wave interaction works independently of all the others, but many such interactions affect each individual wave. The dynamics involves many degrees of freedom (one for each dependent variable or eigenmode at each $\mathbf{k}$) all exchanging energy, and while each three wave interaction tends towards coherence, the fact that there are many of them going on simultaneously results in a weak degree of overall coherence between waves at differing $\mathbf{k}$. When these three wave interactions are as strong as the quasi-linear interactions, they result in an incoherent type of dynamics, which must be treated statistically. Moreover, it has a short time memory: dynamics at the same spatial point but separated in time become uncorrelated beyond a time range called the correlation time. The correlation time is somewhat longer than the time required for individual eddies to turn over, but it is shorter than the time required for the collective disturbance free energy to respond to changes in the background parameters. The dynamics therefore quickly loses memory of any initial state, and somewhat more slowly reaches a state called saturation: the individual waves fluctuate vigorously in amplitude and phase, but their free energy (or amplitude) level remains statistically stationary over time scales of several correlation times. This type of dynamics is what we call turbulence. It is a non-equilibrium, dissipative system with many degrees of freedom, with short correlation times and lengths. It is characterized by its statistical properties, not so much by the evolutionary fate of individual structures. For those familiar with geophysical fluid dynamics, the distinction between individual eddies, or structures, within the turbulence and the statistical properties of the fluctuating dynamics is analogous to the distinction between weather and climate.

The basic picture of turbulence one starts with is an *eddy mitosis* model. Large eddies are driven by some *stirring* process and a balance is formed as they break up into roughly two smaller ones each, on average, then these smaller ones into two smaller ones, each, and then so forth down to scales at which dissipation becomes stronger than the nonlinear dynamics. This energy transfer is called a *cascade* because it occurs in a chain of interactions, each of which is between eddies at not very disparate scales. In terms of the *turbulence spectrum*, the distribution of energy in terms of wave-numbers, the cascade process is *local* in the sense that the ratio of wave-numbers involved in most of the transfer process is rarely larger than two.

The separation between the main energy containing scales and the small scale dissipation range is characterized by the *Reynolds number*, given by the ratio of the product of typical scale length and velocity, and the viscous diffusion coefficient. The turbulence is called *high Reynolds number turbulence* if the cascade process and not the details of the dissipation process controls the balance between drive and dissipation. In terms of an expansion, the small parameter is the inverse Reynolds number, and the turbulence is thought of as random to lowest order, with the energy transfer and any transport processes entering at next order.

Turbulence in fluids is generally three dimensional, but under certain circumstances the vorticity (see Sect. 8.6) of the motion is ordered along a particular axis. This is either the rotation axis in a rapidly spinning system, like a thin accretion disk of material orbiting a central object like a star, or in a thin-atmosphere system with some rotation like geophysical turbulence, or it is the magnetic field in a plasma whose kinetic energy of motion is not sufficient to strongly disturb the magnetic field. Two dimensional turbulence has the special property that its vorticity is strictly perpendicular to its velocity. As a result, a quantity called the *enstrophy* is conserved in addition to the energy. The energy density is one half the square of the velocity, and the enstrophy density is one half the square of the vorticity. With both quantities arising from the single velocity variable conserved, the one with the highest wave-number dependence, namely enstrophy, controls the statistics and effects the direct cascade to small scales, and the other one, namely energy, then cascades to larger scales. This phenomenon is called a *dual cascade* and the transfer of energy to large scales is called an *inverse cascade*.

Turbulence in an idealized sense is incompressible, with all the motion of the shearing type. It is important to note this in light of the common concept of one dimensional nonlinear chaos. Turbulence is not only more statistical than the results of deterministic chaos models, it also has this shearing character. Neutral fluid turbulence is kept incompressible by the pressure, so long as the velocity of the flow eddies is slower than the sound speed. Disturbances in the density can in certain circumstances (e.g., thermal convection) remain, but are merely *passively* advected, with any coupling back to the flows rather weak in comparison to the time scale represented by the action of the eddies upon each other. A quantity whose dynamics is simple incompressible advection by the turbulence but with no back reaction upon the turbulence is called a *passive scalar*. In a magnetized plasma, incompressibility is maintained by the magnetic field if the velocity is slower than the Alfvén velocity, in the usual *low beta* situation in which plasma pressure is much smaller than the effective magnetic field pressure which is part of the magnetohydrodynamic

(MHD) force. The pressure evolves freely, and like the density it is under the MHD approximations a passive scalar.

The above concepts comprise Part A of this chapter. The following Sections will add substance to this Introduction, starting with the eddy mitosis model, continuing with the statistical character, the fluid model, three wave coupling and the cascades, the contrast between two and three dimensional models, and finally MHD turbulence and passive scalar dynamics. Part B will turn to the special case of gradient driven turbulence in a magnetized plasma, whose dynamics parallel to the magnetic field invokes dissipative coupling mechanisms between the fluid vorticity and the transported quantities, such that the latter no longer behave as passive scalars. This type of turbulence is very important for two reasons:

1. it is beyond the realm of the MHD model, and
2. it is found in many instances to be responsible for the dominant transport processes in a confined plasma.

8.2 Eddy Mitosis and the Cascade Model

The basic picture of turbulence in fluids is the one introduced by Kolmogorov in 1941 and independently by von Weizäcker and Heisenberg in different language in subsequent years. At a certain range of spatial scales, energy is supposed to be introduced into the fluid motions. This is called *stirring* and beyond the characterization of a drive rate it is independent of the details. The turbulence is supposed to be homogeneous and isotropic, so that spatial scale and spectral wave-number magnitude, related[1] through $k \leftrightarrow \sqrt{2}\pi/\Delta$, can be used interchangeably. A collection of eddies at the driven spectral range $k \sim k_0$ finds shearing motion tearing each of the eddies apart at about the same rate at which they are driven, so that there is a statistical equilibrium. This *nonlinear de-correlation* process conserves energy, so the energy taken out of the driven range is passed to other scales. Since the dominant instability of a sheared flow has a wavelength slightly shorter than the scale of the flow, most of this energy is passed to the next smallest scale in a logarithmic hierarchy, with a scale ratio of about two. With all of the energy transfer taking place between neighbors in the hierarchy, the picture we have in mind is a *local cascade*. At each scale below the driven range, there is a statistical balance mostly between energy passed in from the next larger scale, and nonlinear de-correlation passing the energy further to the next smaller scale. The shearing rate at each scale is proportional to the vorticity of the flow at that scale, so that the coherent lifetime of a typical eddy is comparable to

---

[1] You can show that for a two-dimensional field of periodic vortices on a wavelength 2Δ in each direction, the Laplacian operator ∇^2 carries an eigenvalue of $-2\pi^2/\Delta^2$.

the *eddy turnover time* the time the eddy takes to transport the flow energy and momentum across itself.

Following these ideas, at each level n with wave-number magnitude k_n a balance is struck between incoming energy due to nonlinear de-correlation at level $n-1$ and stirring at level n, and outgoing energy due to nonlinear de-correlation at level n. This is expressed as

$$(kv)_{n-1} E_{n-1} + \gamma_n E_n = (kv)_n E_n \tag{8.1}$$

where E_n and $(kv)_n$ are the energy and vorticity within level n, and γ_n is the rate of stirring at level n. We can take $E_n = v_n^2/2$.

At the low-k end of the spectrum the turbulence is generated by stirring and hence $\gamma > 0$ there. At the high-k end lies the dissipation range, where $\gamma < 0$. Since the separation of scales is supposed to be arbitrarily large, there is a wide spectral range in between for which γ is negligible compared to (kv). This is called the *inertial range* of the turbulence and is the range in which the turbulence is most pure, that is, the conservative nonlinearity is the only part of the dynamics which is active. Within the inertial range, we then have

$$(kv)_{n-1} E_{n-1} = (kv)_n E_n = T_0 \tag{8.2}$$

where T_0 is a constant giving the energy *throughput* of the cascade. With $E_n = v_n^2/2$ we find the velocity and energy density spectrum,

$$v_n^3 = 2T_0 k_n^{-1} \qquad E_n = (2T_0)^{2/3} k_n^{-2/3} \tag{8.3}$$

To find E_k, the energy per unit range of k, we have to take into account the fact that each level n has a range of k which is proportional to k (i.e., we have a logarithmic spacing). In integral form we have

$$\sum_n E_n = \int d\log k \, E_n = \int dk \, E_k \tag{8.4}$$

which identifies E_k as E_n/k, or $E_n(k_0/k)$ to preserve the dimensions. This yields

$$E_k = E_0 \, k^{-5/3} \tag{8.5}$$

with $E_0 = (2T_0)^{2/3} k_0$. This 5/3 power law is known as the Kolmogorov spectrum. Within the inertial range, this spectrum is *universal* and the dynamics is *self similar*, indicating that the scales of motion in time and space can be arbitrarily reformatted.

If the dissipation is caused by viscosity or some other diffusive process, we have in the dissipation range $\gamma(k) = -\chi k^2$, where χ is called the diffusivity. The Reynolds number is defined by comparing the typical velocity and space scales in the *driven range* with the diffusivity,

$$\mathrm{Re} = LV/\chi \tag{8.6}$$

The *high Reynolds number* regime is the one for which the inertial range involves several levels of the cascade hierarchy, and an important result of that is that the energy balances of both the energy containing range and the inertial range are independent of the strength or form of the dissipation process. For diffusion, the amount of energy consumed by the dissipation should be independent of the diffusion coefficient.

Clearly, the existence of an inertial range, with several levels in the hierarchy for which $\gamma \ll (kv)$, is connected with a large value for Re. It does however take a rather large Reynolds number before the high Reynolds number regime is entered. To see this, we may take a simple situation for which (8.1) can be applied: assume that $\gamma_0 > 0$ only for the first level in the hierarchy, the driven range k_0. For all other scales, let $\gamma_n = -\chi k_n^2$. The amplitude of the driven range is given by the *mixing condition*,

$$\gamma_0 = (kv)_0 \qquad \text{hence} \qquad v_0 = \frac{\gamma_0}{k_0}. \tag{8.7}$$

It is a property in the inertial range that the shearing rate is progressively faster at smaller scale, due to the conservation of energy: with (8.1) satisfied for $\gamma = 0$ we have

$$\frac{(kv)_{n+1}}{(kv)_n} = \left(\frac{k_{n+1}}{k_n}\right)^{2/3} \tag{8.8}$$

and hence with $k_{n+1}/k_n = 2$ the ratio is $2^{2/3}$ or about 1.6. Even with a process like convection for which γ is roughly constant below the scale of inhomogeneity, one does therefore reach an inertial range if χ is small enough, that is, if Re is large enough. But the factor of 1.6 is small enough that Re must be very large in order to have a real inertial range. Typically, a flow is not even turbulent for Re lower than a few thousand, and for flow in three dimensions the high Reynolds number regime is not reached until Re is at least 10^6. In astrophysical situations Re can be in the range 10^{12}–10^{15}. For atmospheric turbulence one finds values some orders of magnitude smaller than that but still large; for air at STP the kinematic viscosity is about $1.5 \times 10^{-5}\,\text{m}^2/\text{s}$, so for $L > 15\,\text{m}$ and $V > 1\,\text{m/s}$ we have $\text{Re} > 10^6$. The highest resolution three dimensional fluid turbulence computations to date are barely into the high Reynolds number regime, although the mere demonstration of a local cascade is possible with a discretized domain of only 64^3 grid nodes. In two dimensions, as discussed below, the cascade dynamics is different and turbulence may set in for Re values rather lower than this; typically, at least two decades between driven and dissipation range in the spectrum are required ($k_N/k_0 \sim 10^2$). For magnetized plasma turbulence the Reynolds number is not as large as this: the dissipative coupling process discussed in Part B of this chapter inhibits the cascade process sufficiently that only one decade of scale separation is needed to reach the high Reynolds number regime.

8.3 The Statistical Nature of Turbulence

One is motivated to ask "nonlinear dynamics, OK, but what makes it turbulence?" The answer is that the motion involved exhibits short space/time correlation and is *ultimately non-periodic* in the sense that no matter how long the time domain in an experiment or computation is allowed to persist, the longest time periods become dominant in a frequency spectrum, the standard deviation of a time series saturates instead of falling with increasing time intervals, and the statistical description of the turbulent flow amplitudes and de-correlation rates in terms of a mean and a standard deviation is independent of how the dynamical system was started. In terms of a computation, the particular initial conditions chosen should have no effect on the statistical result, even though certain initial scenarios can save time by getting the system to statistical equilibrium in a comparatively shorter time than others. Fully developed turbulence is characterized by a *saturated state* in which any relevant statistical measure is stationary in this sense. This is never certain; one should accept that results are only known within error bars characterized by that standard deviation, and in this case one should always quote the standard deviation of the distribution, not that of the mean (the latter falls with the square root of the number of samples if the former is stationary). In this sense turbulence *converges* only statistically, not monotonically as a coherent dissipative process would do.

Current research is occupied with the question of rare but strong events: those well down the tails of the *probability distribution function* (PDF). The PDF is essentially a histogram. The amplitude of a certain variable is sampled and considered as a series. The mean and standard deviation of the series are measured, and the PDF is computed by dividing the abscissa into intervals which are small compared to the standard deviation. The fraction of the events falling into each of these intervals is plotted as the ordinate, normalized in terms of the number of intervals per standard deviation. A purely random process has a Gaussian PDF for any of its measures, but many features of the turbulence are *intermittent* in the sense that the first one or two standard deviations of the PDF appear Gaussian but the tails are different, usually above the Gaussian curve of the same mean and standard deviation. Nevertheless, the basic statistical character of turbulence can be understood in terms of this near-Gaussian PDF. The PDF curves for turbulence are usually compared to the corresponding Gaussian which has the same mean and standard deviation.

Statistical memory in turbulence can be characterized in terms of autocorrelation functions in the principal fluctuating variables. Both time and space correlations are described, with the $1/e$ half-width of these curves defined as the correlation time or correlation length. The time autocorrelation function of a variable ϕ is defined as

$$F(\tau) = \lim_{T\to\infty} \int_{-T/2}^{T/2} \frac{dt}{T}\, \phi(t)\phi(t-\tau) \qquad (8.9)$$

also averaged over space. For a purely wavelike process (e.g., a linear oscillation), $F(\tau)$ is sinusoidal. For a purely random process, it is Gaussian. If the spatial domain is periodic on a length L, it is simple to compute the space autocorrelation function in terms of Fourier components,

$$F(\Delta) = c_0 \sum_{l\neq 0} |\phi_l|^2 \cos(2\pi l \Delta/L) \qquad (8.10)$$

also averaged over time, with the normalizing constant c_0 chosen such that $F(0) = 1$. The curve for $F(\Delta)$ is itself periodic and therefore is only significant for $-L/2 < \Delta < L/2$. In turbulence within a bounded domain (e.g., convection, or tokamak edge turbulence), the correlation length in the relevant direction (e.g., down the gradient) can be comparable to the domain size, but it should be much smaller than the domain size in the *horizontal* direction (e.g., East-West for geophysical fluid turbulence, or perpendicular to the magnetic field lines but within the flux surfaces for gradient driven turbulence within a closed, toroidal MHD equilibrium). Most importantly, the correlation time must be short compared to the time intervals required for the statistical measures to reach saturation. In relevantly scaled units, tokamak turbulence typically has a dynamical scale of order 1, a correlation time of order 10, a saturation time at least of order several hundred, and an transport equilibration time at least of order several thousand. This level of temporal scale separation is what makes plausible the model of thermal transport by turbulent mixing on a slowly varying background.

8.4 Quadratic Nonlinearity and Three Wave Coupling for Small Disturbances

When treating turbulence as such one usually either assumes an homogeneous background, or one with a gradient which drives the turbulence while nevertheless leading to disturbances which are small compared to background quantities. The flow velocity, as in certain circumstances the magnetic field, belongs to the disturbances. The pressure and density belong to the background, and since the disturbances are small these two quantities are treated as constant coefficients except where acted upon by the gradient operator. The scale of motion is also expected to be small compared to the scale of the background, and this simplification is used as a model even when in the case of a thin boundary region (e.g., convection in a thin atmosphere, or tokamak edge turbulence) the scales of inhomogeneity and of the turbulence are comparable but both small compared to the overall size of the equilibrium.

In a situation like this the equations governing the dynamics are linearized except for the terms representing advection or, in the case of turbulence on a magnetized plasma background, the disturbed parallel gradient. These terms have the special character that they are quadratic in the dependent variables. The simplest example of such an effect is $\mathbf{v} \cdot \nabla \mathbf{v}$, the self-advection of the velocity, which is the basis of what gives turbulence its character. The thermodynamic state variables are also advected through quadratic nonlinearities. The simplest example of an equation they satisfy is the continuity equation for the mass density ρ,

$$\frac{\partial \rho}{\partial t} + \nabla \cdot \rho \mathbf{v} = 0 \tag{8.11}$$

expressing mass conservation. This is rearranged by separating the advection and the divergence terms, and linearizing the latter, formally writing

$$\left(\frac{\partial}{\partial t} + \mathbf{v} \cdot \nabla \right) \log \rho + \nabla \cdot \mathbf{v} = 0 \tag{8.12}$$

which we can do with no loss of generality, but then also considering small disturbances so that $\delta \log \rho \to \delta \rho / \rho$, writing

$$\frac{\partial \tilde{\rho}}{\partial t} + \mathbf{v} \cdot \nabla \tilde{\rho} + \rho \nabla \cdot \mathbf{v} = 0 \tag{8.13}$$

and treating the factor of ρ multiplying the divergence as a constant parameter while using the tilde symbol to mark the dependent variable. The advection term $\mathbf{v} \cdot \nabla \tilde{\rho}$ is recognized as a quadratic nonlinearity. The nonlinear part of the divergence term is smaller than this because we are treating situations under the *small scale mixing* condition:

$$\tilde{\rho} \ll \rho \qquad \text{although} \qquad \nabla \tilde{\rho} \sim \nabla \rho. \tag{8.14}$$

In an incompressible model, of course, the advection term is the only effect controlling time evolution in this equation.

The three wave coupling condition follows from the properties of Fourier transforms. We treat the background as homogeneous in understanding the properties of the turbulence. The dependent variables are Fourier decomposed according to

$$\tilde{\rho}(\mathbf{x}) = \sum_{\mathbf{k}} \rho_{\mathbf{k}} \exp(i \mathbf{k} \cdot \mathbf{x}) \qquad \rho_{\mathbf{k}} = \oint \frac{d^n x}{L^n} \exp(-i \mathbf{k} \cdot \mathbf{x}) \, \tilde{\rho}(\mathbf{x}) \tag{8.15}$$

where n is the number of dimensions (usually 2 or 3) and the spatial domain is taken to be periodic on length L in each dimension (this is easily generalized). The incompressible continuity equation (with $\nabla \cdot \mathbf{v} = 0$) then becomes

$$\frac{\partial \rho_{\mathbf{k}}}{\partial t} = - \oint \frac{d^n x}{L^n} \exp(-i \mathbf{k} \cdot \mathbf{x}) \ldots$$
$$\ldots \sum_{-\mathbf{k}'} \sum_{-\mathbf{k}''} \mathbf{v}_{-\mathbf{k}'} \cdot (-i \mathbf{k}'') \rho_{-\mathbf{k}''} \exp(-i \mathbf{k}' \cdot \mathbf{x}) \exp(-i \mathbf{k}'' \cdot \mathbf{x})$$
$$\tag{8.16}$$

where we have used the decomposition in terms of the sum for both dependent variables in the nonlinearity and the integral operation to isolate the single component under the time derivative. Note that the choice of signs in $\mathbf{k}$ is arbitrary due to the condition that $\rho(\mathbf{x})$ is real valued, $\rho_{-\mathbf{k}} = \rho_{\mathbf{k}}^*$, since we sum over all the components. The factors involving $\mathbf{x}$ and the integral can be isolated from the rest of the right hand side,

$$\oint \frac{d^n x}{L^n} \exp\left[-i(\mathbf{k} + \mathbf{k}' + \mathbf{k}'') \cdot \mathbf{x}\right] = \delta_{(\mathbf{k}+\mathbf{k}'+\mathbf{k}''),0} \qquad (8.17)$$

yielding the Kronecker delta. The two components are said to beat against one another (the parlance for waves) to drive component $\mathbf{k}$. The condition under which there is a finite interaction is then

$$\mathbf{k} + \mathbf{k}' + \mathbf{k}'' = 0 \qquad (8.18)$$

which is called the three wave condition: the three wave-numbers must vectorially add to zero. Interactions of this type are called *three wave interactions* and they form the basis for the interactions in turbulence. The three wave condition is therefore a relatively trivial consequence of the quadratic nonlinearity and the orthogonality of Fourier transforms. Even if the spatial domain is not periodic, the basic character of turbulence has this qualitative form.

Three wave models are constructed using a single triplet $\{\mathbf{k}_1, \mathbf{k}_2, \mathbf{k}_3\}$, satisfying the three wave condition, for explanatory purposes. The model consists of an evolution equation for each Fourier component which is a member of the triplet. Turbulence, however, derives its character from the fact that there are very many such triplets and the vast majority of them are mutually incoherent, with the three members having no special relation to each other. To lowest order of approximation (the small parameter is usually Re^{-1} or some equivalent) in the mathematical theories of turbulence the various triplets are supposed to be mutually random, with the PDF of cross correlation having Gaussian form with zero mean. All the energy transfer dynamics comes from the corrections, which are computed according to one or another model. The details are very complicated and are left to the references listed at the end. We will return to the three wave interactions and what they say about the transfer dynamics generally, after introducing the incompressible turbulence models.

8.5 Incompressible Hydrodynamic Turbulence – Energy and Enstrophy

We start with simple hydrodynamic turbulence, following the ideal equation of motion in MHD without the magnetic field,

$$\frac{\partial \mathbf{v}}{\partial t} + \mathbf{v} \cdot \nabla \mathbf{v} = -\frac{1}{\rho} \nabla p \,. \qquad (8.19)$$

We recognize the role of the pressure in maintaining incompressibility by taking the divergence and assuming that $\nabla \cdot \mathbf{v} = 0$,

$$\nabla \cdot \frac{1}{\rho}\nabla p = -\nabla \cdot (\mathbf{v} \cdot \nabla \mathbf{v}) . \tag{8.20}$$

This condition controls the evolution of the pressure, and so even in an inhomogeneous model the only other equation controlling evolution would be an incompressible advection equation for ρ. In an homogeneous model we take ρ to be constant in (8.19) since $\rho \nabla^2 p \gg \nabla \rho \cdot \nabla p$ due to the small fluctuation approximation as in (8.14). Taking the curl, we then have

$$\frac{\partial}{\partial t}(\nabla \times \mathbf{v}) + \mathbf{v} \cdot \nabla(\nabla \times \mathbf{v}) = (\nabla \times \mathbf{v}) \cdot \nabla \mathbf{v} . \tag{8.21}$$

This equation says that the fluid vorticity $(\nabla \times \mathbf{v})$ evolves not only due to advection but also due to a finite component of the velocity gradient along the vorticity. The squared magnitude of the vorticity is the density of a quantity called *enstrophy* in hydrodynamics. Defining it as $W = (1/2)|\nabla \times \mathbf{v}|^2$, we may form its equation by contracting (8.21) with the vorticity,

$$\frac{\partial W}{\partial t} + \nabla \cdot (W \mathbf{v}) = (\nabla \times \mathbf{v})(\nabla \times \mathbf{v}) : \nabla \mathbf{v} \tag{8.22}$$

using the condition $\nabla \cdot \mathbf{v} = 0$. This equation says that the enstrophy is conserved except for a class of motions which involves stretching the velocity component along the vorticity further in the direction of the vorticity. Since the motion is incompressible, the components of the velocity making up most of the vorticity must have a convergence to make up the divergence involved in the stretching.

To find out what this type of motion signifies and what its consequence is, we reconsider the results arising from the turbulent cascade model (Sect. 8.3). Recall that a cascade towards small scale which conserves energy must exhibit a shearing rate which increases towards small scale. Given the energy conservation condition, it is a simple matter to conclude that the enstrophy must grow during this cascade, simply from the extra factors of k in W compared to the energy density $U = v^2/2$. The type of motion which we concluded above allows W to grow may be displayed in a simple sketch: Most of the velocity is in eddies each of which rotates in a local two dimensional plane, to which the vorticity is perpendicular. The third dimension has to have a finite velocity component, along the vorticity vector, and it must stretch in this direction. The eddy motion in the local plane must also contract to maintain incompressibility. This basically involves a vortex tube structure which is stretched in the time it maintains its coherence (not more than a few eddy turnover times), before it is sheared apart by the motion both in the local plane and along the local vorticity vector. We therefore conclude simply on the basis of (1) the incompressibility constraint, (2) energy conservation, (3) the cascade to smaller scales, and (4) the growth of the enstrophy, that the turbulent cascade process behind the eddy mitosis model must involve this *vortex tube stretching* in three dimensions.

The foregoing is significant because in several important applications (e.g., motion at large horizontal scale in a thin atmosphere, *geostrophic* motion in a rapidly rotating background, or motion perpendicular to the magnetic field in a magnetized plasma) the turbulence is quite strictly two dimensional (2D). We will see in Part B how dissipative coupling does this for gradient driven turbulence in a magnetized plasma. In a 2D model, we have the additional constraint that the vorticity is strictly perpendicular to the plane in which the velocity and the gradient operator both lie. In this case we have both $(\nabla \times \mathbf{v}) \cdot \mathbf{v} = 0$ and $(\nabla \times \mathbf{v}) \cdot \nabla = 0$, so that the right hand side of (8.22) is zero. Therefore, the enstrophy as well as the energy is conserved. With W conserved we cannot have the cascade of energy to small scale. To find out how this situation is resolved we may examine three wave coupling in 2D turbulence.

In a 2D incompressible flow, the velocity is given in terms of a stream function, which we label ψ, and the vorticity is a scalar, Ω, times the unit normal to the plane, which we label $\hat{\mathbf{s}}$, so that

$$\mathbf{v} = \hat{\mathbf{s}} \times \nabla \psi \qquad \nabla \times \mathbf{v} = \Omega\,\hat{\mathbf{s}} \qquad \Omega = \nabla_\perp^2 \psi \qquad (8.23)$$

with the perp symbol ($\perp$) used to denote the fact that only the two dimensions in the plane are involved. (8.21) becomes

$$\frac{\partial \Omega}{\partial t} + \mathbf{v} \cdot \nabla \Omega = 0 \qquad (8.24)$$

which is called the 2D Euler equation. Multiplying by $-\psi$ or by Ω, we find

$$\frac{\partial U}{\partial t} + \nabla \cdot \left\{ U\,\mathbf{v} - \left[\psi \left(\frac{\partial}{\partial t} + \mathbf{v} \cdot \nabla \right) \nabla_\perp \psi \right] \right\} = 0 \qquad \frac{\partial W}{\partial t} + \nabla \cdot (W\,\mathbf{v}) = 0 \qquad (8.25)$$

that is, simultaneous energy and enstrophy conservation. (The second term in the first equation is a quantity related to compression and becomes important only if the density is strongly variable, at which point the assumption of incompressibility itself breaks down.)

Three wave interactions set up by Fourier decomposition of (8.24) will also have these conservation properties, therefore representing a conservative mutual transfer of energy and enstrophy from wave to wave. For (8.24), for example, one can easily verify that

$$\frac{\partial}{\partial t} \left(\frac{\Omega_\mathbf{k}^* \Omega_\mathbf{k} + \Omega_{\mathbf{k}'}^* \Omega_{\mathbf{k}'} + \Omega_{\mathbf{k}''}^* \Omega_{\mathbf{k}''}}{2} \right) = 0 \qquad (8.26)$$

for any particular triplet $\{\mathbf{k}, \mathbf{k}', \mathbf{k}''\}$. The equations satisfied by the members of the triplet are

$$\frac{\partial \Omega_{\mathbf{k}}}{\partial t} = C_{\mathbf{k}\mathbf{k}'} \left(\Omega_{-\mathbf{k}''} \psi_{-\mathbf{k}'} - \Omega_{\mathbf{k}'} \psi_{-\mathbf{k}''} \right) \quad (8.27)$$

$$\frac{\partial \Omega_{\mathbf{k}'}}{\partial t} = C_{\mathbf{k}\mathbf{k}'} \left(\Omega_{-\mathbf{k}} \psi_{-\mathbf{k}''} - \Omega_{\mathbf{k}''} \psi_{-\mathbf{k}} \right) \quad (8.28)$$

$$\frac{\partial \Omega_{\mathbf{k}''}}{\partial t} = C_{\mathbf{k}\mathbf{k}'} \left(\Omega_{-\mathbf{k}'} \psi_{-\mathbf{k}} - \Omega_{\mathbf{k}} \psi_{-\mathbf{k}'} \right) \quad (8.29)$$

which we can form by simple permutation among the members of the triplet. The coupling constant is given by

$$C_{\mathbf{k}\mathbf{k}'} = \frac{1}{2} \hat{\mathbf{s}} \cdot (\mathbf{k} \times \mathbf{k}') \quad (8.30)$$

where we use the relative symmetry in the interactions,

$$\hat{\mathbf{s}} \cdot (\mathbf{k} \times \mathbf{k}') = \hat{\mathbf{s}} \cdot (\mathbf{k}' \times \mathbf{k}'') = \hat{\mathbf{s}} \cdot (\mathbf{k}'' \times \mathbf{k}) \quad (8.31)$$

since $\mathbf{k}'' = -\mathbf{k} - \mathbf{k}'$.

Using these forms we can build the energy equations for each of the three components (and their complex conjugates) in the triplet, noting that for each one we have the reality condition, e.g., $\psi_{\mathbf{k}} = \psi^*_{-\mathbf{k}}$. For each energy component given by

$$U_{\mathbf{k}} = |\mathbf{k}\psi_{\mathbf{k}}|^2 = -\psi_{\mathbf{k}}\Omega_{-\mathbf{k}} \quad (8.32)$$

we obtain

$$\frac{\partial U_{\mathbf{k}}}{\partial t} = 2C_{\mathbf{k}\mathbf{k}'} \operatorname{Re}\left[\psi_{\mathbf{k}} \Omega_{\mathbf{k}'} \psi_{\mathbf{k}''} - \psi_{\mathbf{k}} \psi_{\mathbf{k}'} \Omega_{\mathbf{k}''} \right] \quad (8.33)$$

$$\frac{\partial U_{\mathbf{k}'}}{\partial t} = 2C_{\mathbf{k}\mathbf{k}'} \operatorname{Re}\left[\psi_{\mathbf{k}'} \Omega_{\mathbf{k}''} \psi_{\mathbf{k}} - \psi_{\mathbf{k}'} \psi_{\mathbf{k}''} \Omega_{\mathbf{k}} \right] \quad (8.34)$$

$$\frac{\partial U_{\mathbf{k}''}}{\partial t} = 2C_{\mathbf{k}\mathbf{k}'} \operatorname{Re}\left[\psi_{\mathbf{k}''} \Omega_{\mathbf{k}} \psi_{\mathbf{k}'} - \psi_{\mathbf{k}''} \psi_{\mathbf{k}} \Omega_{\mathbf{k}'} \right] . \quad (8.35)$$

Pairs of equal terms appearing with opposite sign give the energy transfer effects. Considering the equations as a permutation among the three modes in the triplet, the second term in each equation gives the energy transfer from the equation below it, and the first term in each equation gives the energy transfer from the equation above it. For example, the energy transfer from mode $\mathbf{k}'$ to mode $\mathbf{k}$ is given by

$$\begin{aligned} T_U(\mathbf{k} \leftarrow \mathbf{k}') &= 2C_{\mathbf{k}\mathbf{k}'} \operatorname{Re}\left[-\psi_{\mathbf{k}} \psi_{\mathbf{k}'} \Omega_{\mathbf{k}''} \right] \\ &= 2C_{\mathbf{k}\mathbf{k}'} \operatorname{Re}\left[(k'')^2 \psi_{\mathbf{k}} \psi_{\mathbf{k}'} \psi_{\mathbf{k}''} \right] . \end{aligned} \quad (8.36)$$

Similarly, for each enstrophy component given by

$$W_{\mathbf{k}} = |\Omega_{\mathbf{k}}|^2 = k^4 |\psi_{\mathbf{k}}|^2 \quad (8.37)$$

we build the enstrophy transfer equations,

$$\frac{\partial W_{\mathbf{k}}}{\partial t} = 2C_{\mathbf{kk'}} \operatorname{Re}\left[\Omega_{\mathbf{k}}\psi_{\mathbf{k'}}\Omega_{\mathbf{k''}} - \Omega_{\mathbf{k}}\Omega_{\mathbf{k'}}\psi_{\mathbf{k''}}\right] \tag{8.38}$$

$$\frac{\partial W_{\mathbf{k'}}}{\partial t} = 2C_{\mathbf{kk'}} \operatorname{Re}\left[\Omega_{\mathbf{k'}}\psi_{\mathbf{k''}}\Omega_{\mathbf{k}} - \Omega_{\mathbf{k'}}\Omega_{\mathbf{k''}}\psi_{\mathbf{k}}\right] \tag{8.39}$$

$$\frac{\partial W_{\mathbf{k''}}}{\partial t} = 2C_{\mathbf{kk'}} \operatorname{Re}\left[\Omega_{\mathbf{k''}}\psi_{\mathbf{k}}\Omega_{\mathbf{k'}} - \Omega_{\mathbf{k''}}\Omega_{\mathbf{k}}\psi_{\mathbf{k'}}\right] . \tag{8.40}$$

Hence, the transfer of enstrophy from mode $\mathbf{k'}$ to mode $\mathbf{k}$ is given by

$$\begin{aligned} T_W(\mathbf{k} \leftarrow \mathbf{k'}) &= 2C_{\mathbf{kk'}} \operatorname{Re}\left[-\Omega_{\mathbf{k}}\Omega_{\mathbf{k'}}\psi_{\mathbf{k''}}\right] \\ &= 2C_{\mathbf{kk'}} \operatorname{Re}\left[-k^2(k')^2 \psi_{\mathbf{k}}\psi_{\mathbf{k'}}\psi_{\mathbf{k''}}\right] . \end{aligned} \tag{8.41}$$

For the three wave system to have an average spectral transfer, the triple correlation $\langle \psi_{\mathbf{k}}\psi_{\mathbf{k'}}\psi_{\mathbf{k''}}\rangle$ will have a definite sign. If only a single triplet is present, we expect the energy or enstrophy to simply redistribute to reach a statistical equilibrium with a triple correlation of zero. But in turbulence, with many triplets present all with differing relationships between the wavenumbers, we expect an asymmetry to develop since there are more available states towards high k than towards low for any physically realizable instantaneous configuration (i.e., with finite total energy content at $k \to \infty$). The triple correlation and hence the average transfer dynamics will be constrained by the simultaneous conservation of both energy and enstrophy. We can consider the effects of interactions which spread energy and enstrophy towards both ends in the spectrum, out of some intermediate spectral range. Examining (8.36) and (8.41), noting that the squared factors are positive definite and the coefficient $C_{\mathbf{kk'}}$ is the same in both, we find that for any definite sign of the triple correlation, the signs of the transfer are opposite. If $k < k' < k''$ then the direction of energy transfer from k'' to k' and from k' to k is the same, and similarly for the oppositely directed enstrophy transfer. For both energy and enstrophy, then, the intermediate range k' sees a throughput.

The overall process is called a *dual cascade* since the two quantities arising from the same dependent variable ψ are transferred in opposite directions in the spectrum. The direction of the cascade dynamics will be determined by which one of these transfer effects controls the overall statistics. It is expected to be a local cascade since the coupling coefficients are largest when the magnitudes $\{k, k', k''\}$ are all comparable. The enstrophy and enstrophy transfer contain extra wave-number factors compared to the energy and energy transfer, so the enstrophy can be expected to mix faster. The statistical behavior of the whole should therefore be controlled by the enstrophy. The enstrophy should be redistributed towards higher k due to the fact that there are more available states there given any physically realizable initial state (i.e., with finite total energy content at $k \to \infty$). Besides, we already found that for the energy cascade to be *direct* (towards smaller scale) the enstrophy must grow. So, with enstrophy also conserved, we expect the energy cascade to be *inverse* (towards larger scale). We therefore expect the dual cascade to send enstrophy preferentially to higher k and energy preferentially to lower k.

The main points of this Section are that

1. a direct energy cascade requires the vortex tube stretching process in three dimensions since it requires a growing enstrophy,
2. two dimensional incompressible hydrodynamics conserves enstrophy as well as energy,
3. the directions of spectral transfer for energy and enstrophy are opposite given a definite sign of the triple correlation, and therefore
4. we expect to find a dual cascade in two dimensional incompressible hydrodynamic turbulence.

We can learn all this simply from the properties of the equations the dynamical model satisfies. We will show that it actually occurs in the computations in Part B of this chapter.

8.6 MHD Turbulence

Magnetohydrodynamic (MHD) turbulence is similar to its neutral fluid counterpart, with two differences: incompressibility is maintained by the magnetic field strength (magnetic rather than plasma pressure), and the additional magnetic nonlinearities cause special effects through which flow eddies and Alfvén waves interact with each other to allow long range spectral transfer.

To examine these effects we start with the MHD force equation under the same assumptions as in Sect. 8.5.

$$\frac{\partial \mathbf{v}}{\partial t} + \mathbf{v} \cdot \nabla \mathbf{v} = -\frac{1}{\rho} \nabla \left(p + \frac{B^2}{8\pi} \right) + \frac{\mathbf{B} \cdot \nabla \mathbf{B}}{4\pi \rho} \tag{8.42}$$

where we have used Ampere's law $\nabla \times \mathbf{B} = (4\pi/c)\mathbf{J}$ to replace the $c^{-1}\mathbf{J} \times \mathbf{B}$ force adding to $-\nabla p$ in the MHD model. In most cases one is interested in the *low beta regime* characterized by

$$\beta = \frac{8\pi p}{B^2} \ll 1 . \tag{8.43}$$

Here, the plasma pressure p is negligible compared to the magnetic pressure $B^2/8\pi$ in the maintenance of incompressibility, which we can see by taking the divergence of (8.42) assuming $\nabla \cdot \mathbf{v} = 0$,

$$\nabla \cdot \frac{1}{8\pi\rho} \nabla B^2 = -\nabla \cdot \left(\frac{1}{\rho} \nabla p + \mathbf{v} \cdot \nabla \mathbf{v} \right) . \tag{8.44}$$

It is therefore possible for p as well as ρ to enter the dynamics of background disturbances while B^2 maintains the dynamical incompressibility as long as $\beta \ll 1$, and the dynamics is slow compared to the compressional Alfvén frequency at the same scale, $\omega \ll k_\perp v_A$, and the velocity is sub-Alfvénic,

$v \ll v_A$ (the last two conditions are different if the motion is driven by the magnetic field). The perp symbol is now reckoned against the background magnetic field. The Alfvén velocity v_A is defined by

$$v_A^2 = \frac{B^2}{4\pi\rho} . \tag{8.45}$$

Most models satisfy these conditions, with the principal exception being the kinematic dynamo process (thermal convection with $\beta \gg 1$, with the magnetic field passively advected).

The equations of incompressible MHD turbulence also include the kinematics of the magnetic field, which assuming $\nabla \cdot \mathbf{v} = 0$ satisfies

$$\frac{\partial \mathbf{B}}{\partial t} + \mathbf{v} \cdot \nabla \mathbf{B} = \mathbf{B} \cdot \nabla \mathbf{v} . \tag{8.46}$$

Shear Alfvén dynamics basically involves a magnetic field whose direction is fluctuating but whose field strength is constant except for the small changes required to maintain the incompressibility. The dependent variable then becomes the unit vector. We can also scale the velocity against the background Alfvén velocity v_{A0}, which is v_A defined in terms of the background field strength B_0. The dependent variables are then

$$\mathbf{u} = \frac{\mathbf{v}}{v_{A0}} \qquad \mathbf{b} = \frac{\mathbf{B}}{B_0} \tag{8.47}$$

in whose terms (8.42) and (8.46) become

$$\frac{1}{v_{A0}} \frac{\partial \mathbf{u}}{\partial t} + \mathbf{u} \cdot \nabla \mathbf{u} = -\nabla I + \mathbf{b} \cdot \nabla \mathbf{b} \tag{8.48}$$

$$\frac{1}{v_{A0}} \frac{\partial \mathbf{b}}{\partial t} + \mathbf{u} \cdot \nabla \mathbf{b} = \mathbf{b} \cdot \nabla \mathbf{u} \tag{8.49}$$

where I is now simply a potential which maintains $\nabla \cdot \mathbf{u} = 0$. On the other hand, $\nabla \cdot \mathbf{b} = 0$ is maintained by the antisymmetry of $\mathbf{ub} - \mathbf{bu}$ as a tensor. We may define the *Elsässer variables*

$$\mathbf{u}_\pm = \mathbf{u} \pm \mathbf{b} \tag{8.50}$$

which satisfy

$$\frac{1}{v_{A0}} \frac{\partial \mathbf{u}_\pm}{\partial t} + \mathbf{u}_\mp \cdot \nabla \mathbf{u}_\pm = -\nabla I . \tag{8.51}$$

Computations of MHD turbulence often use this model, which allows for very simple numerical schemes such as are used for the hydrodynamic model of (8.19). The force potential gradient ∇I need not be computed specifically, though it is often used to subtract off the part of the nonlinearity which has a finite divergence (in computational fluid dynamics this is called the *Laplacian pressure method*). We note that there is only one force potential since

8 Introduction to Turbulence in Magnetized Plasmas   189

$$\nabla \cdot (\mathbf{u}_- \cdot \nabla \mathbf{u}_+) = \nabla \cdot (\mathbf{u}_+ \cdot \nabla \mathbf{u}_-) \tag{8.52}$$

although some numerical schemes apply two of them independently.

The main difference between MHD and fluid turbulence is that there are two equations each describing advection of one of the Elsässer variables by the other. The three wave coupling condition is no longer among different wave-number components of the same variable but is now also among the variables. Several consequences regarding the relationship between the velocity and the magnetic field dynamics follow. One consequence is known as *helicity conservation*, by which the magnetic helicity density $\mathbf{A} \cdot \mathbf{B}$, where $\mathbf{B} = \nabla \times \mathbf{A}$, can be shown to satisfy a transport equation. This exercise, using the divergence free character of all three vector fields $\mathbf{v}$, $\mathbf{B}$, and $\mathbf{A}$, is left to the reader. Helicity conservation leads to a particularly strong inverse cascade for the magnetic energy, due to the fewer factors of the wave-numbers in its three wave coupling terms. Second, the form of the Elsässer variable nonlinearities results in the squared amplitude of each of them being separately conserved, e.g.,

$$\frac{1}{v_{A0}} \frac{\partial}{\partial t} \frac{u_\pm^2}{2} + \nabla \cdot \left(\frac{u_\pm^2}{2} \mathbf{u}_\mp + I \mathbf{u}_\pm \right) = 0 \tag{8.53}$$

due to the linearity of the ∇ operator and the incompressibility of $\mathbf{u}_\pm$. With both of these Elsässer amplitudes conserved, not only the energy but also the *cross helicity* given by $\mathbf{u} \cdot \mathbf{b}$ is conserved. There is therefore some memory of the initial conditions: an initial state with low cross helicity produces a final state with low cross helicity, for example. Third, the spectral transfer of both the energy and the Elsässer amplitudes to high-k dissipation results in an aligning effect: the nonlinear transfer rate of $\mathbf{u}_+$ is given by the vigor of $\mathbf{u}_-$, and vice versa. So if there is an initial inequality in the Elsässer amplitudes, the one which is smaller is transferred to dissipation faster, leaving all the energy in the one which was larger. As a result, $\mathbf{b}$ becomes either aligned or anti-aligned to $\mathbf{u}$. This tendency has been actually observed in the solar wind. Finally, we observe that if one of the Elsässer amplitudes is zero the *mode structure* is that of a shear Alfvén wave, with $\omega = k v_{A0}$, and the nonlinearity vanishes. Therefore, pure shear Alfvén waves have no nonlinear interaction. This can be described as a cancellation between the *Reynolds stress*, from $\mathbf{u} \cdot \nabla \mathbf{u}$, and the *Maxwell stress*, from $\mathbf{b} \cdot \nabla \mathbf{b}$. We will not go further into the details of these points, except to note that their basis is a possibility for a small scale vortex and a small scale magnetic disturbance to drive a rather larger scale magnetic disturbance, or for two similarly small scale magnetic disturbances to drive a rather larger scale flow, and for the flows to be particular sensitive to the direction of the local magnetic field. This affords the possibility of a long range spectral interaction, over about one decade corresponding to $k_1 \sim k_2 \sim 10 k_3$. Interactions like this are thought to be responsible for the generation of large scale magnetic field energy (the dynamo process) or large scale flows (magnetic reconnection) starting with small scale turbulence.

8.7 Part B – Gradient Driven Turbulence in Magnetized Plasmas

Gradient driven turbulence in a magnetically confined plasma is basically $\mathbf{E} \times \mathbf{B}$ fluid turbulence driven by the background pressure gradient. A low beta MHD equilibrium is assumed. The motion is two dimensional, and is kept incompressible by magnetic pressure, which holds since there is not enough energy in the background pressure to drive motions sufficiently vigorous to compress the magnetic field. So far, we have encountered this combination through 2D MHD turbulence. Here, we will go one step further and assume also that there is insufficient energy to bend the magnetic field lines. The motion is fluid-like and two dimensional, with the fluid resistively slipping past the field lines, with dissipative processes acting along the magnetic field to suppress turbulent activity in that direction. This parallel dynamics acts instead to couple the various quantities together as part of the linear forcing effects to which the turbulence is subject. The dynamics is treated as *electrostatic,* which means that although there are finite currents and a finite $\mathbf{E} \times \mathbf{B}$ flow, disturbances in the magnetic field are small enough to be neglected.

We do have to treat the third dimension. The fluid part of the dynamics occurs in what we can call the *drift plane* which is locally perpendicular to the magnetic field. The drift velocities and fluxes occur in this drift plane. The dynamics in the third dimension, along the magnetic field, is shaped principally by the electrons and the parallel current. Although the turbulence is largely two dimensional, its interaction with the parallel dynamics make this a three dimensional phenomenon. The stream function for the $\mathbf{E} \times \mathbf{B}$ velocity is the electrostatic potential. A parallel gradient in that same potential may occur as a result of the turbulence itself or may be set by the magnetic geometry. In ideal MHD the parallel component of the electric field vanishes, to there must be a finite resistivity in order to maintain nonzero disturbances in the potential. The character of this turbulence is therefore always dissipative. In MHD, the balance in the electron parallel dynamics is between the electric field and resistive dissipation acting upon a finite parallel current. Seen another way, finite parallel currents and $\mathbf{E} \times \mathbf{B}$ turbulence, hence finite parallel electric fields, imply each other through the finite resistivity.

But in the usual situation for gradient driven turbulence the electron pressure gradient is not small compared to the electric field, and the MHD approximations are broken. In this situation the principal balance in the parallel dynamics is between these two forces, with parallel currents (through resistivity or electron inertia) equalizing any differences. There is concurrently an energetic coupling between the electron pressure and the $\mathbf{E} \times \mathbf{B}$ flow eddies, mediated by the parallel current. This parallel dynamics is called the *adiabatic response* and in general it has the property of *dissipative coupling* since

1. the pressure and electrostatic potential are thereby coupled, and
2. with the finite resistivity the coupling acts to damp each quantity toward the other.

The corresponding linear eigenmodes are called *drift waves*, and the fully developed nonlinear state is called *drift wave turbulence*. Drift wave turbulence differs from its fluid counterpart principally by this tendency towards coupling between the flow eddies themselves and the quantities they transport. Gradient driven turbulence in magnetized plasmas may also involve other effects, but this drift wave dynamics is never absent.

The concept of the high Reynolds number regime is still relevant to the case of dissipative coupling. Even though the coupling process is active at all scales, there is enough energy and enstrophy being transferred to smaller scales that diffusive processes still have a role in maintaining the statistical equilibrium. Indeed, the mutual coupling allows the direct cascade of what would otherwise be a passive scalar to indirectly affect the entire system. In this type of turbulence the small scale dissipation is collisional diffusion, with a coefficient of $\rho_e^2 \nu_e$, combining the electron gyroradius and collision frequency. With the typical scales of motion we will find below, the Reynolds number has values of order 10^4. The high Reynolds number regime is reached when the energy containing scales are no longer sensitive to the value of the collisional diffusion coefficient. Because the nonlinear transfer processes are moderately inhibited by the dissipative coupling, the high Reynolds number regime is reached for rather small Reynolds number values, about 10^3, in comparison to the pure hydrodynamic 2D fluid case.

This second Part of the lecture treats gradient driven turbulence in the simplest model in which it can be studied. Many complicating effects are left out so that the basic character is in the forefront. The adiabatic response could be electromagnetic, the temperatures could be separately involved, finite ion gyroradius effects could enter, but the basic drift wave dynamics always underlies. The coupling maintains the *mode structure* of the turbulence in the presence of this varying dynamics in each nonlinearity. The basic properties of turbulence and passive scalar cascade dynamics remain active, so that for the overall system both energy cascade tendencies are simultaneously present: direct through the electron pressure, and inverse through the $\mathbf{E} \times \mathbf{B}$ turbulence. The statistical equilibrium in the presence of sources and sinks is called *saturated turbulence* and is regarded as independent of its initial conditions. In this case, saturation occurs when both linear and nonlinear transfer mechanisms are all in statistical balance.

8.8 Passive Scalar Dynamics

In incompressible neutral fluid turbulence the density follows a pure advection equation,

$$\frac{\partial \widetilde{\rho}}{\partial t} + \mathbf{v} \cdot \nabla \widetilde{\rho} = 0 \qquad (8.54)$$

which we derive from (8.13) by taking $\nabla \cdot \mathbf{v} = 0$. The pressure is not free, but is determined by the incompressibility condition in (8.20). In low beta MHD turbulence the incompressibility is maintained by the magnetic field, as in (8.44), so the pressure also follows a pure advection equation. With the rest of the dynamics satisfying the MHD turbulence equations (8.51), we have a closed system in which although the density and pressure disturbances are advected by $\mathbf{v}$ there is no back reaction of the disturbances upon the turbulence. This property is called *passive advection*, and the dependent variable involved is called a *passive scalar*.

A passive scalar also obeys the three wave coupling condition, even though two of the wave-number components apply to the passive scalar and one to the flow. In two dimensions, with the flow itself obeying a dual cascade, the passive scalar conserves only its amplitude. The *energy-like quadratic quantity* conserved by (8.54) is simply $T = \widetilde{\rho}^2/2$, with the factor of 2 chosen to follow the definitions of energy and enstrophy for the flow. Of course, any power of $\widetilde{\rho}$ is also conserved, but there are no extra factors of k for any such quantity. Specifically, the counterpart to flow energy, $|\nabla \widetilde{\rho}|^2$, is *not* conserved by (8.54). In this sense the quantity which is conserved is more like enstrophy than energy. We can call it *free energy* or *entropy*. The three wave equations for a specific triplet $\{\mathbf{k}, \mathbf{k}', \mathbf{k}''\}$ for (8.54) are given by

$$\frac{\partial \widetilde{\rho}_\mathbf{k}}{\partial t} = C_{\mathbf{k}\mathbf{k}'} \left(\widetilde{\rho}_{-\mathbf{k}''} \psi_{-\mathbf{k}'} - \widetilde{\rho}_{-\mathbf{k}'} \psi_{-\mathbf{k}''} \right) \qquad (8.55)$$

$$\frac{\partial \widetilde{\rho}_{\mathbf{k}'}}{\partial t} = C_{\mathbf{k}\mathbf{k}'} \left(\widetilde{\rho}_{-\mathbf{k}} \psi_{-\mathbf{k}''} - \widetilde{\rho}_{-\mathbf{k}''} \psi_{-\mathbf{k}} \right) \qquad (8.56)$$

$$\frac{\partial \widetilde{\rho}_{\mathbf{k}''}}{\partial t} = C_{\mathbf{k}\mathbf{k}'} \left(\widetilde{\rho}_{-\mathbf{k}'} \psi_{-\mathbf{k}} - \widetilde{\rho}_{-\mathbf{k}} \psi_{-\mathbf{k}'} \right) . \qquad (8.57)$$

The important thing to note is that they have the same form as the equations satisfied by the vorticity (8.33)–(8.35). The three wave transfer equations will therefore be the same as in (8.38)–(8.40), becoming

$$\frac{\partial T_\mathbf{k}}{\partial t} = 2C_{\mathbf{k}\mathbf{k}'} \operatorname{Re} [\widetilde{\rho}_\mathbf{k} \psi_{\mathbf{k}'} \widetilde{\rho}_{\mathbf{k}''} - \widetilde{\rho}_\mathbf{k} \widetilde{\rho}_{\mathbf{k}'} \psi_{\mathbf{k}''}] \qquad (8.58)$$

$$\frac{\partial T_{\mathbf{k}'}}{\partial t} = 2C_{\mathbf{k}\mathbf{k}'} \operatorname{Re} [\widetilde{\rho}_{\mathbf{k}'} \psi_{\mathbf{k}''} \widetilde{\rho}_\mathbf{k} - \widetilde{\rho}_{\mathbf{k}'} \widetilde{\rho}_{\mathbf{k}''} \psi_\mathbf{k}] \qquad (8.59)$$

$$\frac{\partial T_{\mathbf{k}''}}{\partial t} = 2C_{\mathbf{k}\mathbf{k}'} \operatorname{Re} [\widetilde{\rho}_{\mathbf{k}''} \psi_\mathbf{k} \widetilde{\rho}_{\mathbf{k}'} - \widetilde{\rho}_{\mathbf{k}''} \widetilde{\rho}_\mathbf{k} \psi_{\mathbf{k}'}] . \qquad (8.60)$$

Hence, the transfer of enstrophy from mode $\mathbf{k}'$ to mode $\mathbf{k}$ is given by

$$T_T(\mathbf{k} \leftarrow \mathbf{k'}) = 2C_{\mathbf{kk'}} \operatorname{Re}\left[-\widetilde{\rho}_{\mathbf{k}}\widetilde{\rho}_{\mathbf{k'}}\psi_{\mathbf{k''}}\right] . \quad (8.61)$$

With only the single conserved quantity, and with the passive scalar triple correlation $\langle \widetilde{\rho}_{\mathbf{k}}\widetilde{\rho}_{\mathbf{k'}}\psi_{\mathbf{k''}}\rangle$ independent of the one for the flow, $\langle \psi_{\mathbf{k}}\psi_{\mathbf{k'}}\psi_{\mathbf{k''}}\rangle$, we expect its cascade dynamics to favor the direction towards small scales simply through statistical mixing.

The interesting property of turbulence with passive scalar mixing is all this variation in the properties of nonlinear spectral transfer. We expect free energy and enstrophy to go towards small scale but flow energy towards large scale. If we start the flow stream function ψ and the passive scalar $\widetilde{\rho}$ with identical scale, we should find them rather different at later times. If dissipative coupling is present, however, they should evolve together. We will show this in Sect. 8.10 covering computations.

8.9 Dissipative Coupling and the Adiabatic Response

In gradient driven turbulence in a magnetized plasma, we have the complication that the MHD approximations, which allow the neglect of pressure gradient effects in the MHD kinematics, are usually broken. We can no longer take the ideal MHD approximation that $E_\parallel$, the component of the electric field parallel to $\mathbf{B}$, to be zero. The parallel electron pressure gradient, $\nabla_\parallel p_e$ is in general not negligible in comparison. The parallel gradient is defined as

$$\nabla_\parallel = \frac{1}{B}\mathbf{B}\cdot\nabla . \quad (8.62)$$

Neglecting special structure of the magnetic equilibrium, $\mathbf{B}$ is taken as homogeneous, its field strength B is constant, and its unit vector $\mathbf{b}$ is a constant coordinate direction. With this, the parallel divergence of the parallel current has the same form as a parallel gradient, simply $\nabla_\parallel J_\parallel$. We will find that the existence of a finite $\nabla_\parallel p_e$ in the electron force balance works together with a finite $\nabla_\parallel J_\parallel$ in the pressure equation. The parallel dynamics is therefore not incompressible.

With massless electrons with finite isotropic resistivity the electron force balance is given by

$$\mathbf{E} + \frac{\mathbf{v}}{c}\times\mathbf{B} = \eta\mathbf{J} - \frac{1}{n_e e}\nabla p_e \quad (8.63)$$

where we note that the velocity here is specifically that of the electrons. Substitution of this electric field into the Maxwell equation for induction $(\partial\mathbf{B}/\partial t)$ we obtain a rather more complicated equation than (8.46). In a constant density model the ∇p_e term vanishes under the curl operation, but with $\mathbf{v}$ different for electrons and ions we have an extra $\mathbf{J}\times\mathbf{B}$ term, often called the *Hall term* in the literature, which gives rise to additional kinematic effects. Under low frequency conditions, the current and pressure are especially closely related, since the MHD equilibrium condition,

$$\frac{\mathbf{J} \times \mathbf{B}}{c} = \nabla(p_e + p_i) \qquad (8.64)$$

will hold *quasi-statically* perpendicular to the field lines: The physical system including pressure and current evolves on slow time scales with the perpendicular balance evolving through *successive equilibria* since the balance is maintained by dynamics on much faster time scales. This holds if the characteristic frequencies of the turbulence satisfy $\omega \ll k_\perp v_A$, which in this context defines the low frequency approximation.

With low frequency motion in quasi-static perpendicular force balance we enter a different regime than MHD. It is related to incompressible MHD, since the concepts of quasi-static perpendicular force balance and incompressibility maintained by the magnetic field basically mean the same thing for perpendicular dynamics. The parallel dynamics is what differs. Here, we enter a situation that in all detail is three dimensional: turbulence very similar to the 2D fluid model of (8.24) with passive scalar advection of densities and pressures as in (8.54), in planes locally perpendicular to the magnetic field, and dynamics satisfying the parallel component of (8.63),

$$E_\| + \frac{1}{n_e e} \nabla_\| p_e = \eta J_\| \qquad (8.65)$$

parallel to the magnetic field. A model like this is called *drift wave* more or less for historical reasons (waves in this physical system *drift* in the direction of $\nabla p_e \times \mathbf{B}$). If the plasma beta is low enough ($\omega \ll k_\| v_A$) we may assume that $E_\|$ is electrostatic,

$$E_\| = -\nabla_\| \phi \qquad (8.66)$$

where ϕ is the electrostatic potential. Substituting this into (8.65), we find the relationship between the parallel current and the parallel forces,

$$J_\| = \frac{1}{\eta} \left(\frac{1}{n_e e} \nabla_\| p_e - \nabla_\| \phi \right) \qquad (8.67)$$

This is called the *adiabatic response*, since dissipation of $J_\|$ towards zero tends to enforce quasi-static parallel force balance in the electrons, with both the pressure gradient and the electric field finite. Strict maintenance of this force balance is called *adiabatic electrons* for historical reasons (note it is not the same thing as an adiabatic equation of state for a fluid). For small disturbances with finite $k_\|$, the difference to MHD is that $\tilde{\phi}$ is dissipated towards $\tilde{p}_e$ by the resistive parallel dynamics, rather than towards zero.

In general the dissipative coupling is greater the larger the gradients in the parallel direction. Ultimately, this is the reason the turbulence is largely two dimensional: the dissipative dynamics involves only the parallel direction, and disturbances and flows in the 2D drift plane are free to evolve in a way similar to high Reynolds number turbulence. In magnetically confined plasmas, the shear of the magnetic field guarantees that the parallel gradients of finite sized disturbances are nonzero (this is a consequence of the combination of magnetic shear and toroidal topology of the closed magnetic flux surfaces which describe the equilibrium). As a result of these properties we also have what is called *flute mode ordering* in the dynamics even with the finite parallel gradient: in terms of wavenumbers,

$$k_\parallel \ll k_\perp \tag{8.68}$$

wherein it is important to note that we still have $k_\parallel \neq 0$. The magnetic field both orders the motion to occur in the drift plane and provides the dissipative mechanisms to couple all the parts of the dynamical system together.

We will examine the consequences of this parallel dynamics, which acts to couple ϕ directly back to p_e in the turbulence, i.e., the electron pressure is no longer a passive scalar. The simplest model in which to study this is a 2D one, since we can model the parallel dynamics through a set of coupling terms. The perpendicular dynamics is still incompressible, but the parallel dynamics for both the vorticity and pressure is influenced by the parallel compression of the current. Here, the vorticity is the $\mathbf{E} \times \mathbf{B}$ vorticity, the pressure is the electron pressure, and we simplify to an isothermal model wherein $p_e = n_e T_e$ with T_e constant. The role of $J_\parallel$, the parallel current, in the pressure dynamics arises from the electron velocity, which since we neglect sound wave effects is given by

$$v_\parallel = -\frac{J_\parallel}{n_e e} \tag{8.69}$$

so that

$$\nabla \cdot \mathbf{v} = \nabla \cdot (\mathbf{b} v_\parallel) = \nabla_\parallel v_\parallel = -\frac{1}{n_e e} \nabla_\parallel J_\parallel \tag{8.70}$$

where due to the small fluctuation rules the term is linearized. The *small fluctuation rules* arise from the ordering described in (8.14). The derivation is regarded as an expansion in the small parameter δ, which we will quantify once we find the characteristic scale of motion. At this point we order the relative amplitude and scale of motion with δ, assuming the ordering

$$\frac{e\widetilde{\phi}}{T_e} \sim \frac{\widetilde{p_e}}{p_e} \sim \frac{k_\parallel}{k_\perp} \sim \frac{\Delta}{L_\perp} \sim \delta \ll 1 \tag{8.71}$$

where Δ is the scale of motion, $L_\perp$ is the scale of the background gradient, and the tilde denotes the disturbances on the equilibrium. The basic rules are that the equations are linearized except for the terms involving $\mathbf{v}_E \cdot \nabla$, the advection by the $\mathbf{E} \times \mathbf{B}$ velocity, since the perpendicular *gradients* of the background and the disturbances are regarded as comparable. We may drop the tilde symbols momentarily, if we recall that ϕ and $J_\parallel$ belong strictly to the disturbances, and that generally second and higher order derivatives are to act solely upon the disturbances (the first derivative carries the scale of $L_\perp$, while higher ones carry the scale Δ).

The role of $J_\parallel$ in the vorticity dynamics arises through the $\mathbf{J} \times \mathbf{B}$ force on the plasma. We start with (8.42), and assume that for the ions $\mathbf{v}$ is $\mathbf{v}_E$, and further following the incompressibility rules in the perpendicular plane, that $\mathbf{v}$ is given by a stream function. This stream function is none other than the electrostatic potential, with an extra factor of c/B:

$$\mathbf{v} = \frac{c}{B^2} \mathbf{B} \times \nabla \phi \qquad \frac{\rho c}{B} \nabla \times \mathbf{v} = \Omega \mathbf{b} \qquad \Omega = \frac{\rho c^2}{B^2} \nabla_\perp^2 \phi \qquad (8.72)$$

which is the same as in 2D hydrodynamics except for the units. Taking the curl of (8.42), treating ρ as constant, and contracting with $\mathbf{b}$, we have

$$\frac{\partial \Omega}{\partial t} + \mathbf{v}_E \cdot \nabla \Omega = \nabla_\parallel J_\parallel \qquad (8.73)$$

This equation is also the one for the *total* divergence free nature of $\mathbf{J}$, so the inertial effects on the left hand side can be regarded as the perpendicular current divergence (from the *polarization current* arising from inertial delays in the $\mathbf{E} \times \mathbf{B}$ drift motion due to the finite gyrofrequency). Although $J_\parallel \mathbf{b}$ itself does not contribute to the cross product with $\mathbf{B}$, its divergence still enters the vorticity in this manner, because $\nabla_\perp \cdot \mathbf{J}_\perp + \nabla_\parallel J_\parallel = 0$, and $\nabla_\perp \cdot \mathbf{J}_\perp$ is what gives rise to the inertia term on the left hand side of (8.73).

Now, following the same rules for the electron pressure, that is, adding the parallel compressibility to a 2D advection equation, as in (8.54), for p_e, we have

$$\frac{\partial p_e}{\partial t} + \mathbf{v}_E \cdot \nabla p_e = \frac{T_e}{e} \nabla_\parallel J_\parallel . \qquad (8.74)$$

Finally, we follow the simplest model for the parallel dynamics. Each of Ω and p_e is forced upon by $\nabla_\parallel J_\parallel$. Taking the parallel divergence of $J_\parallel$ in (8.67), we find

$$\nabla_\parallel J_\parallel = \frac{1}{\eta} \left(\frac{1}{n_e e} \nabla_\parallel^2 p_e - \nabla_\parallel^2 \phi \right) \qquad (8.75)$$

as with the small fluctuation rules the only non-negligible terms are the ones with the highest-order derivatives, as in (8.14). The model which we substitute for this to get a 2D dynamical system is to replace $-\nabla_\parallel^2$ with a positive constant, $k_\parallel^2$, giving the characteristic parallel wavenumber associated with the parallel dynamics. This sets the parallel current divergence to a simple difference between p_e and ϕ, suitably normalized, with a single parameter resulting from all the multiplying factors,

$$\frac{1}{n_e e} \nabla_\parallel \tilde{J}_\parallel = D_\parallel \left(\frac{e\tilde{\phi}}{T_e} - \frac{\tilde{p}_e}{p_e} \right) \qquad (8.76)$$

where we now explicitly treat the dependent variables as small disturbances on an equilibrium. The coupling parameter is given by

$$D_\parallel = \frac{T_e k_\parallel^2}{n_e e^2 \eta} \tag{8.77}$$

which has the units of a frequency, that is, the inverse of a characteristic time constant. Coupling through a constant gives rise to real eigenvalues corresponding to exponential growth or damping. In a physically realistic model this is usually a damping mechanism although in combination with other processes it can lead to instabilities. This process is called *dissipative coupling*.

We rewrite (8.73) and (8.74) explicitly in terms of small disturbances and then substitute (8.76) for the parallel current divergence. We remember to keep the background gradient given by

$$\nabla p_e = -\frac{p_e}{L_\perp} \nabla x \tag{8.78}$$

where in whichever geometry is assumed, x becomes a coordinate whose direction is down the gradient. In deriving these equations we strictly follow the small disturbance ordering in (8.14), and also the flute mode ordering in (8.68). In addition, a characteristic perpendicular scale arises when we compare (8.76) and the form of Ω. Scaling ϕ as $e\tilde{\phi}/T_e$ and allowing the action of the coupling to force $e\tilde{\phi}/T_e$ and $\tilde{p}_e/p_e$ towards each other, we find that the natural scale of $\nabla_\perp^2$ is given by a quantity called the *drift scale*, ρ_s, defined as

$$\rho_s^2 = \frac{c^2 T_e M_i}{e^2 B^2} \tag{8.79}$$

where we have used $n_e = n_i$ and $\rho = n_i M_i$. This has the form of a gyroradius, but it arises from electron mobility and ion inertia, so it is simply the natural scale of the vorticity. Its origin is the combination of the sound speed c_s and the gyrofrequency Ω_i, given by

$$c_s^2 = \frac{T_e}{M_i} \qquad \Omega_i = \frac{eB}{M_i c} \qquad \text{hence} \quad \rho_s = \frac{c_s}{\Omega_i}. \tag{8.80}$$

The turbulence will occupy a range of scales typically of order and larger than ρ_s. We characterize these by Δ or a perpendicular wavenumber $k_\perp$, but remember that $k_\perp$ like the dynamical frequency ω takes a range of values that can span one or two orders of magnitude. The scale of the disturbances will only remain small if $\rho_s \ll L_\perp$. We can now evaluate the small ordering parameter δ, as we are now ordering the scale of motion $\Delta_\perp$ with ρ_s, so that

$$\delta \equiv \frac{\rho_s}{L_\perp} \ll 1. \tag{8.81}$$

It is a result of the dynamics occupying frequencies up to $c_s/L_\perp$, that the gyrofrequency can be ordered as large also only if $\rho_s \ll L_\perp$. In the literature, (8.71, 8.81), the specific application to these problems of (8.14), is called variously *drift ordering* or *gyro-kinetic ordering*. It is a *maximal ordering* since all the small parameters potentially entering, including that in (8.68), are taken together.

Equations (8.73) and (8.74) now become

$$\frac{\partial \widetilde{\Omega}}{\partial t} + \mathbf{v}_E \cdot \nabla \widetilde{\Omega} = D_\parallel \left(\frac{\widetilde{e\phi}}{T_e} - \frac{\widetilde{p}_e}{p_e} \right) \tag{8.82}$$

$$\frac{\partial}{\partial t} \frac{\widetilde{p}_e}{p_e} + \mathbf{v}_E \cdot \nabla \frac{\widetilde{p}_e}{p_e} = -\mathbf{v}_E \cdot \nabla \log p_e + D_\parallel \left(\frac{\widetilde{e\phi}}{T_e} - \frac{\widetilde{p}_e}{p_e} \right) \tag{8.83}$$

where the vorticity is given by

$$\widetilde{\Omega} = \rho_s^2 \nabla_\perp^2 \frac{\widetilde{e\phi}}{T_e}. \tag{8.84}$$

We identify the dissipative coupling mechanism parameterized by $D_\parallel$, explicitly in the form of an inverse time constant, and the background gradient forcing given by $\nabla \log p_e$. These are the linear forcing effects. The rest of the system is just 2D incompressible $\mathbf{E} \times \mathbf{B}$ flow turbulence plus passive advection of the pressure disturbances. Of course, with $D_\parallel$ nonzero, the pressure is no longer passive.

8.10 Computations in the Dissipative Coupling Model for Drift Wave Turbulence

We now develop some physical insight into these phenomena by means of computations. Due to the statistical nature of the turbulence, a deterministic analytical model is not useful. Moreover, due to the fact that it is the non-random part of the turbulence which gives rise to all the energetic phenomena (spectral transfer, free energy drive and damping rates, average transport) a purely statistical calculation which takes the turbulence as random to lowest order in the inverse Reynolds number, and then finds corrections due to the existence of the forcing effects, will have the problem that as soon as the physical process ceases to be self similar and there is more than one nonlinear transfer phenomenon active, that it will be too complicate to solve without additional assumptions that predetermine the results. There is no substitute for direct numerical simulation[2] in this situation: dissipative coupling and the existence of several cascade tendency possibilities, not to mention the fact that we still have to show that the spectral transfer process is in fact

---

[2] This is labelled *DNS* in a segment of the fluid turbulence literature, in contrast to *LES* which is the large eddy simulation already making the assumption that spectral transfer has the effect of small scale, or *subgrid*, dissipation

a local cascade, make it essential that we work from first principles, even in this simplest of gradient driven turbulence models.

We take the dissipative coupling model of Sect. 8.9, given by (8.82) and (8.83), normalizing the equations as suggested by the appearance of the scaled units there. It is possible to assume $\rho_s/L_\perp$ to be small and renormalize it into the dependent variables, so that these are scaled as

$$\phi \leftarrow \frac{L_\perp}{\rho_s} \frac{e\widetilde{\phi}}{T_e} \qquad p \leftarrow \frac{L_\perp}{\rho_s} \frac{\widetilde{p}_e}{p_e} . \tag{8.85}$$

The space coordinates are x and y, with x in the direction down the gradient, and the orientation given by $\nabla x \times \nabla y \cdot \mathbf{b} = 1$. Space and time scales are normalized in terms of

$$t \leftarrow \frac{c_s}{L_\perp} t \qquad \{x, y\} \leftarrow \rho_s^{-1} \{x, y\} . \tag{8.86}$$

This results in the equations being cast in terms of the single dissipative coupling parameter given by

$$D \leftarrow \frac{L_\perp}{c_s} D_\parallel . \tag{8.87}$$

In these terms, (8.82)–(8.83) become

$$\frac{\partial \Omega}{\partial t} + \mathbf{v}_E \cdot \nabla \Omega = D(\phi - p) \tag{8.88}$$

$$\frac{\partial p}{\partial t} + \mathbf{v}_E \cdot \nabla p = \omega_p \mathbf{v}_E \cdot \nabla x + D(\phi - p) \tag{8.89}$$

and the vorticity and $\mathbf{E} \times \mathbf{B}$ advective derivative are

$$\Omega = \nabla_\perp^2 \phi \qquad \mathbf{v}_E \cdot \nabla = \frac{\partial \phi}{\partial x} \frac{\partial}{\partial y} - \frac{\partial \phi}{\partial y} \frac{\partial}{\partial x} . \tag{8.90}$$

The first term on the right side of (8.88) is the gradient drive. The coefficient ω_p allows setting the gradient drive to zero for simple tests. This is the prototypical drift wave turbulence model, and it is a useful model for learning the physical character of gradient driven turbulence, although as can be seen from the list of assumptions used to derive it in Sect. 8.9, it is quite simplified. It is also called the Hasegawa-Wakatani model, after its original authors. The normalisation scheme is called *gyro-Bohm* in the literature.[3]

There is one more thing to say about turbulence computations, in basically any situation one ever faces: subgrid dissipation still has to be treated. In treating high Reynolds number turbulence, we do not try to find the resulting turbulence and transport in the presence of a given level of dissipation,

---

[3] In these units the unit of diffusion is $D_{GB} = \delta \times D_B$, where $D_B = cT_e/eB$ is the *Bohm diffusion coefficient* up to a numerical factor.

and not even in the absence of dissipation, but in the presence of an arbitrarily small level of dissipation. One or more of the nonlinear transfer processes will always send enough free energy to arbitrarily small scales that a computation will not be able to represent a statistically saturated state unless this free energy is absorbed. In the literature there are many ways to do this. One may take an actual physical diffusion (for passive scalar variables) or viscosity (for the vorticity or velocity variables) or resistivity (for the magnetic field in MHD turbulence) and try to achieve enough spatial resolution to make the results independent of the value of the dissipation coefficient. Then the value may be set artificially large, allowing computations to carry moderate resolution and hence a larger set of cases to disassemble the various physical effects. Another method is to use a different operator, for example *hyperviscosity* or ∇^4 instead of ∇^2, which for some situations may reach the high Reynolds number regime with coarser resolution. A final method is to use a sophisticated numerical scheme which incorporates dissipation of structures with large second derivative but switches it off for second derivatives below a certain limit; these schemes are called *high order upwind* in the literature. Which method one chooses may be optimized with regard to computational expense, or can even be a matter of personal taste. Either way, one should apply enough resolution that the high Reynolds number regime is reached: the amount of free energy absorbed at large wavenumber should be independent of the form or size of the subgrid dissipation operator used.

We note that the need for subgrid dissipation is additional to the presence or absence of dissipative coupling. In the model of (8.88) and (8.89) we still require subgrid diffusion of both Ω and p regardless of the value of D. Recall that the presence of a cascade tendency in a particular direction does not rule out the presence of spectral transfer in the other direction. Indeed, even in 2D Euler turbulence (8.24), the enstrophy which tends toward high k contains finite energy and the energy which tends toward low k contains finite enstrophy. The presence of dissipative coupling can indeed diminish the cascade dynamics somewhat, as we will see, but it never completely assumes the overall role of energy sink processes in balancing the gradient source drive.

In the computations presented below, an advanced upwind scheme has been used, the one presented in Sect. I of Colella (1990) [1]. Details are given therein. Both Ω and the combination $(p-\omega_p x)$ are advected by $\mathbf{v}_E$ as passive scalars, with

$$v_E^x = -\frac{\partial \phi}{\partial y} \qquad v_E^y = \frac{\partial \phi}{\partial x} \ . \tag{8.91}$$

This part of the scheme is *explicit* since this part of the right hand side is evaluated with information already to hand, namely, the values of Ω and p on all nodes at the start of a given time step. Subgrid dissipation is effected by this scheme.

The terms involving D are evaluated *implicitly* due to the potentially large damping rates for ϕ towards p, particularly at large scale, since the inverse time constant of the dissipative coupling is $Dk_\perp^{-2}$. That a part of a scheme is implicit means that information at the future time step is required to evaluate the corresponding subset of the terms. Advection of Ω and p by $\mathbf{v}_E$ leads to auxiliaries defined by

$$S_w = \Omega - \tau[\mathbf{v}_E \cdot \nabla\Omega] \qquad S_p = p - \tau[\mathbf{v}_E \cdot \nabla(p - \omega_p x)] \qquad (8.92)$$

where τ is the value of the time step. Then, the dissipative coupling terms are solved together with $\nabla_\perp^2$, according to

$$\left(\frac{\tau D}{1+\tau D} - \nabla_\perp^2\right)\phi = \frac{\tau D\, S_p}{1+\tau D} - S_w \qquad (8.93)$$

and

$$p = \frac{\tau D\,\phi + S_p}{1+\tau D} \qquad (8.94)$$

and then the vorticity is recovered via

$$\Omega = S_w - \frac{\tau D}{1+\tau D}(S_p - \phi) \qquad (8.95)$$

avoiding operation by $\nabla_\perp^2$ directly. In this case τ and D are constants, so that (8.93) is solved by means of Fourier transforms and then (8.94) and (8.95) are applied directly.

To avoid the issue of coupling to large scale *zonal flows*, which is beyond the scope of this lecture, the Fourier components with $k_y = 0$ were set to zero at the end of each time step. The grid was equidistant in both x and y, on a doubly periodic domain of dimensions $L \times L$, with $L = 20\pi$. Three values of the resolution, with grid node counts of 64^2, and 128^2, and 256^2 were used. The time step was 0.05. The turbulence was initialized as such, i.e., at finite amplitude, with $p = \phi = p_0$, with

$$p_0 = a_0 \sum_{k_x, k_y} \left[1 + (k_\perp^2/0.32)^4\right]^{-1/2} \exp(i\Theta)\exp(ik_x x)\exp(ik_y y) \qquad (8.96)$$

where k_x and k_y are the wavenumbers involved in a discrete Fourier transform over the domain, Θ is a random variable on $[-\pi, \pi]$, and the amplitude a_0 was chosen to set the root-mean-square (rms) amplitude to 3.0.

8.11 No Coupling – the Hydrodynamic Limit

The difference between a simple scalar quantity which is passively advected by the flow and the stream function for an incompressible 2D flow is well illustrated by the simplest version of the model, which is $D = 0$ and $\omega_p = 0$. This limit is called *hydrodynamic* since the turbulence is then just the 2D

Euler model of (8.24), for Ω and ϕ, and p just follows the basic passive scalar model of (8.54). Here, the variables ϕ and p are arranged equal, with the same spectrum, so we may watch them evolve apart due to their differing spectral transfer tendencies as indicated by the three wave analyses above.

The diagnostics used are time traces of selected quantities averaged over each node, logarithmic spectra of selected quantities in k_y averaged over x and sometimes also over time, and spatial morphology of the dependent variables displayed as level contours on the spatial domain. The half squared amplitude of ϕ is defined as A_p and the $\mathbf{E} \times \mathbf{B}$ energy density as U_E, according to

$$A_p = \frac{\phi^2}{2} \qquad U_E = \frac{|\nabla \phi|^2}{2} \qquad (8.97)$$

and A_n and A_w are defined similarly for p and Ω, noting that the thermal free energy density is $U_n = A_n$ and the enstrophy density is $W = A_w$. The transport is defined as

$$Q_e = p v_E^x \qquad (8.98)$$

with v_E^x defined in (8.91). These are given as both time traces and k_y-spectra. The spectra are defined as the contribution by each k_y to the total quantity, defined for the transport as

$$Q_e(k_y) = p^*(k_y) v_E^x(k_y) \qquad (8.99)$$

averaged over x, with both k_y and $-k_y$ modes contributing, either as a snapshot at a particular time or averaged over a range of time. The decay rate of the total energy is also given as a time trace, defined as

$$\Gamma_T = \frac{1}{2U} \frac{\partial U}{\partial t} \qquad (8.100)$$

where U is $U_E + U_n$ integrated over the spatial domain. Note that in the description of the computations U and W are given as totals, not as densities.

The amplitudes and decay rate are shown in Fig. 8.1. They show monotonic decay of all quantities, with $\Gamma_T < 0$ at all times. All of the decay in this case is due to the subgrid dissipation at high k. In this case, with the dissipation being switched on for relative second derivatives beyond a certain limit defined by the scheme, the dissipation coefficient is a function of the spatial resolution. A resolution test is therefore the same thing as a Reynolds number test. In the hydrodynamic model, the direct cascade in U_n is the most powerful nonlinear interaction, and we see that in the faster decay of U_n compared to U_E. We see also that the average half squared amplitude of ϕ, shown by A_p, actually rises, as the inverse cascade in U_E moves the $\mathbf{E} \times \mathbf{B}$ energy to larger scales where the factor of $k_\perp^2$ is smaller. The decay rates are shown for each of the three values of the resolution. The initial transient phase, during which the nonlinear transfer dynamics destroys the alignment between p and ϕ, shows a virulent cascade in which free energy is thrown to the highest $k_\perp$

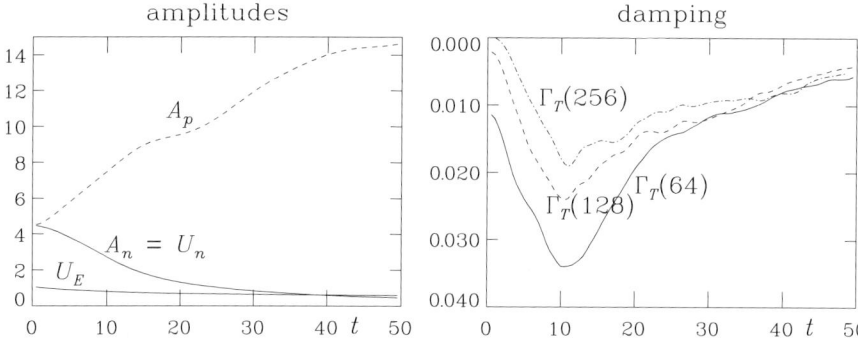

Fig. 8.1. Time evolution of the hydrodynamic model. (*left*) Initial decay of half squared amplitudes of p and ϕ, denoted A_n and A_p, respectively. The free energy density components are also given, U_E for the $\mathbf{E} \times \mathbf{B}$ eddies (ϕ) and U_n for the density (also pressure in this model), noting that U_n is also A_n. (*right*) Energetic losses (mostly in p due to the direct cascade) for three values of the resolution

and hence to the numerical dissipation. At later times, however, the decay rates are comparable, since most of the energy resides at lower $k_\perp$, and there are several factors of two over which the local cascade operates.

A high Reynolds number means that there is enough spectral separation between the largest scales and the dissipation range for the nonlinear interactions to operate unconstrained, whether the dissipation is physical or numerical. The high Reynolds number regime is characterized by an dissipation rate which is independent of the resolution, that is, the Reynolds number. Figure 8.2 shows the evolution of the ϕ and p spectra, and the vorticity spectrum, out of the initial state. Most of the redistribution occurs over the short time interval of $0 < t < 10$. The spectra of A_n and A_w become much flatter,

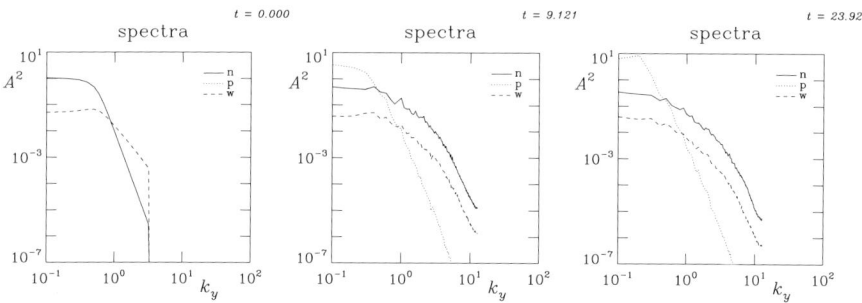

Fig. 8.2. Amplitude spectra in the hydrodynamic model, for p, ϕ, and the vorticity, labelled by "n", "p", and "w", respectively. The times of the snapshots are $t = 0$ (*left*), $t = 9.8$ (*center*), and $t = 24$ (*right*). The spectra evolve rapidly apart due to the differing cascade dynamics for p and ϕ. Note $p = \phi$ at $t = 0$, while at late times p and Ω acquire the same spectrum

but that of A_p steepens. This follows from the cascade dynamics for each of these three quantities. Once most of the $\mathbf{E} \times \mathbf{B}$ energy is at larger scales, the cascade dynamics in the middle range slows down, and then the decaying turbulence enters the high Reynolds number regime.

The evolution of the morphology of the disturbances is shown in Fig. 8.3, in which the top row of frames gives the initial state. The physical nature of the nonlinear interactions is vortex merger for the $\mathbf{E} \times \mathbf{B}$ eddies and shearing apart of the structures for the quantities directly advected by the eddies. As this includes the vorticity, the ϕ disturbances are involved in both the flows

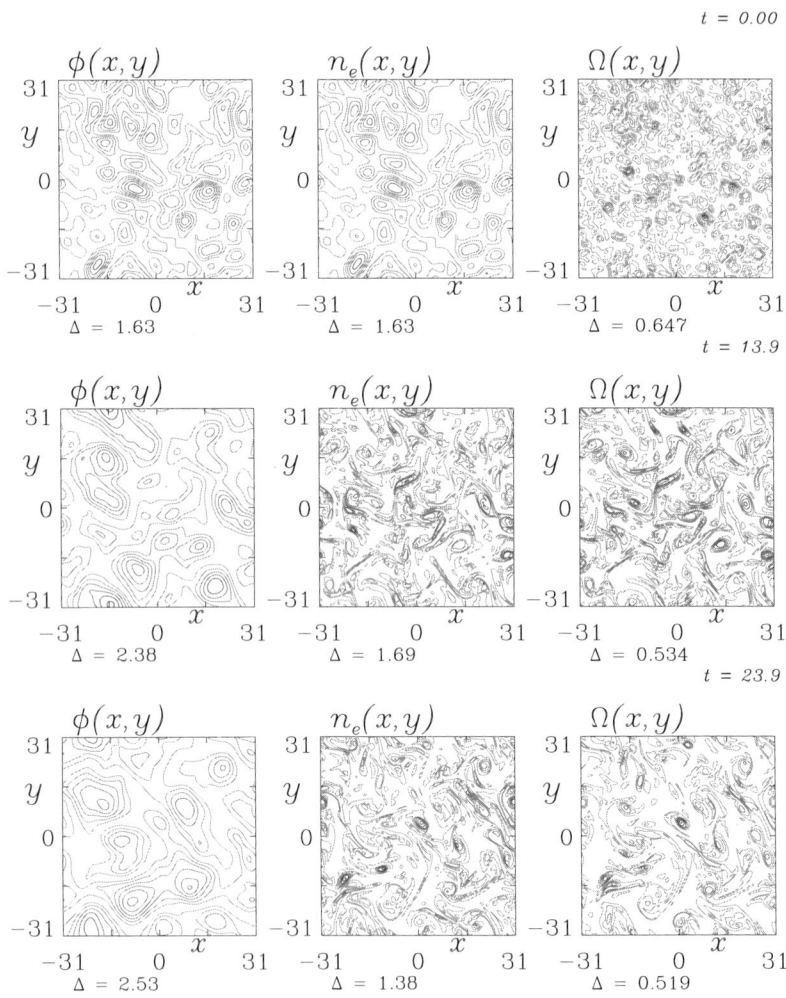

Fig. 8.3. Evolution of the disturbances for the hydrodynamic model. The nonlinear interactions represent $\mathbf{E} \times \mathbf{B}$ vortex merger and shearing apart of advected quantities, including the $\mathbf{E} \times \mathbf{B}$ flow vorticity itself

and the advected quantities. The Euler equation's dynamics is essentially this merger/shearing process, until at late times (if the Reynolds number is high enough) one is left with a shear-free flow field. The fate of a passively advected quantity, usually called a passive scalar, in this case p, is for the disturbances to be sheared apart into sheets, much like the vorticity in the Euler equation, whose narrowness is limited only by the Reynolds number. These interactions represent the physical manifestations of the cascade dynamics in the hydrodynamic model. Hence p evolves away from ϕ and towards Ω, as ϕ and Ω evolve apart, as seen.

8.12 The Effects of Dissipative Coupling

We now examine what a continuous level of driving and dissipation has on the *pure* turbulence we have just seen in the purely decaying hydrodynamic limit. With this continuous forcing, there are always energy sources and sinks available to the turbulence. The free energy will therefore redistribute itself among the available degrees of freedom until these forcing effects and the various transfer effects are all in statistical balance. This balance is called saturation, and in a dissipatively forced system such as gradient driven turbulence one is mostly interested in the maintenance of this saturated state. The approach to saturation is not as important as the properties the turbulence has once it is statistically stationary. This is why we have not been more concerned with the Reynolds number effects encountered in the previous Section. In saturation, they are much less pronounced.

With $D \neq 0$ and $\omega_p = 1$ we can examine the persistence of the basic nonlinear $\mathbf{E} \times \mathbf{B}$ flow and passive scalar dynamics in a situation with dissipative forcing. We highlight an intermediate case with $D = 0.1$ in order to show how the various pieces of the dynamics interact, and then examine the energy transfer dynamics in all limits. This case is referred to as nominal. It was taken to $t = 400$ for the three values of the resolution.

The turbulence evolves initially as it would do in the hydrodynamic limit, but then the dissipative coupling prevents p and ϕ from separating completely. One interesting result is that the cases with various resolution do not differ to the extent they do in the purely hydrodynamic model. Most of the activity is in a driven range of the spectrum, and the action of the dissipation involving D prevents the *long term* or *pure* forms of freely decaying turbulence from establishing themselves. The turbulence approaches saturation after about $t = 100$, and averaged quantities are measured over the interval $200 < t < 400$.

Time traces of the free energy components of the turbulence are shown in Fig. 8.4, and in the right frame the resulting transport is given for the three values of the resolution. The values of Q_e vary within about one standard deviation. It is also noteworthy that the numerical dissipation rate, Γ_E, was 0.028, 0.028, and 0.026, for 64^2, 128^2, and 256^2 grid nodes, respectively. In

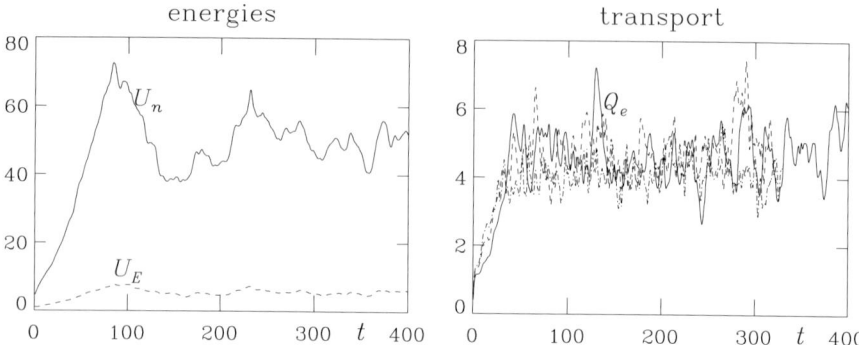

Fig. 8.4. Time evolution of the dissipative coupling model, for the nominal case of $D = 0.1$. *Left*: Half squared amplitudes of p and ϕ, denoted A_n and A_p, respectively, and the free energy density components, U_E for the $\mathbf{E} \times \mathbf{B}$ eddies (ϕ) and U_n for the density, noting that U_n is also A_n. *Right*: The transport caused by the turbulence, for three values of the resolution. For 64^2, 128^2, and 256^2 grid nodes, the values are 4.69 ± 0.80 and 4.89 ± 0.74 and 4.14 ± 0.51, respectively

this case Γ_E is not Γ_T, since ω_p and D are nonzero. Saturation results in $\Gamma_T = 0$ within the uncertainty (the mean is usually less than 0.1 standard deviation). The energy theorem for these equations is

$$\frac{\partial U}{\partial t} = \oint \frac{d^2 x}{L^2} \left(p \frac{\partial p}{\partial t} - \phi \frac{\partial \Omega}{\partial t} \right)$$

$$= \oint \frac{d^2 x}{L^2} \left(\omega_p p v_E^x - D h^2 \right) - \text{numerical dissipation} \quad (8.101)$$

where $h = p - \phi$ and Γ_E is this numerical dissipation normalized as a damping rate (found by measuring Γ_T and the integral over the terms in parentheses divided by $2U$). This shows that the high Reynolds regime has been reached, and so the resolution consideration is not grievous. A grid of 64^2 nodes is sufficient to study phenomenology as we are doing here. The main result from the time traces is that most of the free energy resides in p, and so the main dissipation process is actually the direct cascade of the associated free energy component, U_n, to arbitrarily short wavelengths.

The amplitude spectra and the disturbances themselves are shown in Fig. 8.5. Here we can easily see the varying effectiveness of the dissipative coupling at larger and smaller scales. The spectra look more like the *adiabatic* state of $\phi = p$ at low k_y but then separate at smaller scales for which the nonlinear interactions are more able to compete with the coupling. The high degree of correlation between p and ϕ is clear from the spatial distributions, but a the level of details there is more to see in p. The tendency to form sheets is still evident at the smaller scales, and the dissipative coupling reduces the basic tendency for all the $\mathbf{E} \times \mathbf{B}$ energy to go to larger scales.

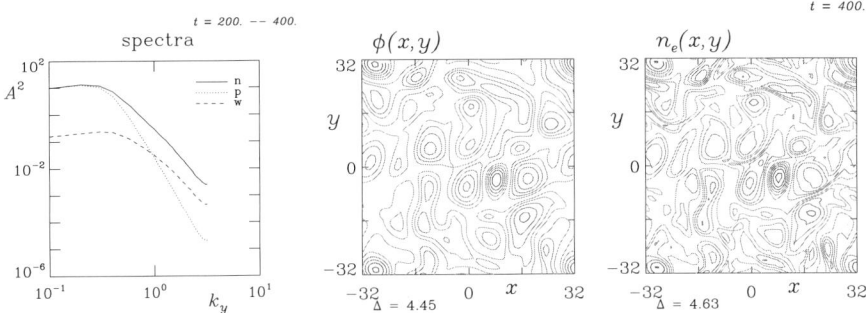

Fig. 8.5. *Left*: Averaged amplitude spectra in the dissipative coupling model, with $D = 0.1$, for p, ϕ, and the vorticity, labelled by "n", "p", and "w", respectively. *Center, right*: Contours of ϕ and $n_e = p$ at $t = 400$, showing the close coupling at larger scales but differences on smaller scales, corresponding to the spectra. The nonlinear interactions affecting p are stronger relative to the coupling at higher $k_\perp$

The energy transfer dynamics is shown in Fig. 8.6. In each case the transfer is defined as positive if the free energy leaves a spectral range k' and enters the spectral range k. The transfer for each three wave triplet is given by (8.36) for the $\mathbf{E} \times \mathbf{B}$ energy U_E, by (8.61) for the thermal free energy U_n, and by (8.41) for the enstrophy W. The spectral range k is defined as the set of all Fourier pairs (k_x, k_y) for which the resulting $k_\perp$ is within $0.5(2\pi/L)$ of k, noting that $k_\perp^2 = k_x^2 + k_y^2$. The symmetry about $k' = k$ is obvious, and so the contours are drawn only where they are positive. Activity above the line $k' = k$ indicates an inverse transfer tendency, while the activity for a direct transfer appears below the line. The quantitative level of each transfer

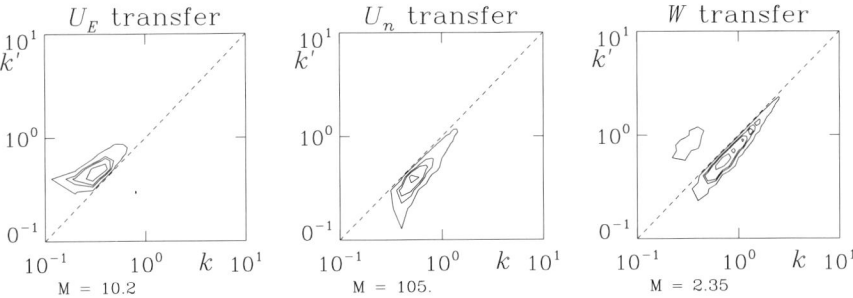

Fig. 8.6. Energy and enstrophy transfer in the dissipative coupling model, with $D = 0.1$. Left to right, the transfer from spectral range k' to k is shown for the $\mathbf{E} \times \mathbf{B}$ energy U_E, the thermal free energy U_n, and the enstrophy W. The maximum value for each frame is given by M. Contours are drawn only where the transfer is positive, noting the symmetry about $k' = k$, so that activity above the dashed line shows an inverse transfer while activity below the dashed line shows a direct transfer. The cascade is local, with most of the activity near $k' = k$

is given by its maximum value, denoted M in the figures. A local cascade is a successive transfer between spectral ranges corresponding to scales within a factor of two of each other, and indeed this is where we find the activity. As argued above, the transfer is clearly an inverse cascade in the $\mathbf{E} \times \mathbf{B}$ energy (energy associated with ϕ, or U_E), and a direct cascade in the mean squared vorticity (fluid enstrophy) and the thermal free energy (energy associated with p, or U_n).

The above introductory results show that the basic properties of the underlying hydrodynamic turbulence are present generally. Most notably, both cascade tendencies, dual for the $\mathbf{E} \times \mathbf{B}$ energy and enstrophy, and direct for the passive scalar quantity, are still active when the turbulence is severely forced by drive and dissipation. It is useful to demonstrate that the simultaneous presence of the nonlinear transfer dynamics for ϕ and p is present for all values of D except the extreme adiabatic limit of $D \to \infty$ at which $\phi = p$ and hence $\mathbf{v}_E \cdot \nabla p = 0$. In Fig. 8.7 we find the transfer dynamics for $D = 0.01, 0.03, 0.3,$ and 1.0, from top to bottom. We find the same basic form in all cases, with the hydrodynamic regime (small D) showing the strongest dynamics along with the strongest turbulence (a wider range of the spectrum is hydrodynamic). For all cases shown the direct transfer in U_n is stronger than the inverse transfer in U_E. Larger values of D are necessary to find the true adiabatic limit. In general, this type of turbulence is either more nonadiabatic than in the $D > 1.0$ cases or it is too weak to be of consequence. We therefore always find that in cases of interest that both Euler fluid and passive scalar nonlinear dynamics is simultaneously present in gradient driven turbulence in magnetized plasmas, and that the nonlinear transfer in all the conserved quantities satisfies the property of a local cascade.

8.13 Summary

Turbulence is a phenomenon involving mostly conservative interactions among a large number of statistically independent degrees of freedom. It involves nonlinear interactions among various accessible states of the disturbances, rather than quasi-linear interactions between disturbances and background (profile) quantities. For low frequency turbulence in fluids and magnetized plasmas the principal nonlinearities are quadratic, with the result that the interactions are three wave triplets each following their own transfer dynamics. The number of such triplets, and the number of triplets of which each degree of freedom is a member, is arbitrarily large, so that each triplet is statistically independent of any other.

Three wave interactions transfer energy mostly locally in the spectrum of scales of motion. For most nonlinearities the tendency is to transfer free energy from larger scales towards smaller scales, where there are more accessible states. A spectrally local transfer chain is called a cascade, termed direct for transfer to smaller scales and inverse for transfer to larger scales.

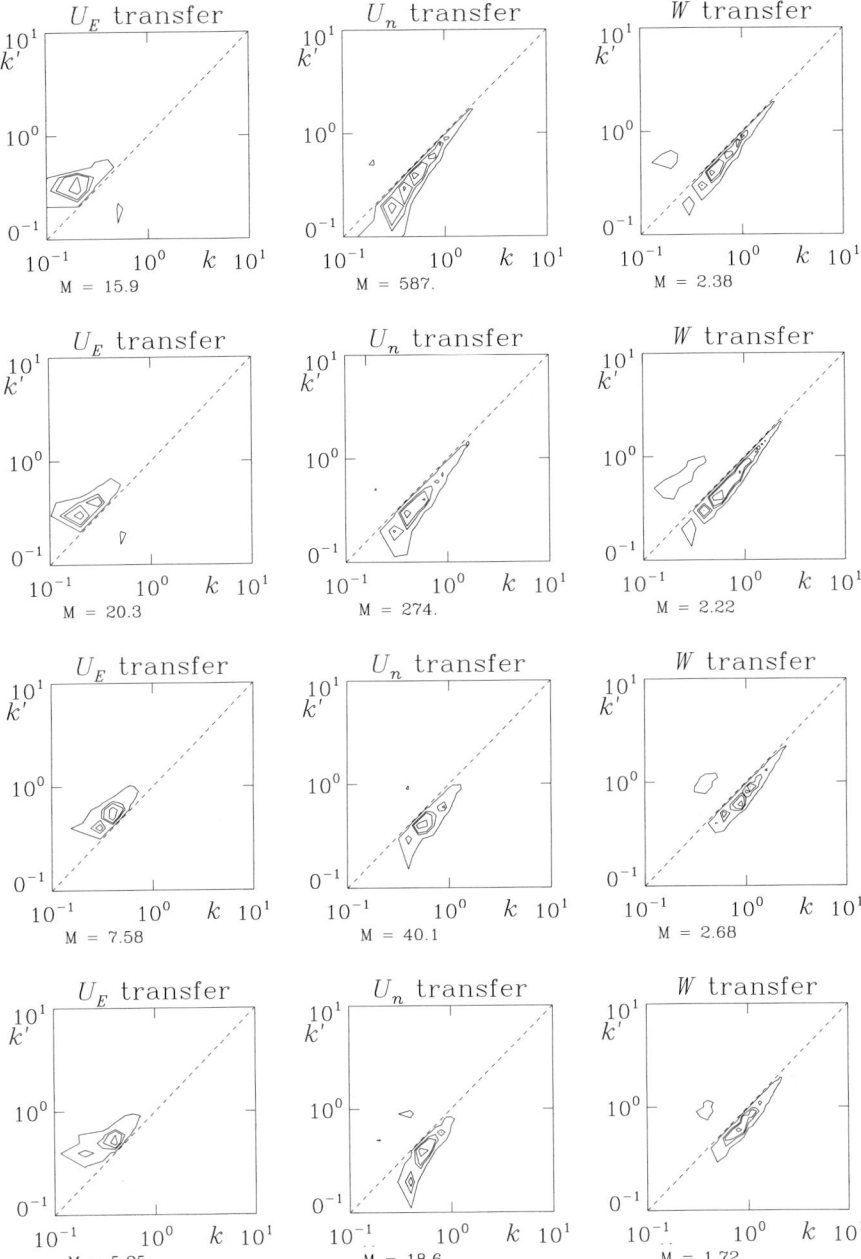

Fig. 8.7. Energy and enstrophy transfer in the dissipative coupling model, from *top* to *bottom* with $D = 0.01$, 0.03, 0.3, and 1.0, respectively. Labelling is as in Fig. 8.6. These cases serve to show that the same basic transfer dynamics is present for all $D < 1.0$, with only the levels changing

A two dimensional, incompressible fluid such as the $\mathbf{E} \times \mathbf{B}$ velocity exhibits a dual cascade due to the simultaneous conservation of energy and enstrophy (mean squared vorticity for an homogeneous fluid), with the enstrophy going to smaller scales and the energy to larger. This dual cascade is a basic consequence of the three wave constraint on the vorticity nonlinearity.

Dissipative coupling reaches limits in which it can be treated as hydrodynamic, with no correlation between the transporting eddies and the transported quantities, or adiabatic, in whose extreme limit the eddy stream function and transported quantity are constrained to be equal. Conventional fluid turbulence in the presence of a thermal gradient is hydrodynamic, and always develops motion at the very largest scales due to the dual cascade. When dissipative coupling is important the dominant scales of motion occur in a range whose lower limit is the drift scale, ρ_s, at which the dynamics loses the self similarity enjoyed by purely hydrodynamic turbulence. In deeply adiabatic dynamics, the direct cascade through the transported quantity is shut off, leaving the vorticity nonlinearity to act alone. Dissipative drift wave turbulence is always somewhere between these two limits, and in all regimes of interest the basic transfer dynamics of each nonlinearity retains its character.

8.14 Further Reading

Basic textbooks on fluid turbulence include those by Frisch [2], Lesieur [3] and Monin and Yaglom [4]. The standard cascade model is the one by Kolmogorov [5] covered in all the basic texts. The dual cascade property of two dimensional turbulence was shown and explained by Kraichnan [6].

MHD turbulence *per se* is covered in a few of the texts on MHD proper, though many of these are mostly concerned with global equilibrium and stability. Recommended in general and for its chapters on MHD turbulence and reconnection in particular is the monograph by Biskamp [7].

On gradient driven turbulence in magnetized plasmas the best sources are still papers in the scientific literature, since most of the texts still suffer greatly from an over-reliance upon linear instabilities and their specific properties. The adiabatic and hydrodynamic limits for drift wave turbulence arise from the earliest standard models [8, 9, 10, 11].

Statistical equilibrium models to investigate the cascade properties analytically were developed for the Hasegawa-Mima model [12] and for the Hasegawa-Wakatani model [13], both of which are known by those names. A general review of the self organizing property of flows with dual cascades is given in [15]. The energy transfer diagnostics presented herein are derived from their use in hydrodynamics [14] and are applied to the dissipative coupling model in [16, 17].

References

1. P. Colella: J. Comp. Phys. **87**, 171 (1990)
2. U. Frisch: *Turbulence: the Legacy of A.N. Kolomogorov* (Cambridge University Press, Cambridge 1995)
3. M. Lesieur: *Turbulence in Fluids*, 3rd ed (Kluwer Academic Publishers, London 1997)
4. A.S. Monin and A.M. Yaglom: *Statistical Fluid Mechanics: Mechanics of Turbulence*, vol 2 (MIT Press, Cambridge, USA 1975)
5. A.N. Komolgorov: J. Fluid Mech. **13**, 82 (1962)
6. R.H. Kraichnan: Phys. Fluids **10**, 1417 (1967)
7. D. Biskamp: *Nonlinear Magnetohydrodynamics* (Cambridge University Press, Cambridge 1993)
8. A. Hasegawa, H. Mima: Phys. Rev. Lett. **39**, 205 (1977)
9. A. Hasegawa, H. Mima: Phys. Fluids **21**, 87 (1978)
10. A. Hasegawa, M. Wakatani: Phys. Rev. Lett. **50**, 682 (1983)
11. M. Wakatani, A. Hasegawa: Phys. Fluids **27**, 611 (1984)
12. D. Montgomery, L. Turner: Phys. Fluids **23**, 264 (1980)
13. F.Y. Gang, B.D. Scott, P.H. Diamond: Phys. Fluids B **1**, 1331 (1989)
14. A.J. Domaradzki: Phys. Fluids **31**, 2747 (1988)
15. A. Hasegawa: Adv. Phys. **34**, 1 (1985)
16. S. Camargo, D. Biskamp, B.D. Scott: Phys. Plasmas **2**, 48 (1995)
17. S. Camargo, D. Biskamp, B.D. Scott: Phys. Plasmas **3**, 3912 (1996)

9 Transport in Toroidal Plasmas

U. Stroth

Abstract. To understand the mechanisms determining confinement of energy and particles in toroidal plasmas, is one of the most prominent and ambiguous objectives of fusion research. There is a wide spectrum of different physical processes involved in plasma transport, ranging from the complexity of particle trajectories in three-dimensional magnetic configurations over kinetic effects related to the particle distribution functions to the wide field of plasma turbulence. This chapter aims to give an overview and an intuitive understanding of the various approaches.

The starting point of this chapter is a plasma confined in a toroidal magnetic field as produced by tokamak or stellarator experiments (see also Chap. 7 in Part II). The plasma pressure is assumed to be low enough in order to guarantee MHD stability (see Chap. 4 in Part I). Hence the zeroth-order plasma flow is directed tangentially to the magnetic flux surfaces and the radial component of it vanishes. Such a plasma is nevertheless not perfectly confined. In order to maintain constant plasma conditions, continuous heating and particle refuelling of the plasma is required. If refuelling and heating cease, the plasma temperature and density decay exponentially. The characteristic time constants of the decay are the particle and energy confinement times.

Particle and energy losses are caused by transport due to two different types of processes, which both go beyond the physics of linear ideal MHD. Resistivity or collisions between particles in the presence of gradients in the thermodynamic variables lead to distortions in the distribution functions. This causes *collisional transport* which is the basic level of transport always present in a plasma. In many cases, collisional transport is only of minor importance for particle and energy confinement. Since it needs not to be ambipolar, however, it determines the plasma potential and the radial electric field. The second and for confinement times more important component is the *turbulent* or so called *anomalous transport*. An introduction to turbulence is given in Chap. 8 in Part II. It is driven by fluctuations in the plasma parameters which are caused due to nonlinearities in the fluid equations.

This chapter covers the basic processes leading to collisional and turbulent transport in toroidal devices. As an introduction, in Sect. 9.1 experimental techniques for investigating confinement times and diffusion coefficients together with key results are presented. Transport can be understood as a random walk problem, which will be revisited. In Sect. 9.2, particle orbits in tokamak and stellarator fields are discussed. The results serve as a basis for the treatment of collisional transport and radial electric fields in Sect. 9.3. Finally, Sect. 9.4 is devoted to turbulent transport in fluids and plasmas and to the physics of transport barriers.

The objective of this chapter is to discuss the physical processes of transport. Therefore understanding of the underlying mechanisms is emphasised and an intuitive deduction of the parameter dependencies of the transport coefficients is favoured over an exact quantitative treatment. In order to view the processes from different viewpoints, the collisional diffusion coefficients are derived in the particle and the fluid picture. Furthermore, turbulent transport processes are set in relation with results from fluid turbulence.

9.1 Experimental Confinement Times and Diffusion Coefficients

The experimental investigation of transport is based on the conservation equations for particles and energy. One distinguishes global (i.e., integrated over the plasma volume) and local (i.e., radially resolved) analyses with confinement times or transport coefficients, respectively, as results. The relevant definitions and some important results are summarised in this section.

9.1.1 Global Confinement Times

The simplest and most common estimates of transport are given in terms of confinement times. In particular, the global *energy confinement time* τ_E is a key parameter in nuclear fusion. It follows from the energy conservation equation in the form

$$\dot{W} = P_{net} - \frac{W}{\tau_E} . \tag{9.1}$$

The temporal change in plasma kinetic energy W is given by the available net heating power P_{net} and a term, which accounts for transport losses. The plasma energy content includes the contributions from electrons and ions. The radial density and temperature profiles $n(r)$ and $T(r)$ are integrated over the plasma volume V:

$$W = \frac{3}{2} \int (n_e T_e + n_i T_i) \, \mathrm{d}^3 r \approx 3 V \bar{n} \bar{T} . \tag{9.2}$$

For simple estimates, definitions of average temperatures $\bar{T}$ and densities $\bar{n}$ are useful. Together with the assumption of a pure hydrogen plasma ($n_e = n_i$) and $T = T_e = T_i$ this yields the handy expression on the right hand side of the equation.[1] In a strict definition, the net heating power P_{net} comprises

---

[1] Temperatures are given in electronvolts.

all energy sources and sinks. The most common sources are ohmic, neutral beam injection (NBI), ion- (ICRH) and electron-cyclotron-resonance heating (ECRH). Losses are mainly due to atomic line and bremsstrahlung radiation and to a smaller extent due to ionisation of and charge exchange with the neutral particle background. The *particle confinement* time τ_p is defined in an analog way. The sources are due to injected hydrogen ice pellets, gas puffing and recycling of plasma ions at the wall, while losses due to electron–ion recombination are negligible.

As illustrated in Fig. 9.1, there are two methods to measure the energy confinement time. The best and commonly used one is to take data in the stationary phase of the plasma discharge, while results from the transient phase, where the plasma decays after switching off the heating power, are dubious, since the plasma conditions and therefore the confinement time change during the decay process:

$$\tau_E = W/P_{net} , \qquad \tau_E = \left| W/\dot{W} \right| . \tag{9.3}$$

For practical use, the loss terms in P_{net} are neglected and the calculated absorbed heating power P is used instead. In many cases this is a reasonable assumption, since the losses are often less than 20% of the input power.

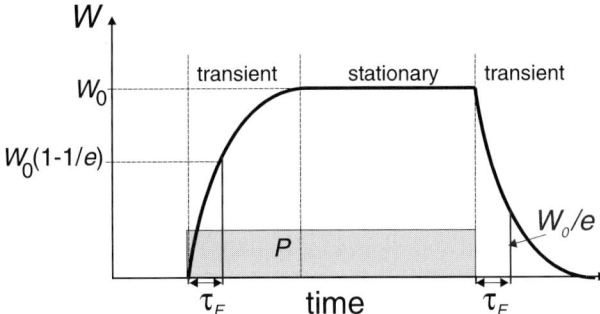

Fig. 9.1. Qualitative evolution of the plasma energy content when a heating pulse is applied. Rise and decay times are characterised by the energy confinement time τ_E

Transport theory aims at the prediction of the local diffusion coefficients. For an order of magnitude comparison of calculated particle D or heat diffusivities χ with measured global confinement times, the following equations are useful: Fick's law relates the radial particle flux density Γ to the radial density gradient and the heat flux density Q to the radial temperature gradient,

$$\Gamma = -D\nabla n ,$$
$$Q = -n\chi \nabla T . \tag{9.4}$$

In a stationary plasma, the heating power input is balanced by losses due to the heat flux across the last closed flux surface (separatrix) at the minor plasma radius a:

$$P = 2Sn\chi\nabla T|_a \approx 2S\bar{n}\bar{\chi}\frac{\bar{T}}{a} \,. \tag{9.5}$$

The factor of 2 accounts for ion and electrons, for which the same profile shape is assumed, and S is the flux surface area, here taken at the separatrix. In the last step, the edge gradient is approximated by the average temperature divided by the minor plasma radius and the bars indicate characteristic values for the various quantities. Using (9.2) and $V/S = a/2$ it is found:

$$\tau_E = \frac{W}{P} \approx \frac{3a^2}{4\bar{\chi}} \,. \tag{9.6}$$

At present, there is no theory which allows the energy confinement time to be deduced from first principles. Hence, empirical scaling expressions for τ_E are used as a normalisation of the confinement quality. This allows one to compare discharges from different experiments and to predict the performance of future devices. The expressions are derived by regression analyses of multi-machine confinement databases in which global plasma parameters are stored such as major R and minor plasma radius a (small half axis of the plasma cross-section), the ellipticity κ of the plasma cross-section, the total plasma current I_p, line-averaged density n_l, and heating power P. An early scaling expression for the energy confinement time in tokamaks has been derived by Goldston in 1984 [1]:[2]

$$\tau_E = 0.0368 a^{-0.37} R^{1.75} P^{-0.50} \bar{n}_l^{0.00} I_p^{1.00} \kappa^{0.50} \,. \tag{9.7}$$

More recent results for the low confinement regime (L-mode) can be found in [2], for the high confinement regime (H-mode) in [3] while the Lackner–Gottardi scaling [4] is a semi-empirical result obtained by constraining the expression to be dimensionally correct.

For *currentless* stellarators, tokamak scaling expressions predict the absence of confinement. Obviously, this is not the case. Therefore the plasma current must be replaced by a more relevant quantity, which is the rotational transform ι. In tokamaks, the rotational transform at the last closed flux surface is related to the plasma current by

$$\iota = \frac{\mu_0 R I_p}{2\pi a^2 B_t} \,, \tag{9.8}$$

with B_t the toroidal magnetic field strength. For scaling studies, ι is taken at $2/3$ of the plasma radius. Furthermore, since the effective plasma radius a_{eff} is used instead of a, no further shaping parameters are needed. Although

---

[2] SI units are used in scaling expressions, but density is in 10^{19} m^{-3}, power in MW and current in MA.

scaling expressions for stellarators and tokamaks work with different parameters, data from both types of devices can be reasonably well reproduced by the ISS95 stellarator scaling expression [5]

$$\tau_E^{ISS95} = 0.079 \times a_{eff}^{2.21} R^{0.65} P^{-0.59} n_l^{0.51} B_t^{0.83} t_{2/3}^{0.4} \ . \tag{9.9}$$

This can be seen in Fig. 9.2. The ISS95 scaling, derived from a database with small-scale stellarator experiments, describes the largest tokamaks rather well. Also confinement of the new large LHD heliotron was rather well predicted. Therefore, the parameters used in stellarator expressions are physically more relevant than those used in tokamak scalings. Common aspects of stellarator and tokamak confinement are the approximate increase of confinement with plasma volume ($\sim a^2 R$), the increase with B_t and the very robust degradation with heating power. Differences are observed in the density scaling (absent in tokamaks) and the scaling with ι, which is not well established in stellarators. To compare stellarator and tokamak expressions, (9.8) has to be used.

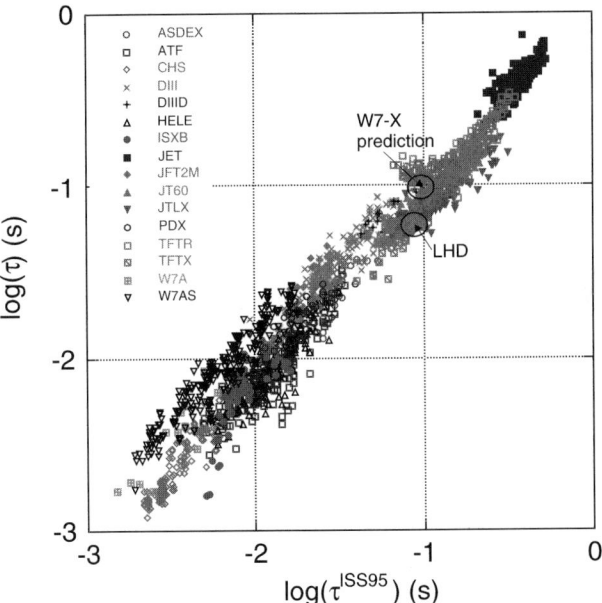

Fig. 9.2. Global energy confinement time versus the prediction from the ISS95 expression for various stellarators and tokamaks (adapted from [5], see also for designations of device) including recent data from the heliotron LHD [6] and the predicted value for the W7-X stellarator

9.1.2 Diffusion Coefficients

The physical mechanisms that determine the confinement times are related to the radial fluxes and according to (9.4) to the particle D and the electron and ion heat diffusivities χ_e and χ_i. Experimentally the diffusivities are deduced from the local particle and energy balance equations. The particle balance reads

$$\frac{\partial}{\partial t} n + \nabla \cdot \mathbf{\Gamma} = G - L . \tag{9.10}$$

A local change in density depends on the divergence of the particle flux Γ and the local sources (G for gain) and sinks (L for loss). Since only the average flux through a magnetic surface is of interest, the relevant equation is obtained by integration over the volume inside the flux surface. The average radial particle flux density $\Gamma(r)$ through the surface S is therefore given by

$$S(r)\Gamma(r) = \int_0^r [G(r') - L(r')] \, \mathrm{d}^3 r' - \frac{\partial}{\partial t} \int_0^r n(r') \, \mathrm{d}^3 r' . \tag{9.11}$$

The experimental particle flux can be calculated, if radially resolved data of all parameters are available. Most important are measurements of the (time dependent) radial density profile. Sources and sinks are the same as for the particle confinement time but they have to be calculated spatially resolved with codes that need temperature and density profiles as input. A typical experimental value for the particle diffusivity in the core of fusion experiments is $D = 0.1 \, \mathrm{m^2 s^{-1}}$ [7].

There are separate energy conservation equations for electrons and ions which both have the form

$$\frac{\partial}{\partial t}\left(\frac{3}{2} n T\right) + \nabla \cdot \left(\frac{3}{2} T \mathbf{\Gamma} + \mathbf{Q}\right) = G_E - L_E - p \nabla \cdot \mathbf{u} . \tag{9.12}$$

The divergence is taken of the heat flux density Q and the particle convection, where each particle carries along its thermal energy. Gain and loss terms are the same as for the energy confinement time with the addition of the electron–ion energy exchange term,

$$Q_{ei} = \pm \frac{m_e^{1/2} Z_i^2 e^4 n^2 \ln \Lambda}{(2\pi)^{3/2} \epsilon_0^2 m_i} \frac{T_i - T_e}{T_e^{3/2}} , \tag{9.13}$$

which has opposite sign for the two species in the form that it is a loss term for the hotter species. Here, $\ln \Lambda$ is the Coulomb logarithm, Z_i the ion charge number and $m_{e,i}$ the electron and ion masses.

The last term in (9.12) accounts for adiabatic compression and is neglected. Integration over a volume inside a flux surface leads to a local heat flux density of the form

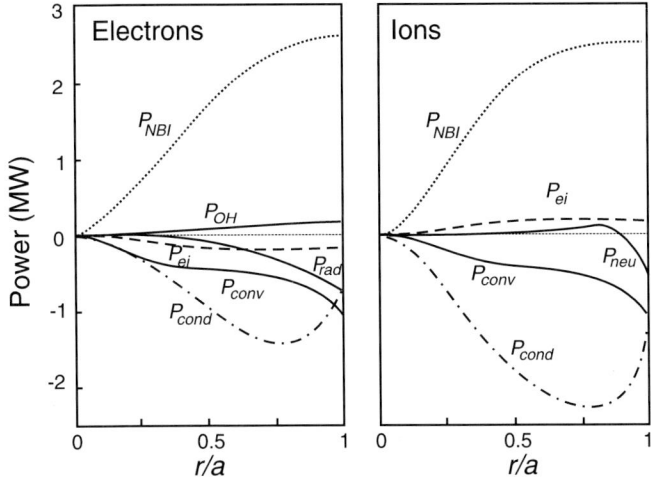

Fig. 9.3. Source and loss terms of a power balance carried out for a DIII-D discharge [8]. The curves represent radially integrated terms for electrons (*left*) and ions (*right*). Dominant is the power input due to neutral beam heating P_{NBI}. A smaller contribution P_{OH} stems from the Ohmic heating from the plasma current. Loss terms for the electrons are due to electron–ion collisions P_{ei} and radiation P_{rad}. For the ions, ionisation of neutrals constitutes an energy gain. For both species, the energy is primarily lost due to conduction P_{cond}, only at the edge, convection $P_{conv} = \frac{3}{2} T \Gamma$ contributes

$$S(r)Q(r) = \int_0^r (G_E - L_E) \mathrm{d}^3 r' - \frac{3}{2} T\Gamma(r) - \frac{\partial}{\partial t} \int_0^r nT \mathrm{d}^3 r'. \quad (9.14)$$

The heat diffusivity follows from (9.4). Again full profile information is required for measurements of the diffusivities. As an example, Fig. 9.3 depicts a power balance analysis of a neutral-beam-heated tokamak discharge [8].

Figure 9.4 shows a power balance analysis of a stationary phase of an ECRH plasma in a stellarator [10]. Here, the power is deposited very localised in the plasma centre. Electron density and temperature profiles were measured by Thomson scattering, ion temperature profile data have been obtained from spectroscopic measurements (see Chap. 13 in Part III) and charge-exchange neutral particle analysis. Due to the high density, electrons and ions are strongly coupled and the single diffusivities cannot be separated. Hence $\chi_e = \chi_i$ was used in the analysis with a value of the order of $1 \, \mathrm{m^2 s^{-1}}$. At the plasma edge for the ions and everywhere for the electrons, the diffusivity is well above the neoclassical predictions (see Sect. 9.3), which are represented by dotted (electrons) and dashed lines (ions).

The experimental investigation of particle diffusion coefficients is much more involved since it is well established that off-diagonal transport coefficients play a dominant role. In particular, it exists a convective inward

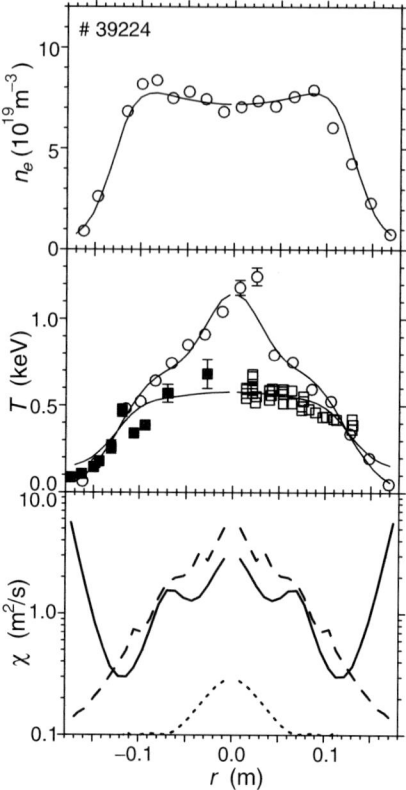

Fig. 9.4. Power balance analysis of a plasma discharge in the W7-AS stellarator with central ECRH [9]. Electron density and temperature from Thomson scattering (*open circles*) and fits (*solid lines*), ion temperature as result from the power balance (*dashed lines*) and from visible spectroscopy (*solid squares*) and neutral particle analysis (*open squares*). Lower graphs: one fluid heat diffusivity from power balance analysis (*solid lines*) compared with the equivalent neoclassical estimates for electrons (*dotted*) and ions (*dashed*)

particle flux $u < 0$ which can lead to density gradients in source-free central regions [see, e.g., [11, 10]]. The particle flux density is then given by

$$\Gamma = -D_{11}\nabla n - n D_{12}\frac{\nabla T}{T} + u\,n\,. \tag{9.15}$$

In the next section it will become clear that off-diagonal elements, here represented by the thermodiffusion coefficient D_{12} and the convective velocity u, are important in transport. In a stationary phase, only one coefficient can be deduced using the particle flux from (9.11). Transient phases, where the parameters of (9.11) change in time, are needed for investigating off-diagonal elements. Figure 9.5 depicts results from a time-dependent analysis

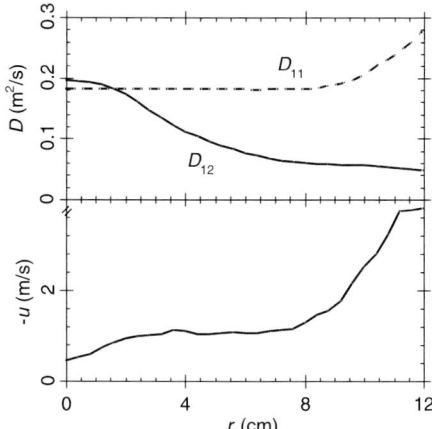

Fig. 9.5. Transport coefficients according to (9.15) deduced from experiments in the W7-AS stellarator with a transient phase in the density profile triggered by a change of the ECRH heating location from on to off-axis [10]

[10], where ECRH was used to modify the electron temperature profile and the subsequent reaction of the density profile was analysed. Emploing (9.15) results in a diffusion coefficient of the order of $0.1\,\mathrm{m^2 s^{-1}}$ and the convective velocity of the order of $-2\,\mathrm{ms^{-1}}$ is inward directed (*inward pinch*) with the typical increase to the plasma edge. Although there exists evidence of a heat pinch in the electron channel [12], the existence of it is still disputed.

9.1.3 The Collisional Transport Matrix

This section gives a brief inside into the technique to calculated transport coefficients from kinetic theory. Later in this chapter, more intuitive arguments will be employed and it will be shown that for the simple cases they lead to the same expressions as the kinetic approach. Kinetic theory is used to calculate collisional transport.

The relations between the fluid fluxes and the microscopic velocity distribution function f are given by

$$\mathbf{\Gamma} = n\mathbf{u} = \int \mathbf{v} f(\mathbf{r},\mathbf{v}) \mathrm{d}\mathbf{v} \,, \tag{9.16}$$

$$\mathbf{Q} = \int \frac{m}{2} v_s^2 \mathbf{v}_s f(\mathbf{r},\mathbf{v}) \mathrm{d}\mathbf{v} \,, \tag{9.17}$$

where $\mathbf{v}$ is the particle velocity and $\mathbf{v}_s$ the velocity in the frame of the moving fluid, $\mathbf{v}_s = \mathbf{v} - \mathbf{u}$. For symmetric distribution functions, as in case of a Maxwellian, both fluxes vanish. Hence non-thermal contributions to the distribution function are responsible for collisional transport.

Starting point for the calculation of diffusion coefficients is the linearised Boltzmann equation

$$\left(\frac{\partial}{\partial t} + \mathbf{v}\cdot\nabla + \frac{q}{m}\mathbf{E}(\mathbf{r})\cdot\nabla_v\right) f_M(\mathbf{r},\mathbf{v}) \approx -\frac{f_1(\mathbf{v})}{\tau_c}, \quad (9.18)$$

which is used to find the non-Maxwellian contribution $f_1 \ll f_M$ with $f = f_M + f_1$. The equation is valid for an unmagnetised plasma including an electric field $E(r)$. τ_c is the collision time and the Maxwellian is given by

$$f_M(\mathbf{r},v) = \frac{n(\mathbf{r})}{\pi^{3/2}[v_{th}(\mathbf{r})]^3} \exp\left(-\frac{v^2}{[v_{th}(\mathbf{r})]^2}\right), \quad (9.19)$$

where $v_{th} = (2T/m)^{1/2}$ is the thermal particle velocity. The Maxwellian is inserted in (9.18) yielding

$$f_1(\mathbf{r},\mathbf{v}) = -\tau_c \mathbf{v}\left(\frac{\nabla n}{n} - \frac{3}{2}\frac{\nabla T}{T} + \frac{v^2}{v_{th}^2}\frac{\nabla T}{T} - \frac{q}{T}\mathbf{E}\right) f_M(\mathbf{r},v). \quad (9.20)$$

The distortion f_1 can now be used to calculate the fluxes defined in (9.16) and (9.17). In case of one spatial dimension the particle flux is given by

$$\Gamma = -4\pi\tau_c \int_0^\infty v^2 f_M \left[\left(\frac{\nabla n}{n} - \frac{3}{2}\frac{\nabla T}{T} - \frac{qE}{T}\right) + \frac{v^2}{v_{th}^2}\frac{\nabla T}{T}\right] v^2 dv \quad (9.21)$$

and the energy flux by

$$\frac{2Q}{m} = -4\pi\tau_c \int_0^\infty v^4 f_M \left[\left(\frac{\nabla n}{n} - \frac{3}{2}\frac{\nabla T}{T} - \frac{qE}{T}\right) + \frac{v^2}{v_{th}^2}\frac{\nabla T}{T}\right] v^2 dv. \quad (9.22)$$

(9.21) and (9.22) are part of a transport matrix and have been quoted to illustrate that off-diagonal elements appear naturally from kinetic theory. For a magnetized plasma, the Lorenz force has to be added to (9.18). Including an equation for the toroidal plasma current density j_φ the full transport matrix for a toroidal plasma can be cast into the form [13]

$$\begin{pmatrix} \Gamma \\ \frac{2Q}{m} \\ j_\varphi \end{pmatrix} = -n \begin{pmatrix} D & D_{12} & D_{13} \\ D_{21} & \chi & D_{23} \\ D_{31} & D_{32} & \sigma_\| \end{pmatrix} \begin{pmatrix} \frac{\nabla n}{n} - \frac{3}{2}\frac{\nabla T}{T} - \frac{qE_r}{T} \\ \frac{\nabla T}{T} \\ -\frac{E_\varphi}{TB} \end{pmatrix}. \quad (9.23)$$

The structure of the driving forces from above can still be recognised. The last row represents the toroidal current. The contribution driven by the radial pressure gradients is known as bootstrap current, while the toroidal electric field E_φ is externally induced to generate the major part of the tokamak current. The transport matrix obeys Onsager symmetry, i.e., $D_{ij} = D_{ji}$ [13].

For comparison with later results, the diffusion coefficient is calculated for the simple case, where only a density gradient is present. Then the integral in (9.21) can be solved analytically and one finds:

$$\Gamma = -\frac{\tau_c v_{th}^2}{2}\nabla n \equiv -D\nabla n \ . \tag{9.24}$$

With the mean-free path $L = v_{th}\tau_c$ between collisions, the diffusion coefficient can be written as

$$D = \frac{\tau_c v_{th}^2}{2} = \frac{L^2}{2\tau_c} \ . \tag{9.25}$$

This is the random walk expression for the diffusion coefficient, as it will be introduced next.

9.1.4 Diffusion as Random-Walk

Simple and intuitive models for diffusion coefficients can be deduced from the random walk ansatz [14]. Since this approach will be extensively use in this chapter, it is worthwhile to recapitulate the basics in one dimension. Starting point is a particle which makes discrete steps either to the left or to the right (see Fig. 9.6). There is one step during the stepping time τ and both directions occur with 50% probability. After N steps, the probability to find the particle at position m is

$$f(m,N) = \frac{N!}{\left[\frac{1}{2}(N+m)\right]!\left[\frac{1}{2}(N-m)\right]!}\left(\frac{1}{2}\right)^N . \tag{9.26}$$

For the example in Fig. 9.6 with $N = 7$ the probability for a specific path is $(1/2)^7$. There are 7! different combinations for 7 independent steps leading, as pathes (**a**) and (**b**) (see Fig. 9.6) to the same final position. The path remains, however, the same, if steps into the same direction are interchanged. The denominator corrects for this double counting.

For large N and $m \ll N$ the Stirling relation

$$\log n! \approx (n + \frac{1}{2})\log n - n + \frac{1}{2}\log 2\pi \ . \tag{9.27}$$

and

$$\log\left(1 \pm \frac{m}{N}\right) \approx \pm\frac{m}{N} - \frac{m^2}{2N} \tag{9.28}$$

can be applied. Simple algebra leads to the form

$$f(m,N) = \left(\frac{2}{\pi N}\right)^{1/2}\exp\left(-\frac{m^2}{2N}\right) . \tag{9.29}$$

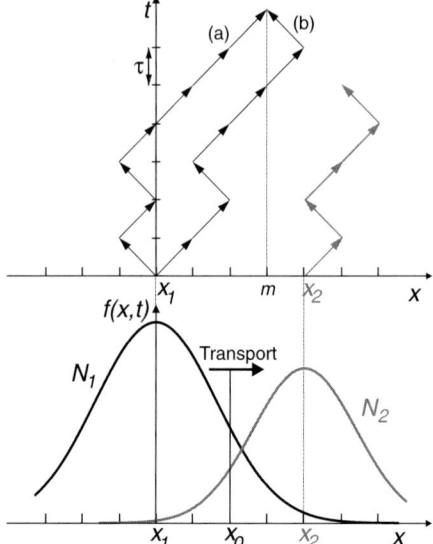

Fig. 9.6. *Upper part*: Illustration of the random-walk problem where particles step with a frequency $\nu = 1/\tau$ either to the *left* or to the *right*. This leads to a probability distribution function for finding the particle at location x as shown in the lower part. If different numbers N_1 and N_2 of particles start from two different locations x_1 and x_2 this leads to particle transport across a position x_0

Hence, after a large number of steps the probability distribution is a Gaussian and its width increases with $N^{1/2}$.

The distribution can be transformed to continuous variables. The position of a particle at m is represented by $x = ml$ and the continuous time is defined by $t = N\tau$. With these definitions, the distribution becomes a probability density to find the particle in a particular cell of the size $2l$:

$$f(x,t) = \frac{f(m,N)}{2l} = (4\pi Dt)^{-1/2} \exp\left(-\frac{x^2}{4Dt}\right). \qquad (9.30)$$

Here the important *random walk diffusion coefficient* has been introduced:

$$D = \frac{l^2}{2\tau} = \frac{1}{2}l^2\nu. \qquad (9.31)$$

The coefficient has the same structure as in (9.25) with the squared of the average step size divided by the stepping time or multiplied by the stepping frequency ν.

Figure 9.6 illustrates how random walk diffusion leads to transport. Assume different numbers N_1 and N_2 of particles starting their random walk at two different locations x_1 and x_2 with possibly different step frequencies ν_1, ν_2. N leads to different amplitudes of the probability distributions and ν to different widths. For the example shown in Fig. 9.6 the probability that particles penetrate a boundary at x_0 to the right is higher than for the opposite process. This leads to a particle flux and therefore to transport from the left to the right.

9.2 Particle Orbits in Toroidal Magnetic Fields

For the understanding of collisional transport in toroidal plasmas a detailed knowledge of the particle orbits is essential. The orbits determine the step size for the random walk process. Therefore the different classes of orbits in stellarators and tokamaks are briefly recapitulated. A review of the orbits and their role in transport can be found in [15] (see also Chap. 7 in Part II).

9.2.1 Particles in a Toroidal Magnetic Mirror

According to the variation of the toroidal and poloidal magnetic field components in an ideal tokamak device,

$$\begin{aligned} B_\varphi &= B_0 R_0/R \; , \\ B_\theta &= B_\theta(r) R_0/R \; , \end{aligned} \qquad (9.32)$$

a helical magnetic field line passes through a magnetic mirror with minimum field strength B_{min} on the outboard or low-field side (LFS) and maximum field B_{max} on the inboard or high-field side (HFS) (Chap. 7 in Part II). The coordinates and relevant parameters of the problem are summarised in Fig. 9.7.

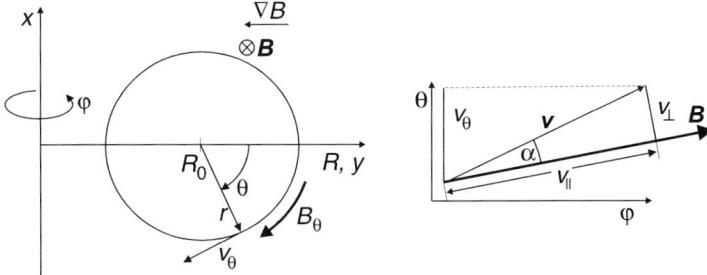

Fig. 9.7. Geometry for the treatment of particle orbits in a tokamak magnetic field. *Left*: poloidal cross-section of a flux surface with minor radius r. The magnetic field lines wind clockwise into the plane. Hence poloidal and toroidal components B_θ and B_φ, respectively, of the field lines are positive. The major radius is $R = R_0 + r\cos\theta$. *Right*: components of a particle velocity vector relative to the magnetic field line. α denotes the pitch angle of the particle

Particles of mass m and charge q traveling along a field line are influenced by two effects, the vertical components of gradient and curvature drifts,

$$\mathbf{v}_D = \mathbf{v}_D^{\nabla B} + \mathbf{v}_D^k = \frac{m}{2qRB_\varphi}\left(v_\perp^2 + 2v_\parallel^2\right)\mathbf{e}_x = \frac{mv^2}{2qRB_\varphi}\left(2 - \sin^2\alpha\right)\mathbf{e}_x \ , \quad (9.33)$$

($\mathbf{e}_x$ is the unit vector) and the mirror effect due to the inhomogeneity parallel to the field, which can lead to magnetic trapping. The condition for particle trapping follows from energy and magnetic moment conservation. According to (9.32), particles are trapped if at the LFS the pitch angle α (see Fig. 9.7) is larger than α_m given by

$$\sin^2\alpha_m = \frac{B_{min}}{B_{max}} = \frac{1-\epsilon}{1+\epsilon} \ ,$$

or

$$\cos\alpha_m \approx 2\epsilon \ , \quad (9.34)$$

where the local inverse aspect ratio is given by $\epsilon = r/R_0$. Integration of the Maxwell's distribution within the limits of particle trapping gives for the *fraction of trapped particles*:

$$f_T = (2\epsilon)^{1/2} \ . \quad (9.35)$$

At an aspect ratio of $A = 1/\epsilon = 4$ approximately 70% of the particles are trapped.

9.2.2 Passing Particles

Particles which are not reflected in the toroidal mirror are called *passing particles*. They have small pitch angles $|\alpha| < \alpha_m$ and circulate freely in toroidal and poloidal direction following the field line. Figure 9.8 shows a qualitative picture of the poloidal projections of passing orbits.

According to the rotational transform $\iota = RB_\theta/rB_\varphi$ the *poloidal transit time* of such a particle is defined as

$$\tau_{tr} = \frac{2\pi}{\omega_{tr}} = \frac{2\pi r}{v_\theta} = \frac{2\pi R_0}{\iota v_\parallel} \ , \quad (9.36)$$

$2\pi R_0/\iota$ being about the length of the field line until it closes after one poloidal turn. During a poloidal turn, the curvature drift (9.33), taken at $\alpha = 0$ ($v_\parallel$), leads to a displacement of the guiding centre from the flux surface of about

$$\delta_P = \left|\frac{v_D}{\omega_{tr}}\right| = \frac{mv}{\iota|q|B_\varphi} = \frac{mv}{|q|B_\theta}\frac{r}{R_0} = \rho_{L\theta}\epsilon \ , \quad (9.37)$$

with the *poloidal Larmor radius* defined as

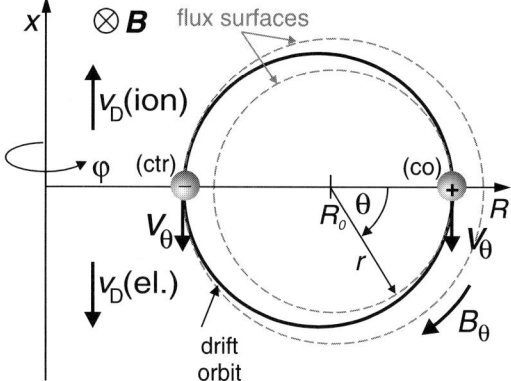

Fig. 9.8. Qualitative picture of poloidal projections of drift orbits of a positively charged particle traveling in co direction (parallel to **B**) and a negatively charged one in counter direction (anti-parallel). For reference, two magnetic flux surfaces are plotted. Note that at the same energy, the shift of electron trajectories is by the mass ratio smaller then that of ion trajectories

$$\rho_{L\theta} = \frac{mv}{|q|B_\theta} \ . \tag{9.38}$$

These orbits define drift surfaces which are deplaced from the flux surfaces.

The direction of the particle displacement from the flux surfaces can be seen in Fig. 9.8. It depends on the charge of the particle and the direction of the velocity with respect to the field direction. Orbits of co electrons and counter ions, meaning parallel or anti-parallel motion with respect to the magnetic field direction, are displaced to the low-field side, orbits of counter ions and co electrons are shifted to the high-field side.

9.2.3 Trapped Particles and Banana Orbits

Particles with larger pitch angles can be reflected on their way from low to high field side. In the case of reflection, the poloidal orbit is not closed. This leads to banana-shaped poloidal projections of the trajectories as shown in Fig. 9.9 and to the name *banana particles*.

For transport processes, the width of the banana orbit, which will be the step size, is of interest. It follows from the integration of the equation of motion with the result

$$\delta_{Ba} = \frac{mv}{qB_\theta} \epsilon^{1/2} = \rho_{L\theta}\, \epsilon^{1/2} = \frac{\rho_L}{\iota\, \epsilon^{1/2}} \ . \tag{9.39}$$

Hence the displacement from the flux surface is increased by a factor $\epsilon^{-1/2}$ with respect to a drift surface (9.37). Furthermore, the *bounce time* τ_{Ba}, i.e., the travel time between two reflections, enters the transport coefficient. It

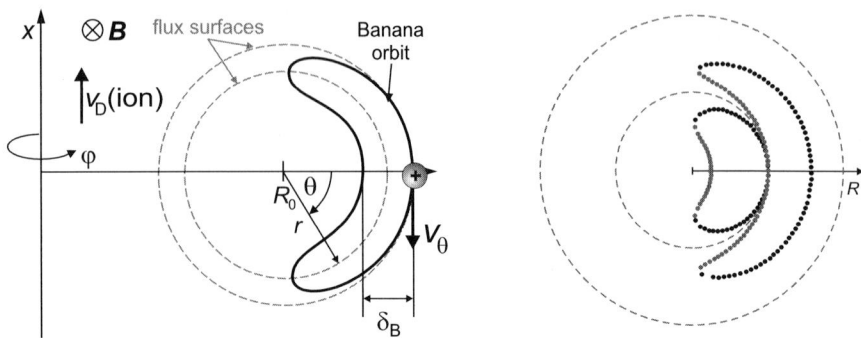

Fig. 9.9. Magnetic flux surface and a qualitative banana orbit (*left*). *Right*: calculated orbits for a co ion, black marks the motion against, grey parallel to the magnetic field

is estimated from the curvature-drift velocity, which needs the time τ_{Ba} to create a displacement of δ_{Ba}:

$$\tau_{Ba} = 2\pi \frac{\delta_{Ba}}{v_D} \approx 2\pi \frac{BR\epsilon^{1/2}}{B_\theta v} = \frac{\tau_{tr}}{\epsilon^{1/2}} \ . \qquad (9.40)$$

It is by a factor of $\epsilon^{1/2}$ longer than the transit time (9.36). The factor 2π has been added since the characteristic times are defined by $\tau = 1/\omega$.

9.2.4 Trajectories in Stellarator Fields

The complex structure of the stellarator magnetic field leads to different classes of particle trajectories. They are determined by the variation of the magnetic field strength on the magnetic flux surfaces, given, e.g., by a mod-B ($|\mathbf{B}|$) plot as in Fig. 9.10.

The example in Fig. 9.10 is for a $l = 1$, $m = 6$ torsatron (see also Sect. 7.5.9 in the Chap. 7 [Part II]). The dominant structure is impressed by the helical field coil. The local maximum of the field (light area) is aligned with this coil. But still visible is also the toroidal inhomogeneity, which is the same as in a tokamak: in average, the field is lower at a poloidal angle of $\theta = 0$ (LFS) than at $\theta = \pi$ (HFS), the local extremes, however, are localised at different positions. The inhomogeneities are characterised by the toroidal and helical ripple $\epsilon_t \approx \epsilon$ and ϵ_h:

$$B = B_0 \left[1 - \epsilon \cos\theta - \epsilon_h \cos(l\theta - m\varphi)\right] \ . \qquad (9.41)$$

As in a tokamak, small pitch angles lead to passing particles. The population of banana particles trapped in the toroidal field ripple is much smaller than in a tokamak, instead there exists a large population of helically trapped particles, which bounce between local field maxima. These particles are responsible for important differences in collisional transport in stellarators with

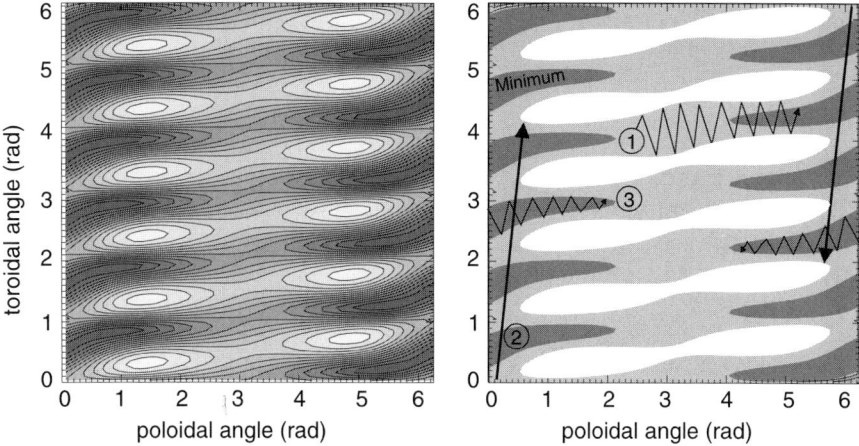

Fig. 9.10. *Left*: mod-B plot of an entire flux surface for a $l = 1$, $m = 6$ torsatron. Plotted is the field strength on the surface spanned by poloidal and toroidal angles. *Dark* relates to low and light to high magnetic fields. The helical structure stems from the helical field coil which winds six times poloidally before it closes. *Right*: Sketch of different classes of particle trajectories, which are from (1) helically trapped particles (2) toroidally trapped banana particles and (3) super-banana particles. A particle on trajectory (2), e.g., is reflected at the arrow tip on a small poloidal angle and travels backward almost an entire toroidal turn until if reaches again the same field value at a large poloidal angle. The $|B|$ structure has been reduced to three contours

respect to that in tokamaks. In addition, helically trapped particles can also be reflected at the toroidal ripple, leading to super-banana orbits, with large excursions from the flux surface. In collisional plasmas, stellarator transport behaves as in tokamaks at the same aspect ratio. In collisionless plasmas, helically trapped particles can be directly lost. Since they are constrained to a fixed poloidal position, the curvature drift leads to a large displacement of the guiding centre from the flux surface. A characteristic time for such particles to get lost due to the curvature drift is given by

$$\tau_{loss} = \frac{a}{v_D} \approx \frac{qaRB}{mv^2} \ . \tag{9.42}$$

Similar as in case of the toroidal ripple (9.35), the *fraction of helically trapped particles* can be estimated by

$$f_H = (2\epsilon_h)^{1/2} \ . \tag{9.43}$$

9.2.5 Influence of a Radial Electric Field

A radial electric field can substantially modify the particle orbits. It induces a drift tangential to the flux surfaces with a poloidal component of

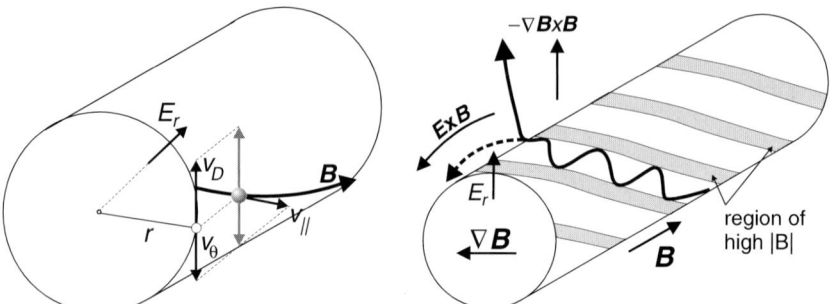

Fig. 9.11. *Left*: superposition of the poloidal components of the parallel thermal velocity and the $E_r \times B$ drift. As indicated on the *right-hand side*, the drift can detrap the helical particles, which bounce between the field maxima indicated in grey

$$u_{D\theta} = \frac{E_r}{B_\varphi} \approx \frac{E_r}{B} . \tag{9.44}$$

As depicted in Fig. 9.11, this drift adds to the poloidal component of the parallel velocity to a total poloidal particle velocity of

$$v_\theta^* = v_\| \frac{B_\theta}{B} - \frac{E_r}{B} . \tag{9.45}$$

When both components cancel, the condition for the *toroidal resonance* is fulfilled. The resonant electric field is then given by

$$E_{res} = v_\| B_\theta = \epsilon B_\varphi \iota v_\| \approx \epsilon B \iota v_\| . \tag{9.46}$$

This is a critical condition for confinement since resonant particles stay localised at a fixed poloidal position and behave the same as helically trapped particles, which can be rapidly lost due to the curvature drift.

On the other hand, the $E_r \times B$ drift can help to confine helically trapped particles. Detrapping happens if the typical loss time of the particle is long compared to the poloidal transit time due to the $E_r \times B$ drift, or

$$\tau_{loss} = \frac{a}{v_D} \gtrsim \frac{\pi a}{v_{E \times B}} = \frac{\pi a B}{E_r} . \tag{9.47}$$

The mechanism is depicted on the right hand side of Fig. 9.11. The poloidal drift detraps the particles and subsequently the poloidal orbit is closed again and the net radial displacement due to the curvature drift cancels after one poloidal turn. According to (9.47), the radial electric field sufficient to reduce losses of helically trapped particles is given by

$$|E_r| \geq \frac{\pi m v^2}{|q| R_0} . \tag{9.48}$$

In Fig. 9.12, calculated orbits of counter and co passing as well as helically trapped particles in a $l = 2$ stellarator are depicted.

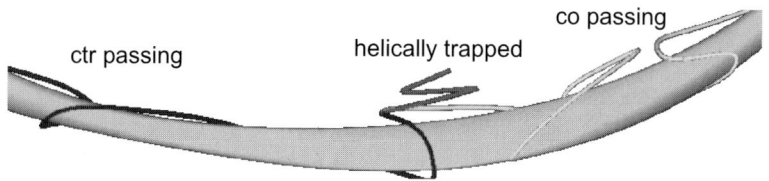

Fig. 9.12. Calculated orbits of counter and co passing as well as helically trapped particles with respect to a flux surface of an $l = 2$ stellarator

9.3 Collisional Transport

In this section the transport aspects which rely on particle–particle collisions will be discussed. In order to give insight into the basic physical processes, the semi-quantitative derivations will be carried out in the particle and the fluid picture, both yielding the same parametrical dependencies. In general one distinguishes *classical* and *neoclassical transport* if the toroidal topology of the magnetic field is neglected or taken into account, respectively. Neoclassical theory was developed in the 60ies by Galeev and Sagdeev [16]. Reviews of collisional transport theory can be found in [17, 18]. Stellarator effects of neoclassical transport in the important low-collisionality regime have been treated, e.g., in [19, 20]. Comparisons with experiments can be summarised as follows: in discharges with moderate ion heating, in both stellarators and tokamaks the ion heat diffusivity can be of the order of the neoclassical prediction [7, 9]. At the plasma edge, theory predicts too small values. Only in tokamaks, strong ion heating has been well explored showing ion diffusivities which are higher than neoclassical. Electron heat transport in tokamaks and also in most stellarator regimes is well above theory [21]. Only for collisionless electrons, the diffusivity can be comparable with the neoclassical prediction [22, 23].

9.3.1 Classical Transport in the Particle Picture

In the particle picture, the diffusion coefficient of classical transport is obtained from the random-walk ansatz (9.31) with the Larmor radius as characteristic step size and the electron–ion collision time as step time. The electron diffusion coefficient therefore is

$$D_{kl} = \frac{\rho_e^2}{2\tau_{ei}} \approx \left(\frac{m_e v}{eB}\right)^2 \frac{1}{2\tau_{ei}} . \tag{9.49}$$

This represents the mono-energetic coefficient, which increases linear with collision frequency. In order to obtain characteristic values for a thermal plasma, the velocity can be replaced by the thermal velocity $v_{th} = (2T/m)^{1/2}$. Correct coefficients are obtained by integration over velocity space weighted by a Maxwellian.

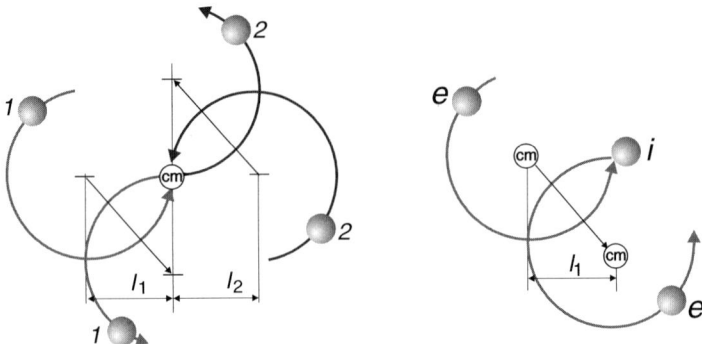

Fig. 9.13. Particle gyro motion in like (*left*) and unlike (*right*) particle collisions. Only in collisions of unlike particles, the centre of mass (cm) of the electron population is displaced

As indicated in Fig. 9.13, only collisions of unlike particles contribute to transport. In like particle collisions the centre of mass stays in place. The proper quantity to be inserted for the step time is the electron–ion collision time

$$\tau_{ei} = \frac{2\pi \epsilon_0^2 m_e^2 v_e^3}{e^4 Z_i n \ln \Lambda_e}, \tag{9.50}$$

with Z_i the ion charge number and $\ln \Lambda_e$ the electron Coulomb logarithm (see also Chap. 1 in Part I). The diffusion coefficient for the ion species has the same form. The same numerical value is obtained, if the ion Larmor radius and the ion–electron collision time is used instead of the electron parameters. Hence the classical transport in the particle picture is ambipolar.

9.3.2 Classical Transport in the Fluid Picture

The starting point in the fluid picture is Ohm's law and the objective is to calculate the radial component of the fluid velocity **u**. Vector multiplied with **B** Ohm's law reads:

$$\frac{\mathbf{j} \times \mathbf{B}}{\sigma} = \mathbf{E} \times \mathbf{B} + (\mathbf{u} \times \mathbf{B}) \times \mathbf{B}. \tag{9.51}$$

σ is the electric conductivity. Combined with the force balance $\nabla p = \mathbf{j} \times \mathbf{B}$, where p is the plasma pressure and $\mathbf{j}$ the electric current density, a simple transformation leads to

$$\frac{1}{\sigma}\nabla p = \mathbf{E} \times \mathbf{B} - B^2 \mathbf{u} + (\mathbf{B} \cdot \mathbf{u})\mathbf{B} = \mathbf{E} \times \mathbf{B} - \mathbf{u}_\perp B^2. \tag{9.52}$$

The component of the fluid velocity perpendicular to **B** is therefore:

$$\mathbf{u}_\perp = -\frac{\nabla_\perp p}{\sigma B^2} + \frac{\mathbf{E} \times \mathbf{B}}{B^2}. \tag{9.53}$$

In general, this expression contains poloidal and radial components. In cylindrical geometry, however, the electric field and the pressure gradient are purely radial. If a constant temperature is assumed, the radial particle flux is given by:

$$\Gamma_r = nu_r = -\frac{nT}{\sigma B^2}\nabla n = -D_{kl}\nabla n \ . \tag{9.54}$$

Since the electric conductivity has the form

$$\sigma = \frac{ne^2 \tau_{ei}}{m_e} \ , \tag{9.55}$$

the same coefficient as in the particle picture is recovered [see (9.49) with $v = (2T/m)^{1/2}$].

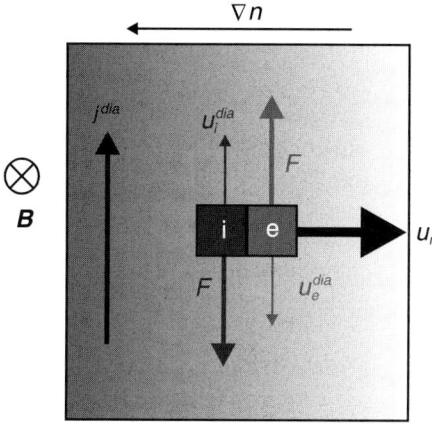

Fig. 9.14. Classical transport in the fluid picture is caused by the diamagnetic drift $\mathbf{u}^{dia}$ and the friction between electron and ion fluids. The friction force $\mathbf{F}$ causes an ambipolar radial drift of the plasma. The shading indicates a density gradient to the left

Figure 9.14 gives an intuitive explanation of the origin of classical transport in the fluid picture. A diamagnetic current is needed to stabilise the pressure gradient. The current is generated by opposite diamagnetic drifts of ion and electron fluids. Hence a opposite friction force acts on the fluids, which is of course caused by electron–ion collisions. In the same way as single particles, also a plasma fluid reacts to an arbitrary force $\mathbf{F}$ with a drift perpendicular to the force and the magnetic field,

$$\mathbf{u}_D = \frac{\mathbf{F} \times \mathbf{B}}{qnB^2} \ . \tag{9.56}$$

The direction of the fluid drift depends on the charge of the fluid and is therefore directed down the density gradient for both electrons and ions. Hence in the fluid picture, too, classical diffusion is ambipolar.

9.3.3 Pfirsch–Schlüter Transport in the Particle Picture

The first correction to transport caused by toroidal geometry takes into account the fact, that passing particles follow drift surfaces rather than magnetic flux surface. Trapped particles are neglected in this approximation. In the particle picture the random walk ansatz is used with the displacement of the drift surface (9.37) as step width and again the collision time as step time. In case of a rather collisionless plasma, $\tau_{ei} > \tau_{tr}$, where particles can freely circulate around the torus, the displacement assumes its maximum value δ_p. With (9.37) it follows for the contribution of passing particles to the diffusion coefficient:

$$D_{ps} = \frac{\delta_p^2}{2\tau_{ei}} = \left(\frac{mv}{\iota q B_\varphi}\right)^2 \frac{1}{2\tau_{ei}} . \tag{9.57}$$

Combined with the classical contribution (9.49), the *diffusion coefficient in the Pfirsch–Schlüter regime* is:

$$D_{neo} = D_{kl}\left(1 + 2/\iota^2\right) , \tag{9.58}$$

where the factor two has been added to be conform with more refined calculations. At a typical value of $\iota = 1/3$, the classical coefficient is enhanced by about a factor of 20.

In a more collisional plasma with $\tau_{ei} < \tau_{tr}$, the particles cannot complete the poloidal orbit and the displacement from the flux surface reduces to $v_D \tau_{ei}$. Therefore, the diffusion coefficient drops at high collisionalities and resumes the form

$$D_{ps} = \frac{(v_D \tau_{ei})^2}{2\tau_{ei}} = \left(\frac{mv^2}{qB_\varphi R}\right) \frac{\tau_{ei}}{2} . \tag{9.59}$$

The dependence of the diffusions coefficient on collisionality will be is summarised in Fig. 9.15.

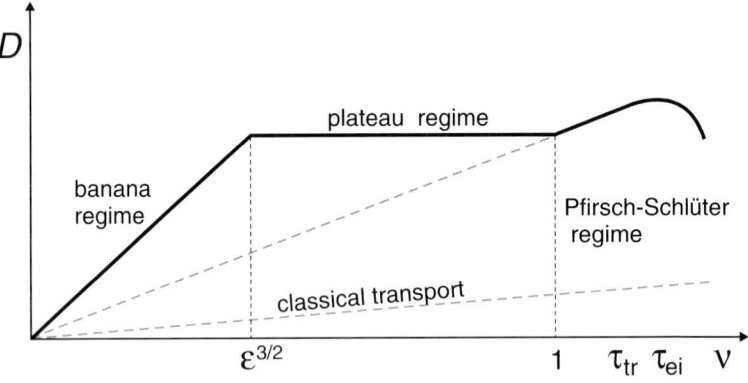

Fig. 9.15. Qualitative dependence of the neoclassical diffusivity on collisionality $\nu \sim \tau_{tr}/\tau_{ei}$ in a tokamak magnetic field

9.3.4 Pfirsch–Schlüter Transport in the Fluid Picture

In the fluid picture, the first correction to classical transport caused by toroidal geometry accounts for two effects: the averaging of the now poloidally asymmetric radial flux (9.54) and an additional contribution due to a vertical electric field component in (9.53). As depicted in Fig. 9.16, this comes from the fact that the helical field lines cannot completely cancel the charge up-down separation caused by the curvature drifts (Chap. 7 in Part II). In toroidal symmetry without an externally induced field from an Ohmic transformer, the toroidal component of the electric field vanishes.

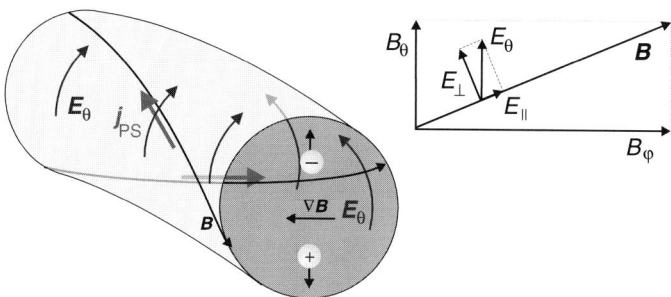

Fig. 9.16. Poloidal electric field E_θ and the Pfirsch–Schlüter current generated by charge separation due to the curvature drift in a segment of a torus. The small figure introduces the relation between poloidal, parallel and perpendicular electric field components

With the local radial flow velocity component from (9.53), the flux surface average radial particle flux is given by

$$\Gamma_r = \frac{n}{2\pi} \int_0^{2\pi} d\theta \, (1 + \epsilon \cos\theta) \left(\frac{E_\perp}{B} - \frac{1}{\sigma_\perp B^2} \frac{\partial p}{\partial r} \right) . \quad (9.60)$$

The first term in parentheses in (9.60) accounts for integration in toroidal co-ordinates.

Here one has to distinguish between the conductivity parallel and perpendicular to the magnetic field and, as indicated in Fig. 9.16, $E_\perp$ is the component tangential to the flux surface. Both contributions, the $E \times B$ drift and the one due to a resistive diamagnetic current (second term), depend through the magnetic field (9.32) on the poloidal coordinate.

The objective is to calculate the electric field from the Pfirsch–Schlüter current $j_{ps} = j_\parallel$ according to

$$j_\parallel = \sigma_\parallel E_\parallel = \sigma_\parallel E_\perp \frac{B_\theta}{B_\varphi} . \quad (9.61)$$

The Pfirsch–Schlüter current follows from the quasi-neutrality condition

$$\nabla \cdot \mathbf{j} = \nabla \cdot \left(j_\| \frac{\mathbf{B}}{B} + \mathbf{j}_\perp \right) = \nabla \cdot \left(j_\| \frac{\mathbf{B}}{B} - \frac{\nabla p \times \mathbf{B}}{B^2} \right) = 0 , \quad (9.62)$$

where the perpendicular current is substituted by the radial force balance vector multiplied with $\mathbf{B}$. Using toroidal symmetry, the first term is converted to

$$\nabla \cdot \left(j_\| \frac{\mathbf{B}}{B} \right) = \frac{1}{rR} \frac{B_{\theta 0}}{B_0} \frac{\partial}{\partial \theta} (Rj_\|)$$

and the second one to ($p' = \frac{\partial p}{\partial r}$ is now used)

$$\nabla \cdot \left(\nabla p \times \frac{\mathbf{B}}{B^2} \right) = -\frac{p'}{rR} \frac{\partial}{\partial \theta} \left(R \frac{B_\varphi}{B^2} \right) = 2 \frac{p'}{rR} \frac{R_0}{B_0} (1 + \epsilon \cos \theta) \epsilon \sin \theta .$$

Equating the two terms yields the equation

$$\frac{\partial}{\partial \theta} (Rj_\|) = 2 \frac{p' R_0}{B_{\theta 0}} (1 + \epsilon \cos \theta) \epsilon \sin \theta ,$$

which can be integrated in the large-aspect-ratio approximation ($\epsilon^2 \to 0$) resulting in an analytic expression for the *Pfirsch–Schlüter current*:

$$j_\| = -\frac{2p'}{B_{\theta 0}} \epsilon \cos \theta . \quad (9.63)$$

The expression is used to calculate $E_\perp$ from (9.61), which is then inserted into (9.60) to calculate the radial transport. In leading order of ϵ and for constant temperature, the *Pfirsch–Schlüter transport* in the fluid picture is therefore given by

$$\Gamma_r = \frac{n}{\sigma_\perp B_\varphi^2} \left(1 + \frac{2\sigma_\perp}{\sigma_\|} \frac{1}{t^2} \right) \nabla n , \quad (9.64)$$

which for $\sigma_\| = \sigma_\perp$ has the same form as (9.58), calculated in the particle picture.

9.3.5 The Toroidal Resonance

As discussed in Sect. 9.2.5, does a radial electric field modify the orbits of passing particles and thus also the step width used in the random-walk argument. The poloidal thermal velocity component of the particle has now to be corrected by the $E \times B$ drift. According to (9.45) it is:

$$v_\theta^* \approx v_\| \frac{B_\theta}{B} - \frac{E_r}{B} = \frac{v_\| r t}{R} \left(1 - \frac{E_r}{v_\| \epsilon t B} \right) . \quad (9.65)$$

Consequently, the transit time (9.36) is now given by

$$\tau_{tr}^* = \frac{2\pi}{\omega_{tr}^*} = \frac{2\pi r}{v_\theta^*} = \tau_{tr}\left(1 - \frac{E_r}{E_{res}}\right)^{-1}, \qquad (9.66)$$

with the resonant electric field defined in (9.46). For passing particles, the displacement from the magnetic surface and therefore the step width is reduced to [see (9.37)]

$$\delta_p^* = \left|\frac{v_D}{\omega_\theta^*}\right| = \delta_p\left(1 - \frac{E_r}{E_{res}}\right)^{-1}, \qquad (9.67)$$

and the random-walk diffusion coefficient including a radial electric field has the form:

$$D_{ps}^* = \frac{\delta_p^{*2}}{2\tau_{ei}} = \frac{D_{ps}}{\left(1 \pm \frac{E_r}{E_{res}}\right)^2}. \qquad (9.68)$$

At the resonant field, the poloidal component of the thermal particle velocity is canceled by the drift motion. In this case passing particles behave like helically trapped ones and are directly lost due to the curvature drift leading to a strongly increased diffusivity. As it will be discussed in Fig. 9.20, this feature shows up in full calculations of neoclassical transport including radial electric fields.

9.3.6 Neoclassical Transport in the Particle Picture

The missing element, which remains to be considered, is the influence of trapped particles. In the particle picture, the large excursions of the banana orbits discussed in Sect. 9.2.3 enhance the diffusivity. As illustrated in Fig. 9.17, collisions detrap banana particles and transform them to passing ones. This can occur randomly on the inner and outer part of the orbit, which corresponds to a random displacement of the banana particle with the step

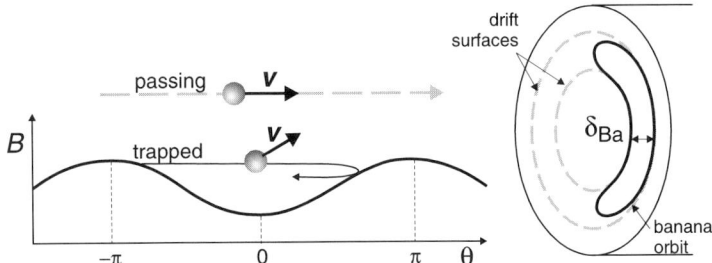

Fig. 9.17. Depending on the perpendicular velocity component, particles can be trapped in the toroidal magnetic mirror. A transition from a banana to a passing orbit can be caused by turning the velocity vector by less then 90°. Wherever the collision occurs, the particle then follows a smaller or larger drift surface as indicated on the *right*

width δ_{Ba}. To detrap a particle, the pitch angle has to be turned by less than 90°. Therefore the relevant step time is shorter than the 90° collisions time, namely

$$\tau_{eff} \approx \tau_{ei}\epsilon. \tag{9.69}$$

First collisionless particles are considered, where the effective collision time is longer than the bounce time (9.40):

$$\tau_{eff} > \tau_{Ba} = \tau_{tr}\epsilon^{-1/2}, \tag{9.70}$$

or

$$\frac{\tau_{tr}}{\tau_{ei}} < \epsilon^{3/2}. \tag{9.71}$$

While τ_{eff} sets the step time for collisionless particles, the step width is the banana width (9.39) in the random-walk expression. Since only a ratio of $\epsilon^{1/2}$ of the particles is trapped, the *diffusivity in the banana regime* is

$$D_{Ba} \approx \epsilon^{1/2}\frac{\delta_{Ba}^2}{2\tau_{eff}} \approx \frac{1}{\epsilon^{3/2}\iota^2}\frac{\rho_L^2}{2\tau_{ei}} = \frac{1}{\iota^2\epsilon^{3/2}}D_{kl} = \frac{1}{\epsilon^{3/2}}D_{ps}. \tag{9.72}$$

Hence in a typical tokamak, the diffusivity is enhanced by a factor of typically fifty above the classical value.

With increasing collisionality at $\tau_{eff} \approx \tau_{Ba}$ the particles cannot complete their banana orbit anymore, the step width becomes smaller and the diffusivity stops to increase. For approximately $\tau_{ei} = \tau_{tr}$ the banana particle population vanishes completely and the Pfirsch–Schlüter value has to be recovered. This transitional region from banana to Pfirsch–Schlüter transport with

$$\epsilon^{3/2} < \frac{\tau_{tr}}{\tau_{ei}} < 1 \tag{9.73}$$

is called *plateau regime* and is approximated by a constant value calculated with the effective collisionality equal to the bounce frequency, $\tau_{eff} = \tau_{Ba}$. Under this condition, particles make exactly one collision after completing one banana orbit. This results in the *diffusivity for the plateau regime*:

$$D_{Pl} \approx \epsilon^{1/2}\frac{\delta_{Ba}^2}{2\tau_{Ba}} = \frac{1}{\iota^2}\frac{\rho_L^2}{2\tau_{tr}} = \frac{1}{4\pi\iota R_0 v}\left(\frac{mv^2}{qB}\right)^2. \tag{9.74}$$

Figure 9.15 summarises the qualitative characteristics of the neoclassical diffusion coefficient as function of the normalised collisionality $\nu \sim \tau_{tr}/\tau_{ei}$. In the banana and Pfirsch–Schlüter regimes, the diffusivity increases linear with collisionality. At the transition from Pfirsch–Schlüter to plateau regime, banana particles contribute more and more to transport, since the time trapped particles have between two collisions and to complete their orbit increases until it is long enough to allow for a full banana orbit. This corresponds to the largest step width possible. At lower collisionalities the step width is constant but the step frequency decreases and so does the diffusivity.

9.3.7 Elements of Stellarator Transport

In stellarators, helically trapped particles introduce new effects which are related to the large radial excursions of their orbits as discussed in Sect. 9.2.4. The step width of the random-walk process $\delta r = v_D \tau_{eff}$ is given by the curvature drift multiplied with the effective collision time $\tau_{eff} = \tau_{ei}\epsilon_h$, needed to detrap the particles (ϵ_h is the helical ripple). Without a radial electric field and for collisionless particles, this yields a simple random-walk expression for the *diffusivity in the $1/\nu$ regime*

$$D_{1/\nu} = \epsilon_h^{1/2}\frac{(v_D\tau_{eff})^2}{2\tau_{eff}} \approx \frac{1}{2}\epsilon_h^{3/2}\left(\frac{mv^2}{2qR_0B}\right)^2 \tau_{ei} . \qquad (9.75)$$

Again the fraction of trapped particles $\epsilon_h^{1/2}$ has to be taken into account. The name comes from the $1/\nu = \tau_{ei}$ dependence. With the collisionality inserted and the velocity replaced by the temperature, the expression reads

$$D_{1/\nu} \approx \frac{3\pi \, \epsilon_0^2 \, (3m_e)^{1/2} \, \epsilon_h^{3/2}}{e^6 \, n_e \, \ln\Lambda \, R_0^2 \, B^2} T_e^{7/2} . \qquad (9.76)$$

The characteristic feature also visible in Fig. 9.18 is the linear increase of the diffusivity with $1/\nu$ which converts to a $T_e^{7/2}$ dependence. This behaviour is opposite to that in a tokamak and would be a severe problem for a stellarator reactor if it could not be healed by a radial electric field. At very small collisionalities the diffusivity drops again, since the particle losses create a depletion of the Maxwell distribution at large pitch angles which cannot be sufficiently fast replenished by electron–electron or ion–ion collisions.

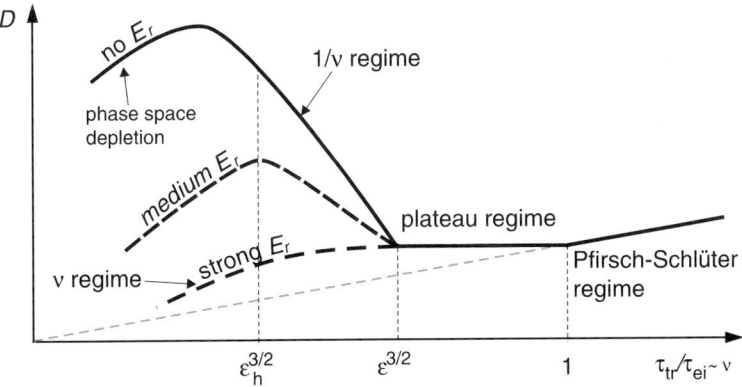

Fig. 9.18. Qualitative dependence of the neoclassical diffusivity on collisionality $\nu \sim \tau_{tr}/\tau_{ei}$ and radial electric field E_r in a stellarator

Strong radial electric fields radically change the course of the diffusivity at low collisionalities. As indicated in Fig. 9.11, the $E_r \times B$ drift drives a poloidal motion of the trapped particles and lets them behave as passing particles. Therefore instead of the transit time, the expression $\omega_{E \times B} = 2\pi/\tau_{E \times B}$ with $\tau_{E \times B} = 2\pi r B/E_r$ has now to be used in (9.37) to calculate the step width of passing particles. This leads to a *diffusion coefficient in the ν regime* of the form

$$D_\nu \approx \left(\frac{v_D}{2\omega_{E \times B}}\right)^2 \frac{1}{2\tau_{ei}} \approx \epsilon^2 \left(\frac{mv^2}{qE_r}\right)^2 \frac{1}{8\tau_{ei}}. \tag{9.77}$$

For strong electric fields, the $E \times B$ drift is the dominant poloidal motion and therefore all particles contribute to this kind of transport. The coefficient decreases with collisionality as in case of a tokamak. In terms of the electron temperature, the much weaker $T_e^{1/2}$ dependence is found:

$$D_\nu \approx \epsilon^2 \frac{e^2 \pi n_e \ln \Lambda_e}{12\epsilon_0^2 (3m_e)^{1/2} E_r^2} (T_e)^{1/2}. \tag{9.78}$$

In Fig. 9.18, the qualitative effect of the radial electric field on the diffusivity can be seen.

Although the effect of particles trapped in the magnetic field ripple is strongest in stellarators, also the toroidal inhomogeneity of the tokamak field due to discrete coils can have implications for transport [24]. The particles trapped in this field ripple can lead to bipolar radial fluxes and therefore to changes in the radial electric field [25].

9.3.8 Neoclassical Transport in the Fluid Picture

The starting point for the derivation of neoclassical transport in the two-fluid picture is the stationary equation of motion valid for electrons and ions:

$$qn(\mathbf{E} + \mathbf{u} \times \mathbf{B}) - \nabla p - \nabla \Pi^{neo} = 0. \tag{9.79}$$

Electron–ion collisions can be neglected. The physics is hidden in the neoclassical viscosity. While the perpendicular viscosity related to shear flows can be important for momentum transport and transport barrier physics, neoclassical transport is related to the parallel viscosity $\nabla_\parallel \Pi^{neo}$ given by the diagonal elements of the tensor $\nabla \Pi^{neo}$. The radial component of the parallel viscosity can be neglected and due to the symmetry of a tokamak, the toroidal one is zero. Hence the complicated expression is reduced to the poloidal parallel viscosity caused by the momentum loss, which can occur in the compression–expansion cycle when the fluid is convected in poloidal direction from the low–field side to the high–field side and back (*magnetic pumping*):

$$\nabla \Pi^{neo} \approx (\nabla_\parallel \Pi^{neo})_\theta \mathbf{e}_\theta \approx -mn\mu_{\parallel\theta} u_\theta \mathbf{e}_\theta \approx -n\hat{\mu}_{\parallel\theta} m^{1/2} u_\perp \mathbf{e}_\perp. \tag{9.80}$$

In order to keep the mathematics simple, the poloidal component of the flow velocity was replaced by the one perpendicular to $\mathbf{B}$ and $\mathbf{e}_r$ and the explicit mass dependence was extracted from the viscosity coefficient $\mu_\parallel$.

Next, the radial and perpendicular components of (9.79) are regarded:

$$qn(E_r + u_\perp B) - p' = 0 , \tag{9.81}$$

$$qnu_r B + n\hat{\mu}_{\parallel\theta} m^{1/2} u_\perp = 0 . \tag{9.82}$$

The perpendicular component leads to an expression for the neoclassical radial particle flux of the form

$$u_r = -\frac{1}{qB}\hat{\mu}_{\parallel\theta} m^{1/2} u_\perp = -\frac{1}{q^2 n B^2}\hat{\mu}_{\parallel\theta} m^{1/2} (p' - qnE_r) , \tag{9.83}$$

where (9.81) was used to replace $u_\perp$. This equation is valid for electrons and ions, therefore ambipolarity imposes a condition on the radial electric field. A simple expression is found when the ion flux (which dominates due to the mass dependence) is set to zero. Then the *ambipolar radial electric field* follows from the bracket in the equation set to zero:

$$E_r^{amb} = -\frac{p_i'}{q_i n} = -\frac{T_i}{q}\left(\frac{n'}{n} + \frac{T_i'}{T_i}\right) . \tag{9.84}$$

The situation which corresponds to this approach is explained in Fig. 9.19. In the left part, the ions move poloidally according to their diamagnetic drift velocity. The friction force F_{vis} caused by parallel viscosity counteracts this flow leading to ion transport in the $q\mathbf{F}_{vis} \times \mathbf{B}$ direction, hence outward. For the electrons the same arguments are valid resulting again in outward directed transport. Since, however, the ion viscosity is by a factor $(m_i/m_e)^{1/2}$ larger, ion transport is by the same factor higher then electron transport. This violates ambipolarity and the plasma charges up negatively on the fast

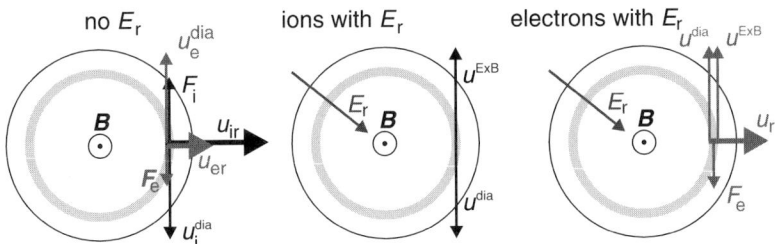

Fig. 9.19. Without a radial electric field (*left*), the viscose forces $F_{e,i}$ related to the poloidal diamagnetic flows create more ion than electron transport. The plasma charges up negatively until the $E_r \times B$ drift cancels the diamagnetic drift of the ions (*middle*). The electron transport (*right*) is then twice the one related to the diamagnetic drift

Alfvénic time scale. This process stops when the $E_r \times B$ drift cancels the diamagnetic drift of the ions. For the electrons the $E_r \times B$ drift points into the same direction as the diamagnetic drift, leading to a perpendicular velocity which is in case of $p_i = p_e = p$ twice the diamagnetic one:

$$u_{e\perp} \approx -2\frac{p'}{enB}. \quad (9.85)$$

The radial electron transport caused by the friction force related to the perpendicular flow sets now the approximate value for *neoclassical transport*. It is given by

$$\Gamma^{neo} \approx n u_{er} \approx -\frac{\hat{\mu}_{\|\theta} m_e^{1/2}}{enB^2}\frac{p'}{p}. \quad (9.86)$$

Only a small deviation from E_r^{amb} is needed to increase the ion transport from zero to the value of the electron transport. Since the electron transport will be only little affected by the small electric field change, (9.86) gives a rather good approximation of the ambipolar neoclassical transport. The calculation of the viscosity coefficient is complex [26]. It turns out that it has the same dependence on collisionality as it has been derived for the diffusivity in the particle picture and hence qualitatively this approach leads to the same diffusivities.

9.3.9 The Ambipolar Electric Field

The previous section has shown that the radial electric field plays the key role in order to make the radial fluxes ambipolar. In numerical calculations, the ambipolar field is determined by equating the neoclassical fluxes (9.23) for electrons and ions by iterating E_r. For this purpose, the full transport matrix is required with the coefficients $D_{ij}(\nu, E_r, r)$ depending on collisionality, electric field and radial position. Since these calculations, which include folding of the mono-energetic coefficients with a Maxwellian and averaging over flux surfaces, are, like in the case of the DKES (drift kinetic equation solver) code [27], very time consuming, the results are fitted by analytic functions [28] to be readily available for all parameters. Figure 9.20 shows the complexity of the physics related to the radial electric field. Using the DKES code for the geometry of the stellarator Wendelstein 7-AS, the particle fluxes were calculated at a fixed position of $r = 13\,\text{cm}$ as function of E_r. Density and ion temperature were kept fixed while the electron temperature has been varied.

First the ion transport is addressed, which exhibits three maxima that can be understood on the basis of the previous sections. The central peak corresponds to losses of helically trapped particles. Collisionless particles in the tail of the Maxwellian are in the $1/\nu$ regime and consequently lost. The peak is, however, very small since already small E_r values are sufficient to confine these particles. The two other peaks are due to the *toroidal resonance*

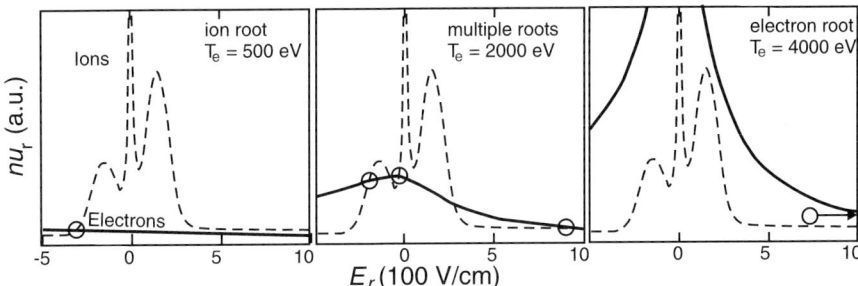

Fig. 9.20. Neoclassical electron and ion flux in the stellarator Wendelstein 7-AS for different electron temperatures at a fixed ion temperature of $T_i = 500\,\text{eV}$ calculated at $r = 13\,\text{cm}$ with the DKES code. The *circles* mark possible stable solutions for the ambipolar electric field

of co and counter circulating particles, which of course both exist in the Maxwellian. The width of the peaks is related to the ion temperature.

The same structure can also be found for the electrons. Due to the higher thermal velocity, the toroidal resonances are shifted to fields outside the plot range. The peak from the helically trapped particles is broader as for the ions and shows up as soon as the electron temperature is increased. The intersections in Fig. 9.20 of the curves for electrons and ions correspond to possible solutions for the ambipolar electric field. Only the stable solutions are marked by circles. At a stable solution, a deviation of the field from the value of the intersection leads to bipolar fluxes which bring the field back to the ambipolar value. At unstable intersection this is not the case. In the left part of Fig. 9.20, electrons and ions have the same temperature and only one negative field value is possible. This is the normal case for a fusion plasma and is called the *ion root*. Experimental results in the ion root agree in general rather well with the simple expression (9.84). This is demonstrated in Fig. 9.21 for three different discharges in the W7-AS stellarator, namely (from left to right) a high confinement NBI discharge [9], one with combined ERCH and NBI heating [23] and a pure ECRH discharge. Typical is the negative radial electric field due to the ion root in the outer plasma region. It coincides with the region of the steepest pressure gradient. The spectroscopically measured data coincide reasonably well with both models, the full neoclassical calculation and the simple model of (9.84). Similar results are also reported in [29] and from tokamak discharges [30].

At low densities and strong electron heating, situations as in the middle and right part of Fig. 9.20 can be realised. Collisionless helically trapped electrons make the electron losses superior to the ion ones. At very high electron temperatures (right), only one solution related to a strong positive field is found, called the *electron root*. The strong field converts the $1/\nu$ *regime*

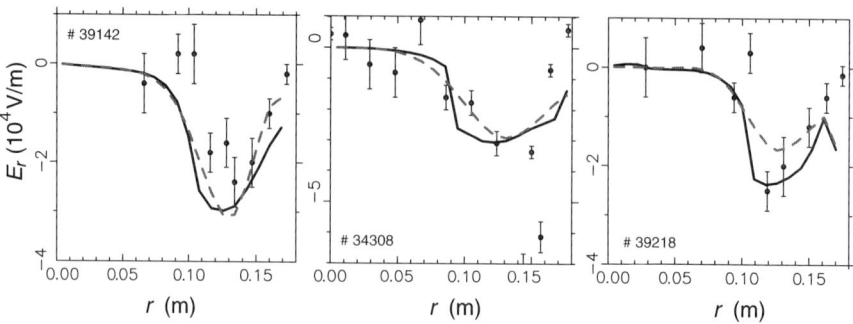

Fig. 9.21. Radial electric field for discharge in W7-AS heated with (from *left* to *right*) NBI, combined NBI and ECRH and pure ECRH. Simulations of the ambipolar electric field are from the DKES code (*solid line*) and from the simple model of (9.84) (*dashed line*)

into the ν *regime*. In-between the two extremes, multiple solutions exist and interesting transitions between electron and ion root solutions can occur.

As an example an experiment is quoted, where one could continuously increase the heating power and therefore the electron temperature. This can lead to fast transitions in the radial electric field. Starting from the ion root solution, increasing electron temperatures will move the intersection of the curves to the central ion peak visible in Fig. 9.20 where it stays at small E_r values until the electron flux exceeds the ion flux. Than a fast transition occurs and the electric field has to jump from small to large positive values. The possibility of fast transitions in the plasma potential were observed in the CHS torsatron with a heavy ion beam probe [31, 32].

Figure 9.20 shows the situation at a fixed radial position. In order to describe the radial profile of the electric field, the calculation has to be carried out at all radii. Figure 9.22 shows the comparison of such calculations with spectroscopically measured data for the electric field. The results are from an ECRH discharge in W7-AS, the theoretical curves were calculated with the DKES code. In the central plasmas the electron root is present. In the region were multiple roots are possible, the plasma chooses, as also described by an minimum energy argument related to the $E \times B$ flow [34], the smallest field possible. At the plasma edge the ion root is realised. At the transition from large positive to small fields, a layer of strong $E \times B$ flow shear is present in the plasma, which has the potential to suppress plasma turbulence (see next section). Such an interaction of neoclassical with turbulent transport can lead to transport barriers as found in the W7-AS stellarator [35].

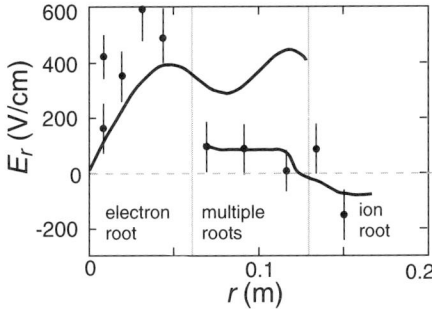

Fig. 9.22. Radial electric field for an ECRH discharge in W7-AS. Simulations are from the DKES code and data from the poloidal plasma rotation measured by active charge-exchange spectroscopy (adapted from [33])

9.4 Turbulent Transport

Although toroidal effects strongly increase the value of the collisional diffusion coefficients, they are not strong enough to account for the level found experimentally for particle, electron and in general also ion-heat transport. It is well established, that fluctuations in the plasma parameters are responsible for the observed transport losses [36]. The fluctuations in density, temperature and potential are caused by plasma turbulence and the related losses have been termed *anomalous transport*, although *turbulent transport* should be preferred. Turbulence is called to be *electrostatic* if radial transport is caused by the $E \times B$ drift in the fluctuating electric field. If magnetic fluctuations are large enough to locally destroy magnetic flux surfaces, one speaks of *electromagnetic turbulence*. Reviews of the theoretical progress in simulating turbulent transport can be found in [37, 38, 39], experimental results will be summarised in Sect. 9.4.5. An introduction to turbulence given in Chap. 8 in Part II.

For a better understanding of the elementary processes, in the next section fluid turbulence is briefly discussed, followed by introductions to the phenomenology of plasma turbulence, to linear instabilities which drive the turbulence, to a simple drift-wave model and the physics of transport barriers. The final section gives some experimental results on turbulence.

9.4.1 Fluid Turbulence

Turbulence in fluids is characterised by eddies which appear with random sizes and life times. The physical principles that govern the transition from laminar to turbulent flows and the statistical properties of the eddies is studied since centuries. An introduction to the main concepts can be found in [40] while more recent reviews are given in the textbooks [41, 42].

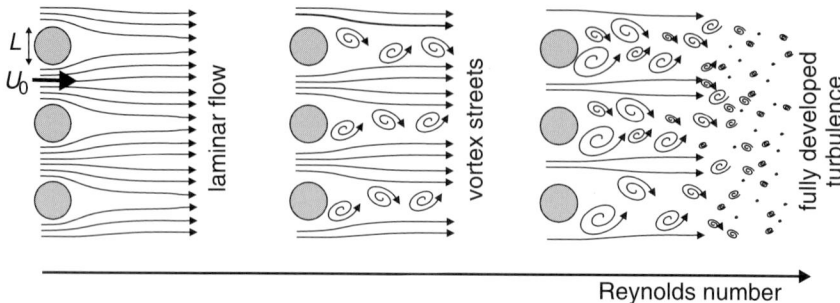

Fig. 9.23. A neutral fluid flows with the velocity U_0 through obstacles of size L. Shown is the qualitative dependence of the flow pattern on Reynolds number. The structure changes from laminar over vortex streets to turbulent

In incompressible fluids ($\nabla \mathbf{u} = 0$), turbulence is described by the *Navier–Stokes equation*

$$mn \left(\frac{\partial}{\partial t} + \mathbf{u} \cdot \nabla \right) \mathbf{u} = -\nabla p + \mu \Delta \mathbf{u} , \qquad (9.87)$$

with the hydrodynamic derivative on the left and the pressure force and viscous damping on the right; μ being the viscosity. Figure 9.23 depicts a fluid running through obstacles of the characteristic size L, with a flow velocity U_0 and a related time constant $T = L/U_0$. These parameters can be used to make the Navier–Stokes equation dimensionless:

$$\frac{d\hat{\mathbf{u}}}{d\hat{t}} = -\hat{\nabla}\hat{p} + \frac{1}{R_e} \hat{\Delta}\hat{\mathbf{u}} , \qquad (9.88)$$

with $\hat{t} = t/T$, $\hat{\mathbf{u}} = \mathbf{u}L/T$ and $\hat{p} = p(T/L)^2/mn$. The *Reynolds number* R_e is a dimensionless parameter, which, at given geometry, determines the state of the fluid. Starting from a laminar flow, with increasing Reynolds number the fluid first develops streets of regular eddies and than goes over into a fully turbulent state. The Reynolds number can also be understood as the ratio of the non-linear to the viscous term, namely

$$R_e = \frac{u\nabla u}{\mu \Delta u} = \frac{U_0^2}{L} \bigg/ \frac{\mu U_0}{L^2} = \frac{U_0 L}{\mu} . \qquad (9.89)$$

The non-linear term drives and the viscous one damps the turbulence. This can be demonstrated with the one-dimensional version of the Navier–Stokes equation, where all constants were set to 1 and the pressure term is neglected:

$$\frac{\partial u}{\partial t} = -(u)\frac{\partial u}{\partial x} + \frac{\partial^2 u}{\partial x^2} . \qquad (9.90)$$

Figure 9.24 shows numerical solutions of this equation, with the initial conditions $u = u_0 \sin kx$ and $y(x) = 0$. The left part shows the results without

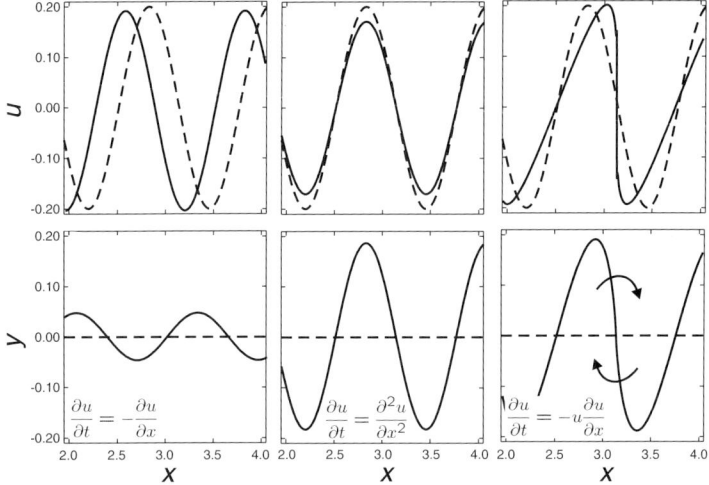

Fig. 9.24. Numerical solutions of the equations given in the lower figures. *Dashed line*: initial conditions of the velocity (*top*) and the position of a fluid line (*bottom*). *Solid lines*: simulation results. From *left* to *right* the effect of wave propagation, damping and creation of higher harmonics can be seen. The arrows indicate that these structures given the velocity profile from above will transform into eddies

viscosity and without the u of the non-linear term, where a propagating wave is found. Viscosity alone (middle plot) leads to damping and the non-linearity alone (right) to the creation of higher harmonics and wave-breaking, which creates eddies. Eddies are characterised by non-zero circulation Z,

$$Z = \oint \mathbf{u} \cdot \mathrm{d}\mathbf{l} = \int (\nabla \times \mathbf{u}) \cdot \mathrm{d}\mathbf{S} = \int \boldsymbol{\Omega} \cdot \mathrm{d}\mathbf{S} \neq 0 \;, \tag{9.91}$$

or by their vorticity $\boldsymbol{\Omega} = \nabla \times \mathbf{u}$, which is conserved in ideal fluids.

Further quantities used to characterise turbulent fluids are the volume averaged *energy density*

$$E = \frac{1}{2V} \int u^2 \mathrm{d}^3 r \;, \tag{9.92}$$

and the *enstrophy*

$$\Omega^* = \frac{1}{2}\left\langle \boldsymbol{\Omega}^2 \right\rangle = \frac{1}{2V} \int \boldsymbol{\Omega}^2 \mathrm{d}^3 r \;. \tag{9.93}$$

In the seminal *K41 theory*, Kolmogorov has investigated energy transfer between different spatial scales of the turbulence (for a review see [41] and references therein). The result of this study is summarised in the left part of Fig. 9.25. Starting point was three-dimensional isotropic turbulence where the scales on which energy input takes place (the *injection range*) is well separated from the *dissipation range*, where energy is transferred into heat. The theory describes the *inertial range* in-between, where energy is transferred on

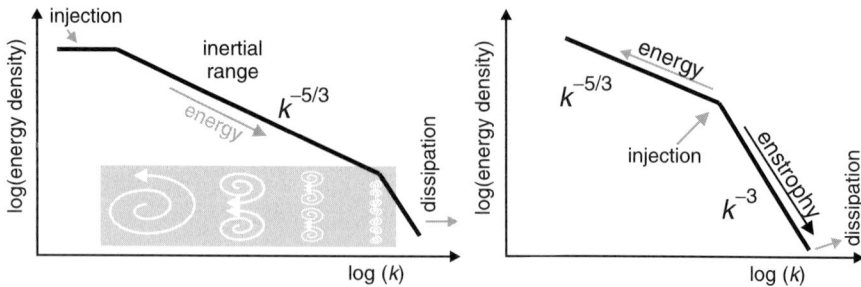

Fig. 9.25. Power spectra of three-dimensional (*left*) and two-dimensional turbulence (*right*) in neutral fluids

a direct cascade from larger to smaller scales. This leads to a distribution of the turbulent energy over the different spatial scales (power spectrum) that decays as a function of wave number as $E_k \sim k^{-5/3}$. Experimentally such spectra have been found, e.g., in low-temperature Helium gas flows [43].

Since in magnetised plasmas the fast dynamics parallel to the magnetic field is separated from the slow drifts perpendicular to it, turbulence cannot be expected to be isotropic in three dimensions. Therefore a brief summary of the modification in the turbulent cascade introduced by the two-dimensional geometry is helpful. The changes which occur are best seen at the Navier–Stokes equation written in terms of the vorticity:

$$\left(\frac{\partial}{\partial t} + \mathbf{u} \cdot \nabla\right)\mathbf{\Omega} = (\mathbf{\Omega} \cdot \nabla)\mathbf{u} + \mu\Delta\mathbf{\Omega} \ . \tag{9.94}$$

The first term on the right hand side stands for a process called *vorticity stretching*. It produces vorticity while the second term destroys vorticity due to viscosity. The creation of vorticity is due to angular momentum conservation. If a vortex in an incompressible fluid is stretched in the direction of the angular momentum vector, it becomes at the same time narrower in the other directions and the rotation spins up. In a 2D fluid, the vorticity equation becomes particulary simple. The flow velocity has only two components $u_x(x,y)$, $u_y(x,y)$ and the vorticity has only a component in z direction. Consequently vorticity stretching vanishes and the two-dimensional Navier–Stokes equation reads:

$$\left(\frac{\partial}{\partial t} + \mathbf{u} \cdot \nabla\right)\mathbf{\Omega} = \mu\Delta\mathbf{\Omega} \ . \tag{9.95}$$

Therefore in an ideal ($\mu = 0$) 2D fluid, both energy and enstrophy are conserved which introduces an additional constraint to the turbulence. A consequence of this constraint is shown on the right hand side of Fig. 9.25. Instead of one direct energy cascade one now finds a *dual cascade* [44]. In case of a localised injection range, an inverse energy cascade with a slope in the power spectrum given by $k^{-5/3}$ is separated from a direct enstrophy cascade with a slope k^{-3}. Energy is also transferred to smaller scales but the majority goes into the large scales. Since there is no dissipation at large scales this would lead to a so-called infrared catastrophe if friction with the boundaries of the fluid did not dissipate energy. Indications of the dual cascade were found in experiments with soap films running through a comb of obstacles [45].

9.4.2 Phenomenology of Turbulent Plasma Transport

There exists overwhelming evidence that electrostatic turbulence is responsible for the bigger part of radial transport in fusion experiments. This means that transport perpendicular to the ambient field is caused by $\tilde{E} \times B$ drifts due to the fluctuating electric field $\tilde{E}$. Figure 9.26 illustrates how a local potential fluctuation $\tilde{\phi}$ creates a circular flow, also called eddy. In the left figure, the eddy acts on the background density, which has a gradient in radial direction, and advects the same amount of density inward at the top and outward at the bottom of the perturbation. Hence this process does not create any net particle transport. If the potential perturbation stays in place, the motion will create a density perturbation $\tilde{n}$ with positive values at the bottom and negative ones at the top (middle part of Fig. 9.26). Under this condition, net outward transport would occur. This shows that both density and potential fluctuations are necessary to create net outward transport. It also shows that the phase between both fluctuations is an important quantity. Only if both perturbations are out of phase, the transport is non-zero, but if the potential perturbation propagates to be in phase with the density perturbation (right part of Fig. 9.26) the net-transport is again zero.

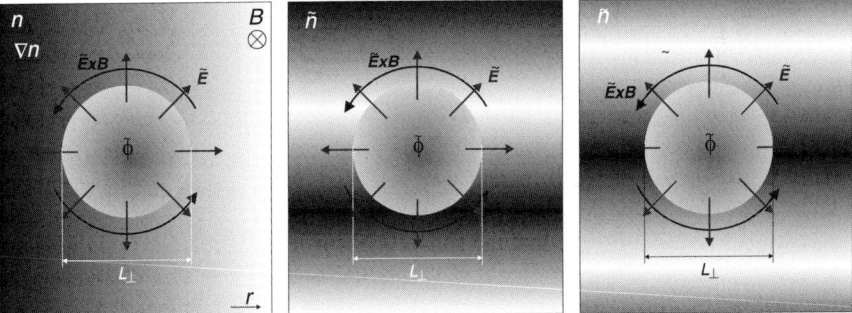

Fig. 9.26. The mechanism of radial transport created by the $E \times B$ drift due to an electrostatic perturbation. *Left*: the eddy acts on the background density with a gradient in radial direction indicated by the shading. No net-transport in radial direction (*across a vertical line*) occurs. *Middle*: the situation changes if a density perturbation exists, that is out of phase with the potential perturbation. The causes radial net-transport. *Right*: If the perturbations are in phase, the net-transport is again zero

For the study of turbulence, the plasma parameters are decomposed into a constant background and a small fluctuating contribution:

$$\phi(\mathbf{r},t) = \phi_0(\mathbf{r}) + \tilde{\phi}(\mathbf{r},t) \; ; \qquad n(\mathbf{r},t) = n_0(\mathbf{r}) + \tilde{n}(\mathbf{r},t) \; . \tag{9.96}$$

The fluctuations are Fourier transformed. For turbulent transport averaged over a flux surface and in time it then follows

$$\tilde{\Gamma} = \langle \tilde{n}\tilde{u} \rangle_{F,t} = \sum_{k=-\infty}^{\infty} \tilde{n}_k \tilde{u}_k^* \; , \tag{9.97}$$

where the sum extents over all complex Fourier coefficients. Of course, it would be desirable to have information on all Fourier coefficients,

$$\tilde{\phi}(\mathbf{r},t) = \sum_{k,\omega} \tilde{\phi}_{k,\omega} \exp\left[\mathrm{i}(\mathbf{k}\mathbf{r} - \omega t)\right] \; ; \qquad \tilde{n}(\mathbf{r},t) = \sum_{k,\omega} \tilde{n}_{k,\omega} \exp\left[\mathrm{i}(\mathbf{k}\mathbf{r} - \omega t)\right] \; . \tag{9.98}$$

Experimentally in general, however, only frequency spectra measured at one radial location $\mathbf{r}_0$ are available. So experimental data are often frequency spectra:

$$\tilde{\phi}_\omega = \int_{-\infty}^{\infty} \tilde{\phi}(\mathbf{r},t) \exp(\mathrm{i}\omega t)\mathrm{d}t = \sum_k \tilde{\phi}_{k,\omega} \exp(\mathrm{i}\mathbf{k}\mathbf{r}_0) \; . \tag{9.99}$$

From this follow characteristic quantities as, e.g., correlation functions. If both parameters, potential and density, are measured at the same location, the *cross-correlation function*,

$$C_{\phi n}(\tau) = \int \tilde{\phi}(t)\tilde{n}(t+\tau)\mathrm{d}t = \int \tilde{\phi}_\omega \tilde{n}_\omega^* \exp(\mathrm{i}\omega\tau)\mathrm{d}\omega$$

is used to find similar events which occur in both signals with a possible delay or time lag τ. The second part of the equation stands for the *Wiener–Khintchine theorem*. The auto-correlation function $C_{\phi\phi}(\tau)$, which is $C_{\phi\phi}(\tau) = 1$ for $\tau = 0$, gives a measure of the duration for which characteristic structures persist in the signal. This defines characteristic quantities of the fluctuations as the *correlation time* τ_{corr} or, in space, the *correlation length* L_{corr}. They are defined as the time and distance, respectively, after which the auto-correlation decays to $1/e$. This also defines characteristic wave numbers and frequencies of the fluctuations:

$$\bar{k} = 2\pi/L_{corr} \; ; \qquad \bar{\omega} = 2\pi/\tau_{corr} \; . \tag{9.100}$$

These experimentally accessible quantities are used to calculate estimates of the turbulent diffusion coefficient. To this end, one assumes that the fluctuations can be represented by one Fourier harmonic at the characteristic wave number and frequency. An estimate is obtained by considering the $E \times B$ drift in radial direction given by

$$\tilde{u}_r = \left.\frac{\mathbf{B}\times\nabla\tilde{\phi}}{B^2}\right|_r \approx i\frac{\bar{k}_\theta}{B}\tilde{\phi}\,, \tag{9.101}$$

where $\bar{k}_\theta$ is the characteristic poloidal wave number. If the perturbation exists for the time τ_{corr}, a fluid element is radially displaced by $\delta r = |\tilde{u}_r|\tau_{corr}$, and from the random walk ansatz (9.31) follows a diffusion coefficient of the form

$$D \approx \frac{\delta r^2}{\tau_{corr}} = |\tilde{u}_r|^2 \tau_{corr} = \left(\frac{\bar{k}_\theta}{B}\right)^2 |\tilde{\phi}|^2 \tau_{corr}\,. \tag{9.102}$$

Hence, in the simplest case, the diffusion coefficient increases quadratically with fluctuation amplitude. According to the Boltzmann relation ($\tilde{n}/n_0 \approx e\tilde{\phi}/T_{e0}$) this can also be translated into the density fluctuation amplitude. In the case of strong turbulence, the expression turns into a linear dependence on amplitude.

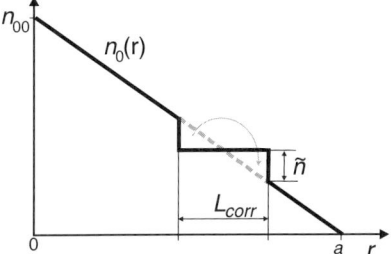

Fig. 9.27. Mixing-length model. An eddy of the size L_{corr} flattens the density gradient and causes the density fluctuation $\tilde{n}$

Equation (9.102) also follows from the *mixing-length model* as illustrated in Fig. 9.27. According to this model, an eddy leads to a flattening of the density gradient in the range of L_{corr}. This results in a density fluctuation amplitude of

$$\tilde{n} = L_{corr}\nabla n_0\,. \tag{9.103}$$

Using L_{corr} for δr and again $\delta r = |\tilde{u}_r|\tau_{corr}$ this leads to the same diffusion coefficient as in (9.102):

$$\tilde{\Gamma} = \tilde{n}\tilde{u}_r = \delta r \nabla n_0 \tilde{u}_r = |\tilde{u}_r|^2 \tau_{corr}\nabla n_0 = D\nabla n_0\,. \tag{9.104}$$

In these simple considerations the cross-phase between density and potential fluctuations is neglected. It enters if both quantities are measured. With $\tilde{n} = |\tilde{n}|\exp(i\delta_n)$, $\tilde{\phi} = |\tilde{\phi}|\exp(i\delta_\phi)$ and (9.101) it follows for the turbulent transport:

$$\tilde{\Gamma} = 2k_\theta|\tilde{n}||\tilde{\phi}|\sin\delta_{n\phi}/B\,, \tag{9.105}$$

with the cross-phase $\delta_{n\phi} = \delta_n - \delta_\phi$. Hence the level of transport depends on the phase relation between density and potential fluctuations. The highest transport is found if the cross-phase is $\pi/2$. This situation is represented in the middle part of Fig. 9.26. The right part of the figure represents the situation where both quantities are in phase and transport is zero.

9.4.3 Two Fundamental Linear Instabilities

There are two fundamental linear instabilities, which are relevant for plasma turbulence, the interchange and the drift-wave instability. As a major signature, the two cases lead to different cross-phases. It has been shown [46] that the characteristics of the linear instabilities are also present in the state of fully developed turbulence if the dominant instability mechanism remains the same. On the basis of Fig. 9.28, which follows the ones in [47], the mechanisms of these instabilities and their characteristics are discussed.

The *drift wave*, in the left part of Fig. 9.28, occurs in arbitrary magnetic fields. In the simplest case, the field is constant with a density gradient

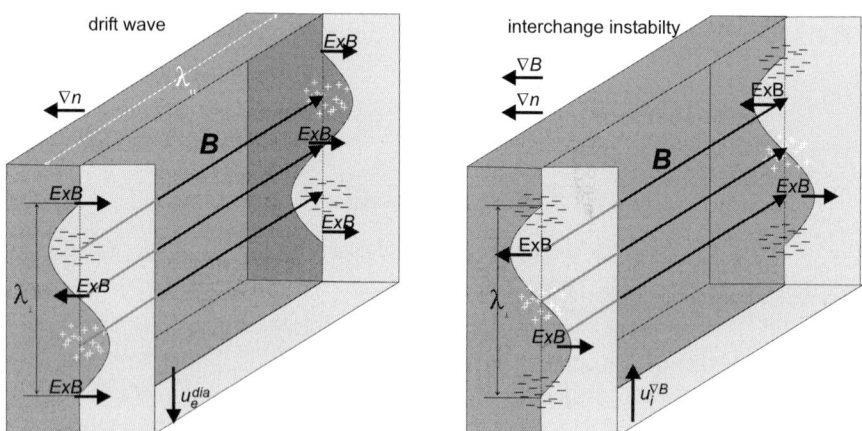

Fig. 9.28. Microscopic mechanisms and properties of the linear drift-wave (*left*) and the interchange instability (*right*). Dark and light areas represent high and low density, respectively. The equilibrium profile is perturbed by a sinusoidal function. The structure of the drift wave is three-dimensional, the electrons respond to the parallel pressure gradient and create electric charges in phase with the density perturbation. The resulting $E \times B$ drift results in a propagation of the perturbation into the electron diamagnetic direction u_e^{dia}. The structure of the interchange mode is two-dimensional and the dynamics restricted to the drifts perpendicular to **B**. Here charges are generated by the curvature drifts $u_{e,i}^{\nabla B}$ at the limit between high and low density (see text), hence out of phase, and the $E \times B$ drift acts to amplify the initial perturbation

perpendicular to it. Essential for the occurrence of a drift wave is a three-dimensional perturbation of the pressure equilibrium represented by the parallel and perpendicular wave numbers $k_\parallel \neq 0$ and $k_\perp \gg k_\parallel$ (with respect to **B**), respectively. Hence drift waves consist of density perturbations which are elongated parallel to the magnetic field.

For the dynamic of the drift wave, the high electron mobility parallel to the field is crucial. The electrons respond to the parallel pressure gradient, which creates positive charges in the region of positive density perturbations and vice versa. The resulting electric field leads to the $E \times B$ drift that advects, as discussed in Fig. 9.26, the background density. Details of the dynamics will be discussed in Sect. 9.4.4. In case of adiabatic electrons, i.e., instantaneously responding electrons, the phase between density and potential perturbation is zero. In this case, the $E \times B$ drift leads to a simple displacement of the perturbation into the *electron-diamagnetic* direction ($\sim \nabla p \times \mathbf{B}$). The drift wave becomes unstable if the parallel response of the electrons to the parallel pressure gradient is delayed. In this case, the potential perturbation lags behind the density perturbation and the region of the outward directed $E \times B$ drift shifts to the region of positive density perturbations, which now are amplified. There are a number of effects, which contribute to the non-adiabaticity of the electrons: the resistivity due to electron–ion collisions, magnetic induction, particle-wave interaction due to Landau damping and, at high frequencies, electron inertia.

The *interchange instability* as on the right side of Fig. 9.28 only exists in regions of bad magnetic field-line curvature, i.e., on the low–field side of the torus. On the high-field side, the regions of low and high density are interchanged (good curvature) and the same mechanism as discussed below has a stabilising effect. In contrast to the drift wave, the density perturbation is now two-dimensional ($k_\parallel = 0$). The dynamics is restricted to the plane perpendicular to **B** and is driven by the vertically directed and charge dependent curvature drift, which transports, e.g., ions from a region of high density upward into a region of low density. Hence an ion excess and positive charges appear at this border. The electrons drifting downward from a region of low to one of high density support this effect, since their density is too low to cancel the charges of the present ions. Halve a period further away, the same process cause a negative net-charge. Hence the curvature drift creates potential perturbations which are by $\pi/2$ out of phase with the density perturbations. In this case the resulting $E \times B$ drift amplifies the original density perturbation. If this process is maintained it leads to a lamination of the plasma density with regions of high density streaming outward and low density streaming inward (so-called *streamers*). Kelvin–Helmholtz-like instabilities than lead to a decay of the perturbation into smaller scales and to a transition into turbulence.

In the edge of a fusion plasma, drift-wave turbulence is expected to make the dominant contribution to transport [46, 48]. The instabilities which are most likely to explain transport in the core of fusion plasmas are related to the interchange instability. The drive, however, does not come from the density gradient alone. For an illustration of ion-temperature gradient (ITG) driven turbulence [49] the shaded areas in Fig. 9.28 have to be identified with hotter and colder ion fluid. Since the curvature drift (9.33) depends on the thermal particle energy, the hotter fluid drifts faster than the colder one, which leads to a compression of ions at the upper side of the dark-shaded area and to positive charges and to a dilution and negative charges on the lower side. Hence the charge pattern shown in the figure is recovered. ITG is the best candidate to explain the ion thermal transport in tokamaks. The equivalent electron temperature gradient (ETG) driven turbulence [50] functions in the same way. The contribution of ETG turbulence to transport is not settled yet, since it produces small-scale fluctuations on the scale of the electron gyro radius, which leads to a low level of transport. For both ion and electron thermal transport in tokamaks, also trapped particle modes (TEM) [51] might be of importance. Here electric charge is accumulated due to the toroidal precession of banana electrons. Since these particles are located on the low-field side of the torus their effect is concentrated on the bad-curvature region and therefore very efficient.

Negative magnetic shear reduces the interchange drive [52]. This is illustrated in Fig. 9.29. A perturbation that is constant on a magnetic field line is tilted on its ways around the torus. In the same way as on the right hand side of Fig. 9.28 the curvature drift generates charges at the limit between high and low-density regions. For the present geometry, the upper limit becomes positively and the lower one negatively charged. In case of negative

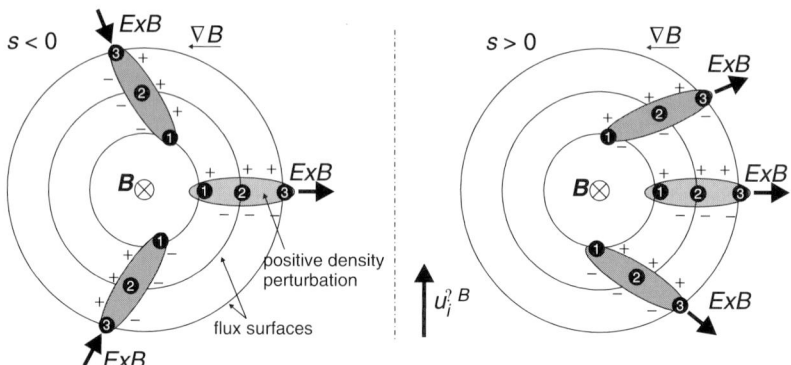

Fig. 9.29. Influence of negative (*left*) and positive (*right*) magnetic shear on the interchange instability. The numbers indicate the positions of three field lines at three different toroidal angles. The charges are generated by the curvature drift by the same mechanism as discussed in Fig. 9.28. In case of negative shear, the $E \times B$ drift is for a larger poloidal section stabilising (inward directed) than in case of positive shear

shear ($s < 0$), i.e., the field lines on the outer flux surfaces are twisted more strongly than on inner ones, the tilt of the perturbation changes in such a way that the $E \times B$ drift is radially inward directed in a larger portion of the poloidal cross-section than in case of positive shear. This has a stabilising effect. Positive shear (right) is the normal case for a tokamak, while stellarators exist with small to strong negative shear.

9.4.4 Elements of a Drift Wave Model

In this section, the elements of a simple drift-wave model are discussed. The basis are the processes illustrated in Fig. 9.30. Starting point is an elongated positive density perturbation with the dimension parallel to the magnetic field much larger than perpendicular to it: $L_\perp \ll L_\parallel$. As in the linear model shown on the left hand side of Fig. 9.28, due to their small mass electrons react first to the parallel pressure gradient and create an increasing positive potential $\dot\phi$ inside the density perturbation. The time constant of this process is set by the polarisation drift, which drives the ions perpendicular to the magnetic field out of the density perturbation and thus counteracts the charge built-up. The background density is then, as already discussed in Fig. 9.26, advected by the perpendicular electric field $\mathbf{E}_\perp$. Important for the understanding of a drift wave turbulence model are three points:

1. the perturbations are three-dimensional,
2. the thermal motion of the electrons electrons determines the parallel and
3. the drift motion of the ions the perpendicular dynamics.

In the simplest case, the magnetic field is constant and the ions are cold ($T_i = 0$). The background density gradient is characterised by the density decay length $L_n = -n_0/\nabla n_0$.

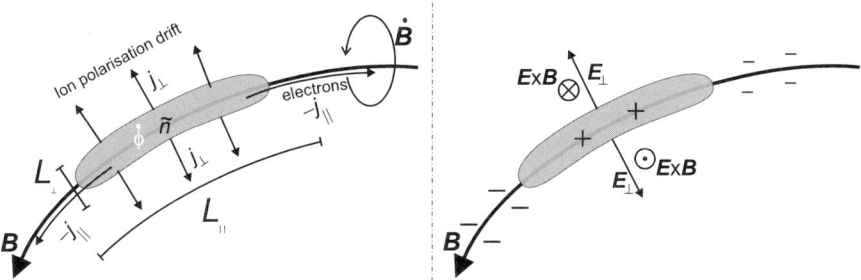

Fig. 9.30. Three-dimensional dynamics of drift-wave turbulence. *Left*: a density perturbation elongated along the magnetic field creates a parallel electron current $j_\parallel$, which is hindered by self-induction. The current creates a positive potential ϕ inside the density perturbation leading to an perpendicular ion polarisation current $j_\perp$. *Right*: The electric field leads to advection of the background density

The generalised Ohm's law governs the dynamics parallel to the magnetic field which is driven by the parallel pressure gradient $\nabla_\| p_e$ (the ion pressure is zero due to cold ions):

$$\frac{m_e}{e}\frac{\partial j_\|}{\partial t} = enE_\| + \nabla_\| p_e - en\frac{j_\|}{\sigma}, \qquad (9.106)$$

with the electric conductivity σ. The electric field can be replaced by

$$E_\| = -\nabla_\|\phi - \frac{\partial A_\|}{\partial t}, \qquad (9.107)$$

with the vector potential related to the parallel current through

$$\nabla_\perp^2 A_\| = \mu_0 j_\| . \qquad (9.108)$$

Altogether this leads to an *equation for the parallel current*:

$$en\frac{\partial A_\|}{\partial t} + \frac{m_e}{e}\frac{\partial j_\|}{\partial t} = -en\nabla_\|\phi + \nabla_\| p_e - 0.511\frac{m_e\nu}{e}j_\| . \qquad (9.109)$$

The driving terms of the current are the parallel electrostatic force, the pressure gradient and resistivity; on the left hand side one finds the effects of induction and inertia.

The quasi-neutrality condition $\nabla_\perp \mathbf{j}_\perp + \nabla_\| j_\| = 0$ couples the parallel dynamics to the perpendicular one, which is determined by ion drifts of the form

$$\mathbf{u}_{i\perp} = \frac{\mathbf{E}\times\mathbf{B}}{B^2} + \frac{m_i}{eB^2}d_t^{E\times B}\mathbf{E}_\perp \equiv \mathbf{u}^{E\times B} + \mathbf{u}_i^{pol} . \qquad (9.110)$$

For zero ion temperature, the diamagnetic term is not present and the perpendicular ion flow $\mathbf{u}_{i\perp}$ is given by the $E\times B$ and the polarisation drift. The latter contains as nonlinearity once more the $E\times B$ drift:

$$d_t^{E\times B} = \frac{\partial}{\partial t} + \mathbf{u}^{E\times B}\nabla . \qquad (9.111)$$

The $E\times B$ drift is the same for electrons and ions and no current arises from it. Due to the mass dependence, the electron polarisation drift is, however, negligible and a current perpendicular to $\mathbf{B}$ arises from the ions:

$$\mathbf{j}_\perp = \frac{m_i n}{B^2}d_t^{E\times B}\mathbf{E}_\perp . \qquad (9.112)$$

With $E_\perp = -\nabla_\perp\phi$ it follows the *vorticity equation*

$$\frac{m_i n}{B^2}d_t^{E\times B}\nabla_\perp^2\phi = \nabla_\| j_\| . \qquad (9.113)$$

As a third element the *continuity equation* describes the advection of the density. The electron version of it has the form

$$\left(\frac{\partial}{\partial t} + \mathbf{u}^{E\times B} \cdot \nabla\right) n = -n\nabla_\| u_{e\|} \approx \nabla_\| j_\|/e \,. \tag{9.114}$$

The density changes due to advection of the background density by to the $E \times B$ drift and due to parallel electron losses, which are accompanied by an equal amount of ion losses due to the polarisation drift. The coupled set of (9.109),(9.113) and (9.114) contain the main elements to perform drift-wave simulations. If field-line curvature is introduced, the electron diamagnetic drift will contribute to the perpendicular current. The resulting equations are solved in modern turbulence codes such as DALF3 [46].

In the *electrostatic limit*, the left hand side of (9.109) vanishes and $j_\|$ can be determined and inserted into (9.113) and (9.114). The resulting coupled equations have the form

$$\left(\frac{\partial}{\partial t} + \mathbf{u}^{E\times B} \cdot \nabla\right) n = \frac{1}{m_e \nu_e} \nabla_\|(\nabla_\| p_e - en\nabla_\| \phi) \,, \tag{9.115}$$

$$\frac{m_i n}{B^2}\left(\frac{\partial}{\partial t} + \mathbf{u}^{E\times B} \cdot \nabla\right) \nabla_\perp^2 \phi = \frac{e}{m_e \nu_e} \nabla_\|(\nabla_\| p_e - en\nabla_\| \phi) \,. \tag{9.116}$$

The linearised and normalised form of these equations are know as Hasegawa–Wakatani equations [53] and a reduction to a collisionless plasma with a constant for the parallel gradient leads to the Hasegawa–Mima equations [54].

9.4.5 Experimental Results

In fusion plasmas, the edge region is the best accessible to fluctuation measurements. First extensive experimental studies of turbulence in fusion experiments were carried out in the 80ies on small and medium size tokamaks. A review of the early experiments and comparisons with linear models can be found in [55] and of more recent data in [56]. The experiments, which have been in the majority carried out with Langmuir probes, revealed the existence of broad turbulent fluctuation spectra and it became evident that fluctuations contribute considerably to transport. The observed normalised fluctuation amplitudes $\tilde{n}/n_0$ are up to a few tenth of percent and the Boltzmann relation is approximately fulfilled: $\tilde{n}/n_0 \approx e\tilde{\phi}/T_{e0}$. As a general trend, the relative fluctuation amplitudes where found to drop when density is increased [57, 58, 21]. The values for the cross-phase between potential and density fluctuations was found to be in the range 0.2–0.5π. Figure 9.31 summarises results from these early studies with some more recent data added. It shows that the fluctuation amplitudes approximately follow the relation $\tilde{n}/n_0 \sim (kL_n)^{-1}$, which is equal to the prediction (9.103) from the mixing-length model in Fig. 9.27 when the spatial scale length is set by the correlation length ($k = 2\pi/L_{corr}$, $L_n = -n_0/\nabla n_0$). A survey on fluctuation measurements is given in Chap. 14 in Part III.

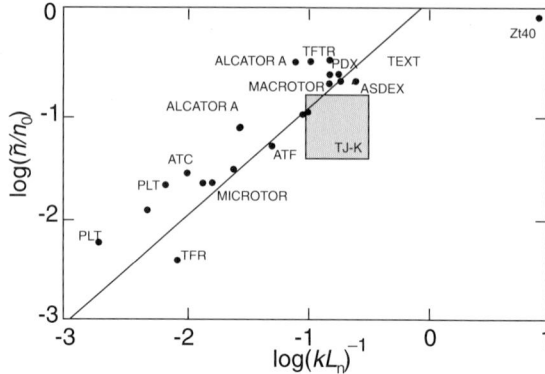

Fig. 9.31. Relative fluctuation amplitude vs. mixing length estimate for different toroidal confinement experiments [55] with more recent added data from TJ-K [59] and ASDEX [60]

In so called gas-puff experiments, the structure of the fluctuations has been made visible by observing H_α-light emission, from which in the edge plasma region the density fluctuations can be inferred [60]. The data revealed structures elongated along magnetic field lines with poloidal dimensions of the order of a centimetre. More recently, the perpendicular turbulent structure has been investigated by gas puff imaging with fast cameras and compared to numerical simulations [61]. Correlation measurements along the field lines indicated that the parallel size of the structures is several metres [62, 63]. Hence, these results were consistent with a separation of parallel and perpendicular dynamics as discussed in context of Figs. 9.28 and 9.30. The cross-phases between potential and density fluctuations have been determined from dominant structures by conditional-averaging techniques to be $\delta_{n\phi} \lesssim \pi/2$, [60, 64, 65]. On the other hand, simulations predict for this regime drift-wave turbulence with cross-phases $\delta_{n\phi} \approx 0$ [46, 66]. Since most of the experimental data stem from regions outside the separatrix, i.e., where the field lines are open (scape–off layer), the interchange character might be related to the sheath dynamics at the intersection of the field line with the limiters or target plates.

Statistical properties of the fluctuations have also been studied in detail [67, 68]. In many devices it was found that transport fluctuation have an intermittent nature, i.e., large amplitude fluctuation appear more frequently than a Gaussian distribution would predict and contribute considerably to transport [69, 70, 71]. This feature leads to a non-gaussian shape of the probability density function (PDF) and can partly be explained by the fact that transport results out of a product of two fluctuating quantities [72]. It was, however, also pointed out that the large-scale fluctuations could be a sign of an avalanche-type of a transport mechanism [73] as it is know from sand piles at a critical gradient [74]. This raises the question, whether critical gradient models such as ion-temperature-gradient driven turbulence [49] could be responsible for these features. Figure 9.32 shows as an example the transport fluctuations $\tilde{E} \times B$ with the corresponding PDF from the toroidal low-temperature plasma in TJ-K. The large spikes in the fluctuations are clearly visible and they constitute the wings in the PDF shown on the right.

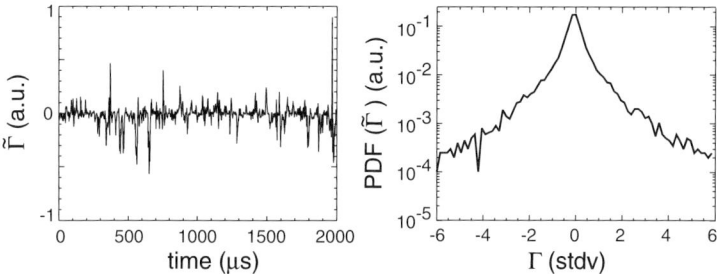

Fig. 9.32. Time trace of turbulent transport fluctuations $\tilde{E} \times B$ (*left*) and the probability density function of this signal from the torsatron TJ-K [71]

In order to get information on turbulence in the core of a fusion plasma, more sophisticated diagnostics than Langmuir probes have to be used due to high plasma temperatures. Examples are beam emission spectroscopy [57] or heavy ion beam probes [31]. Since these diagnostics are quite complex, systematic studies are sparse. Alternatively, toroidal low-temperature plasmas with dimensionless parameters similar to those in the edge of fusion plasmas [59] can help to gather information on turbulence on closed magnetic field lines. The reduced plasma parameters allow the use of multi-probe arrays with up to 64 tips to measure also the spatial structure of the fluctuations.

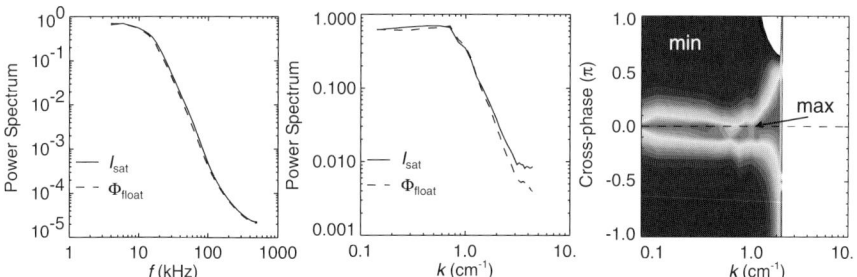

Fig. 9.33. Frequency and poloidal wavenumber resolved spectra of potential and density fluctuations and the cross-phase wavenumber spectrum. *Grey* relates to high, *black* to low probability to find the specific value for the cross-phase. Data are from the toroidal low-temperature plasma in TJ-K [78]

Such probe arrays have been already used in the 80ies in fusion experiments [75] and in the 90ies in linear devices to investigate the initial transition from drift-waves with a few mode numbers to a turbulent state [76, 77]. Recently poloidal Langmuir probe arrays with 64 tips have been applied to a toroidal low-temperature plasma [59, 78, 79]. Some results of these studies are presented in Fig. 9.33. Shown are frequency and wave-number spectra of the density and potential fluctuations, which all are broad as expected for a turbulent system. The figure on the right represents the cross-phase spectrum. For each wave number, the distribution of the phases between density and potential fluctuations is plotted. The plot shows that the phases are centred around zero for all wavenumbers. This is consistent with drift-wave turbulence and agrees with numerical simulations [66].

Up to now there is no clear evidence that magnetic fluctuations contribute measurably to transport. The influence from coherent magnetic turbulent fluctuations on transport was shown to be weak [80]. There is some evidence, that small-scale fluctuations are correlated with confinement properties [81, 82]. It is too early, however, to draw any final conclusions about the importance of electro-magnetic turbulence for transport.

9.4.6 Transport Barriers

Transport barriers play an important role in fusion plasmas. They consist of a radially limited area in which turbulence and transport are substantially reduced. They show up as a region where the electron and/or ion temperature and/or density profile gradient is considerably steeper than in the vicinity. The first observation of a transport barrier was made in the edge of the ASDEX plasma [83] and the emerging improved confinement regime was doped H-mode in contrast to the low confinement L-mode. As depicted in Fig. 9.34, the H-Mode is characterised by a transport barrier close to the separatrix, where a steep density gradient develops. Concurrently, the density fluctuations drop in the same radial region, where the density gradient steepens. Later on, the H-mode was observed in many stellarators and tokamaks. Although extensive studies of the processes which trigger the transition have been carried out [84, 85], a final explanation is still lacking.

In the meanwhile, also in the plasma core transport barriers have been generated and the physics of the local transport suppression remains a field of intensive research. For recent results related to internal transport barriers see [87, 88].

The mechanism of turbulence suppression has been attributed to sheared $E \times B$ flows [89, 90, 91]. On the right hand side of Fig. 9.34, this mechanism is illustrated. A gradient in the radial electric field leads to sheared poloidal plasma flows, which can tear apart perturbations with long radial correlation lengths. In computer simulations, it has been demonstrated [92] that this process can suppress turbulence efficiently. Reviews of experimental and

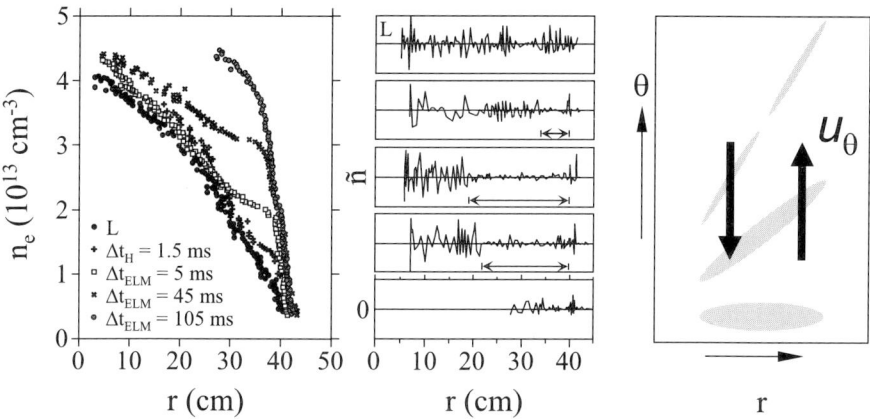

Fig. 9.34. *Left*: temporal evolution of the radial density profile after transition from L to H-mode and (*middle*) the concurrent drop in the radial profile of the density fluctuations (adapted from [86]) The fluctuation profiles from *top* to *bottom* correspond to the density profiles starting from L-mode with increasing time delay after the transition as indicated in the left plot. *Right*: Schematic drawing to illustrate the de-correlation mechanism of large scale fluctuations (*in grey*) due to sheared plasma flow (*arrows*)

theoretical work on this topic can be found in [93, 94], respectively. As a rule of thumb, suppression occurs when the shearing rate

$$S_v = r\frac{RB_\theta}{B}\frac{\mathrm{d}}{\mathrm{d}r}\left(\frac{E_r}{B_\theta R}\right) \tag{9.117}$$

exceeds the turbulence growth rate [90]. A simple estimate for the growth rate of drift waves is given by the ratio of sound velocity to major radius c_s/R [95].

Still unresolved is, how the radial electric field is generated that is responsible for the poloidal flows. The mechanism of the radial electric field generation is determined by the two fluid equations. For both electrons and ions the radial, poloidal and toroidal components of the momentum balance equations have the form

$$nm\frac{\mathrm{d}u_r}{\mathrm{d}t} = qn(E_r + u_\theta B_\varphi - u_\varphi B_\theta) - \nabla p, \tag{9.118}$$

$$nm\frac{\mathrm{d}u_\theta}{\mathrm{d}t} = -qnu_r B_\varphi - nm^{1/2}\hat{\mu}_{\|\theta}u_\theta, \tag{9.119}$$

$$nm\frac{\mathrm{d}u_\varphi}{\mathrm{d}t} = qnu_r B_\theta - nm^{1/2}\hat{\mu}_{\|\varphi}u_\varphi + F_\varphi. \tag{9.120}$$

The radial flow u_r is driven by the electrostatic force, the Lorentz force and the pressure gradient. Poloidal and toroidal flows are influenced by torques exerted by the Lorentz force due to a radial flow of charge and by viscous damping $\hat{\mu}$ due to the inhomogeneous magnetic field. In toroidal direction an external force F_φ can be applied through neutral beam heating. In absence of turbulence, these equations describe neoclassical electric field and transport [see (9.81) and (9.82)].

The radial momentum balances for electrons and ions determine the neoclassical radial electric current

$$J_r^{neo} \sim nq(u_r^i - u_r^e) \,. \tag{9.121}$$

But in general there exists a variety of other mechanisms to drive a radial current, which can be written as [96]

$$J_r = J_r^{neo} + J_r^{NBI} + J_r^{edge} + J_r^{v\nabla v} + J_r^{bias} \,. \tag{9.122}$$

For the different contributions see below. The total current then determines the radial electric field

$$\epsilon_\perp \frac{\partial E_r}{\partial t} = J_r \,, \tag{9.123}$$

which feeds back into the momentum balance equations where the flows have to adjust to the modified force balance. Hence the radial current is the key quantity to modify the poloidal flows. The evolution to a new force balance in (9.118)–(9.120), which have to be adjusted in order to fulfill the equilibrium condition characterised by $J_r = 0$, takes place on the fast Alfvénic time scale. If sheared flows are generated by this mechanism, they can, as discussed above, act to suppress turbulence. A transport reduction leads through the energy (9.12) to a change in the pressure gradient. This change takes place on the slow energy-confinement time scale and again modifies the force balance. On the left hand side of Fig. 9.35, this interaction is illustrated.

The complex interaction between electric field and transport discussed above creates a link between the section on neoclassical transport to the present one on turbulent transport. The same processes which cause under stationary conditions ambipolar neoclassical transport also create, in transient phases, neoclassical currents. According to (9.83), neoclassical transport is related to the parallel viscosity $\mu_\parallel$. Since $\mu_\parallel$ depends on collisionality, this bears a source of confinement transitions: if the collisionality varies in a plasma, e.g., due to varying temperature, $\mu_\parallel$ changes and brings the momentum equations in an unbalance. This drives a radial current and E_r adjusts. If this happens close to a threshold for turbulence suppression, a transition into a transport barrier can be triggered. This would be an example, where increased heating initiates a transition into improved confinement (see, e.g., [25]).

Stellarator plasmas have additional degrees of freedom. Ripple-trapped particles can cause bipolar flows. Furthermore, as discussed in Fig. 9.22, stellarator plasmas can generate a natural shear layer with a potential to suppress

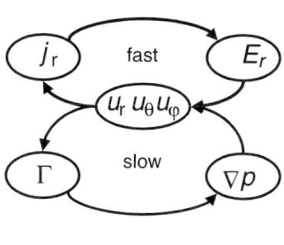 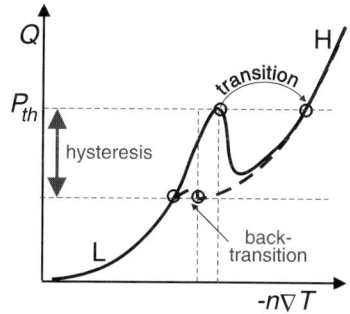

Fig. 9.35. *Left*: illustration of the interaction of the fast processes determining the radial electric field through the momentum balance (9.118)–(9.120) and the slow time scale, on which the pressure gradient can be changed according to (9.12) due to changes in turbulent transport. *Right*: Hysteresis in the energy flux as function of the temperature gradient in a transition from L to H-mode. The line represents the solution of the diffusion (9.4). By increasing the heating power and hence the power flux, the temperature gradient is determined by a point on the line which moves to the right. When the threshold power P_{th} is reached, a transition occurs with a steeper gradient at the same power flux, hence a reduced transport coefficient. The back transition is indicated by the *dashed line*

turbulence at the transition from electron to ion-root confinement. This feature could be at the origin of the internal transport barriers observed in CHS [32] and W7-AS [35].

According to (9.122), further possibilities to drive radial currents and to modify E_r exist:

1. The toroidal torque due to tangential neutral beam injection enters in F_φ in (9.120). It is accounted for by J_r^{NBI} and might have relevance for H-mode and internal transport barrier formation.
2. Ion losses at the plasma edge due to orbits colliding with the wall introduce J_r^{edge}. These losses are of importance for H-mode transitions.
3. Reynolds stress is a candidate for a turbulent drive of poloidal flows [97], which than also cause radial currents $J_r^{v\nabla v}$ and a modification of E_r. This could lead to spontaneous transport barrier formation and might possibly happen preferably in the vicinity of rational values of the rotational transform, where transport barriers were found in tokamaks [98] and stellarators [99].
4. Transitions can also be forced externally by radial currents J_r^{bias} drawn through a biased limiter or special probe. It has been shown on many devices, that transitions into the H-mode can be triggered this way [100, 85].

Finally, the right hand side of Fig. 9.35 depicts, as a general feature, the bifurcation of the plasma state at transitions into transport barriers [89, 101].

The signature is a hysteresis in the flux-gradient diagram. If, e.g., the heating power is increased during a plasma discharge, the temperature gradient and according to (9.4) the power flux increase. When a critical gradient is reached, the radial electric field determined from the mechanism just discussed suffices to suppress turbulence and a new steady state is reached, which due to a lower diffusivity has a steeper gradient at the same power flux. For the back transition obtained by a reduction in heating power, the condition is set from a different process. Now flow shear must be sufficient to maintain a quiescent state and not to suppress fluctuations in an already turbulent state. For this, less flow shear is sufficient and therefore, the transition occurs at a lower temperature gradient.

References

1. R.J. Goldston: Plasma Phys. Contr. Fusion **26**, 87 (1984)
2. S. Kaye, M. Greenwald, U. Stroth et al: Nucl. Fusion **37**, 1303 (1997)
3. ITER Physics Expert Groups on Confinement and Transport and Confinement Modelling and Database, ITER Physics Basis Editors: Nucl. Fusion **39**, 2175 (1999)
4. K. Lackner, N.A.O. Gottardi: Nucl. Fusion **30**, 767 (1990)
5. U. Stroth, M. Murakami, R.A. Dory et al: Nucl. Fusion **36**, 1063 (1996)
6. H. Yamada, S. Murakami, K. Yamazaki et al: Nucl. Fusion **43**, 749 (2003)
7. F. Wagner, U. Stroth: Plasma Phys. Contr. Fusion **35**, 1321 (1993)
8. R.J. Groebner, W. Pfeiffer, F.P. Blau et al: Nucl. Fusion **26**, 543 (1986)
9. U. Stroth, J. Baldzuhn, J. Geiger et al: Plasma Phys. Contr. Fusion **40**, 1551 (1998)
10. U. Stroth, T. Geist, J.P.T. Koponen et al: Phys. Rev. Lett. **82**, 928 (1999)
11. K.W. Gentle, O. Gehre, K. Krieger: Nucl. Fusion **32**, 217 (1992)
12. T.C. Luce, C.C. Petty: Nucl. Fusion **34**, 121 (1994)
13. S.P. Hirshman, W.I. van Rij: Comput. Phys. Commun. **43**, 143 (1987)
14. S. Chandrasekhar: Rev. Mod. Phys. **15**, 1 (1943)
15. B.B. Kadomtsev, O.P. Pogutse: Nucl. Fusion **11**, 67 (1971)
16. A.A. Galeev, R.Z. Sagdeev: Theory of neoclassical diffusion. In: *Reviews of Plasma Physics*, vol 7, ed by M.A. Leontovich (Consultants Bureau, New York 1979) pp 257–343
17. F.L. Hinton, R.D. Hazeltine: Rev. Mod. Phys. **48**, 239 (1976)
18. R. Balescu: *Neoclassical Transport*, vol 2 (North-Holland, Amsterdam, Netherlands 1988)
19. K.C. Shaing, S.A. Hokin: Phys. Fluids **26**, 2136 (1983)
20. C.D. Beidler, W.N.G. Hitchon, W.I. van Rij et al: Phys. Rev. Lett. **58**, 1745 (1987)
21. U. Stroth: Plasma Phys. Contr. Fusion **40**, 9 (1998)
22. H. Maaßberg, R. Burhenn, U. Gasparino et al: Phys. Fluids, B **5**, 3627 (1993)
23. M. Kick, H. Maaßberg, M. Anton et al: Plasma Phys. Contr. Fusion **41**, A549 (1999)
24. J.W. Connor, R.C. Grimm, R.J. Hastie: Nucl. Fusion **13**, 221 (1973)

25. K.C. Shaing, A.Y. Aydemir, W.A. Houlberg et al: Phys. Rev. Lett. **80**, 5353 (1998)
26. S.P. Hirshman, N.J. Sigmar: Nucl. Fusion **21**, 1071 (1981)
27. W.I. van Rij, S.P. Hirshman: Phys. Fluids B **1**, 563 (1989)
28. H. Maaßberg, R. Brakel, R. Burhenn et al: Plasma Phys. Contr. Fusion **35**, B319 (1993)
29. H. Ehmler, Y. Turkin, C.D. Beidler et al: Nucl. Fusion **43**, L11 (2003)
30. X.Z. Yang, B.Z. Zhang, A.J. Wootton et al: Phys. Fluids, B **3**, 3448 (1991)
31. A. Fujisawa, H. Iguchi, H. Sanuki et al: Phys. Rev. Lett. **79**, 1054 (1997)
32. A. Fujisawa, H. Iguchi, T. Minami et al: Phys. Rev. Lett. **82**, 2669 (1999)
33. R. Brakel, M. Anton, J. Baldzuhn et al: Plasma Phys. Contr. Fusion **39**, B273 (1997)
34. H. Maassberg, C.D. Beidler, U. Gasparino et al: Phys. Plasmas **7**, 295 (2000)
35. U. Stroth, K. Itoh, S.I. Itoh et al: Phys. Rev. Lett. **86**, 5910 (2001)
36. C.P. Ritz, R.V. Bravenec, P.M. Schoch et al: Phys. Rev. Lett. **62**, 1844 (1989)
37. W. Horton: Rev. Mod. Phys. **71**, 735 (1999)
38. X. Garbet: Plasma Phys. Contr. Fusion **43**, A251 (2001)
39. A. Yoshizawa, S.-I. Itoh, K. Itoh et al: Plasma Phys. Contr. Fusion **43**, R1 (2001)
40. L.D. Landau, E.M. Lifschitz: *Course of Theoretical Physics, Fluid Mechanics*, vol 6 (Pergamon Press, Oxford 1959)
41. U. Frisch: *Turbulence* (Cambridge University Press, New York 1995)
42. B. Pope: *Turbulent Flows* (Cambridge University Press, Cambridge 2000)
43. J. Maurer, P. Tabeling, G. Zocchi: Europhys. Lett. **26**, 31 (1994)
44. R.H. Kraichnan: Phys. Fluids **10**, 1417 (1967)
45. M.A. Rutgers: Phys. Rev. Lett. **81**, 2244 (1998)
46. B. Scott: Plasma Phys. Contr. Fusion **39**, 1635 (1997)
47. F. Chen: Phys. Fluids **8**, 912 (1965)
48. B. Scott: Plasma Phys. Contr. Fusion **39**, 471 (1997)
49. F. Romanelli: Phys. Fluids B **1**, 1018 (1989)
50. W. Horton, B.G. Hong, W.M. Tang: Phys. Fluids **31**, 2971 (1988)
51. B.B. Kadomtsev, O.P. Pogutse: Turbulence in toroidal systems. In: *Reviews of Plasma Physics*, vol 5, ed by M.A. Leontovich (Consultants Bureau, New York 1970) pp 249–400
52. J.F. Drake, Y.T. Lau, P.N. Guzdar et al: Phys. Rev. Lett. **77**, 494 (1996)
53. A. Hasegawa, M. Wakatani: Phys. Rev. Lett. **50**, 682 (1983)
54. A. Hasegawa, K. Mima: Phys. Fluids **21**, 87 (1978)
55. P.C. Liewer: Nucl. Fusion **25**, 543 (1985)
56. A.J. Wooton, B.A. Carreras, H. Matsumoto et al: Phys. Fluids B **2**, 2879 (1990)
57. R.J. Fonck, N. Bretz, G. Cosby et al: Plasma Phys. Contr. Fusion **34**, 1993 (1992)
58. L. Giannone, R. Balbín, H. Niedermeyer et al: Phys. Plasmas **1**, 3614 (1994)
59. C. Lechte, S. Niedner, U. Stroth: New J. Phys. **4**, 34 (2002)
60. M. Endler, H. Niedermeyer, L. Giannone et al: Nucl. Fusion **35**, 1307 (1995)
61. S.J. Zweben, D.P. Stotler, J.L. Terry et al: Phys. Plasmas **9**, 1981 (2002)
62. A. Rudyi, R.D. Bengtson, A. Carlson et al: Investigation of Low-Frequency Fluctuations in the Edge Plasma of ASDEX. In: *Proc. of the 16th EPS Conference on Controll. Fusion and Plasma Phys., Venice, Italy, 13.–17. March 1989*, vol. 13b, part 1, ed by S. Segre, H. Knoepfel, E. Sindoni (The European Physical Society, Geneva, 1989) pp 27–30

63. H. Thomsen, M. Endler, J. Bleuel et al: Phys. Plasmas **9**, 1233 (2002)
64. J. Bleuel, M. Endler, H. Niedermeyer et al: New J. Phys. **4**, 38 (2002)
65. O. Grulke, T. Klinger, M. Endler et al: Phys. Plasmas **8**, 5171 (2001)
66. S. Niedner, B.D. Scott, U. Stroth: Plasma Phys. Contr. Fusion **44**, 397 (2002)
67. M. Pedrosa, M.A. Ochando, J.A. Jiménez et al: Plasma Phys. Contr. Fusion **38**, 365 (1995)
68. B.A. Carreras, R. Balbin, B. van Milligen et al: Phys. Plasmas **6**, 4615 (1999)
69. C. Hidalgo: Plasma Phys. Contr. Fusion **37**, A53 (1995)
70. B.A. Carreras, B.P. van Milligen, M.A. Pedrosa et al: Phys. Plasmas **5**, 3632 (1998)
71. N. Mahdizadeh, M. Ramisch, U. Stroth et al: Phys. Plasmas **11**, 3932 (2004)
72. B.A. Carreras, C. Hidalgo, E. Sánchez et al: Phys. Plasmas **3**, 2664 (1996)
73. P.H. Diamond and T.S. Hahm: Phys. Plasmas **2**, 3640 (1995)
74. P. Bak, C. Tand, K. Wiesenfeld: Phys. Rev. A **38**, 364 (1988)
75. S.J. Zweben, R.W. Gould: Nucl. Fusion **25**, 171 (1985)
76. A. Latten, T. Klinger, A. Piel et al: Rev. Sci. Instrum. **66**, 3254 (1995)
77. T. Klinger, A. Latten, A. Piel et al: Plasma Phys. Contr. Fusion **39**, B145 (1997)
78. C. Lechte: Microscopic Structure of Plasma Turbulence in the Torsatron TJ-K. Ph.D. thesis, Christian-Albrechts-Universität, Kiel (2003)
79. U. Stroth, F. Greiner, C. Lechte et al: Phys. Plasmas **11**, 2558 (2004)
80. G. Fiksel, S.C. Prager, P. Pribyl et al: Phys. Rev. Lett. **75**, 3866 (1995)
81. C. Laviron, F. Clairet, L. Colas, L. et al: Local Analysis of Transport and Turbulence in Tore Supra. In: *Proc. of the 16th IAEA Conference on Plasma Physics and Controlled Nuclear Fusion Research, Montreal, Canada, 7.–11. October 1996*, vol 1, IAEA–CN–64/A6–3 (IAEA, Vienna, 1997) pp 535–546
82. L. Colas, X.L. Zou, M. Paume et al: Nucl. Fusion **38**, 903 (1998)
83. F. Wagner, G. Becker, K. Behringer et al: Phys. Rev. Lett. **49**, 1408 (1982)
84. K.H. Burrell: Phys. Plasmas **4**, 1499 (1997)
85. R.R. Weynants, S. Jachmich, G. Van Oost: Plasma Phys. Contr. Fusion **40**, 635 (1998)
86. F. Wagner, F. Ryter, A.R. Field et al: Recent Results of H-Mode Studies on ASDEX. In: *13th IAEA Conference on Plasma Physics and Controlled Nuclear Fusion Research, Washington, DC, 1.–6. October 1990*, vol 1, IAEA–CN–53/A–IV–2 (IAEA, Vienna 1991) pp 277–290
87. R.C. Wolf: Plasma Phys. Contr. Fusion **45**, R1 (2003)
88. J.W. Connor, T. Fukuda, X. Garbet et al: Nucl. Fusion **44**, R1 (2004)
89. S.I. Itoh, K. Itoh: Phys. Rev. Lett. **60**, 2276 (1988)
90. T.S. Hahm, K.H. Burrell: Phys. Plasmas **2**, 1648 (1995)
91. K. Itoh, S.I. Itoh: Plasma Phys. Contr. Fusion **38**, 1 (1996)
92. Z. Lin, T.S. Hahm, W.W. Lee et al: Science **281**, 1835 (1998)
93. K.H. Burrell: Phys. Plasmas **6**, 4418 (1999)
94. P.W. Terry: Rev. Mod. Phys. **72**, 109 (2000)
95. J.W. Connor, H.R. Wilson: Plasma Phys. Contr. Fusion **42**, R1 (2000)
96. K. Itoh, S.I. Itoh, A. Fukuyama: *Transport and Structural Formation in Plasmas*, chapt 12 (Insitute of Physics Publishing, London 1999)
97. P.H. Diamond, Y.P. Kim: Phys. Fluids B **3**, 1626 (1991)
98. N.J. Lopes Cardozo, G.M.D. Hogeweij, M. de Baar et al: Plasma Phys. Contr. Fusion **39**, B303 (1997)

99. C. Hidalgo, M.A. Pedrosa, E. Sánchez et al: Plasma Phys. Contr. Fusion **42**, A153 (2000)
100. R.J. Taylor, M.L. Brown, B.D. Fried et al: Phys. Rev. Lett. **63**, 2365 (1989)
101. N. Kasuya, K. Itoh, Y. Takase: *Bifurcation Phenomena in Plasmas*, ed by S.I. Itoh and Y. Kawai (Kyusho University, Fukuoka, Japan 2002)

10 Non-Neutral Plasmas and Collective Phenomena in Ion Traps

G. Werth

Abstract. Single components non-neutral plasmas confined in static or radio-frequency traps show oscillations and instabilities as basic manifestations of collective effects. They are discussed for the weak and strong coupling regime. This chapter covers introductory aspects to chapters on applications and is related to the chapter on Strongly Coupled Plasmas (Chap. 6, Part I) as well as on Collective Effects in Dusty Plasmas (Chap. 11, Part II).

10.1 Introduction

A cloud of ions or electrons represents an example of a single component non-neutral plasma. Such plasmas can be confined by simple arrangements of electric and magnetic fields, known as ion traps. Many examples exist where confinement times of days or even month are routinely achieved. This allows to observe properties of such plasmas which are not easily available with neutral plasmas. Non-neutral plasmas have proven to be excellent subjects for a wide range of plasma phenomena. Particularly it is possible to vary the temperature and density of such a plasma over a wide range. Transport phenomena, plasma oscillations, or plasma instabilities can be investigated under different conditions. The temperature of a plasma can be reduced to such low values that strong correlations between the trapped particles occur allowing the formation of Coulomb crystals.

In this contribution we will first sketch the basic properties of ion traps, operated with static or time-varying electromagnetic fields. We restrict ourselves to so-called Paul and Penning traps since they are most commonly used for low energy plasma experiments. We then turn to the discussion of some phenomena which are characteristic for non-neutral plasmas.

10.1.1 Basics of Ion Traps

Confinement of charged particles by electromagnetic fields requires a force acting on the particles pointing into the direction of the trap center in all three dimensions. According to Ehrenfest's theorem it is not possible to create a three dimensional potential minimum by electrostatic fields. The solution of this problem is found by either applying a time varying electric field for confinement, leading to what is commonly called the "Paul trap", or superimposing a magnetic field to the static electric potential. This is called the "Penning trap". A large variety of electrode configurations have been used to create such traps. Analytical description of the motional properties of charged particles is most easily found when the confining forces depend linearly on the distance of the particle from the trap center. The trap potential Φ then depends on the square of the coordinates. Taking into account Laplace equation $\Delta\Phi = 0$ and allowing for rotational symmetry of the trap we arrive at the quadrupolar form of the potential

$$\Phi = \Phi_0 \left(r^2 - z^2/2\right), \qquad (10.1)$$

where r and z are the radial and axial coordinates. At the surface of electrodes which create such a potential by application of a voltage the potential is constant. This leads to the classical form of ion traps with a ring electrode and two electrically connected end-caps of hyperbolic shape as shown in Fig. 10.1. The properties of these traps have been extensively described in the literature [1, 2]. Although this type of traps has been used successfully in numerous experiments it has become more common in recent years to replace the classical trap by simpler structures which are more easy to machine and to align as will be outlined in more detail below. These traps, of course, do not produce a pure quadrupole potential. It can, however, be well approximated: A series expansion of any potential in a rotational symmetric electrode arrangement starts near its minimum with a quadratic dependence on the distance from this point. When a particle moves in an area close to the potential minimum the higher order contributions to the potential are small and can in many cases be neglected.

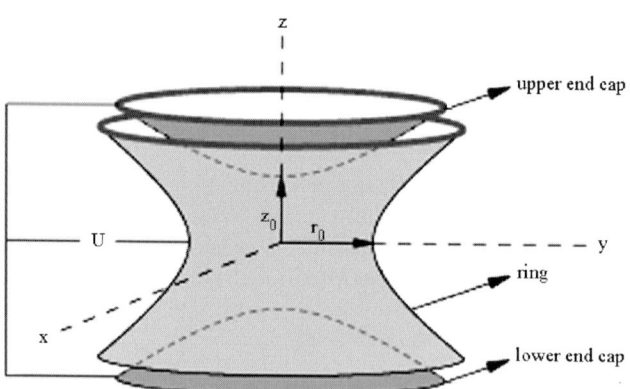

Fig. 10.1. Standard electrode configuration of ion traps. For Paul traps we apply $U = U_{dc} + V\cos(\Omega t)$, for Penning traps $U = U_{dc}$ and $B = B_z$

Paul Traps

Paul traps apply an alternating voltage $U = U_0 \cos(\Omega t)$ between the trap electrodes. A dc voltage U_0 may be added. A particle moving in this field experiences a force, whose time average is finite because of the inhomogeneity of the field [3]. It is directed toward the region of weaker field and leads to a three dimensional time-averaged potential minimum which serves for confinement. By symmetry reasons the electric field is zero at the center of the trap. A particle moving outside the center always experience the force of the electric field and thus oscillates at its frequency Ω. The energy in this so-called micro-motion can be substantial, particularly when storing large clouds of charged particles where many of the ions move in a region far away from the trap center. It represents a significant obstacle when the temperature of the ion cloud shall be reduced by some cooling mechanism and so-called rf heating processes limit the attainable temperature.

A variant of the classical Paul trap reduces this difficulty to a certain degree: The linear Paul trap consists of four metallic rods arranged as shown in Fig. 10.2. An rf voltage and a superimposed dc voltage are applied between adjacent rods in the same way as known from the linear mass-filter. The confinement in the plane perpendicular to the rods follows the same dynamical procedure as described for the classical Paul trap. Along the axis the ions are enclosed by a static voltage applied at additional electrodes at the end of the rods or by use of segmented rods as in Fig. 10.2. The advantage of this trap is that a zero field region appears along the axis of the trap. Therefore many more ions can be placed in a low field region compared to the classical Paul trap and rf heating processes are less severe. Other advantages are that the structure is very simple and easy to produce, and that the open space between the electrodes allows for easy optical access to a stored ion cloud.

The potential of the linear Paul trap in the plane perpendicular to the rod direction (x-y plane) is given by

$$\Phi = \frac{U(t)}{2r_0^2}\left(x^2 - y^2\right), \qquad (10.2)$$

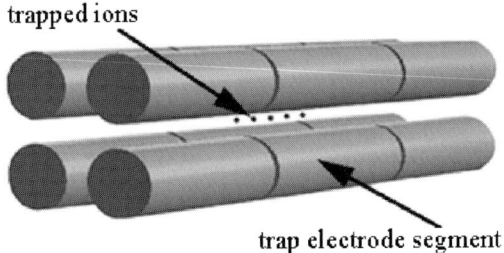

Fig. 10.2. Geometry of a linear Paul trap

where $U(t) = U_0 + V_0 \cos(\Omega t)$. r_0 is the distance of the rods from the center. The equations of motion for a single particle of charge e and mass m in the x-y plane are then

$$\ddot{u} = \frac{e}{mr_0^2}[U_0 + V_0 \cos(\Omega t)]\, u \tag{10.3}$$

$u = x, y$. Using the dimensionless parameters $a = -(4eU_0)/(mr_0^2\Omega^2)$, $q = (2eV_0)/(mr_0^2\Omega^2)$ and $\tau = 1/2\,\Omega t$ one obtains the Mathieu differential equations

$$\frac{d^2x}{d\tau^2} + [-a + 2q\cos(2\tau)]\,x = 0, \tag{10.4}$$

$$\frac{d^2y}{d\tau^2} + [+a - 2q\cos(2\tau)]\,y = 0. \tag{10.5}$$

Depending on the size of the parameters a, q the solution of these equation leads to trajectories which are finite in time or go to infinity. Figure 10.3 sketches the set of parameters for stable and unstable solutions. For stable solutions the ion motion can be described as a sum of harmonic oscillators in each direction. For small values of the stability parameters a and q it can well be approximated by

$$u(t) = u_0\left[1 + \frac{q}{2}\cos(\Omega t)\right]\cos(\omega_u t). \tag{10.6}$$

This is a harmonic oscillation of a slow frequency ω_u (macro-motion) given by

$$\omega_u = \frac{\Omega}{2}\left(\frac{q^2}{2} + a\right)^{1/2}, \tag{10.7}$$

whose amplitude u_0 is modulated by the driving frequency Ω (micro-motion).

The oscillation of the micro-motion can be considered as arising from a harmonic pseudo-potential. The depth D of this potential is given by

$$D_x = \frac{1}{2}mr_0^2\omega_x^2 = \frac{e^2 V_0^2}{4m\,r_0^2\,\Omega^2} + \frac{eU_0}{2}, \tag{10.8}$$

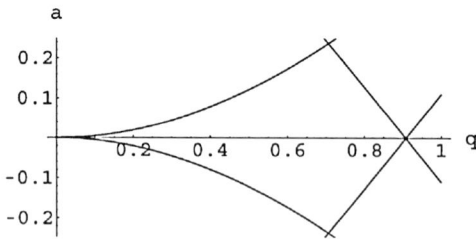

Fig. 10.3. Stability diagram of a linear Paul trap

$$D_y = \frac{1}{2}mr_0^2\omega_y^2 = \frac{e^2 V_0^2}{4m\, r_0^2\, \Omega^2} - \frac{eU_0}{2}. \tag{10.9}$$

Typical values of D for a trap radius of 1 cm, an ion mass of 100 atomic units, a driving frequency of $\Omega/2\pi = 1\,\mathrm{MHz}$ and $V_0 = 1000\,\mathrm{V}$, $U_0 = 10\,\mathrm{V}$ are several tens of eV.

Equations (10.8) and (10.9) hold for a single particle. When clouds of ions are stored the potential is modified by the Coulomb interaction between the ions. Then, of course, the motion is no longer analytically solvable. For low densities and high temperatures, however, the motion can still be considered as that of a single particle with small changes caused by the additional potential. The potential will be no longer harmonic resulting in a shift of the ions oscillation frequencies. These shifts can be estimated by space charge models, considering, e.g., the ions as a cloud with a Maxwellian distribution of charge density [4] as confirmed experimentally [5, 6]. The corresponding fractional frequency shifts can be as large as 50% at charge densities of $10^6\,\mathrm{cm}^{-3}$. This is a typical density for the space charge limit when the traps potential depth is of the order of several eV. The average kinetic energy of the ion cloud depends on the trapping parameters and can be estimated to about 1/10 of the pseudo-potential depth.

Penning Traps

Penning traps use a static voltage U_0 applied to the trap electrodes. The polarity is chosen such that a potential minimum occurs in the axial direction in which particle can be stored. Radial confinement is provided by a superimposed magnetic field along the symmetry axis. The potential well in the axial direction leads to a harmonic oscillation with frequency

$$\omega_z = \left(\frac{eU_0}{2mr_0^2}\right)^{1/2}, \tag{10.10}$$

where $z_0^2 = (r_0^2)/2$ has been assumed.

Without the presence of the electric quadrupole field an ion would perform a cyclotron oscillation in the magnetic field B with frequency

$$\omega_c = \frac{e}{m}B. \tag{10.11}$$

The defocusing force of the electric field modifies this motion: The sum of the electric force F_E and the Lorentz force F_L is balanced by the centrifugal force:

$$-m\frac{v^2}{\rho} = -evB + e\frac{U_0}{2r_0^2}\rho, \tag{10.12}$$

where v is the ions velocity and ρ the radius of the cyclotron orbit. With $v = \omega\rho$ this can be written as

$$\omega^2 = \frac{e}{m}B\omega - \frac{e}{m}\frac{U_0}{2r_0^2} = \omega_c\omega - \frac{\omega_z^2}{2}. \quad (10.13)$$

This equation has two solutions for ω:

$$\omega_\pm = \frac{\omega_c}{2} \pm \left(\frac{\omega_c^2}{4} - \frac{\omega_z^2}{2}\right)^{1/2} \quad (10.14)$$

which are the eigenfrequencies in the radial plane.

In order to produce a quadrupole potential the shape of the classical Penning trap electrodes are hyperboloids of revolution as in the case of Paul traps. It has become common practice, however, to use a cylindrical trap with open end-caps, because they are much easier to machine and to align and they offer easy access to the inside (Fig. 10.4).

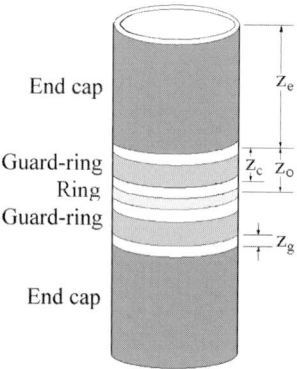

Fig. 10.4. Open end-cap Penning trap

The potential now, of course, is no longer harmonic. A convenient way to deal with this geometry is by series expansion of the potential in terms of Legendre polynomials $\mathcal{P}_n$:

$$\Phi = U_0 \sum_{n=2}^\infty c_n \left(\frac{\varrho}{r_0}\right)^n \mathcal{P}_n\left(\frac{z}{\rho}\right), \quad (10.15)$$

where $\rho = (x^2 + y^2 + z^2)^{1/2}$. For small values of ϱ the higher order components in the expansion are small and the potential can still be considered as a quadrupole potential. The existence of higher order parts in the trapping potential shifts the stored particle's eigenfrequencies which in addition depend on the oscillation amplitudes. Moreover it leads to instabilities in the ion motion at operating points which otherwise would be stable. This will be discussed in detail in a later section.

Ion Loading and Detection

Loading of ions into the trap is most easily performed when the ions are created at low energies inside the trap. The standard way is to ionize an neutral atom beam passing through the trap by electron impact. More recently photo-ionization of neutral atoms has been used to create ions [7, 8]. This has the advantage that the number of created ions can be well controlled and only one specific element or even isotope can be ionized, while electron-ionization is less specific and bears the risk of creating unwanted ions from neutral background molecules.

Injection of ions from outside requires some kind of friction. This can most easily provided by collisions with neutral background atoms. As a rule of thumb the density of the gas should be such that the mean free path between the ion–neutral collisions is of the order of the trap size. This leads to pressures in the 10^{-4}–10^{-5} hPa range as experimentally verified [9]. Alternatively friction can be provided by a laser beam counter-propagating the injected ions and Doppler tuned to a resonant wavelength of the ion.

A dynamical way of loading ions from outside has been developed for Penning traps: When ions are travelling along the magnetic field lines, the first end-cap electrodes is held on ground potential while the second one is on a high potential. When the ions energy is lower than the seconds end-cap potential they are reflected. During the transit time of the ions through the trap, typically a few microseconds, the first end-cap is switched on and the ions are trapped. When some cooling mechanism for the trapped ions is applied the trap can be opened again to some extend, the procedure can be repeated and ions can be stacked.

For ion detection several methods are available. Most easily the ions are dumped to a particle detector outside the trap. Apart from the fact that this is a destructive technique it gives little information on internal properties of a stored ion cloud. Nondestructive detection techniques relay on the induced image charges of the oscillating ion cloud in the trap electrodes. Pick-up of the induced voltages and a Fourier transform gives information on the position and amplitude of these frequencies. The most detailed information is obtained by Doppler imaging: When the trapped ions have a level scheme suited for laser excitation the fluorescence photons from the decay of the excited level can be detected by a position sensitive device such as a CCD camera. When the laser spectral bandwidth is narrower than the Doppler width of the optical transition, the ion velocity profile can be obtained by scanning the laser across the Doppler width: Different velocity classes appearing at different positions in the trap are selectively excited and can be spatially resolved. An example is given in Fig. 10.5 where the velocity distribution of a stored Ca^+ ion cloud in a Paul trap is measured: For large de-tuning of the laser from the ionic resonance only fast ions at the edge of the ion cloud are Doppler shifted into resonance with the laser and emit fluorescence light. Slower ions near the trap center become visible only for small laser de-tuning. The result of

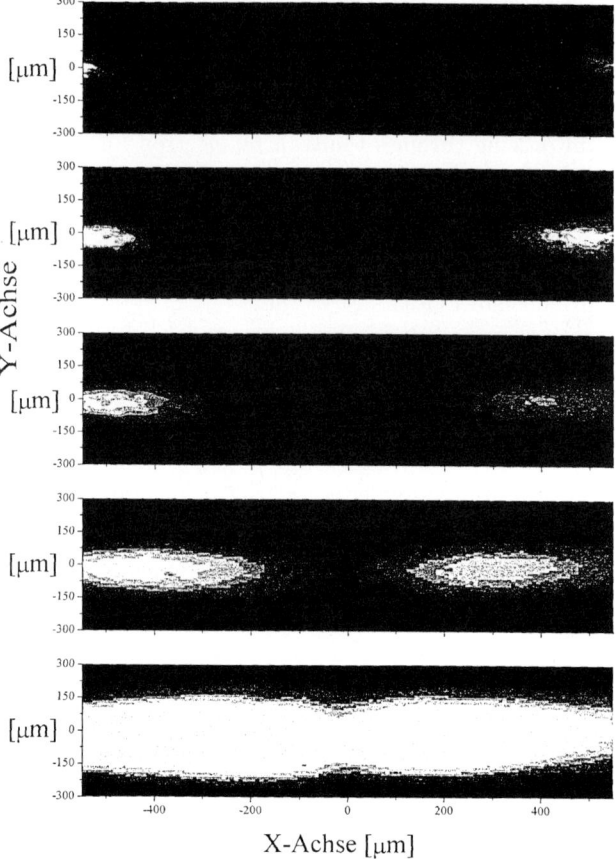

Fig. 10.5. Spatially resolved fluorescence from a stored Ca$^+$ ion cloud in a Paul trap for different de-tunings of the exciting laser from resonance. For large de-tuning (*top*) only hot ions at the edge of the cloud are excited because they are Doppler shifted into resonance. When the de-tuning is decreased ions closer to the trap center are excited and emit fluorescence. The laser beam is in the horizontal direction, the trap's axis perpendicular to the plane

these measurement is that the mean ion velocity increases linearly with the distance from the trap center.

Laser Doppler Cooling of Ions

As mentioned above a stored ion cloud in thermal equilibrium has a temperature which depends on the operating parameters of the trap and the initial conditions and is under typical conditions of the order of 10 000 K. An important parameter to characterize the behavior of a stored ion or electron cloud is the coupling parameter Γ_c, the ratio of the Coulomb interaction energy to the thermal energy (see also Chap. 6 in Part I and Chap. 11 in Part II):

$$\Gamma_c = \frac{1}{4\pi\varepsilon_0} \frac{e^2}{a\, k_B T}\,. \tag{10.16}$$

Here k_B is the Boltzmann constant. The Wigner–Seitz radius a, defined as $(3/4\pi n)^{1/3}$, where n denotes the ion density, is essentially the inter-particle spacing. For $\Gamma_c \gg 1$ the ion trajectories can to a good approximation be considered as those of single particles with small deviations caused by the space charge potential. When, however the ion temperature is reduced, Γ_c may become much larger than one and correlation effects become dominant as will be discussed later.

Among the different ways to reduce the ion temperature by far the most effective one is laser cooling. If the stored ions have a level scheme which allows laser excitation from the electronic ground state to an excited state, a laser tuned to the low frequency side of the resonance transition excites only those ions which are Doppler shifted into resonance when they move in the direction towards the laser beam. The recoil momentum from the absorbed photon reduces the velocity component in this direction. The net momentum transfer after re-emission of the photon is zero since the emission is isotropic when averaged over many absorption-emission cycles. When the resonance transition is a strong electric dipole transition, as in the case of alkali-like ions, an absorption-emission cycle lasts typically 10^{-7} s. The velocity reduction per cycle is

$$\Delta v = \frac{\hbar k}{m}, \tag{10.17}$$

where k is the photons wave number and m the ions mass. When the laser is operated at saturation intensity this is sufficient to reduce the ions velocity to almost at rest in a time of several ms, depending on the wavelength of the radiation and the mass of the ion. The velocity components in all three spatial components are reduced at the same time when the laser has a tilt angle with respect to the trap axis or by Coulomb interaction in an ion cloud. Also Coulomb collisions serve for distribution of the reduced velocity in all directions. The final temperature obtainable by this method ("Doppler cooling") is reached when the Doppler profile of the resonance absorption equals the natural linewidth, given by the lifetime τ of the excited state:

$$k_B T = \frac{\hbar}{2\tau}\,. \tag{10.18}$$

For typical cases $T \approx 1\,\text{mK}$. Details of the Doppler cooling process in traps can be found in [10].

Only few ions fulfill the requirement of a two level system suited for laser cooling on strong electric dipole transitions. Most noticeable are Be^+ and Mg^+ which are most often used for experiments with cooled ions. Other ions of the earth-alkalines such as Ca^+, Sr^+, or Ba^+ have low lying long lived metastable states into which the laser excited state may decay, removing the ion for a long time from the cooling cycle. This can be overcome by an additional laser which re-pumps the ions from the metastable state into the cooling cycle, thus creating an effective two-level system.

When a specific ion of interest can not be cooled directly by laser radiation it may be cooled indirectly by Coulomb interaction with laser cooled ions which are simultaneously confined ("sympathetic cooling"). The final temperature is similar to the one in case of direct cooling.

It should be noted that the effectivity of cooling is different for Paul- and Penning-traps. Since the potential of a Penning trap is static and no heating effects occur it is rather simple to cool large cloud of ions to low temperatures. Care has to be taken for the magnetron motion since a reduction in magnetron energy leads to an increase in radius. By proper spatial and spectral laser tuning and by coupling the magnetron oscillation to other modes this problem can be overcome[11, 12]. In Paul traps only the energy on the macro-motion can be reduced by laser cooling since the micro-motion is constantly driven by the time varying electric trapping field. Of course, since the ion cloud diameter shrinks upon macro-motion cooling, also the micro-motion amplitude will be reduced when the ion move on the average in a smaller electric field. The final total temperature, however, depends on the number of ions. Because of the rf heating effects caused by phase changing Coulomb collisions in the Paul-trap it is much more difficult to cool large ion clouds as compared to Penning-traps and generally higher laser powers are required.

Apart from Doppler cooling other techniques of energy reduction exist such as resistive, collisional, or evaporative cooling. Itano et al. [13] have given an overview of different methods. They are, however, not as effective as laser cooling and play no significant role in plasma physics in traps.

10.2 Ion Cloud as Non-Neutral Plasma

Specific plasma properties in an ion cloud of temperature T, density n, and charge e show up when a characteristic dimension, the Debye length

$$\lambda_D = \left(\frac{k_B T}{4\pi n e^2}\right)^{1/2} \tag{10.19}$$

is small compared to the dimensions of the cloud. This is in general the case when we deal with ion clouds which fill a significant fraction of the trap volume. For an ion cloud at a temperature of 300 K and a density of the order of 10^6 cm^{-3}, λ_D is about 0.5 millimeters while the cloud dimensions usually are several mm. When laser cooling or any other method of energy reduction is applied plasma effects show up even at smaller ion numbers.

As mentioned above an important parameter to characterize a plasma is the plasma parameter Γ_c which relates the Coulomb interaction energy between adjacent ions to the thermal energy. Γ_c is a measure of the coupling strength. For $\Gamma_c \ll 1$ the plasma is called weakly coupled and shows a cloud-like behavior, while molecular dynamics calculations suggest that for $\Gamma_c > 175$ the correlation is sufficiently strong to allow the formation of Coulomb crystals. This obviously requires very low temperatures which may be achieved by laser cooling.

Here we will first describe some observations in weakly coupled plasmas which resemble in many aspects those observed in neutral plasmas. We then turn to the strong correlation regime where crystal formation is the main issue. General properties of weakly coupled non-neutral plasmas are extensively discussed in the literature [14, 15, 16].

10.3 Weakly Coupled Non-Neutral Plasmas

A cloud of charged particles of mass m and charge e confined in a trap of rotational symmetry around the z-axis arranges itself in thermal equilibrium as a ellipsoid having approximately a constant charge density. The aspect ratio $\alpha = a_p/b_p$ of the ellipsoid, where a_p and b_p are the major and minor axes, is given by the expression

$$\omega_z^2 = \omega_p^2 \frac{Q_1^0(\beta_L)}{\alpha^2 - 1}, \qquad (10.20)$$

where ω_z is the ions' oscillation frequency along the z-axis in the traps potential and ω_p is the plasma frequency

$$\omega_p = \left(\frac{4\pi n e^2}{m}\right)^{1/2}. \qquad (10.21)$$

$Q_1^0(\beta_L)$ is the associated Legendre function of the second kind, and $\beta_L = \alpha/(\alpha^2 - 1)^{1/2}$. Figure 10.6 shows an example of a stored Be$^+$ ion cloud in a Penning trap containing about 10^4 ions where laser induced fluorescence serves to monitor the ion density [17].

There is a fundamental limit in single component non-neutral plasmas of the density of particles that can be magnetically confined, the Brillouin density [18]:

$$n_{max} = \frac{\varepsilon_0 \, m \, \omega_c^2}{2e^2}. \qquad (10.22)$$

It corresponds to a plasma which rotates without shear at the frequency $\omega_c/2$ around the trap axis where $\omega_c = (q/m)B$ is the particles cyclotron frequency. Then the Lorentz force acting on the ions is compensated by the force from the space charge potential. In a frame of reference rotating with the plasma, according to Larmor's theorem, the magnetic field vanishes and the particles pursue straight line orbits in the interior of the plasma. The effect of a magnetic field on the properties of a plasma is often characterized by a parameter

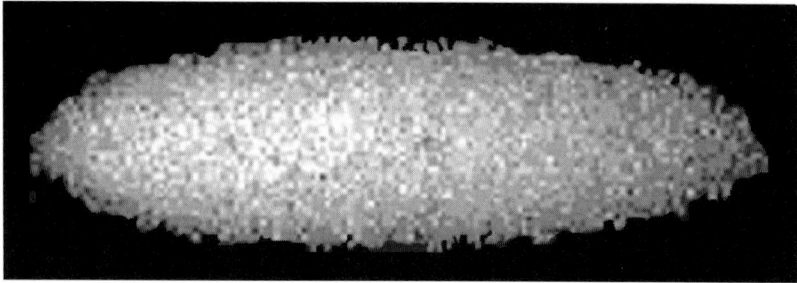

Fig. 10.6. Sideview image obtained by laser induced fluorescence from a Be$^+$ plasma in thermal equilibrium confined in a Penning trap. The magnetic field and the trap axis are in the horizontal direction [17]

$S = 2\omega_p^2/\omega_c^2$. The Brillouin condition can then be written as $S = 1$, defining what is termed Brillouin flow in a plasma. In a field $B = 0.1\,\mathrm{T}$ the maximum density of singly ionized atomic particles of mass around 50 is approximately 5×10^6 ions/cm^3.

10.3.1 Plasma Oscillations

The result of small displacements of the ions from their positions of equilibrium is a subsequent motion dominated by the coupling between them. This requires a description in terms of normal modes of oscillations of the system as a whole, rather than in terms of individual ions. A plasma ellipsoid confined in an ion trap can sustain oscillations in many different types of modes; a thorough treatment of linear normal modes is given by Bollinger et al. [19]. The normal modes are distinguished as having all particles move with the same frequency (but of course not the same phase). The mode in which all ions move together in the same phase is the *center-of-mass mode*, whose axial frequency ω_z is the same as a single particle would have in the trap. The mode in which the displacement of each ion is proportional to the distance of its equilibrium position from the trap center is called the *breathing* mode at $\omega = \sqrt{3}\omega_z$. A plasma containing N ions has of course $3N$ normal modes of oscillation. In the special case of $N = 2$ ions having equilibrium positions along the z–axis, but free to make small oscillations in three dimensions, there are six degrees of freedom: the mode frequencies are $\omega_x, \omega_y, \omega_z$ for the center of mass along the three coordinate axes, $\sqrt{3}\omega_z$ "breathing" mode along the z–axis, $(\omega_y^2 - \omega_z^2)^{1/2}$ and $(\omega_x^2 - \omega_z^2)^{1/2}$ "rocking" in the $y - z$ and $x - z$ planes. Other modes include azimuthally propagating waves (*diocotron waves*) close to harmonics of the plasma rotation frequency arising from shears in the rotational drift velocity.

If we approximate the boundary of the confined plasma by an infinite circular cylinder of radius a, the oscillating potential modes, as derived by

Trivelpiece and Gould have the form [20]

$$\varphi = A\, J_m\left(\frac{p_{mn} r}{a}\right) \exp\left[i\left(m\theta + k_z z - \omega t\right)\right], \tag{10.23}$$

with eigenfrequencies given by

$$k_z a = \pm p_{mn} \left[\frac{\omega^2(\omega^2 - \omega_p^2 - \omega_c^2)}{(\omega^2 - \omega_p^2)(\omega^2 - \omega_c^2)}\right]^{1/2}. \tag{10.24}$$

A is the amplitude of the potential, p_{mn} the n-th root of the m-th order Bessel function J_m, and k_z is the axial wave number. The modes are characterized by two integers, (m, n) with $m \geq 1$ and $n \geq 0$. The index m denotes an azimuthal dependence $\exp(i\, m\theta)$ and the index n describes the variation along the spheroidal surface. Figure 10.7 sketches some of the low order oscillation modes. The $(1, 0)$ mode represents a dipole oscillation of the plasma ellipsoid [Fig. 10.7 (a)], the $(2, 0)$ mode is a quadrupole oscillation where the plasma density remains uniform but the aspect ratio oscillates in time [Fig. 10.7 (b)], in the $(2, 1)$ mode the spheroid has a tilt angle with respect to the trap axis (produced, e.g., by trap misalignments) and precesses around the axis [Fig. 10.7 (c)]. In the $(2, 2)$ mode [Fig. 10.7 (d)] the plasma is distorted in the radial plane forming a triaxial ellipsoid which rotates around the z-axis.

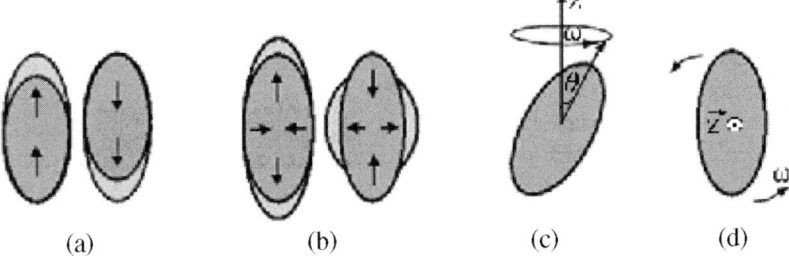

Fig. 10.7. Some low order modes of plasma oscillations of non-neutral plasmas in a trap. (**a**) Dipole oscillation, (**b**) quadrupole oscillation, (**c**) tilt rotation, (**d**) asymmetry rotation

Dubin [21] has published an non-perturbative analytical solution to the problem of the eigenmodes of a cold finite spheroidal plasma. The assumptions are made that both λ_D and $n_0^{-1/3}$ are small compared with the size of the plasma, and that the effects of a finite temperature and correlation are negligible. Then the plasma may be regarded as a cold–fluid ellipsoid of revolution of uniform density.

Studies of modes in confined ions in a quadrupole Penning trap at the Brillouin limit [22] have confirmed that the azimuthally propagating mode frequencies agree with the predictions of a simple fluid model. Plasma modes have also been studied by a Fourier analysis of the induced voltage from the oscillating plasma in the walls of the trap [23], and by laser Doppler imaging using phase coherent detection [17]. The Doppler images provide a direct measurement of the mode's axial velocity (Fig. 10.8), while also yielding an accurate measurement of the mode eigenfrequency.

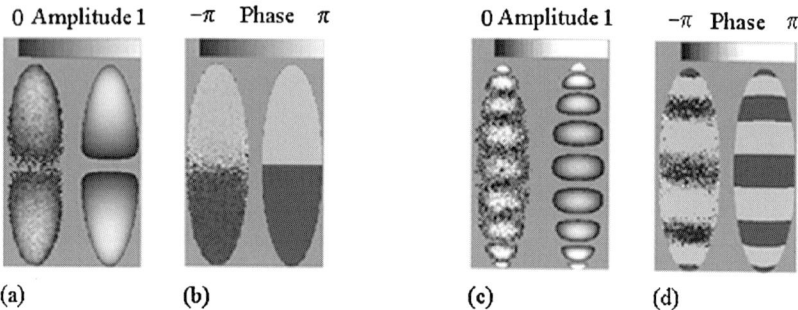

Fig. 10.8. Experimental (*left*) and simulated (*right*) amplitude distribution analyzed into amplitudes of (2,0) mode (**a**) and (9,0) mode (**c**); Experimental (*left*) and simulated (*right*) phase distribution (**b**) and (**d**) in terms of the same modes [17]

10.3.2 Rotating Walls

The crossed electric and magnetic field of a Penning trap causes a charged particle to rotate around the trap center at the magnetron frequency. In a plasma cloud the space charge potential which adds to the trapping potential causes a change in the rotation frequency. Additionally the cloud may be forced to rotate at different frequencies by an external torque. This can arise from momentum transfer by resonant laser radiation when the laser beam crosses the ion cloud outside the trap center in the radial plane [24, 25, 26]. Obviously it can be applied only to ions with atomic transitions accessible with available light sources.

A different way of exerting a torque on the plasma is to apply an electric field rotating around the symmetry axis as introduced in 1997 at the University of California in San Diego [27]. The ring electrode of a Penning trap is split into different segments. A sinusoidal voltage V_{wj} of frequency ω_w is applied to the segments at $\theta_j = 2\pi j/k$ when k is the number of segments, with $V_{wj} = A_w \cos[m(\theta_j - \omega_w t)]$. Figure 10.9 illustrates an 8-segment configuration.

The plasma rotation frequency caused by the torque is in general somewhat less than ω_w. The slip decreases with increasing amplitude A_w and

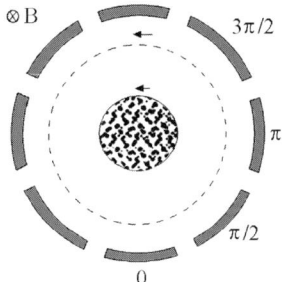

Fig. 10.9. Schematic of $m = 2$ rotating field wall sector signals with rf phases applied to 8 segments of a cylindrical trap electrode [27]

scales approximately with the square root of the plasma temperature [28]. For steady state confinement the additional centrifugal force from the plasma rotation leads to a change in plasma density and spatial profile. Thus variation of the rotating wall frequency allows to compress or expand the plasma when the field is co- or counter-rotating with the rotation of the plasma, respectively (Fig. 10.10). Density compression by an order of magnitude up to 13% of the Brillouin density has been obtained [27]. If the rotation frequency matches one of the plasma mode frequencies, energy from the field is coupled into the plasma, and heating occurs. By monitoring the plasma temperature, the spectrum of the different modes can be displayed (Fig. 10.11).

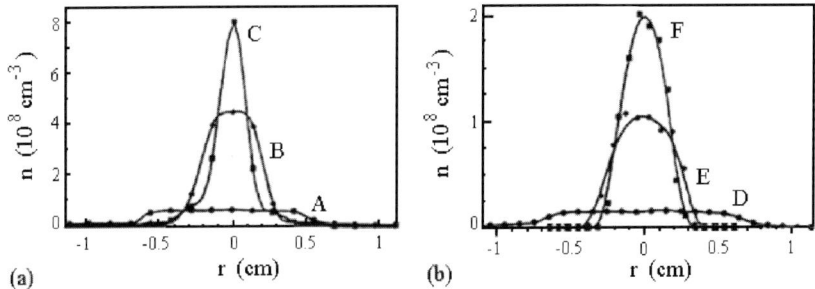

Fig. 10.10. Measured density profiles, demonstrating rotating wall compression and expansion. (a) for electron; (b) for Mg^+ ions. Profiles B and E describe the density distribution without a rotating wall, A and D expansion by a field rotating backward with respect to the ion cloud, C and F compression by forward rotating field [29]

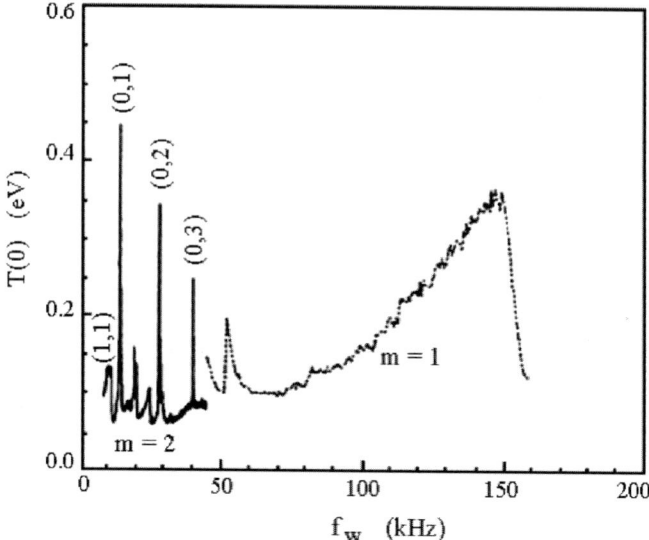

Fig. 10.11. Evolution of central ion temperature during gradual ramp of rotating field frequency. Heating resonances due to excitation of $k \neq 0$ plasma modes are observed [27]

10.4 Collective Effects

10.4.1 Individual and Center-of-Mass Oscillations

As described above, for a single ion in a Penning trap the solution of the equation of motion leads to three independent harmonic oscillations: The axial frequency

$$\omega_z = \left(\frac{eV}{mr_0^2}\right)^{1/2}, \tag{10.25}$$

where V is the applied trap voltage and r_0 the radius of the ring electrode (and $z_0^2 = r_0^2/2$), the reduced cyclotron frequency

$$\omega_+ = \omega_c/2 + \left(\frac{\omega_c^2}{4} - \frac{\omega_z^2}{2}\right)^{1/2} \tag{10.26}$$

and the magnetron frequency

$$\omega_- = \omega_c/2 - \left(\frac{\omega_c^2}{4} - \frac{\omega_z^2}{2}\right)^{1/2}. \tag{10.27}$$

$\omega_c = (e/m)B$ is the free ion's cyclotron frequency. These frequencies can be experimentally observed when the ion motion is resonantly excited by an external drive field.

10 Non-Neutral Plasmas and Collective Phenomena in Ion Traps 285

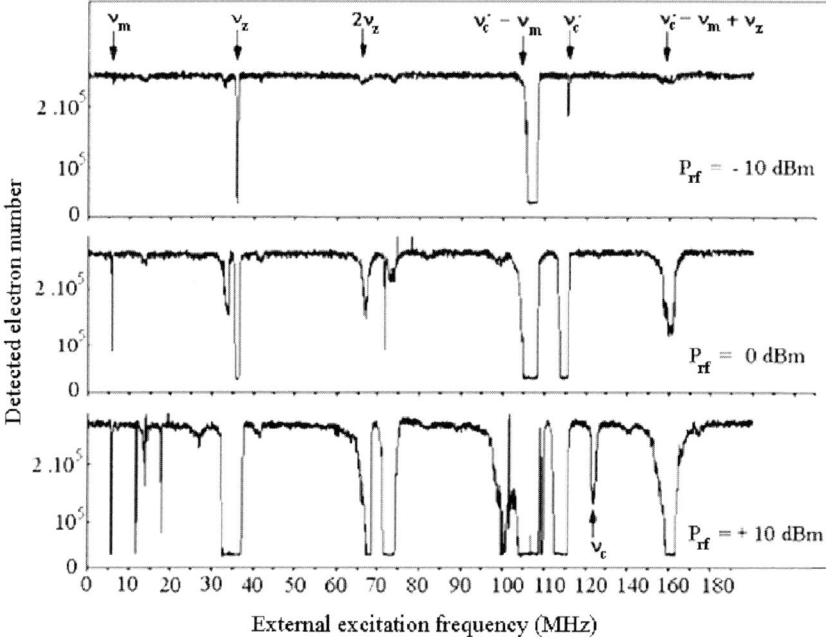

Fig. 10.12. Motional spectrum of a stored electron cloud in a Penning trap observed at different amplitudes of the exciting field

In case of an ion cloud additionally linear combinations of these frequencies become visible (Fig. 10.12).

A closer inspection of the axial frequencies ω_z and $2\omega_z$ shows that the resonance line has two components (Fig. 10.13): A broad asymmetric line and a more narrow feature at the high frequency side of the resonance.

The determination of the dependence of the center frequencies of these resonances on the stored ion number shows that the broad line shifts linearly to lower frequencies with increasing ion number while the position of the narrow feature remains unchanged at the value for a single ion. This suggests the interpretation that the broad resonance represents the excitation of the individual ion oscillations, shifted and broadened by the space charge potential of the neighboring particles, while the narrow feature is the excitation of the center-of-mass motion of the ion cloud. Similar features have been observed in Paul traps [30]. The asymmetry of the resonances reflects the fact that the effective trapping potential deviates from the ideal quadrupolar shape and thus becomes inharmonic. In case of the individual ion motion the potential is mainly modified by the space charge potential of the remaining ions. For the center-of-mass motion space charge plays no role but trap imperfections like misalignments or truncation of the electrodes represent the dominant effect. These deviations can be quantified by a series expansion of the potential in

Legendre polynomials $\mathcal{P}_n(\cos\theta)$

$$\Phi(r) = \Phi_0 \sum_{n=2}^{\infty} c_n \left(\frac{r}{r_0}\right)^n \mathcal{P}_n(\cos\theta) . \qquad (10.28)$$

The coefficients c_n represent the strength of the higher order components in the potential. For $n = 2$ we have the pure quadrupole potential. The asymmetry of the observed resonances allows to determine the size of the higher order contributions. If we restrict ourselves to the most important part, characterized by c_4 and c_6, the equation of motion for the z-direction including the external drive field can be solved with reasonable assumptions [31]. A fit of the observed data as shown in Fig. 10.13 to the solution yields values for c_4 and c_6. In our case we obtain for the individual resonance the values $c_4 = -9.8 \times 10^{-3}$, $c_6 = 3.8 \times 10^{-5}$, normalized to $c_2 = 1$. For the center-of-mass resonance the values are about one order of magnitude larger: $c_4 = -7.5 \times 10^{-2}$, $c_6 = 2.8 \times 10^{-4}$. The excitation of the center-of-mass resonance requires a threshold amplitude of the exciting field [30, 32]. This suggest some damping mechanism for the motion which has to be overcome to excite the oscillation of the ion cloud. Ion collision with background molecules and induced image currents have been identified as source of this damping [31].

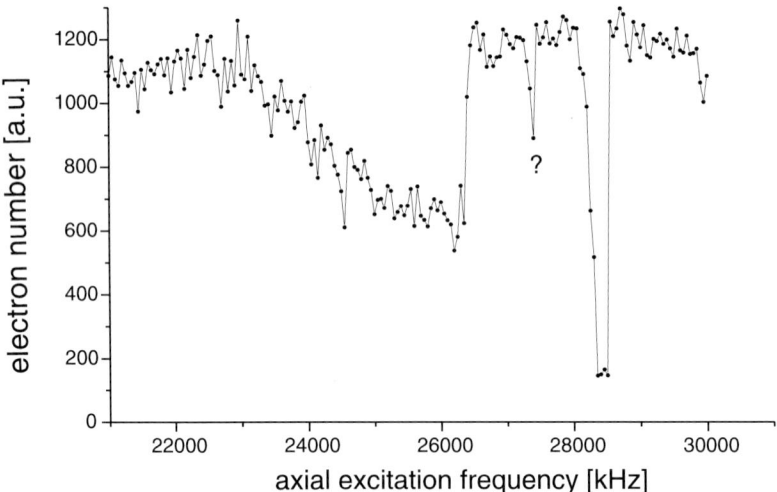

Fig. 10.13. Excited axial resonance of stored electrons in a Penning trap showing an individual (*left*) and center-of-mass resonance. The narrow feature in between comes from a linear combination of other trap eigenfrequencies

10.4.2 Instabilities in the Ion Motion

The higher order components in the trapping potential have a significant effect on the stability of confinement. The limit of stability for a Penning trap is given by the relation

$$\omega_c^2 > \frac{\omega_z^2}{2} \tag{10.29}$$

or

$$\frac{e}{m} B^2 > \frac{V}{r_o^2} \ . \tag{10.30}$$

This means that for a given ion mass and a given magnetic field strength stability is assured up to a maximum value for the applied trapping voltage. Experimentally, however, one finds that for certain values of the voltage ions can not be confined for extended periods of time (Fig. 10.14). The unstable regions become wider when the time between ion creation and detection is increased.

Kretzschmar [33] has developed a theory of anharmonic perturbations in a Penning trap using the series expansion of the potential as in (10.28) and showed that the ion motion becomes unstable when the eigenfrequencies are linearly dependent:

$$n_c \omega_+ + n_z \omega_z + n_m \omega_- = 0 \ , \tag{10.31}$$

n_c, n_z, n_m are integers. As shown in [34] the observed instabilities can be assigned to the predicted values for certain combinations of n_c, n_z, n_m. The assignment is indicated in Fig. 10.14.

This is equivalent to instabilities observed earlier in Paul traps. At operating points where the radial and axial ion oscillation frequencies ω_r and ω_z, respectively, are related as

$$n_r \omega_r + n_z \omega_z = n_\Omega \Omega \ , \tag{10.32}$$

n_r, n_z, n_Ω are integers, rapid ion loss from the trap is observed. It should also be noted that a similar relation describes instabilities in storage rings and accelerators, known as stop-bands, where the frequencies appearing in (10.31) are replaced by the betatron oscillation frequencies and the period of revolution [35].

10.5 Strongly Coupled Non-Neutral Plasmas

Numerical simulations have shown that ordering characteristics of liquids appear when the plasma parameter Γ_c as defined in (10.16) has a value of approximately two [36]. At values of about $\Gamma_c > 175$, a phase transition into a crystal occurs [37, 38]. It requires, as evident from (10.16) cooling of the stored ion cloud to low temperatures. Such low temperatures can be efficiently provided by laser cooling as described above.

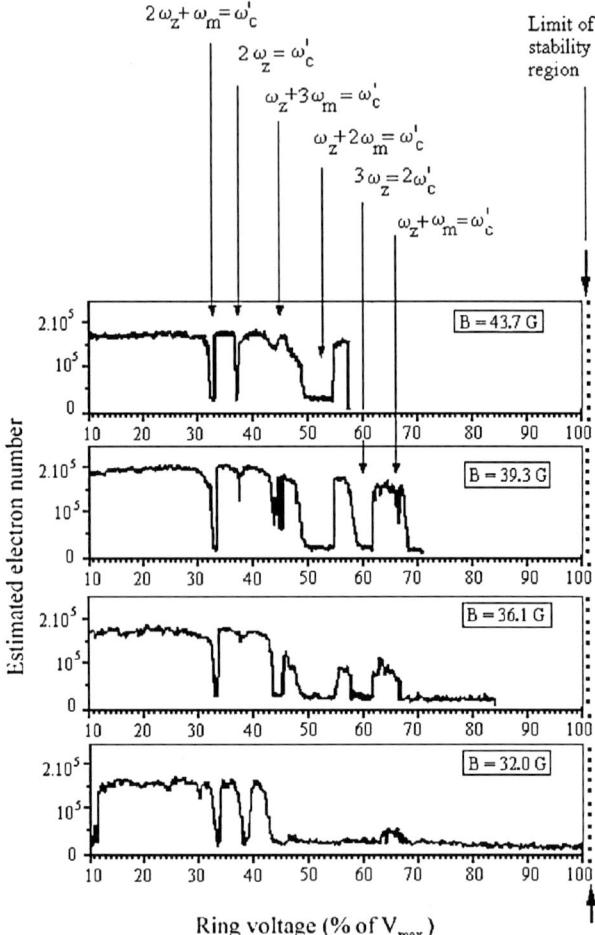

Fig. 10.14. Detected electron number at different trapping voltages of a Penning trap. The voltages are normalized to the maximum allowed voltage for stable confinement according to (10.30)

Experimentally, phase transitions from a gaseous or liquid behavior of a confined ion cloud to a crystalline structure are detected by a characteristic kink in the observed fluorescence spectrum, when a cooling laser is swept from below the optical resonance frequency towards the center frequency. (Fig. 10.15). It indicates a spatial redistribution of the ions leading to a suddenly reduced Doppler width of the transition.

The first observations of such phase transitions were made at the Max-Planck Institute for Quantum Optics in Garching in 1987 [39] on Mg^+ ions and in the same year on Hg^+ ions at NIST, Boulder [40]. Both groups used Paul traps for confinement. The interpretation as the formation of an ordered

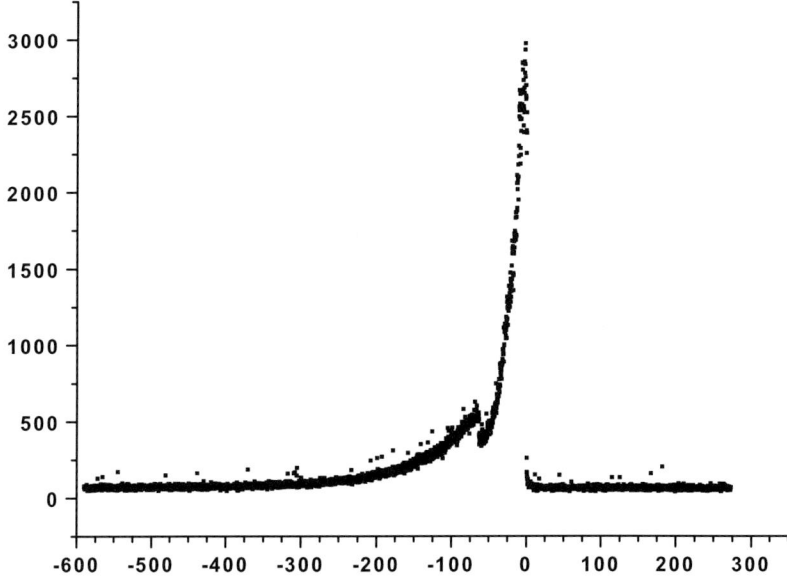

Fig. 10.15. Fluorescence (ordinate in arbitrary units) from a stored Ca^+ ion cloud when a cooling laser is swept from the low frequency (abscissa, arbitrary units) side towards the resonance center. The kink in the fluorescence intensity indicates the onset of crystallization. The fluorescence vanishes when the laser frequency crosses the resonance center because then the ion cloud is heated and expands

state was confirmed by direct observation with a photon counting imaging system. Small clusters of ions were seen to form regular figures in the plane perpendicular to the symmetry axis of the trap [39, 41] (Fig. 10.16). They are similar in structure as configurations observed many years ago by Wuerker, Shelton and Langmuir [42] on charged Al particles trapped in air at a pressure of about a mPa in a three dimensional device similar to a Paul trap.

The formation of large crystals in three dimensional Paul traps becomes increasingly difficult when the ions are placed outside the trap center, since they are heated up by the rf trapping field, and high cooling laser power is required. This effect is somewhat reduced in linear Paul traps, see Fig. 10.2.

In this device many ions can be placed in the field free region along the axis where they experience little heating from the rf field. The formation of Coulomb crystals in linear Paul traps appears for few ions as a string along the axis and turns into a zig-zag or a helix when the axial confinement potential is increased, see Fig. 10.17.

For larger ion numbers the ions arrange in concentric shells [43, 46] (Fig. 10.18) as predicted by theory [44, 45].

Studies of the formation process as function of the cooling power reveal details of the formation and melting process [46]. When different ion species

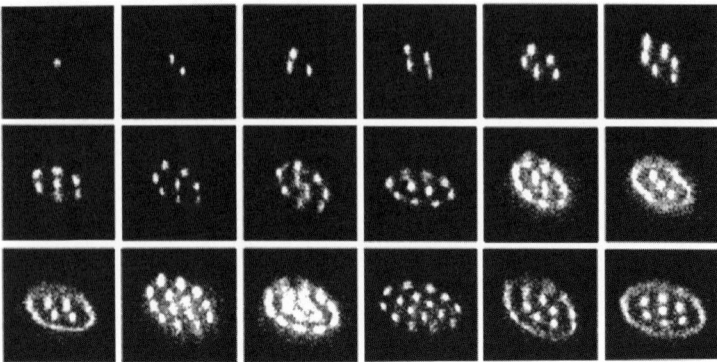

Fig. 10.16. CCD-camera picture of ions confines in a two dimensional rf trapping potential [41]

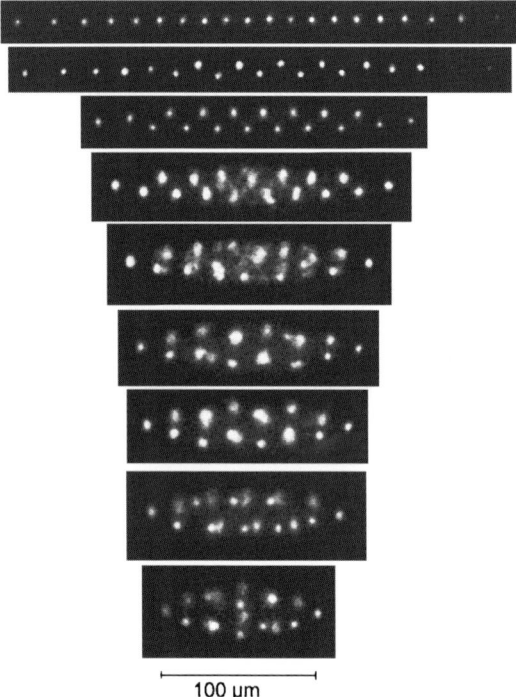

Fig. 10.17. Ion string in a linear Paul trap squeezed into a zig-zag and a helix when the axial confinement strength is increased from *top* to *bottom* [41]

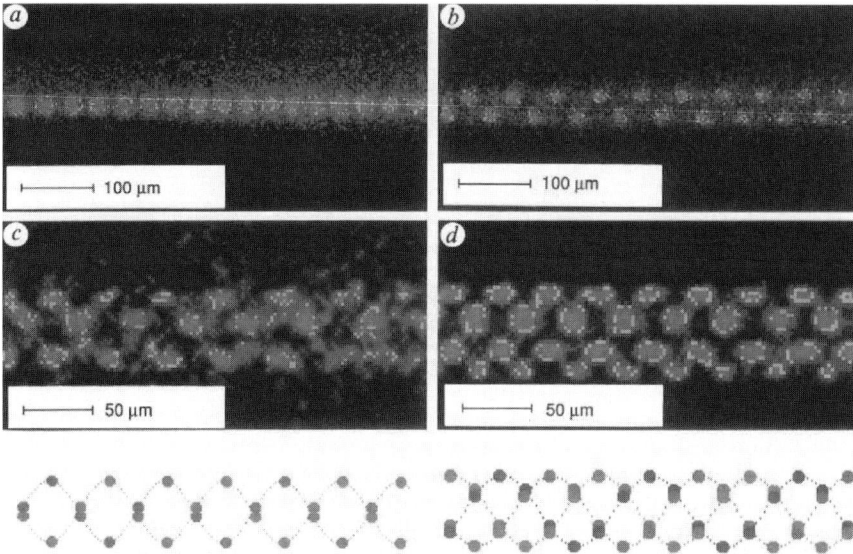

Fig. 10.18. CCD camera images of crystalline structures of laser cooled Mg$^+$ ions with increasing ion density. (**a**): String, (**b**): Zig-zag, (**c**): Two interwoven helices, (**d**): Three interwoven helices. Visualisations of the helices are shown in the lower traces [43]

are stored simultaneously in the trap and one of them is laser cooled and crystallized, the other one will cool by Coulomb interaction to the same temperature. Since the effective radial potential of the ions depends on their masses, lighter ions are confined more strongly than the heavier ones. This leads to spatial separation of the two species as shown in experiments at the university of Åarhus [47], see Fig. 10.19.

The absence of time varying electric fields and the associated rf heating in Penning traps allows to obtain crystals containing much larger numbers of particles. They also arrange in shell structures as first observed at NIST [48].

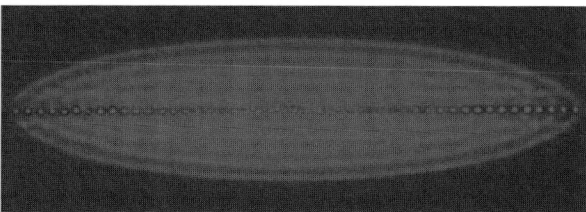

Fig. 10.19. Bicrystal containing 47 ^{24}Mg$^+$ ions, and approximately 1300 ^{40}Ca$^+$ ions. The Mg$^+$ ions are arranged in an equidistant string structure on the trap axis. The distance between the Mg$^+$ ions is 15.5 mm [47]

They have been investigated in detail using Bragg diffraction patterns created by the regular array of scattering ions [49, 50]. Since the plasma rotates, the pattern observed in a plane perpendicular to the axis will consist of concentric rings. They may, however, be observed by a stroboscopic technique when the CCD-camera used for observation is triggered in phase with the rotation frequency [28], see Fig. 10.20.

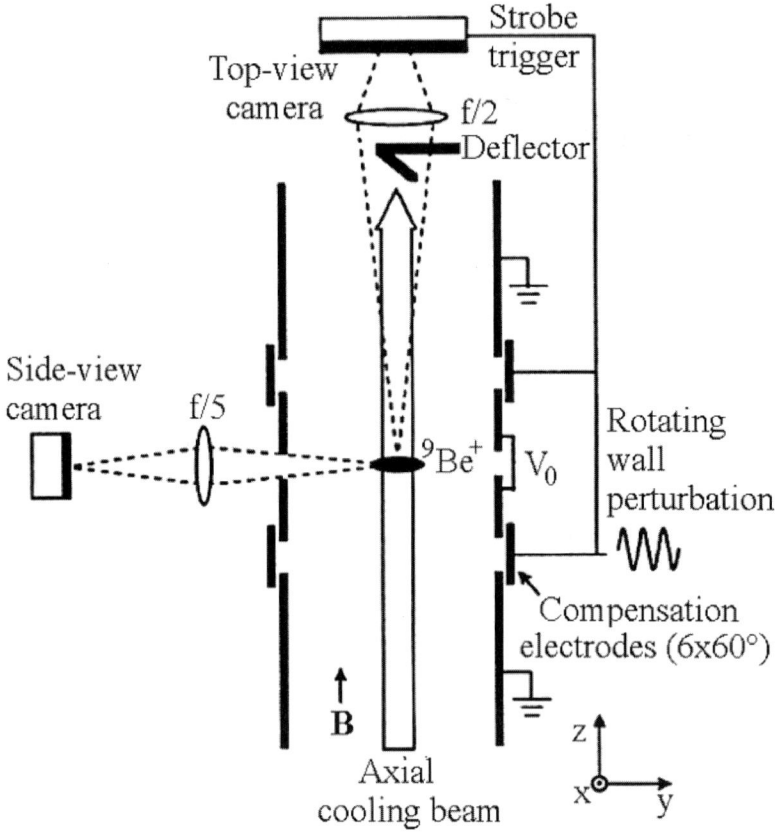

Fig. 10.20. Cylindrical Penning trap used to obtain Bragg diffraction pattern of ion crystals, using phase lock of plasma rotating field to stroboscopically "stop" the rotation [17]

On the basis of these diffraction images it has been established that three dimensional long range order in form of a bbc lattice begins to appear when the ion number exceeds about $N \approx 5 \cdot 10^4$. In some oblate plasmas, a mixture of bbc and fcc orders was seen [51], see Fig. 10.21.

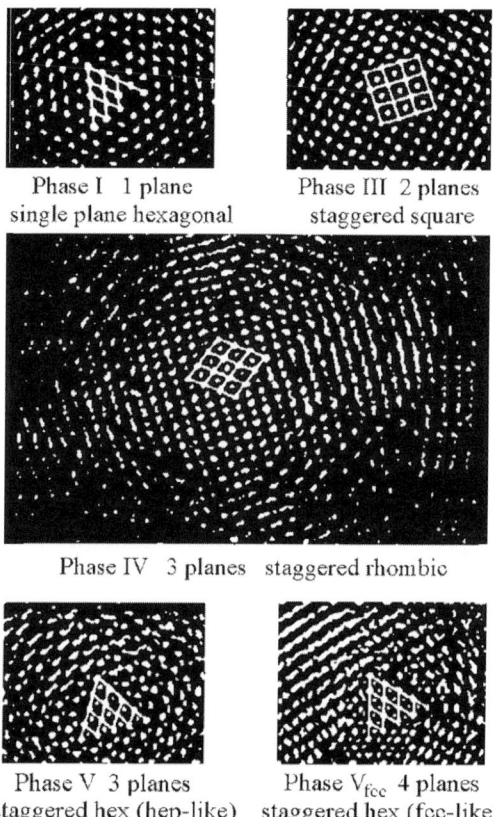

Fig. 10.21. Images of five different structural phases observed on Be$^+$ ions in a Penning trap. The trap axis is perpendicular to the observation plane. The lines show fits to the indicated structure [51]

Also structural phase transitions have been observed when the axial density is changed in a two dimensional extended plane of confined ions [52] in accordance with theoretical predictions [53].

10.6 Summary

Ion traps are often considered as mere containers which keep charged particles in place and makes them available for high precision experiments such as hyperfine structure measurements, lifetimes of long-lived excited states or g-factor determinations. In many cases details of the ion motion inside the trap are of little importance. For extreme precision often operation with only a single stored particle is required to avoid perturbations by ion–ion interaction. When, however, many ions are confined simultaneously new effects show up which are of interest in themselves. They can be explained when we treat the ion cloud as a single component non-neutral plasma. Plasma oscillations, instabilities and collective motions can be experimentally observed. Investigation of the plasma behavior by induced image charges in the trap electrodes or by spatially resolved imaging of scattered laser light reveals many details. Cooling the plasma by laser radiation leads to strong coupling where the formation of Coulomb crystals starts. A great variety of crystalline structures have been observed and the investigation of crystal formation may not only be of interest for plasma physics but also for solid state physics. In this respect we can consider ion traps as miniature laboratories to study the behavior and the properties of such plasmas under different conditions.

Acknowledgements

Our experiments on collective behavior and instabilities of confined electron and ion clouds were supported by the Deutsche Forschungsgemeinschaft. Most of the work from our laboratory has been performed by my students R. Alheit, G. Tommaseo, P. Paasche, and C. Angelescu. I thank J. Alonso for help with the manuscript.

References

1. P.K. Ghosh: *Ion Traps* (Clarendon Press, Oxford 1995)
2. F.G. Major, V. Gheorghe, G. Werth: *Charged Particle Traps* (Springer, Heidelberg Berlin New York 2004)
3. H. Dehmelt: Radiofrequency spectroscopy of stored ions. I: Storage. In: *Adv. At. Mol. Phys.*, vol 3, ed by R.D. Bates and I. Estermann (Academic Press, New York 1967) pp 53–72
4. C. Meis, M. Desaintfuscien, M. Jardino: Appl. Phys. B-Lasers Opt. **45**, 59 (1988)
5. R.D. Knight, M. Prior: J. Appl. Phys. **50**, 3044 (1979)
6. H. Schaaf, U. Schmeling, G. Werth: Appl. Phys. B-Lasers Opt. **25**, 249 (1981)
7. N. Kjaergaard, L. Hornekaer, A.M. Thommesen et al: Appl. Phys. B-Lasers Opt. **71**, 207 (2000)
8. S. Gulde, D. Rotter, P. Barton et al: Appl. Phys. B-Lasers Opt. **73**, 861 (2001)
9. J. Coutandin, G. Werth: Appl. Phys. B-Lasers Opt. **29**, 89 (1982)
10. W.M. Itano, D.J. Wineland: Phys. Rev. A **25**, 35 (1982)
11. G. Horvart, R.C. Thompson: Phys. Rev A **59**, 4530 (1999)
12. R.F. Powell, D.M. Segal, R.C. Thompson: Phys. Rev. Lett. **89**, 93003 (2002)
13. W.M. Itano, J.C. Bergquist, J.J. Bollinger et al: Phys. Scr. **T59**, 106 (1995)
14. D. Davidson: *Physics of Nonneutral Plasmas* (Addison-Wesley, Redwood City Cal 1990)
15. D.H.E. Dubin, T.M. O'Neil: Rev. Mod. Phys. **71**, 87 (1999)
16. T.M. O'Neil: Phys. Scr. **T59**, 341 (1995)

17. T.B. Mitchell, J.J. Bollinger, X.P. Huang et al: Mode and transport studies of laser-cooled ion plasmas in a Penning trap. In: *Trapped Charged Particles and Fundamental Physics, AIP Conf. Proc. 457*, ed by D.H.E. Dubin, D. Schneider (AIP Press, Melville 1999), pp 309–318
18. L. Brillouin: Phys. Rev. **367**, 260 (1945)
19. J.J. Bollinger, D.J. Heinzen, F.L. Moore et al: Phys. Rev. A **48**, 525 (1993)
20. A.W. Trivelpiece, R.W. Gould: J. Appl. Phys.**30**, 1784 (1959)
21. D.H.E. Dubin: Phys. Rev. Lett. **66**, 2076 (1991)
22. R.G. Greaves, M.D. Tinkle, C.M. Surko: Phys. Rev. Lett. **74**, 90 (1995)
23. F. Anderegg, N. Shiga, J.R. Danielson et al: Phys. Rev. Lett. **90**, 115001 (2003)
24. D.J. Wineland, J.J. Bollinger, W.M. Itano et al: J. Opt. Soc Am. B **2**, 1721 (1985)
25. J.J. Bollinger, D.J. Wineland: Phys. Rev. Lett. **53**, 348 (1984)
26. D.J. Heinzen, J.J. Bollinger, F.L. Moore et al: Phys. Rev. Lett. **66**, 2080 (1991)
27. X.-P. Huang, F. Anderegg, E.M. Hollmann et al: Phys. Rev. Lett. **78**, 875 (1997)
28. X.-P. Huang, J.J. Bollinger, T.B. Mitchell et al: Phys. Plasmas **5**, 1658 (1998)
29. E.M. Hollmann, F. Anderegg, C.F. Driscoll: Phys. Plasmas **7**, 2776 (2000)
30. R. Alheit, X.Z. Chu, M. Hoefer et al: Phys. Rev. A **56**, 4023 (1997)
31. G. Tommaseo, P. Paasche, C. Angelescu et al: Eur. Phys. J. D **28**, 39 (2004)
32. P. Paasche, T. Valenzuela, D. Biswas et al: Eur. Phys. J. D **18**, 295 (2002)
33. M. Kretzschmar: Z. Naturf. **45a**, 965 (1990)
34. P. Paasche, C. Angelescu, S. Ananthamurthy et al: Eur. Phys. J. D **22**, 183 (2003)
35. F. Hinterberger: *Physik der Teilchenbeschleuniger und Ionenoptik* (Springer, Heidelberg Berlin New York 1997)
36. P. Hansen: Phys. Rev. A **8**, 3096 (1973)
37. E.L. Pollock, J.P. Hansen: Phys. Rev. A **8**, 3110 (1973)
38. W.L. Slattery, G.D. Doolen, H.E. DeWitt: Phys. Rev. A **21**, 2087 (1980)
39. F. Diedrich, E. Peik, J.M. Chen et al: Phys. Rev. Lett. **59**, 2931 (1987)
40. D.J. Wineland, J.C. Bergquist, W.M. Itano et al: Phys. Rev. Lett. **59**, 2935 (1987)
41. M. Block, A. Drakoudis, H. Leuthner: J. Phys. B–At. Mol. Opt. Phys. **33**, L375 (2000)
42. R.F. Wuerker, H. Shelton, R.V. Langmuir: J. Appl. Phys. **30**, 342 (1959)
43. G. Birkl, S. Kassner, H. Walther: Nature **357**, 310 (1992)
44. J.P. Schiffer, M. Drewsen, J.S. Hangst et al: Phys. Rev. Lett. **82**, 3964 (1999)
45. J.P. Schiffer, M. Drewsen, J.S. Hangst et al: Proc. Natl. Acad. Sci. **897**, 10697 (2001)
46. M. Drewsen, C. Brodersen, L. Hornekær et al: Phys. Rev. Lett. **81**, 2878 (1998)
47. L. Hornekaer, N. Kjaergaard, A.M. Thommesen et al: Phys. Plasmas **8**, 1371 (2000)
48. S.L. Gilbert, J.J. Bollinger, D.J. Wineland: Phys. Rev. Lett. **60**, 2022 (1988)
49. J.N. Tan, J.J. Bollinger, B. Jelenkovik: Phys. Rev. Lett. **75**, 4198 (1995)
50. X.-P. Huang, J.J. Bollinger, T.B. Mitchell et al: Phys. Rev. Lett. **80**, 73 (1998)
51. W.M. Itano, J.J. Bollinger, J.N. Tan et al: Science **279**, 686 (1998)
52. T.B. Mitchell, J.J. Bollinger, D.H.E. Dubin et al: Science **282**, 1290 (1998)
53. J.P. Schiffer: Phys. Rev. Lett. **70**, 818 (1993)

11 Collective Effects in Dusty Plasmas

A. Melzer

Abstract. Dusty plasmas, i.e. plasmas containing particles of nanometer to micrometer size, allow the observation of kinetic and collective effects on a microscopic scale with high temporal resolution. Collective effects in dusty plasmas manifest in various new wave modes, in linear and nonlinear particle oscillations and in unexpected particle-particle interactions. These collective phenomena are discussed in this chapter after an introduction into the fundamental properties of particle charging and the forces on the particles.

11.1 Introduction

Dust-containing plasmas have been known for decades in astrophysical situations, like comet tails, the rings of Saturn and Jupiter or interstellar clouds [1]. In the end of the 1980's, dust particles have been found to grow in reactive gases in plasma processing devices [2]. Then, after the discovery of the plasma crystal in 1994 [3, 4], dusty plasmas have become one of the most active and growing fields of plasma physics where a number of new collective effects arise from the presence of the dust and it has become possible to investigate dynamic processes in strongly coupled systems on a microscopic, kinetic level.

Dusty plasmas consist of macroscopic particles of nanometer to micrometer size immersed in a gaseous plasma environment. The particles typically attain several hundred or thousand elementary charges due to the inflow of plasma electrons and ions. The charge-to-mass ratio of these particles is, however, by orders of magnitude smaller than for ions, not to mention electrons. The relevant time scales of dust dynamics are of the order of milliseconds to seconds which allows direct observation by video microscopy. Moreover, new types of dynamic phenomena arise from the fact that the dust charge itself becomes a dynamic variable that depends on the local plasma properties.

Due to the high charges the electrostatic potential energy of the particles by far exceeds their thermal energy: the system is said to be strongly coupled (see Chap. 6 in Part I). Dusty plasmas enable to study a vast variety of phenomena, like fluid and crystalline plasmas, phase transitions, strong-coupling effects, waves, Mach cones and many more, on a microscopic kinetic level. Thus, dusty plasmas are a unique system bridging the fields of plasma physics, condensed matter and material science.

Dusty plasmas share a number of physical concepts and similarities with non-neutral plasmas, like pure ion plasmas in Paul or Penning traps (see also Chap. 10 in Part II) [5, 6], as well as with colloidal suspensions [7, 8], where charged plastic particles are immersed in an aqueous solution. To stress the analogy to complex fluids or colloidal suspensions, the names "complex plasmas" or "colloidal plasmas" are frequently used when referring to strongly coupled dusty plasmas.

In the following first the basic concept and properties of dusty plasmas, like charging, forces and interaction potentials, are introduced. After that, different type of collective effects, like waves and normal modes in weakly and strongly coupled dusty plasmas are described.

11.2 Particle Charging

The charging of the dust particles is the most fundamental, but also one of the most difficult questions. Here, we will focus on the main charging mechanisms. For a complete quantitative description, a number of charging processes in non-uniform and non-equilibrium situations have to be taken into account. Measurements of the dust charge are presented in Sect. 11.3.7.

11.2.1 Orbital Motion Limit Currents

Generally, the dust particles are treated as spherical probes (Chap. 5 in Part I) at floating potential ϕ_{fl}, where the sum of all currents to the particle vanish

$$\sum_q I_q(\phi_{\mathrm{fl}}) = \frac{dQ}{dt} = 0 \ . \tag{11.1}$$

In the ideal case of a Maxwellian plasma, the currents to the particle can be described by the orbital motion limit (OML) model first derived by Mott-Smith and Langmuir in 1926 [9]. There, it is assumed that electrons and ions move towards the dust particle from infinity on collisionless orbits subject to their electrostatic interaction [see Fig. 11.1 (a)]. Usually, the ions are the attracted and the electrons are the repelled species and the particle potential is negative with respect to the plasma potential $\phi_p < 0$. Then, ions with an initial velocity $v_{i,0}$ and an impact parameter $b < b_c$ will fall onto the dust particle, where

$$b_c^2 = a^2 \left(1 - \frac{2e\phi_p}{m_i v_{i,0}^2}\right) \tag{11.2}$$

is the critical impact parameter determined from conservation of energy and angular momentum. This describes an increased ion collection cross section due to attraction ($\phi_p < 0$). Electrons are repelled from the dust and their

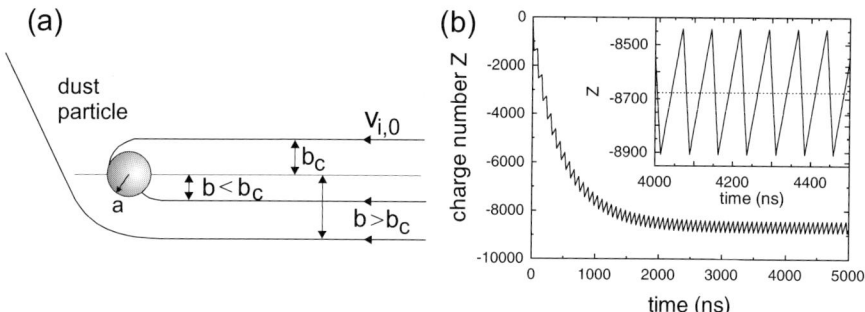

Fig. 11.1. (a) Ion trajectories around a dust particle with different impact parameters. (b) Temporal evolution of the dust charge in an rf sheath [10]

density in the vicinity of the dust is reduced according to a Boltzmann distribution. Finally, the ion and electron OML currents to the dust are given by the following expressions

$$I_i = \pi a^2 n_i e \sqrt{\frac{8k_B T_i}{\pi m_i}} \left(1 - \frac{e\phi_p}{k_B T_i}\right),$$

$$I_e = -\pi a^2 n_e e \sqrt{\frac{8k_B T_e}{\pi m_e}} \exp\left(\frac{e\phi_p}{k_B T_e}\right), \quad (11.3)$$

where a is the particle radius, n_i, n_e are ion and electron densities, T_i, T_e the respective temperatures and m_i, m_e their masses.

The assumption of collisionless ion trajectories is an essential ingredient with respect to conservation of angular momentum. However, this condition is often violated in plasma discharges. Recent investigations [11] have shown that the presence of collisions might considerably modify the currents onto the dust, since they will lead to ions trapped in the electrostatic potential well performing Keplerian orbits around the dust [12].

An issue that was recognized only recently is that in the OML approach ions having an angular momentum within a certain range cannot reach the dust, although their impact parameter is smaller than b_c (angular momentum barrier) [13, 14]. However, for particles much smaller than the Debye length ($a \ll \lambda_D$) that fraction of ions is negligible and the OML results are still valid.

In many cases, like in the sheath environment of a discharge, the ions have a drift velocity u_i that can be much larger than the thermal ion velocity $v_{th,i}$. The ion current to a dust particle can then be written in the form [15]

$$I_i = \pi a^2 n_i e u_i f(u_i), \quad (11.4)$$

where $f(u_i)$ is a rather complicated function of the ion drift velocity. For $u_i \gg v_{th,i}$, the above (11.4) reduces to [10]

$$I_\mathrm{i} = \pi a^2 n_\mathrm{i} e u_\mathrm{i} \left(1 - \frac{2e\phi}{m_\mathrm{i} u_\mathrm{i}^2}\right), \tag{11.5}$$

which is obtained by replacing the thermal ion energy $k_\mathrm{B} T_\mathrm{i}$ in (11.3) by the kinetic energy of the drifting ions $m_\mathrm{i} u_\mathrm{i}^2/2$.

11.2.2 Other Charging Currents

Two other possible charging mechanisms with electron emission should be mentioned, here, which are considered important mostly for astrophysical situations. These are charging by UV radiation and secondary electron emission. Other possible charging mechanisms such as thermal emission, field emission, sputtering, ion-induced electron emission are neglected since they are usually not important for dusty plasmas in the laboratory or in space.

The absorption of UV radiation releases photoelectrons and hence constitutes a positive charging current. The magnitude of the current depends on the photoemission yield of the dust material. Photoemission by UV radiation might be important for dust grains in astrophysical situations near stars. This effect is, e.g., made responsible for dust layers floating above the surface of the Moon [16].

Secondary electron emission from the impact of energetic electrons might also be important under certain conditions in astrophysical situations. Since the maximum yield for secondary electron emission can be much larger than 1, many more electrons can be released for each incoming electron even leading to positively charged dust grains [17].

11.2.3 Particles as Floating Probes

As mentioned above, the floating potential of the particle is determined by the condition that all currents on the particle vanish (11.1). In laboratory discharges usually only the OML collection currents are considered. This results in

$$1 - \frac{e\phi_\mathrm{fl}}{k_\mathrm{B} T_\mathrm{i}} = \sqrt{\frac{m_\mathrm{i} T_\mathrm{e}}{m_\mathrm{e} T_\mathrm{i}}} \frac{n_\mathrm{e}}{n_\mathrm{i}} \exp\left(\frac{e\phi_\mathrm{fl}}{k_\mathrm{B} T_\mathrm{e}}\right). \tag{11.6}$$

This equation can be solved numerically for ϕ_fl for given values of the plasma parameters.

In Table 11.1 the resulting floating potentials are calculated for typical conditions. For the astrophysically important case of the isothermal ($T_\mathrm{e} = T_\mathrm{i}$) hydrogen plasma the well-known Spitzer value $\phi_\mathrm{fl} = -2.5\, k_\mathrm{B} T_\mathrm{e}/e$ is found. Under typical laboratory discharges in heavier gases with $T_\mathrm{e} \gg T_\mathrm{i}$ a good rule-of-thumb approximation is $\phi_\mathrm{fl} \approx -2\, k_\mathrm{B} T_\mathrm{e}/e$. However, for electron energy distributions with even a small suprathermal electron component the floating potential will be decisively different. If the electron density is reduced with respect to the ion density, e.g. in the sheath of a discharge, the electron charging current is reduced and the particle attains a more positive floating potential.

Table 11.1. Normalized floating potentials $e\phi_{\mathrm{fl}}/k_B T_e$ in quasineutral plasmas ($n_e = n_i$) in hydrogen, helium and argon for different electron-to-ion temperature ratios

T_e/T_i	1	2	5	10	20	50	100
H	−2.504	−2.360	−2.114	−1.909	−1.700	−1.430	−1.236
He	−3.052	−2.885	−2.612	−2.388	−2.160	−1.862	−1.645
Ar	−3.994	−3.798	−3.491	−3.244	−2.992	−2.660	−2.414

To determine the dust charge from the floating potential on the particle the dust particle is considered as a spherical capacitor of capacitance C. The particle charge is then given by

$$Q_d = Z_d e = C\phi_{\mathrm{fl}} , \tag{11.7}$$

where Z_d is the number of elementary charges on the dust. In vacuum the capacitance is

$$C = 4\pi\varepsilon_0 a , \tag{11.8}$$

and in an ambient plasma with shielding length λ_D the capacitance is

$$C = 4\pi\varepsilon_0 a \left(1 + \frac{a}{\lambda_D}\right) , \tag{11.9}$$

which in the typical case of $a \ll \lambda_D$ reduces to the vacuum value.

That means that a particle of $a = 1\,\mu\mathrm{m}$ radius attains $Z_d = 695$ elementary charges per volt floating potential. With the above mentioned rule-of-thumb estimation of the floating potential an approximate formula of the dust charge is given by

$$Z_d = 1400\, a_{\mu\mathrm{m}}\, T_{e,\mathrm{eV}} \tag{11.10}$$

with $T_{e,\mathrm{eV}}$ being the electron temperature in electron volts and $a_{\mu\mathrm{m}}$ the particle radius in microns.

If secondary electron emission is important, as for cosmic grains, the floating potential ϕ_{fl} is not a unique value, but might be multivalued under certain conditions [17]: the total current to the dust vanishes for two stable values of the floating potential, one is positive and the other is negative. That means that in the *same* plasma negatively as well as positively charged dust grains can exist. The oppositely charged particles can then immediately coagulate, which might have an enormous influence on the growth mechanism of planetesimals in astrophysical dusty plasmas.

11.2.4 Charging in the RF Sheath

In laboratory dusty plasmas the particles are usually trapped in the sheath of rf discharges. Electrons are able to follow the applied rf voltage (typically at a frequency of 13.56 MHz), whereas ions and dust grains only react to the time averaged fields. The electron component of the plasma can be described as oscillating back and forth between the electrodes leading to a periodically increasing and collapsing sheath at the electrodes (see also Chap. 5 in Part I). A dust particle trapped in sheath therefore periodically "sees" a quasineutral plasma environment, when the sheath is flooded by electrons, and a pure ion sheath, when the sheath is expanded. Hence, the electron charging current to the dust particle will also be modulated by the rf frequency. The temporal behavior of a dust particle in an rf sheath is shown in Fig. 11.1 (b). The charging time is of the order of a few microseconds. In equilibrium, the particle charge is modulated by few hundred elementary charges around a mean particle charge of a few thousand.

11.3 Forces on Particles

In this part the main forces acting on dust particles in a plasma discharge will be briefly discussed. These forces are gravity, electric field force, ion drag force, thermophoresis and neutral drag.

11.3.1 Gravity

The gravitational force simply is

$$\mathbf{F}_\mathrm{g} = m_\mathrm{d}\mathbf{g} = \frac{4}{3}\pi a^3 \varrho_\mathrm{d} \mathbf{g} , \qquad (11.11)$$

where $\mathbf{g}$ is the gravitational acceleration and $m_\mathrm{d}, \varrho_\mathrm{d}$ are the mass and the mass density of the dust grains, respectively. Since this force scales with a^3 it is the dominant force for micrometer particles and becomes negligible for particles in the nanometer range.

11.3.2 Electric Field Force

Obviously, the force due to an electric field $\mathbf{E}$

$$\mathbf{F}_\mathrm{E} = Q_\mathrm{d}\mathbf{E} = 4\pi\varepsilon_0 a \phi_\mathrm{fl} \mathbf{E} \qquad (11.12)$$

is the governing force for charged particles. With the applied capacitor model the force scales linearly with the particle size. One might argue that the electric field force on the (negative) particle is reduced due to a counterforce on the (positive) shielding cloud, thus the external electric field would be shielded. Hamaguchi and Farouki [18] have rebutted that reasoning: the shielding cloud is only a response of the plasma environment to the charged particle and the cloud is not "attached" to the particle. Therefore, there is no counterforce and the full force (11.12) is effective.

In addition, dipole moments on the particles **p** might exist. These forces can then result in a polarization force

$$\mathbf{F}_{\text{dip}} = \nabla(\mathbf{p} \cdot \mathbf{E}) \,. \tag{11.13}$$

Dipole moments can be induced by an external electric field or by directional charging processes (for dielectric particles) where due to an ion flow the front side of the particle is charged more positively than the back side. Polarization forces are usually considered negligible, except for very large particles [19].

11.3.3 Ion Drag Force

Drag forces arise from the motion of a plasma species relative to the dust particle, i.e. ions in the case of the ion drag and neutral gas atoms/molecules in the case of the neutral drag. The force is given by the number of particles interacting with the dust per unit time $n\sigma \mathbf{v}_{\text{rel}}$ and the corresponding momentum transfer Δp, yielding

$$\mathbf{F}_{\text{drag}} = \Delta p n \sigma \mathbf{v}_{\text{rel}} \,, \tag{11.14}$$

where n is the density of the streaming species, σ is the cross section for interaction and $\mathbf{v}_{\text{rel}}$ is the relative velocity.

The ion drag is due to a streaming ion motion with velocity $\mathbf{u}_i$ by ambipolar diffusion or acceleration in the sheath. The ion drag consists of two parts, the collection force $\mathbf{F}_{\text{coll}}$ due to ions directly hitting the dust particle and the Coulomb force $\mathbf{F}_{\text{Coul}}$ due to Coulomb scattering of the ions in the field of the dust particle [20, 21, 22, 23], see Fig. 11.1 (a). The collection force is just due to those ions which are also responsible for the ion charging of the dust and is thus given by [compare (11.2) and (11.3)]

$$\mathbf{F}_{\text{coll}} = m_i v_s n_i \mathbf{u}_i \pi a^2 \left(1 - \frac{2e\phi_{\text{fl}}}{m_i v_s^2}\right) \,. \tag{11.15}$$

Here, $v_s = (u_i^2 + v_{\text{th},i}^2)^{1/2}$ is the mean velocity of the ions and $\sigma = \pi b_c^2$ is the collection cross section.

The Coulomb force is due to those ions which are not collected by the dust, but are deflected in the electric field of the dust grain. It is given by

$$\mathbf{F}_{\text{Coul}} = m_i v_s n_i \mathbf{u}_i 4\pi b_{\pi/2}^2 \ln \left(\frac{\lambda_D^2 + b_{\pi/2}^2}{b_c^2 + b_{\pi/2}^2}\right)^{1/2} \,, \tag{11.16}$$

where $b_{\pi/2} = Q_\mathrm{d} e/(4\pi\varepsilon_0 m_\mathrm{i} v_\mathrm{s}^2)$ is the impact parameter for 90° deflection and $\sigma = 4\pi b_{\pi/2}^2 \ln\left[(\lambda_\mathrm{D}^2 + b_{\pi/2}^2)/(b_\mathrm{c}^2 + b_{\pi/2}^2)\right]^{1/2}$ is the cross section for Coulomb scattering [24]. In this Coulomb cross section only ion trajectories within one Debye length λ_D around the dust are considered. For the highly charged dust grains also Coulomb collisions outside the Debye sphere might contribute to the ion drag force [21, 25, 26].

The total ion drag force is then given by

$$\mathbf{F}_\mathrm{ion} = \mathbf{F}_\mathrm{coll} + \mathbf{F}_\mathrm{Coul} \qquad (11.17)$$

and is directed along the ion streaming motion. In this description, ion-neutral collisions are neglected. However, recent investigations [27] have indicated that ion-neutral collisions can substantially modify the force and can even lead to a force opposite to the ion motion.

11.3.4 Neutral Drag Force

The neutral drag is due to friction of the dust particle with the neutral gas background. The force on a moving dust grain with velocity $\mathbf{v}_\mathrm{d}$ is readily given as [28]

$$\mathbf{F}_\mathrm{n} = -\delta \frac{4}{3}\pi a^2 m_\mathrm{n} v_{\mathrm{th,n}} n_\mathrm{n} \mathbf{v}_\mathrm{d}, \qquad (11.18)$$

where $m_\mathrm{n}, n_\mathrm{n}$ and $v_{\mathrm{th,n}}$ are the mass, the density and the thermal velocity of the neutral gas atoms, respectively. The parameter δ lies in the range between 1 and 1.44 and takes into account how the neutral gas atoms are deflected from the particle surface (diffuse, reflecting, isotropic etc.). We will often use the force in the form

$$\mathbf{F}_\mathrm{n} = -m_\mathrm{d}\beta_E \mathbf{v}_\mathrm{d} \quad \text{with} \quad \beta_E = \delta\frac{8}{\pi}\frac{p}{a\varrho_\mathrm{d} v_{\mathrm{th,n}}}. \qquad (11.19)$$

Here, β_E is the so-called Epstein friction coefficient. It depends linearly on the gas pressure p and is inversely proportional to the particle radius a which means that in relation to their mass smaller particles experience stronger damping than larger particles.

11.3.5 Thermophoresis

The thermophoretic force acts on a dust particle due to a temperature gradient in the neutral gas. In a simplified picture, neutral gas atoms hitting the dust grain from the "hotter" side have a larger momentum and thus exert a stronger force than atoms from the "colder" side, which leads to a force towards colder gas regions. From gas kinetic theory this force is found to be

$$\mathbf{F}_\mathrm{th} = -\frac{16}{15}\sqrt{\pi}\frac{a^2 k_\mathrm{n}}{v_{\mathrm{th,n}}}\nabla T_\mathrm{n} \qquad (11.20)$$

with T_n being the temperature of the neutral gas and k_n the thermal conductivity of the gas. The thermophoretic force has been intentionally applied for levitation of particles using strong temperature gradients by Rothermel et al. [29].

11.3.6 Dust Levitation and Trapping

After the description of the relevant forces on the dust particles now the question is how these forces lead to particle levitation and trapping.

For large particles in the micrometer range the dominant forces are gravity and electric field force. Ion drag and thermophoresis can usually be neglected. The neutral drag is of interest only for moving particles and not for the identification of a stable equilibrium. For large, negatively charged particles, a force balance is only obtained in the sheath of the lower electrode, where the upward electric field force is strong enough to balance gravity [see Fig. 11.2 (c)]. Since the electric field strength in the sheath increases towards the electrode there typically is a single position where electric field force and gravity balance. There, horizontally extended, but vertically restricted dust arrangements are possible [compare Fig. 11.5 (b)].

For smaller particles in the nanometer range or for large particles under microgravity conditions gravity is not important. Smaller electric fields are sufficient to levitate and trap the particles. The electric field force is pointing into the plasma bulk for negatively charged particles, whereas ion drag and thermophoresis usually point outward. Thus, trapping of dust particles should be possible in the entire plasma volume and three-dimensionally extended dust clouds should be formed. However, it is found that for nanometer particles or under microgravity conditions large regions without dust particles, so-called "voids", exist [30, 31, 32] which are assumed to be due to the interaction of the ion drag and electric field force [33]. Thus, the study of 3D effects will be one of the challenging problems for the ongoing and planned investigations on the International Space Station [32].

11.3.7 Vertical Oscillations and Dust Charges

Here, various types of vertical oscillations will be discussed. These oscillations have been used, for example, to measure the dust charge.

Linear Resonances

In laboratory experiments with large particles in rf discharges [see Figs. 11.2(a) and 11.2(b)] the gravitational force on the microspheres is balanced by the electric field force with a spatially dependent electric field $E(z)$. Thus a unique equilibrium position z_0 exists where the particles are trapped [see Fig. 11.2 (c)], i.e.

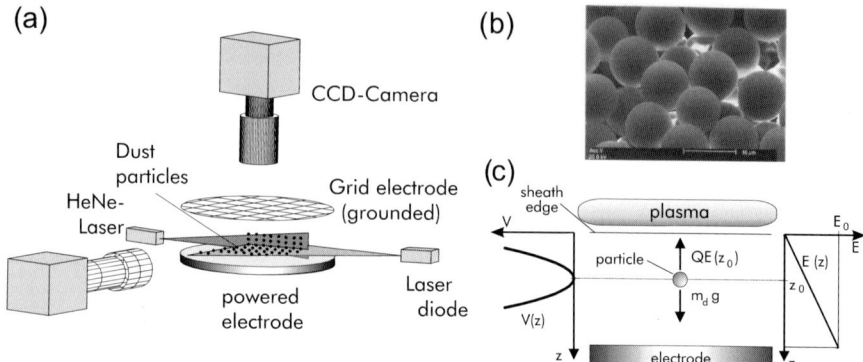

Fig. 11.2. (a) Scheme of the experimental setup in a typical experiment on dusty plasmas. The particles are illuminated by vertical and horizontal laser sheets. The particle motion is recorded from *top* and from the *side* with video cameras. (b) Electron micrograph of the melamine formaldehyde (MF) particles typically used in the experiments. (c) Trapping of the particles in the sheath of an rf discharge. See text for details

$$Q(z_0)E(z_0) = m_\mathrm{d} g \ . \tag{11.21}$$

The equation of motion for a particle in the vertical direction (relative to the equilibrium position) is then given by

$$m_\mathrm{d}\ddot{z} + m_\mathrm{d}\beta_E \dot{z} + Q(z)E(z) = F_\mathrm{ext} \ , \tag{11.22}$$

where also the neutral gas drag is considered and F_ext are other external forces applied to the particle. For a general equation of motion, one has to consider that the dust charge depends on the plasma conditions and therefore is itself a dynamic variable.

For the simpler case when a constant particle charge $Q(z) = Q_0$ and a linearly increasing electric field $E(z) = E_0 + E_1(z - z_0)$ is assumed [34, 35] the microspheres are trapped in a harmonic potential well [36, 37]

$$\frac{1}{2}m_\mathrm{d}\omega_0^2(z - z_0)^2 = \frac{1}{2}Q_0 E_1 (z - z_0)^2 \tag{11.23}$$

with a resonance frequency of

$$\omega_0^2 = \frac{Q_0}{m_\mathrm{d}} E_1 \ . \tag{11.24}$$

This equation of motion is just that of a damped harmonic oscillator. By measuring the resonance frequency the charge-to-mass ratio and thus the particle charge Q_0 can be derived [36, 37].

The charge measurements have been performed using monodisperse MF microspheres [see Fig. 11.2 (b)], which are perfectly spherical and have a very

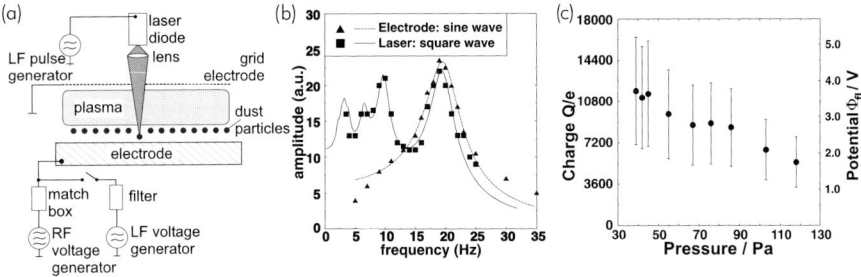

Fig. 11.3. Measuring the charge on MF microspheres. (a) Experimental setup for excitation of resonances by rf voltage modulation and laser manipulation. (b) Resonance curves obtained for a 9.47 μm MF particle for both excitation techniques. (c) Measured dust charge as a function of discharge pressure [38, 39]

low mass dispersion. The vertical oscillations were driven by applying a very low-frequent modulation of the electrode rf voltage, see Fig. 11.3 (a). In doing so, the sheath width is modulated and the particle is forced to oscillate in the trapping potential well. From a frequency scan, a resonance in oscillation amplitude was obtained at about 20 Hz, see Fig. 11.3 (b). The corresponding particle charge is found to be about 10 000 elementary charges from (11.24) and the floating potential is about 3 V [Fig. 11.3 (c)]. It is seen that the dust charge slowly increases from about 6 000 to 11 000 elementary charges with decreasing neutral gas pressure (120 to 40 Pa).

The width of the measured vertical resonance peak is determined by the neutral gas drag on the particle and is in quantitative agreement with the friction coefficient β_E in (11.19).

Besides electrode modulation, a non-invasive laser manipulation technique has been used to excite vertical resonances [39]. There, a laser beam is focused onto a single dust particle which is then pushed by the radiation pressure of the beam. By periodically switching "on" and "off" the laser beam vertical oscillations are driven. Comparing the main resonance frequency ω_0 measured from laser manipulation with that from the voltage modulation no difference within experimental errors is found. By laser excitation additional spurious resonances at $\omega_0/2$, $\omega_0/3$ etc. are excited due to the square wave laser excitation (laser "on" and "off").

In ex-situ charge measurements [40, 41] particles are dropped through a discharge into a Faraday cup where their charge is measured. In these experiments the influence of different charging mechanisms like electron beams or UV radiation was investigated. The particles were, e.g., irradiated by a strong UV source where the particles are found to be charged positively. With an additional photoemitting cathode nearby, the particles are charged negatively due to the electrons released by this cathode.

Nonlinear Oscillations

In recent experiments [42, 43, 44] also nonlinear vertical oscillations have been demonstrated.

Parametric resonances have been observed by Schollmeyer et al. [42]. There, a wire was placed close to the dust particles in the sheath and a sinusoidal electric potential was applied to the wire. At low excitation voltages, a vertical resonance at ω_0 was observed like in the case of the electrode modulation described above. At higher voltages, however, a second resonance at $2\omega_0$ was observed [see Fig. 11.4 (a)]. The presence of the second harmonic at $2\omega_0$ is a clear indication of parametric resonance. They are a consequence of a periodic modulation of the external confining potential. The equation of motion then reads

$$\ddot{z} + \beta_E \dot{z} + \omega_0^2 (1 + h \cos \omega t) = 0 , \qquad (11.25)$$

where h is the modulation depth and ω the modulation frequency of the confining potential frequency ω_0. This equation is known as Mathieu's equation in mechanics. When friction is present a threshold in modulation depth h must be overcome to excite the second resonance which also has been demonstrated in the experiment. It was concluded, that the presence of a driven wire in the plasma disturbs the sheath environment and thus the vertical confinement which leads to the excitation of parametric resonances. Thus, wires have to be used with great care in dusty plasma experiments.

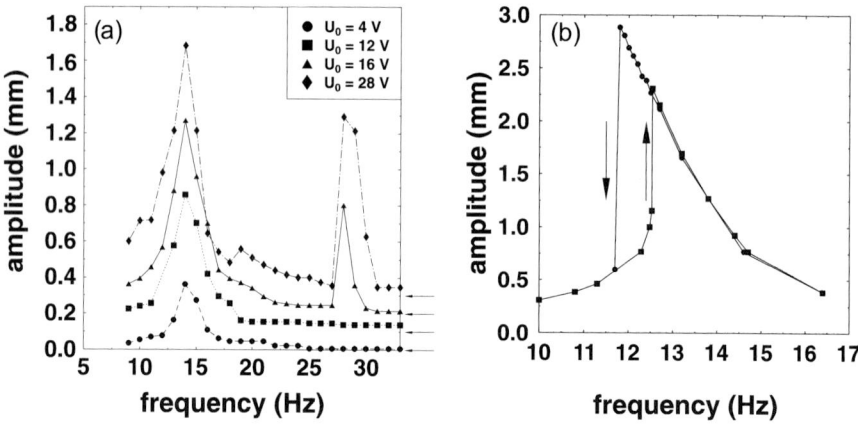

Fig. 11.4. (a) Parametric resonances observed at $2\omega_0$ for high excitation voltages on a wire close to the dust particles [42]. (b) Nonlinear resonance for high voltages using the electrode modulation technique [44]. A hysteric behaviour is observed where a jump in amplitude occurs at about 12.5 Hz when increasing the frequency and at about 11.5 Hz when decreasing

A different type of nonlinearity has been observed in experiments of Ivlev et al. [43] and Zafiu et al. [44] where large amplitude sinusoidal voltages have been applied to a wire [43] or to the lower electrode [44]. Besides the excitation of a second harmonic, the main resonance was found to exhibit hysteretic behavior accompanied by a strong asymmetry [see Fig. 11.4 (b)]. This can be explained when the confining potential contains higher than parabolic terms. The equation of motion in this case is

$$\ddot{z} + \beta_E \dot{z} + C_1 z + C_2 z^2 + C_3 z^3 \ldots = 0 , \qquad (11.26)$$

where the coefficients $C_{1,2,3}$ determine the nonlinear potential well: $C_1 = \omega_0^2$ describes the linear resonance frequency, C_2 the upwards-downwards asymmetry and C_3 the weakening or strengthening of the confining potential for large oscillation amplitudes.

In Fig. 11.4 (b) the resonance is bent towards smaller frequencies reflecting the weakening of the potential well with larger amplitudes, i.e. $C_3 < 0$. The coefficients $C_{1,2,3}$ can be obtained from a comparison of the measured resonance curve and a numerical solution of (11.26). As can be seen from (11.22) these coefficients are directly related to the spatial dependence of the dust charge and the electric field in the sheath. From the analysis of the nonlinear resonance, Zafiu et al. [44] were able to relate the nonlinear coefficients to a position dependent dust charge due to the reduction of electron density deep in the sheath.

Finally, the effect of a position dependent dust charge and finite charging times can lead to the onset of self-excited vertical oscillations [45]. Energy can be gained during an oscillation when the actual dust charge is different from the equilibrium charge at each point due to delayed charging. When that energy gain can compensate energy loss due to friction growing oscillations can be observed.

11.4 Particle–Particle Interaction

After the discussion of the plasma forces on the dust particles, now the question will be addressed how the charged dust particles interact with each other under equilibrium conditions.

11.4.1 Strongly Coupled Systems and Plasma Crystals

Highly charged dust particles in a plasma environment can be described as strongly coupled systems. A detailed overview on strong coupling is given in [46]. Here, only the key points are briefly summarized. Strong coupling occurs when the electrostatic energy between neighboring particles exceeds their thermal energy which is described by the Coulomb coupling parameter

$$\Gamma = \frac{Z_d^2 e^2}{4\pi\varepsilon_0 b k_B T_d}, \quad (11.27)$$

where T_d is the dust temperature and the Wigner–Seitz radius $b = (3/4\pi n_d)^{\frac{1}{3}}$ is of the order of the interparticle distance (see Chap. 6 in Part I). The one-component plasma (OCP) describes the simple case of point charges interacting via pure Coulomb interaction immersed in a homogeneous neutralizing background. There, a fluid-solid phase transition is found when the coupling parameter exceeds the critical value $\Gamma_c = 168$ [47]. For larger values $\Gamma > \Gamma_c$ the point charges arrange in an ordered crystalline structure, for smaller the charges are in an unordered fluid-like state.

When screening of the point charges by the ambient plasma background is taken into account the interaction is described by a Debye–Hückel or Yukawa potential

$$\phi(r) = \frac{Z_d e}{4\pi\varepsilon_0 r} \exp\left(-\frac{r}{\lambda_D}\right) \quad (11.28)$$

with the screening length λ_D. Thus, the Yukawa system is described by the additional parameter screening strength $\kappa = b/\lambda_D$ which is the Wigner–Seitz radius in units of the screening length. For $\kappa \to 0$ the OCP limit is retrieved. In Yukawa systems the critical coupling parameter for the fluid-solid transition increases almost exponentially with the screening strength κ [see Fig. 11.5 (a)].

Dust particles trapped in the sheath of an rf discharge form horizontally extended crystalline arrangements with 1 up to 10 vertical layers, the so-called plasma crystal [3, 4, 36, 49], see Fig. 11.5 (b). Due to the high particle charge and low dust temperature the coupling parameter Γ of these systems typically is of the order of a few thousand corresponding to an ordered crystalline state. Although also other crystal structures have been observed [50, 51], typically,

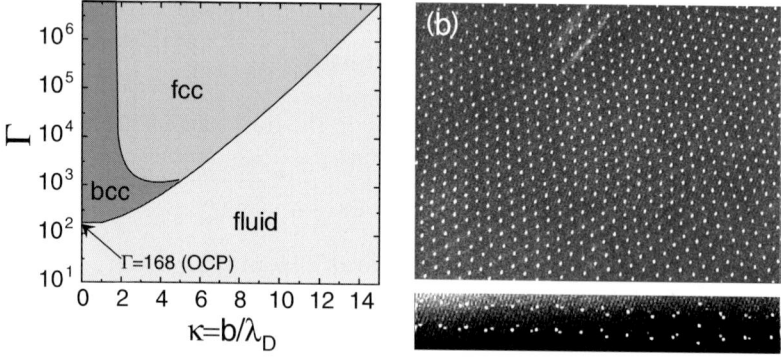

Fig. 11.5. (a) Phase diagram of the 3D Yukawa system [48]. (b) *Top* and *side view* of a two-layer plasma crystal in an rf discharge [38]. The horizontal interaction describes what goes on in the horizontal plane (*top view*), the vertical interaction can be seen in the *side view*

a hexagonal structure is found within a given layer in which one particle is surrounded by six nearest neighbors. In the vertical direction, however, the dust particles of different layers are found to be arranged directly on top of each other [37]. The horizontal hexagonal crystal structure is compatible with a repulsive Yukawa interaction, whereas the vertical arrangement cannot be understood by purely repulsive forces.

11.4.2 Horizontal Interaction

Konopka et al. [52] have collided two dust particles at different relative speeds. The horizontal interaction has been derived from the measured collision dynamics accounting for an external confining potential and friction by the neutral gas. The interaction is indeed found as a purely repulsive Yukawa potential with dust charges around 15 000 elementary charges and a screening length of the order of the electron Debye length [Fig. 11.6 (a)].

11.4.3 Vertical Interaction

The vertical interaction has been studied extensively in experiments [53, 54, 55] and in simulations [56, 57]. They have shown that the force between vertically aligned dust particles is very peculiar: the dust particles experience a nonreciprocal, attractive force which means that the dust particle of the lower layer is attracted by the upper particle, but the upper particle experiences only the Coulomb repulsion from the lower.

The origin of this nonreciprocal attraction lies in the ion streaming motion in the sheath. Ions enter the sheath with Bohm velocity (see Chap. 5 in Part I) and move downwards through the plasma crystal. In the electric field of the upper dust particle the ions are deflected below the upper particles. There they form a region of enhanced positive space charge ("ion cloud") which provides the attraction for the lower particles. Since the ions move at supersonic velocity in the sheath the attractive force can only be communicated downwards and not upwards, which results in the peculiar situation that only the lower dust particle experiences attraction, but there is no reaction on the upper particle.

The nonreciprocal attraction has been directly demonstrated using laser manipulation techniques in a system with two vertically aligned particles [53, 54] [see Fig. 11.6 (b)]. When the upper particle was pushed by a laser beam the lower particle was found to closely follow the motion of the upper one which demonstrates the attractive force on the lower particle [see Fig. 11.6 (c)]. When, however, the lower particle is pushed, the aligned pair is separated and the upper particle moves away from the lower which directly shows the repulsive force on the upper particle [see Fig. 11.6 (d)]. The strength of the attractive force on the lower particle has been measured from a balance of laser pressure and attractive force. The attractive force was found to be larger than the repulsive force which explains the vertical alignment.

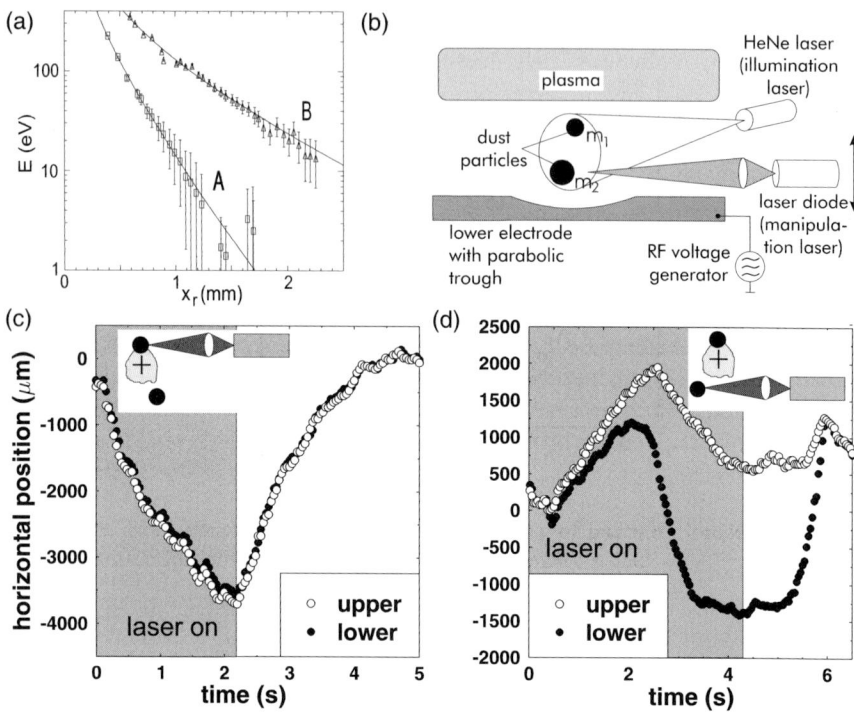

Fig. 11.6. (a) Measured horizontal interaction potential from collisions of two dust particles. The curves A and B reflect two different discharge conditions [52]. The *solid lines* represent best fits of a Yukawa potential with $Z_d = 13\,900$ and $\lambda_D = 0.34$ mm for curve A and $Z_d = 17\,100$ and $\lambda_D = 0.78$ mm for curve B. (b) Measurement of the attractive force between two dust particles. Scheme of the experimental setup. (c) Horizontal position of upper and lower particle when the upper particle is pushed and (d) when the lower particle is pushed [53, 54]

11.4.4 Phase Transitions

Phase transitions of plasma crystals from an ordered, solid state to a fluid and gas-like state have been observed when the gas pressure in the discharge is reduced [38, 58]. The phase transition has been investigated in view of correlation functions and defect organization [38, 59]. The phase transitions occur due to an enormous increase of the dust temperature from essentially room temperature at high gas pressures to about 50 eV at low gas pressures [38]. This dramatic dust heating cannot be explained by changes of the plasma parameters with gas pressure.

In detailed simulations [60, 61] Schweigert et al. have shown that the particle heating and the phase transitions are driven by an instability due to the nonreciprocal, attractive force. This is also substantiated by the fact that single layer crystals, which are not subject to the nonreciprocal attraction, do not experience such a melting transition. Many characteristic features of the plasma crystal melting have been explained in great detail and in good agreement with the simulations where the nonreciprocal attraction is identified as the origin of the melting transition.

11.5 Waves in Weakly Coupled Dusty Plasmas

We now turn to the topic of waves in dusty plasmas. There is a vast amount of literature on waves in dusty plasmas. Here, we focus on those types of waves which have been observed in experiments. For a more detailed overview the reader is referred to recent monographs [1, 62].

In general, two categories of waves can be identified, namely those which do not require strong coupling of the dust particles and those which rely on the strong coupling. In the first category, we find, e.g., the dust-acoustic (DAW) and dust ion-acoustic wave (DIAW). The dust lattice wave (DLW) with its different "polarizations" requires an ordered dust arrangement on lattice sites and thus belongs to the second category. Here, we start with the discussion of the weakly coupled waves, the DAW and the DIAW.

11.5.1 Dust-Acoustic Waves

The dust acoustic wave [63, 64] is a very low-frequent wave with wave frequencies of the order of the dust plasma frequency $\omega_{\rm pd}$ which, due to the high dust mass, is much less than the ion plasma and electron plasma frequency $(\omega_{\rm pi}, \omega_{\rm pe})$

$$\omega_{\rm pd} = \left(\frac{Z_{\rm d}^2 e^2 n_{\rm d0}}{\varepsilon_0 m_{\rm d}}\right)^{1/2} \ll \omega_{\rm pi}, \omega_{\rm pe} \,, \tag{11.29}$$

where $n_{\rm d0}$ is the equilibrium (undisturbed) dust density. The DAW is driven by electrons and ions and the inertia is provided by the massive dust particles. Thus, the DAW is a complete analog to the ion-acoustic wave, where the dust particles take the role of the ions and the ions and electrons take the role of the electrons in the ion-acoustic wave (see Chap. 2 in Part I).

For the derivation of the DAW the equation of continuity, the momentum equation and Poisson's equation for the dust species are used:

$$\frac{\partial n_{\rm d}}{\partial t} + \frac{\partial}{\partial x}(n_{\rm d} v_{\rm d}) = 0 \,, \tag{11.30}$$

$$\frac{\partial v_{\rm d}}{\partial t} + v_{\rm d} \frac{\partial v_{\rm d}}{\partial x} + \gamma_{\rm d} \frac{k_{\rm B} T_{\rm d}}{m_{\rm d} n_{\rm d}} \frac{\partial n_{\rm d}}{\partial x} = \frac{Z_{\rm d} e}{m_{\rm d}} \frac{\partial \phi}{\partial x} - \beta_E v_{\rm d} \,, \tag{11.31}$$

$$\frac{\partial^2 \phi}{\partial x^2} = -\frac{e}{\varepsilon_0}(n_i - n_e - Z_{\rm d} n_{\rm d}) \,. \tag{11.32}$$

There are a few small differences to the ion-acoustic wave: the momentum equation (11.31) includes friction with the neutral gas, and Poisson's equation (11.32) includes all three charged species, electrons, ions and dust. To solve these equations, the dust density and velocity as well as the electron and ion densities are considered as fluctuating quantities where a Boltzmann distribution for electrons *and* ions is assumed

$$n_e = n_{e0} \exp\left(\frac{e\phi}{k_B T_e}\right) \qquad n_i = n_{i0} \exp\left(-\frac{e\phi}{k_B T_i}\right). \qquad (11.33)$$

Here, n_{e0} and n_{i0} denote the equilibrium (undisturbed) values of the electron and ion density. The undisturbed densities are considered quasineutral, i.e. $n_{i0} = n_{e0} + Z_d n_{d0}$, where the dust is assumed to be negatively charged and adds to the electron charge density. The full dispersion relation of the DAW is obtained as [65]

$$\omega^2 + i\beta_E \omega = \left\{ \gamma_d \frac{k_B T_d}{m_d} + \epsilon Z_d^2 \frac{k_B T_i}{m_d} \frac{1}{[1 + T_i/T_e(1 - \varepsilon Z_d) + q^2 \lambda_{Di}^2]} \right\} q^2, \qquad (11.34)$$

where $\epsilon = n_{d0}/n_{i0}$ is the relative dust density in units of the ion density and q is the wave vector. For cold dust ($T_d = 0$) and cold ions ($T_i \ll T_e$) the dispersion relation simplifies to

$$\omega^2 + i\beta_E \omega = \frac{\omega_{pd}^2 q^2 \lambda_{Di}^2}{1 + q^2 \lambda_{Di}^2}. \qquad (11.35)$$

The dispersion relation of the DAW is shown in Fig. 11.7 (a). For large wave numbers $q^2 \lambda_{Di}^2 \gg 1$ the wave is not propagating and oscillates at the dust plasma frequency ω_{pd}. For small wave numbers $q^2 \lambda_{Di}^2 \ll 1$ the wave is acoustic $\omega = q C_{DAW}$ with the dust-acoustic wave speed

$$C_{DAW} = \left(\frac{k_B T_i}{m_d} \epsilon Z_d^2\right)^{1/2}. \qquad (11.36)$$

As for the ion-acoustic wave, the wave speed is determined by the temperature of the lighter species (T_i) and the mass of the heavier (m_d) (see Chap. 2 in Part I). The wave speed also includes the contribution of the dust charge Z_d and the relative dust concentration ϵ. It is also interesting to note that the governing shielding length is the *ion* Debye length λ_{Di} as the ions are the oppositely charged fluid that shields the repulsion between the dust particles.

In the experiment, the wave motion is influenced by friction with the neutral gas. When waves are excited, e.g. by a voltage on a wire [66], the wave frequency ω has to be taken as a real value and, consequently, the wave vector has to be treated as complex $q = q_r + iq_i$, where the real part $q_r = 2\pi/\lambda$ is related to the wave length λ and the imaginary $q_i = 1/L$ to the damping length L in the system. Figures 11.7 (b) and 11.7 (c) show the

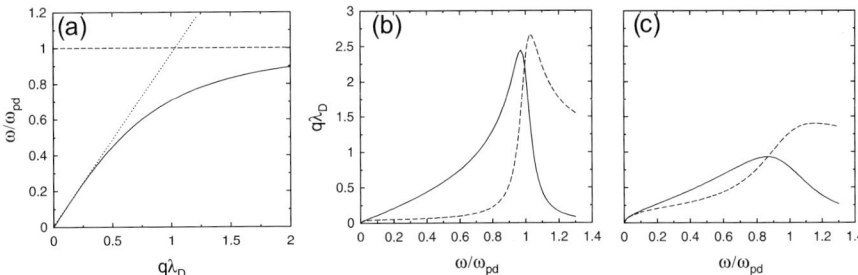

Fig. 11.7. (a) Dispersion relation of the dust-acoustic wave without damping. The *solid line* is the full dispersion relation, the *dotted line* indicates the acoustic limit with the dust-acoustic velocity. (b) Dispersion relation with small friction ($\beta_E = 0.1\,\omega_{\mathrm{pd}}$) and (c) with large friction ($\beta_E = 0.5\,\omega_{\mathrm{pd}}$). Here, the *solid line* refers to the real part of the wave vector and the *dashed line* to the imaginary part. Note, that in (b) and (c) the axes have been exchanged compared to (a)

DAW dispersion for small and large values of the friction coefficient β_E. For small friction the real part of the wave vector behaves similarly to the case of no damping. Close to $\omega = \omega_{\mathrm{pd}}$ the wave vector turns over and decreases dramatically towards zero again. In this range the imaginary part of the wave vector jumps from small values, i.e. low damping, to a large value. For $\omega > \omega_{\mathrm{pd}}$ an overcritically damped DAW is found. For larger friction constants [see Fig. 11.7 (c)] the wave speed ω/q increases and the maximum observable wave number decreases drastically. Moreover, the real and imaginary part of the wave vector are comparable over the entire range: the DAW is found to be strongly damped throughout.

Dust acoustic waves have been observed experimentally in weakly [67, 68] and strongly coupled dusty plasma systems [66]. In the weakly coupled system [67, 68], a dc discharge is driven between an anode disk and the chamber walls. The dust particles are accumulated from a dust tray placed below the anode region. The dust is found to form dust density waves with a certain wavelength and frequency [see Fig. 11.8 (a)]. By applying a sinusoidal voltage on the anode the wave can be driven and the dispersion relation is obtained [see Fig. 11.8 (b)]. The wave shows a linear, acoustic dispersion in agreement with the DAW at long wavelengths.

In a different experiment [66], dust-acoustic waves have been driven in a plasma crystal by a sinusoidal voltage on a wire close to the crystal. The propagation of the wave in the crystal was observed by video cameras and the corresponding wave length and damping length are derived. The measured dispersion relation was found to be in close agreement with a damped DAW, although the system is strongly coupled.

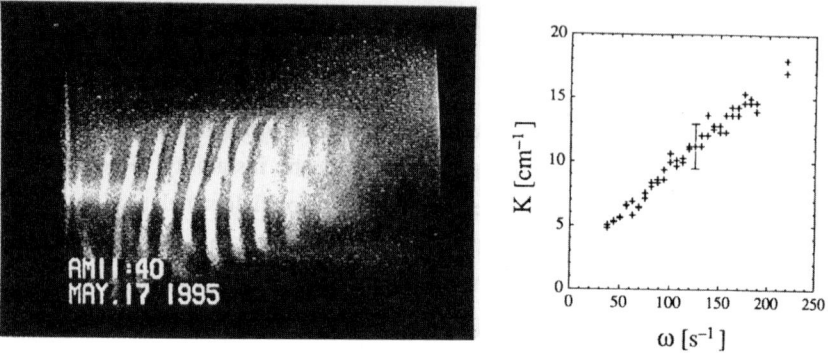

Fig. 11.8. Observation of the DAW in a DC discharge. (**a**) The DAW is seen as regions of high and low dust density in scattered light. (**b**) Measured dispersion relation of the DAW [67, 68]

11.5.2 Dust Ion-Acoustic Wave

The dust ion-acoustic wave has been already introduced in Chap. 2 in Part I. In contrast to the DAW where the dust is the moving species, the dust particles are considered immobile in the DIAW since the wave frequencies are of the order of the ion plasma frequency $\omega_{\rm pi} \gg \omega_{\rm pd}$. The influence of the dust lies only in the reduction of the free electron density since a certain fraction of the electrons are attached to the dust. The dispersion relation of the DIAW is then given as

$$\omega^2 = \frac{\omega_{\rm pi}^2 \lambda_{\rm De}^2 q^2}{1 + q^2 \lambda_{\rm De}^2} = \left(\frac{n_{\rm i0}}{n_{\rm e0}}\right) \frac{k_{\rm B} T_{\rm e}}{m_{\rm i}} \frac{q^2}{1 + q^2 \lambda_{\rm De}^2}\,, \tag{11.37}$$

which is that of the pure ion-acoustic wave with the additional factor of $n_{\rm i0}/n_{\rm e0} > 1$. With increasing dust charge density and thus reduced electron density the speed of the DIAW will increase in comparison to the pure ion-acoustic wave. That increase of the DIAW wave speed has been observed by Merlino et al. [65] and is shown in Fig. 11.9.

11.6 Waves in Strongly Coupled Dusty Plasmas

In this section, the dust lattice wave with its different "polarizations", compressional, shear and transverse mode, will be discussed. As the name suggests, the dust lattice wave requires the particles to be ordered in a crystal lattice. Here, we will deal with lattice waves in 2D systems.

For the compressional (longitudinal) mode, the particle motion is along the wave propagation direction leading to compression and rarefaction of the

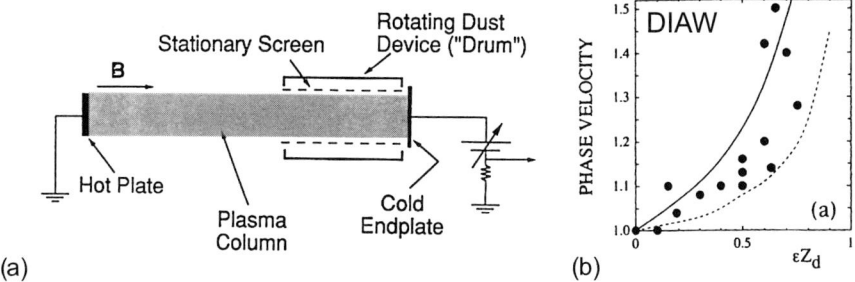

Fig. 11.9. (a) Experimental setup for the observation of the DIAW. The dust is immersed into the plasma by a rotating dust "drum". (b) Measured velocity of the DIAW with increasing dust charge density ϵZ_D [65]

dust. In the shear mode, the dust motion is perpendicular to the wave propagation but inside the 2D crystal plane. The transverse mode also describes particle motion perpendicular to the wave propagation, but here the dust motion is an out-of-plane motion and thus requires the consideration of the vertical confinement of the dust. These three wave types have been observed in the experiment and will be presented in the following.

11.6.1 Compressional Mode in 1D

The dispersion relation of the dust lattice wave (DLW) will be illustrated using the simpler model of a 1D chain of dust particles (see Fig. 11.10). On a linear chain the dust particles have equidistant equilibrium positions $X_n = nb$, where b is the interparticle distance. Neighboring dust particles are considered to be connected by springs of spring constant k. The equation of motion for the n-th particle then is

$$m_d \ddot{x}_n - m_d \beta_E \dot{x}_n = k(x_{n-1} - 2x_n + x_{n+1}) , \qquad (11.38)$$

where x_n is the elongation of the n-th particle from its equilibrium position X_n. Here, also friction with the neutral gas is included. Using the ansatz for

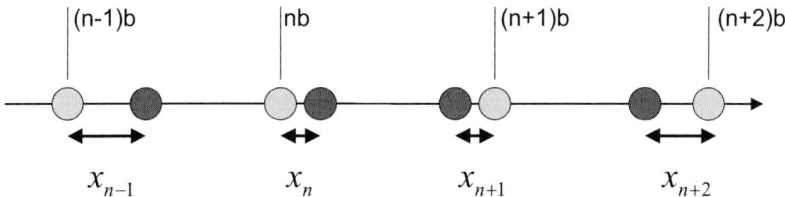

Fig. 11.10. Sketch of a linear dust arrangement with longitudinal particle displacements. The *light grey circles* indicate the equilibrium positions of the dust particles at fixed lattice positions. The *darker circles* represent a snapshot of a wave – like compressional disturbance

waves on a linear chain $x_n = A\exp(inqb - i\omega t)$ with wave vector q and wave frequency ω the equation of motion becomes

$$-m_\mathrm{d}\omega^2 - im_\mathrm{d}\beta_E\omega = k\left[\exp(iqb) + \exp(-iqb) - 2\right]$$
$$= 2k\left[\cos(qb) - 1\right] \quad (11.39)$$

and the dispersion relation

$$\omega^2 + i\beta_E\omega = 4\frac{k}{m_\mathrm{d}}\sin^2\left(\frac{qb}{2}\right) \quad (11.40)$$

is obtained. Now the ad-hoc introduction of the spring constant k has to be related to the repulsive interaction between the dust particles. The spring constant is just the second derivative of the interaction potential which yields for a Debye–Hückel interaction (see Sect. 11.4.1)

$$k = \left.\frac{\mathrm{d}^2\phi}{\mathrm{d}x^2}\right|_{x=b} = \frac{Z_\mathrm{d}^2 e^2}{4\pi\varepsilon_0 b^3}\exp(-\kappa)\left(2 + 2\kappa + \kappa^2\right), \quad (11.41)$$

where the screening strength $\kappa = b/\lambda_\mathrm{D}$ has been used. Finally, the dispersion relation can be extended to include also the influence of many neighbors. Therefore, simply the "springs" to all other neighbors at distance ℓb have to be considered yielding [69]

$$m_\mathrm{d}\ddot{x}_n - m_\mathrm{d}\beta_E\dot{x}_n = \sum_{\ell=1}^{\infty} k(\ell b)(x_{n-\ell} - 2x_n + x_{n+\ell}). \quad (11.42)$$

The full 1D dispersion relation then is given by

$$\omega^2 + i\beta_E\omega = \frac{1}{\pi}\omega_\mathrm{pd}^2\sum_{\ell=1}^{\infty}\frac{e^{-\ell\kappa}}{\ell^3}\left(2 + 2\ell\kappa + \ell^2\kappa^2\right)\sin^2\left(\frac{\ell qb}{2}\right), \quad (11.43)$$

where the dust plasma frequency $\omega_\mathrm{pd}^2 = Z_\mathrm{d}^2 e^2/\varepsilon_0 m_\mathrm{d} b^3$ has been introduced for the strongly coupled case by identifying $n_\mathrm{d} \approx b^{-3}$ in (11.29).

This dispersion relation can be extended to the case of a two-dimensional lattice in a straightforward manner for the compressional mode as well as for the shear mode [70, 71]. The computed dispersion relation of a 2D dust lattice wave is shown in Fig. 11.11. The compressional mode has a form that reflects the sine dependence: For long wavelengths $qb \ll 1$ the dispersion is acoustic. For shorter wave lengths the compressional mode becomes dispersive and attains a maximum near $qb = \pi$. In contrast, the shear mode is nearly acoustic for all wavelengths. It should be noted, here, that the sound speed of the compressional mode is much larger than that of the shear mode. This dispersion holds for finite values of the screening strength κ. For pure Coulomb interaction $\kappa = 0$ the sum in the compressional dispersion relation (11.43) would diverge. Instead for pure Coulomb interaction it is found that $\omega \propto \sqrt{q}$ for long wavelengths (and thus $\omega/q \to \infty$ for $q \to 0$).

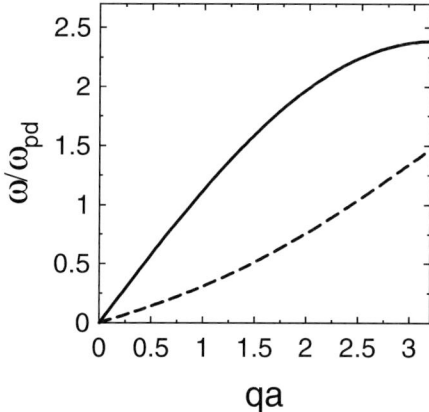

Fig. 11.11. Dispersion relation of a 2D dust lattice wave without damping. The *solid line* is the compressional mode and the *dashed line* is the shear mode for $\kappa = 2$

11.6.2 Compressional Dust Lattice Waves

For compressional dust lattice waves in 1D and 2D systems the excitation of the lattice wave using a focused laser beam has been demonstrated as a powerful technique [71, 72]. In the 2D case, the laser beam of an argon ion laser was expanded into a line focus and directed onto the first row of particles in a 2D plasma crystal. By periodic modulation of the laser power a plane wave was launched in the plasma crystal [see Fig. 11.12 (a)]. The wave motion of the dust was analyzed in terms of the phase and amplitude [see Figs. 11.12 (b) and 11.12 (c)] as a function of distance from the excitation region. The phase dependence directly reflects the wavelength λ and the amplitude decrease the damping length L for a given excitation frequency. The real part of the wave vector is derived from the wavelength as $q_\mathrm{r} = 2\pi/\lambda$ and the imaginary part from the damping length as $q_\mathrm{i} = 1/L$ [see Figs. 11.12 (d) and 11.12 (e)]. In that way the dispersion relation of the wave has been measured. The shape of the measured dispersion relation is found to be quite different from that of the DAW. In contrast, it shows good agreement with the 2D DLW relation. From this comparison the screening strength is determined to be $\kappa = 1 \pm 0.3$. This means that interparticle distance b and shielding length λ_D are comparable and are found to be close to the electron Debye length. However, the shielding mechanism is not fully understood so far: either electrons, which however are present in the sheath only for a limited fraction of the rf period, could be responsible for shielding or ions that enter the sheath with Bohm velocity and that thus could have an "effective" temperature close to the electron temperature.

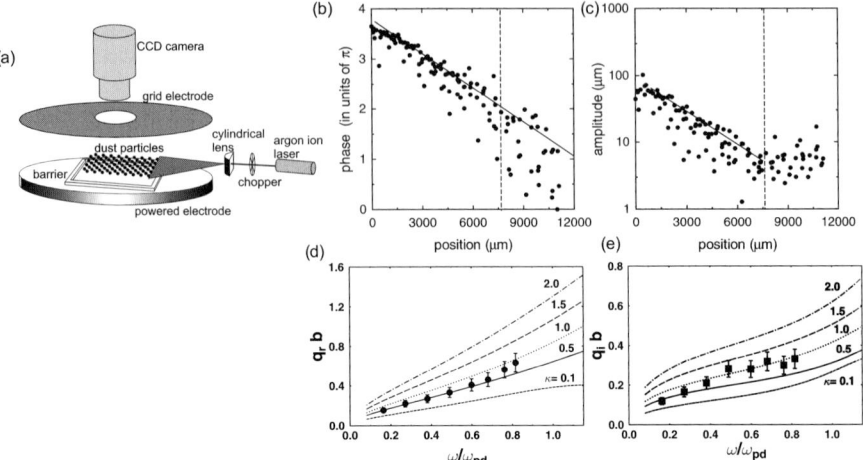

Fig. 11.12. (a) Scheme of the experimental setup for the excitation of 2D dust lattice waves. (b) Phase and (c) amplitude of the dust particle motion as a function of distance from the excitation region for an excitation frequency of 2.8 Hz. (d, e) Real and imaginary wave vector as a function of frequency. The symbols denote the experimental data. The lines indicate the dispersion relation of the 2D DLW for various values of the screening strength κ [71]

11.6.3 Shear Dust Lattice Waves

For shear dust lattice waves a short pulse of a narrow laser beam excites a row of particles along the direction of the beam (see Fig. 11.13) [73]. The velocity pulse created by the beam spreads in a direction perpendicular to the beam. The dust particle motion and pulse travel direction are perpendicular, thus a shear propagation is observed here. The outward velocity of the beam is much smaller than for a compressional pulse and in agreement with the acoustic velocity of the shear wave.

Recently, Nunomura et al. [74] have studied the wave propagation along different lattice orientations and found reasonable agreement with the theoretical DLW dispersion relation. The same authors have developed a method to derive the dispersion relation of the compressional and shear mode from the pure thermal Brownian motion of the dust particles [75]. This powerful technique allows the measurement of the entire dispersion from a single video sequence (see Sect. 11.6.6).

11.6.4 Mach Cones

When an object moves through a medium with a velocity faster than the wave speed in that medium a V-shaped disturbance, the Mach cone, is excited. This phenomenon is well known, e.g. from the sonic boom behind a plane at

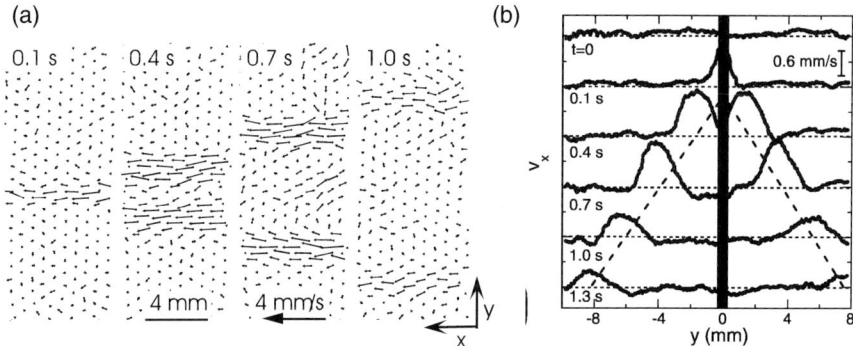

Fig. 11.13. Shear dust lattice waves. (**a**) Dust particle velocity vectors at certain time steps after a laser beam pulse. The initial laser beam pushed the particles in the central region from *right* to *left*. (**b**) Velocity profiles perpendicular to the beam direction. The central bar indicates the excitation region [73]

supersonic velocity. Similarly, Mach cones can be observed in dusty plasmas using objects faster than the acoustic speed of the DLW. Mach cones in dusty plasmas have first been observed by Samsonov et al. [76, 77]. There, dust particles which accidentally are trapped below the actual 2D plasma crystal are found to move at large, supersonic, speeds at low gas pressure. The disturbance by these lower particles excites a Mach cone in the upper plasma crystal.

Mach cones have also been generated in plasma crystals using the focal spot of a laser beam that was moved at supersonic speeds V through the crystal using a moving galvanometer scanning mirror (Fig. 11.14) [78]. The laser technique allows the formation of Mach cones in a repetitive and controllable manner. The Mach cone has an opening angle μ that satisfies the relation

$$\sin \mu = \frac{c}{V} \quad \text{with} \quad c = \lim_{q \to 0} \frac{\partial \omega}{\partial q} \tag{11.44}$$

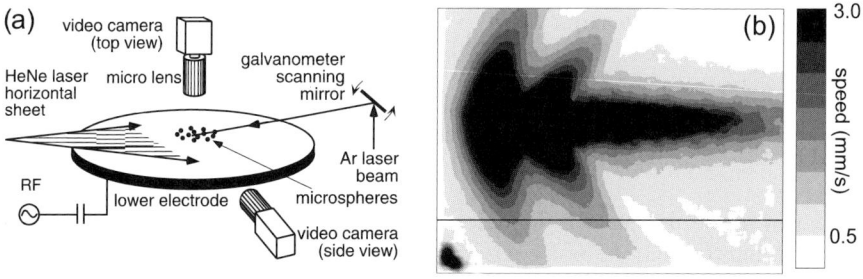

Fig. 11.14. Mach cones in dusty plasmas. (**a**) Scheme of the experimental setup. (**b**) Grey scale map of the particle velocities [78]

322   A. Melzer

being the sound speed of the DLW. Thus from the opening angle μ the sound speed of the DLW is readily obtained. Figure 11.14 (b) shows the Mach cone observed by the laser manipulation technique. A strong first Mach cone is easily seen. Additional secondary and tertiary Mach cones are also observable. These additional features arise from the dispersive characteristics of the DLW at shorter wavelengths [79]. Like the wave pattern of a moving ship, these multiple Mach cones can be interpreted as interference patterns of the wave packages launched by the moving laser beam.

The Mach cone in Fig. 11.14 is a compressional Mach cone due to excitation of compressional waves. Shear Mach cones have been demonstrated by Nosenko et al. [80]. Shear Mach cones are observed at much lower velocities V due to the much smaller acoustic velocity of the shear waves.

Mach cones are assumed to be observable in the rings of Saturn by the Cassini spacecraft after its arrival at Saturn in 2004 [81]. In Saturn's rings, large boulders moving in Keplerian orbits are likely to have supersonic speeds relative to the smaller dust particles which move at speeds determined by their electrostatic interactions with Saturn's plasma environment. The observation of Mach cones would allow detailed studies of the plasma conditions in the rings.

11.6.5 Transverse Dust Lattice Waves

The last wave type discussed here is the transverse dust lattice wave with an out-of-plane (vertical) particle motion perpendicular to the wave propagation is expected. Such vertical displacements are stabilized against the Coulomb repulsion of the particles by the vertical confinement potential. For the dispersion of the transverse DLW, vertical displacements z_n in a 1D chain of particles are considered (see Fig. 11.15). The equation of motion then reads

$$m_d \ddot{z}_n - m_d \beta_E \dot{z}_n + m_d \omega_0^2 z_n = k_z(z_{n-1} - 2z_n + z_{n+1}) , \qquad (11.45)$$

where ω_0 is the strength of the vertical confinement [compare (11.24)] and k_z is the vertical "spring" constant, using $z \ll b$

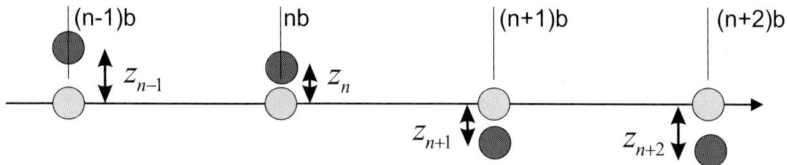

Fig. 11.15. Sketch of a linear dust arrangement with transverse particle displacements

$$F_z = F(r)\frac{z}{r} = F(\sqrt{z^2+b^2})\frac{z}{\sqrt{z^2+b^2}}$$

$$\approx F(b)\frac{z}{b} = \frac{Z_d^2 e^2}{4\pi\varepsilon_0 b^3}\exp(-\kappa)(1+\kappa)z = k_z z \quad (11.46)$$

and F_z is the vertical component of the Coulomb force $F(r)$. Following the above procedure, the dispersion relation of the transverse DLW is given by [82]

$$\omega^2 + i\beta_E\omega = \omega_0^2 - \frac{1}{\pi}\omega_{\rm pd}^2\exp(-\kappa)(1+\kappa)\sin^2\left(\frac{qb}{2}\right). \quad (11.47)$$

One can see that the influence of the vertical confinement ω_0^2 is necessary to yield a real dispersion relation. It is interesting to note that this wave is a backward wave ($\partial\omega/\partial q < 0$) and $\omega \to \omega_0$ for $q \to 0$ ("optical" wave).

Transverse dust lattice waves have been observed by Misawa et al. In their experiment, a linear chain of dust particles shows vertical oscillations (see Fig. 11.16) which propagate along the chain [83]. From the time traces it is immediately seen that the wave is a backward wave (negative slope in the space-time diagram). The authors have determined part of the dispersion relation where a finite frequency is found for $q \to 0$ and the dispersion also has a negative slope, as expected for the transverse DLW. However, the overall agreement of the measured and the theoretical dispersion is not very satisfying.

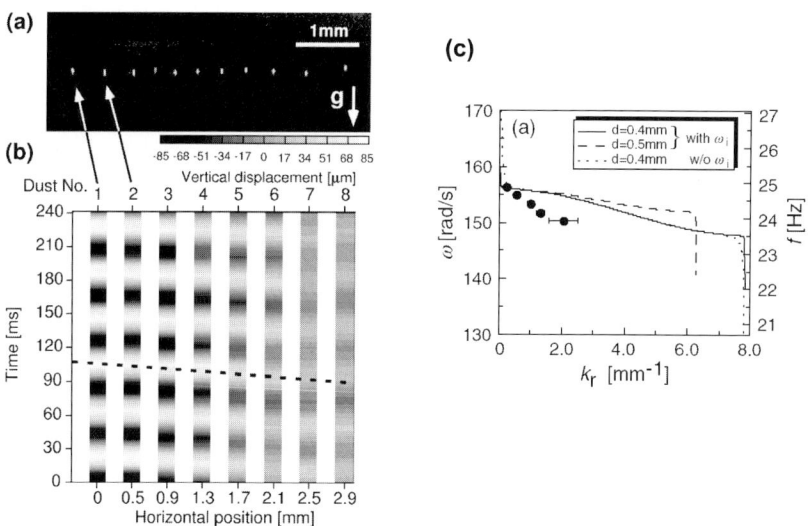

Fig. 11.16. Transverse dust lattice waves. (a) Still image of a 1D particle chain, (b) Grey scale image of the vertical displacement of the dust particles in the chain. The wave is seen to propagate backwards. (c) Measured dispersion relation of the transverse DLW [83]

324   A. Melzer

Recently Liu et al. [84] have measured the transverse DLW dispersion relation using laser excitation techniques with very good agreement with the theoretical dispersion relation.

11.6.6 Normal Modes in Finite Clusters

So far, waves in extended 1D and 2D systems have been presented. Now, we would like to draw the attention to collective effects in finite systems. Therefore, only a small number of dust particles $N = 1$ to 500, say, are trapped in the sheath above the lower electrode. In the horizontal plane an additional weak parabolic confinement (of strength $\omega_{\rm h}$) is applied. Under the interplay of the horizontal parabolic confinement and Coulomb repulsion the particles arrange in concentric shells [see Fig. 11.17 (b)], the so-called finite Coulomb clusters. The structure and their dynamic properties dramatically depend on the particle number N. Like in atomic and nuclear physics there exist "magic" particle numbers of high stability against perturbations, e.g. the N = 19 (1,6,12) cluster. The notation $(N_1, N_2, N_3, \ldots)$ refers to N_1 particles in the inner ring, N_2 in the second and so on. Such structures have indeed been considered as a possible model of the atom by J.J. Thomson in 1904 [86]. Finite Coulomb clusters are also observed in colloidal suspensions [87], in quantum dots [88] and electrons on liquid helium [89].

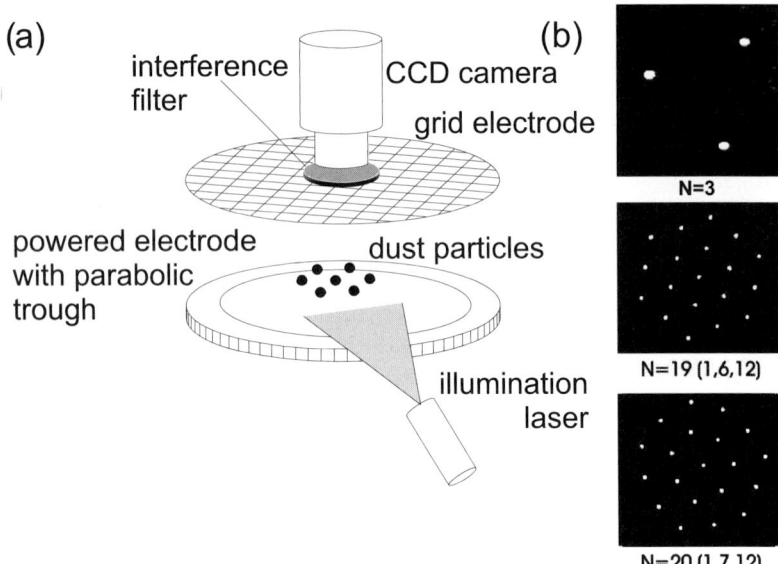

Fig. 11.17. (a) Scheme of the experimental setup for the investigation of finite Coulomb clusters. (b) Finite clusters with $N = 3$, 19 and 20 particles [85]

The finite clusters can be described in terms of their total energy

$$E = \frac{1}{2}m_d\omega_h^2 \sum_{i=1}^{N} r_i^2 + \frac{Z_d^2 e^2}{4\pi\varepsilon_0} \sum_{i>j}^{N} \frac{\exp(-r_{ij}/\lambda_D)}{r_{ij}} , \qquad (11.48)$$

where $\mathbf{r}_i = (x_i, y_i)$ is the position of the i-th particle in the horizontal plane and $r_{ij} = |\mathbf{r}_i - \mathbf{r}_j|$. The first term is the potential energy due to the confinement and the second is the Coulomb repulsion of the particles.

The equilibrium structure of these systems is derived from the minimum of the total energy [90, 91]. Experimentally they have been observed by Juan et al. [92], Klindworth et al. [93] and Goree et al. [94]. The observed cluster structures are in perfect agreement with the theoretical predictions.

The dynamic properties of finite clusters are described in terms of their normal modes which replace the dispersion relation of infinite systems [95]. The normal modes are obtained from the dynamical matrix

$$\mathsf{A} = \begin{pmatrix} \dfrac{\partial^2 E}{\partial x_i \partial x_j} & \dfrac{\partial^2 E}{\partial x_i \partial y_j} \\ \dfrac{\partial^2 E}{\partial y_i \partial x_j} & \dfrac{\partial^2 E}{\partial y_i \partial y_j} \end{pmatrix} , \qquad (11.49)$$

where the second derivatives are themselves $N \times N$ matrices that contain the possible combinations of i and j. The eigenvalues and eigenvectors of A describe the normal mode oscillations of the finite clusters. The eigenvalues are the oscillation frequencies and the eigenvectors describe the mode oscillation patterns.

This is demonstrated for the simple case of $N = 3$ particles in Fig. 11.18 (a) where the $2N$ eigenmodes are presented. Modes that occur in any cluster are the two sloshing modes (i.e. oscillations of the entire cluster in the horizontal confining potential, modes number 5 and 6), the rotation of the entire cluster (mode number 2) and the breathing mode (i.e. coherent, purely radial motion of all particles, mode number 1). For the three particle cluster also two "kink" modes are found (mode number 3 and 4).

These modes can be extracted experimentally from the thermal motion of the dust particles for infinite systems [73] and for finite Coulomb clusters [85]. First, the velocity of the Brownian motion of all particles $\mathbf{v}_i(t)$ is determined from a video sequence. Then, the contribution of the thermal motion to each of the eigenmodes is determined from the projection of the thermal velocities onto the eigenmode pattern by

$$f_\ell(t) = \sum_{i=1}^{N} \mathbf{v}_i(t) \cdot \mathbf{e}_{i,\ell} , \qquad (11.50)$$

where $\mathbf{e}_{i,\ell}$ is the eigenvector for particle i in mode number ℓ. Finally, the spectral power density of each mode ℓ

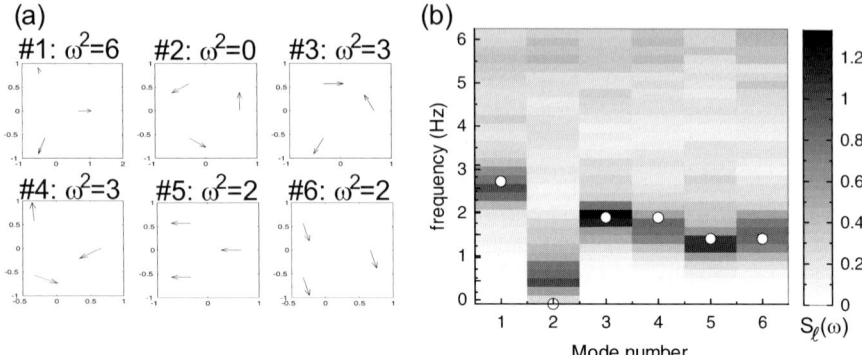

Fig. 11.18. (a) The six eigenmodes of a $N = 3$ cluster with the corresponding mode frequencies ω^2 (in units of $\omega_{\rm h}^2/2$). (b) Spectral power density of the three particle cluster derived from the thermal motion of the particles. The *white dots* indicate the best-fit theoretical values of the mode frequencies [85]

$$S_\ell(\omega) = \frac{2}{T} \left| \int_0^T f_\ell(t) \exp(i\omega t)\, dt \right|^2 \qquad (11.51)$$

of f_ℓ is calculated. The power spectrum contains the contribution of the thermal motion to each of the $2N$ eigenmodes. For the three particle cluster the power spectrum is shown in Fig. 11.18 (b). The observed power spectrum is in very good agreement with the expected mode frequencies. The power spectrum contains the full dynamic properties of the system.

This technique can be applied to clusters of any size and has been demonstrated for clusters up 145 particles [85]. From the analysis of the mode spectra the fundamental properties like particle charge and screening strength can be directly extracted. The particle charge was found to be about 10 000 elementary charges on a $2a = 9.5$ µm particle and the screening strength κ was found in the range between 0.5 and 2 [85].

Moreover, also the stability of the cluster ("magic number" configurations) can be judged from the power spectrum. The mode with the lowest eigenfrequency (besides cluster rotation) determines the stability of the cluster: when its eigenfrequency is close to zero this mode is easily excited and can lead to changes in cluster configuration. For the 19-particle cluster, which is a "magic number" configuration due to its high symmetry [see Fig. 11.17 (b)], the lowest eigenmode frequency is of the order of 1.2 Hz in the experiment. This gap demonstrates the stability of the 19-particle cluster. For the 20-particle cluster, the lowest eigenmode frequency (corresponding to differential rotation of inner and outer ring) is below 0.5 Hz which indicates that this cluster is only weakly stable against this intershell rotation [85, 93].

11.7 Summary

An overview over the collective effects in dusty plasmas has been given together with an introduction to the fundamental properties of dusty plasma like particle charging, interaction potentials and forces.

To summarize, the main properties of dusty plasmas compared to "usual" plasmas are the following:

– Dusty plasmas are at least three-component plasmas (electrons, ions and dust). In this sense, dusty plasmas are comparable to negative ion plasmas.
– The typical charge on the charge carriers (dust) is of the order of 10 000 elementary charges which leads to strong coupling with Γ of the order of several thousand.
– The dust charge is variable and depends on the local plasma parameters. Thus, the charge becomes a dynamic variable.
– The dust mass is by orders of magnitudes larger than that of electrons and ions. Thus, the dust plasma frequency $\omega_{\rm pd}$ is by orders of magnitude smaller than that of electrons and ions leading to convenient time scales for the observation of dynamic processes.
– The slow time scales allow that electrons *and* ions contribute to shielding which should result in different shielding scales.
– The dust size is not negligibly small leading to surface phenomena and forces which are unimportant in "usual" plasmas (see Chap. 5 in Part I).

All of these unique properties of dusty plasmas lead to a number of new collective effects, like new types of waves (DAW, DIAW, DLW), plasma crystallization and phase transitions.

References

1. F. Verheest: *Waves in Dusty Space Plasmas* (Kluver Academic Publishers, Dordrecht 2000)
2. A. Bouchoule (Ed.): *Dusty Plasmas* (John Wiley & Sons, Chichester 1999)
3. J.H. Chu and Lin I: Phys. Rev. Lett. **72**, 4009 (1994)
4. H. Thomas, G.E. Morfill, V. Demmel et al: Phys. Rev. Lett. **73**, 652 (1994)
5. D.J. Wineland, J.C. Bergquist, W.M. Itano et al: Phys. Rev. Lett. **59**, 2935 (1987)
6. F. Diedrich, E. Peik, J.M. Chen et al: Phys. Rev. Lett. **59**, 2931 (1987)
7. N.A. Clark, A.J. Hurd and B.J. Ackerson: Nature **281**, 57 (1979)
8. P. Pieranski: Contemp. Phys. **24**, 25 (1983)
9. H.M. Mott-Smith and I. Langmuir: Phys. Rev. **28**, 727 (1926)
10. T. Nitter: Plasma Sources Sci. Technol. **5**, 93 (1996)
11. M. Lampe, V. Gavrishchaka, G. Ganguli et al: Phys. Rev. Lett. **86**, 5278 (2001)
12. J. Goree: Phys. Rev. Lett. **69**, 277 (1992)
13. J.E. Allen, B.M. Annaratone, U. de Angelis: J. Plasma Phys. **63**, 299 (2000)
14. M. Lampe, G. Joyce, G. Ganguli: Phys. Plasmas **7**, 3851 (2000)

15. E.C. Whipple: Rep. Prog. Phys. **44**, 1197 (1981)
16. T. Nitter, O. Havnes and F. Melandsø: J. Geophys. Res. **103**, (A4), 6605 (1998)
17. N. Meyer-Vernet: Astronom. Astrophys. **105**, 98 (1982)
18. S. Hamaguchi, R.T. Farouki: Phys. Rev. E **49**, 4430 (1994)
19. U. Mohideen, H. Rahman, M.A. Smith et al: Phys. Rev. Lett. **81**, 349 (1998)
20. M.S. Barnes, J.H. Keller, J.C. Forster et al: Phys. Rev. Lett. **68**, 313 (1992)
21. M.D. Kilgore, J.E. Daugherty, R.K. Porteous et al: J. Appl. Phys. **73**, 7195 (1993)
22. J. Perrin, P. Molinàs-Mata, P. Belenguer: J. Phys. D–Appl. Phys. **27**, 2499 (1994)
23. X. Chen: J. Phys. D–Appl. Phys. **29**, 995 (1996)
24. F.F. Chen: *Introduction to Plasma Physics and Controlled Fusion, Vol. 1: Plasma Physics* (Plenum Press, New York London 1984)
25. S. Khrapak, A.V. Ivlev, G. Morfill et al : Phys. Rev. E **66**, 046414 (2002)
26. S. Khrapak, A.V. Ivlev, G. Morfill et al: Phys. Rev. Lett. **90**, 225002 (2003)
27. I.V. Schweigert, A.L. Alexandrov, F.M. Peeters: IEEE Trans. Plasma Sci. **32**, 623 (2004)
28. P.S. Epstein: Phys. Rev. **23**, 710 (1924)
29. H. Rothermel, T. Hagl, G. Morfill et al: Phys. Rev. Lett. **89**, 175001 (2002)
30. J.L. Dorier, C. Hollenstein, A. Howling: J. Vac. Sci. Technol. A **13**, 918 (1995)
31. G. Praburam, J. Goree: Phys. Plasmas **3**, 1212 (1996)
32. G.E. Morfill, H. Thomas, U. Konopka et al: Phys. Rev. Lett. **83**, 1598 (1999)
33. J. Goree, G. Morfill, V. Tsytovich et al: Phys. Rev. E **59**, 7055 (1999)
34. P. Belenguer, J.P. Blondeau, L. Boufendi et al: Phys. Rev. A **46**, 7923 (1992)
35. E. Tomme, D. Law, B.M. Annaratone et al: Phys. Rev. Lett. **85**, 2518 (2000)
36. A. Melzer, T. Trottenberg, A. Piel: Phys. Lett. A **191**, 301 (1994)
37. T. Trottenberg, A. Melzer, A. Piel: Plasma Sources Sci. Technol. **4**, 450 (1995)
38. A. Melzer, A. Homann, A. Piel: Phys. Rev. E **53**, 2757 (1996)
39. A. Homann, A. Melzer, A. Piel: Phys. Rev. E **59**, R3835 (1999)
40. B. Walch, M. Horanyi, S. Robertson: IEEE Trans. Plasma Sci. **22**, 97 (1994)
41. A. Sickafoose, J. Colwell, M. Horanyi et al: Photoelectric charging of dust particles. In: *Physics of Dusty Plasmas II* ed by J. Nakamura, P. Shukla (Elsevier Science, Amsterdam 2000) pp 367–372
42. H. Schollmeyer, A. Melzer, A. Homann et al: Phys. Plasmas **6**, 2693 (1999)
43. A. Ivlev, R. Sütterlin, V. Steinberg et al: Phys. Rev. Lett. **85**, 4060 (2000)
44. C. Zafiu, A. Melzer, A. Piel: Phys. Rev. E **63**, 066403 (2001)
45. S. Nunomura, T. Misawa, N. Ohno et al: Phys. Rev. Lett. **83**, 1970 (1999)
46. R. Redmer: Z. Phys. Chemie-Int. J. Res. Phys. Chem. Chem. Phys. **204**, 135 (1998)
47. S. Ichimaru: Rev. Mod. Phys. **54**, 1017 (1982)
48. S. Hamaguchi, R. Farouki, D.H.E. Dubin: Phys. Rev. E **56**, 4671 (1997)
49. Y. Hayashi, K. Tachibana: Jap. J. Appl. Phys. **33**, L804 (1994)
50. J. Pieper, J. Goree, R. Quinn: Phys. Rev. E **54**, 5636 (1996)
51. M. Zuzic, A.V. Ivlev, J. Goree et al: Phys. Rev. Lett. **85**, 4064 (2000)
52. U. Konopka, G. Morfill, L. Ratke: Phys. Rev. Lett. **84**, 891 (2000)
53. A. Melzer, V. Schweigert, A. Piel: Phys. Rev. Lett. **83**, 3194 (1999)
54. A. Melzer, V. Schweigert, A. Piel: Phys. Scr. **61**, 494 (2000)
55. K. Takahashi, T. Oishi, K. Shimomai et al: Phys. Rev. E **58**, 7805 (1998)
56. V.A. Schweigert, I.V. Schweigert, A. Melzer et al: Phys. Rev. E **54**, 4155 (1996)

57. A. Melzer, V.A. Schweigert, I.V. Schweigert et al: Phys. Rev. E **54**, 46 (1996)
58. H. Thomas, G.E. Morfill: Nature **379**, 806 (1996)
59. R.A. Quinn, C. Cui, J. Goree et al: Phys. Rev. E **53**, 2049 (1996)
60. V.A. Schweigert, I.V. Schweigert, A. Melzer et al: Phys. Rev. Lett. **80**, 5345 (1998)
61. I.V. Schweigert, V.A. Schweigert, A. Melzer et al: Jetp Lett. **71**, 58 (2000)
62. P.K. Shukla, A.A. Mamun: *Introduction to Dusty Plasma Physics* (Institute of Physics Publishing, Bristol 2002)
63. N.N. Rao, P.K. Shukla, M.Y. Yu: Planet Space Sci. **38**, 543 (1990)
64. P. Shukla: Phys. Plasmas **8**, 1791 (2001)
65. R.L. Merlino, A. Barkan, C. Thompson et al: Phys. Plasmas **5**, 1607 (1998)
66. J.B. Pieper, J. Goree: Phys. Rev. Lett. **77**, 3137 (1996)
67. A. Barkan, R.L. Merlino, N. D'Angelo: Phys. Plasmas **2**, 3563 (1995)
68. C. Thompson, A. Barkan, N. D'Angelo et al: Phys. Plasmas **4**, 2331 (1997)
69. F. Melandsø: Phys. Plasmas **3**, 3890 (1996)
70. X. Wang, A. Bhattacharjee, S. Hu: Phys. Rev. Lett. **86**, 2569 (2001)
71. A. Homann, A. Melzer, R. Madani et al: Phys. Lett. A **242**, 173 (1998)
72. A. Homann, A. Melzer, S. Peters et al: Phys. Rev. E **56**, 7138 (1997)
73. S. Nunomura, D. Samsonov, J. Goree: Phys. Rev. Lett. **84**, 5141 (2000)
74. S. Nunomura, J. Goree, S. Hu et al: Phys. Rev. E **65**, 066402 (2002)
75. S. Nunomura, J. Goree, S. Hu et al: Phys. Rev. Lett. **89**, 035001 (2002)
76. D. Samsonov, J. Goree, Z. Ma et al: Phys. Rev. Lett. **83**, 3649 (1999)
77. D. Samsonov, J. Goree, H. Thomas et al: Phys. Rev. E **61**, 5557 (2000)
78. A. Melzer, S. Nunomura, D. Samsonov et al: Phys. Rev. E **62**, 4162 (2000)
79. D.H.E. Dubin: Phys. Plasmas **7**, 3895 (2000)
80. V. Nosenko, J. Goree, Z.W. Ma et al: Phys. Rev. Lett. **88**, 135001 (2002)
81. O. Havnes, T. Aslaksen, T.W. Hartquist et al: J. Geophys. Res. **100**, 1731 (1995)
82. S.V. Vladimirov, P.V. Shevchenko, N.F. Cramer: Phys. Rev. E **56**, R74 (1997)
83. T. Misawa, N. Ohno, K. Asono et al: Phys. Rev. Lett. **86**, 1219 (2001)
84. B. Liu, K. Avinash, J. Goree: Phys. Rev. Lett. **91**, 255003 (2003)
85. A. Melzer: Phys. Rev. E **67**, 115002 (2003)
86. J.J. Thomson: Philos. Mag. **39**, 237 (1904)
87. S. Neser, T. Palberg, C. Blechinger et al: Optical Methods and Physics of Colloidal Dispersions. In: *Progress in Colloid and Polymer Science 104* ed by G. Lagaly (Springer, Berlin Heidelberg New York 1997) pp 194–197
88. M.A. Reed and W.P. Kirk (Eds.): *Nanostructure Physics and Fabrication* (Academic, Boston 1989)
89. P. Liderer, W. Ebner, V.B. Shikin: Surf. Sci. **113**, 405 (1982)
90. V.M. Bedanov, F. Peeters: Phys. Rev. B **49**, 2667 (1994)
91. Y.J. Lai, Lin I: Phys. Rev. E **60**, 4743 (1999)
92. W.T. Juan, Z.H. Huang, J.W. Hsu et al: Phys. Rev. E **58**, R6947 (1998)
93. M. Klindworth, A. Melzer, A. Piel et al: Phys. Rev. B **61**, 8404 (2000)
94. J. Goree, D. Samsonov, Z.W. Ma et al: in Y. Nakamura, T. Yokota and P.K. Shukla (Eds.): *Advances in Dusty Plasmas* (Elsevier, Amsterdam 2000)
95. V.A. Schweigert, F. Peeters: Phys. Rev. B **51**, 7700 (1995)

12 Plasmas in Planetary Interiors

R. Redmer

Abstract. We give an overview of the properties of matter in planetary interiors. Of special interest are the equation of state and the conductivities. This field has gained much more interest because of the detection of Jupiter-like, extrasolar planets since 1995.

12.1 Introduction

Some astrophysical problems are closely related to plasma physics. For instance, matter in stars is probably the standard example for a plasma, i.e. systems consisting of free charge carriers such as electrons and ions. The temperature and density in the Sun rise from $2\,000$ K and $10^{-8}\,\mathrm{g\,cm^{-3}}$ in the photosphere to about 16×10^6 K and $156\,\mathrm{g\,cm^{-3}}$ in the central core region according to the standard model [1].

The equation of state (EOS) relates pressure, density, and temperature and is needed for the solution of the hydrodynamic equations in order to determine their variation along the radius. The optical properties such as the opacity are important for the energy transport from the central region to the surface of stars. As a further example, the network of nuclear reactions, e.g. the p-p chain and the CNO cycle (see also Chap. 17 in Part III), describes the energy source of stars. The corresponding reaction rates have to be known for a dense plasma medium. The physical properties of stars and their evolution are, in principle, understood [2] and not the subject of this contribution.

Matter inside planets is also exposed to extreme conditions. Although the temperature is not as high as in stars, it can reach several thousand Kelvin and, simultaneously, pressures in the megabar range. For instance, planetary models predict about $6\,000$ K and 3.5 Mbar in the centre of the Earth and $20\,000$ K and 50 Mbar in that of Jupiter. Matter under those conditions, between normal condensed matter at low temperatures and fully ionized plasmas at high temperatures, is also called *warm dense matter*. In this context, the physics of dense fluids and partially ionized plasmas is relevant for planetary interiors, especially for the giant planets in our solar system and extrasolar planets around other stars.

In this contribution, we give a review of the thermophysical properties of partially ionized plasmas. We start with an overview of the solar system and the extrasolar planets known until now. We then focus on the EOS of hydrogen-helium mixtures up to megabar pressures which is needed to model Jupiter-like planets. Furthermore, the electrical and thermal conductivity are discussed which determine the magnetic field configuration and the heat transport inside such planets. We finish with a summary and an outlook.

12.2 Solar System

Our solar system consists of the Sun and nine planets; their main properties are given in Table 12.1. Besides, numerous smaller bodies are orbiting the Sun. For instance, the asteroid belt between Mars and Jupiter at distances of 2–4 AU contains about 10^6 objects having a radius greater than 1 km. The Kuiper belt with about 35 000 objects greater than 100 km is located beyond the orbit of Pluto at distances of 30–100 AU. Sometimes the system Pluto with its moon Charon is already considered a Kuiper belt object. This region is the source of short-period comets. The Oort cloud still further out extends to distances of about 50 000 AU at the edge of our solar system. It is the

Table 12.1. Overview of the solar system, see [3]. G_{surf} is the gravitation and T_{surf} the temperature on surface. For gaseous planets, the *surface* is usually defined by the surface of the clouds in the atmosphere or by the radius at which the pressure is 1 bar. Units: 1 AU= 1.495×10^6 km, $M_E = 5.974 \times 10^{24}$ kg, $g_E = 9.81$ ms^{-2}, h: hours, d: days, y: years. The number of moons refers to the known satellites by June 2003

Planet	Distance (AU)	Mass (M_E)	Density (g cm^{-3})	G_{surf} (g_E)	T_{surf} (K)	Rotational Period	Orbital Period	Moons
Sun	–	333×10^3	1.41	28	5 800	25.4 d	–	–
			Inner planets:					
Mercury	0.387	0.0553	5.43	0.378	440	59 d	88 d	–
Venus	0.724	0.8152	5.20	0.907	730	243 d	224.7 d	–
Earth	1.000	1.0000	5.52	1.000	287	23.934 h	365.26 d	1
Mars	1.524	0.1075	3.93	0.377	218	24.623 h	686.98 d	2
		Asteroid belt between 2–4 AU						
			Outer planets:					
Jupiter	5.203	317.88	1.33	2.364	120	9.925 h	11.856 y	61
Saturn	9.555	95.162	0.69	0.916	88	10.656 h	29.424 y	31
Uranus	19.204	14.535	1.32	0.889	59	17.24 h	83.75 y	22
Neptune	30.087	17.141	1.64	1.125	48	16.11 h	163.7 y	14
Pluto	39.505	0.0022	2.06	0.067	37	6.387 d	248.02 y	1
		Kuiper belt between 30–100 AU						
		Oort cloud up to 50 000 AU						

source of long-period comets and contains about 10^{12} bodies with a total mass of Jupiter.

For a better understanding of the formation of the universe in general and our solar system in particular, more information about the planets and their evolution is needed. For instance, the internal structure and composition of the giant planets are still poorly known so that several models have been developed [4, 5, 6]. We have to solve the general equations of hydrostatic equilibrium in order to determine the mass M, the pressure P, and the temperature T along the radius r of the planet under consideration:

$$\frac{\mathrm{d}M(r)}{\mathrm{d}r} = 4\pi r^2 \varrho(r) , \qquad (12.1)$$

$$\frac{\mathrm{d}P(r)}{\mathrm{d}r} = -\frac{GM(r)\varrho(r)}{r^2} , \qquad (12.2)$$

$$\frac{\mathrm{d}T(r)}{\mathrm{d}r} = \frac{\mathrm{d}M(r)}{\mathrm{d}r}\left(\frac{\mathrm{d}P}{\mathrm{d}M}\right)\frac{T}{P}\nabla_T . \qquad (12.3)$$

with $M = 4\pi \int \varrho(r) r^2 \mathrm{d}r$ and $\nabla_T = \partial \ln T / \partial \ln P$. These equations are the same as for stars but without the energy source terms for the nuclear reactions. For simplicity, we neglect the rotation of the planet as well as its evolution in time, i.e. we discuss the (present) static case. It is obvious that we need a relation between the density ϱ, pressure P, and temperature T, i.e. an EOS as input. Furthermore, the temperature gradient ∇_T is usually determined for adiabatic conditions, i.e. along isentropes. This illustrates the importance of accurate EOS data for conditions relevant for planetary interiors.

The physical properties of planets vary along the radius, from cold matter at normal densities in the outer regions to warm dense matter in the centre. For instance, shell models as shown in Fig. 12.1 for Jupiter and Saturn have been studied extensively by Guillot [7] and Gudkova and Zharkov [8].

The outer shell consists mainly of molecular hydrogen and helium according to their mixing ratio Y. Other components such as H_2O, NH_3 or CH_4 are of minor importance. The pressure and temperature increase towards the centre of the planet and, most interestingly, the molecular insulating phase of hydrogen undergoes a transition to an atomic conducting phase at about 2 Mbar (1 Mbar=100 GPa) and several thousand Kelvin. This transition from nonmetallic to metallic-like behavior is one of the key problems in studies of dense hydrogen. The transition of molecular solid hydrogen to a monoatomic (alkali) metal at high pressures has already been proposed by Wigner and Huntington in 1935 [9]. Until now the metallic phase could not be verified in diamond anvil cells exposing solid hydrogen at room temperature to ultra-high pressures of 3.5 Mbar [10]. However, the transition to metallic-like hydrogen has been found already at 1.4 Mbar in the warm dense fluid at about 2 500 K by measuring the electrical conductivity in multiple shock-wave experiments [11]. This finding has strong implications on current

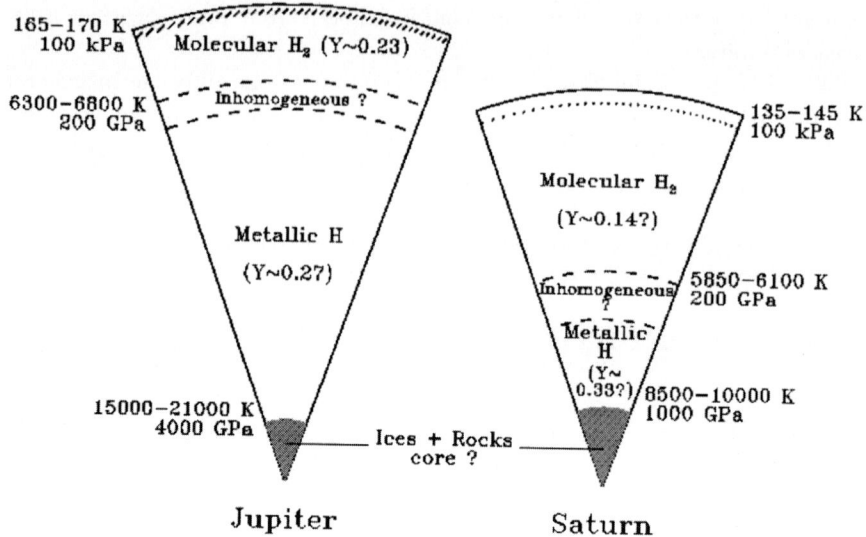

Fig. 12.1. Model of the interior of giant planets according to [7]

models for interiors of Jupiter-like planets, especially for their magnetic field configuration [12].

The central region of giant planets consists probably of a rocky core with a mass fraction of about 5% (Jupiter) and 10% (Saturn) of their total mass. Uranus and Neptune are smaller and colder than Jupiter and Saturn and contain a lesser fraction of hydrogen and helium. Their core region accumulates more than half of the planetary mass so that they are also called "icy giants".

12.3 Extrasolar Planets

A special problem in astrophysics is the understanding of the early stages in the evolution of stars and the regularities of the formation of planetary companions out of protoplanetary clouds. For instance, the existence of planets around other stars, possibly of earth-like planets, and their probability is of central importance in the search for life in the universe.

The quest of extrasolar planets has a long history [13] and various detection methods have been proposed [14]. The first extrasolar planet was detected only 1995 at 51 Pegasi [15], a main–sequence star of spectral type G5V in a distance of 50 Ly (light years) with a mass of 1.06 $M_{\rm Sun}$. The planet of about half the Jupiter mass has an orbital period of only 4.23 d and a semi-major axis of 0.0512 AU. The orbital parameter of a planet can be derived from the periodical Doppler shift of the light emitted from the star which reflects its orbit around the common centre of mass. This periodical movement of a star is sometimes called "stellar wobble" and an indirect prove of a planetary companion. With a given accuracy of 10^{-8} of the spectrographs, this method can prove minimum radial velocities of about $3\,\mathrm{m\,s^{-1}}$. For instance, Jupiter causes a radial movement of the Sun of about $12.5\,\mathrm{m\,s^{-1}}$, while the much lighter Earth generates a "wobble" of only $10\,\mathrm{cm\,s^{-1}}$. Therefore, this radial velocity method is capable of detecting giant planets like Jupiter around other stars, while earth-like companions remain invisible.

The mass of the extrasolar planet $M_P \sin i$ and its orbital semimajor axis a can be derived via Kepler's laws by using the mass of the star known from its spectral type, the measured orbital period T and the radial velocity v_R. The mass of the planet can only be determined with respect to the inclination i of the orbital pole to the line-of-sight. By the end of 2003, 119 extrasolar planets were known in 104 planetary systems, among them 11 systems with two planets and 2 systems with three planets, see [16, 17]. Most of these planets were detected with the radial velocity method mentioned above. Other promising methods are, e.g., transit photometry (measuring the small change of the brightness of the star when a planet passes across), astrometry (watching the periodical changes of the position of the star), or pulsar timing (proving periodical oscillations in the signal of a pulsar due to a planetary companion). We show the distribution of the masses and semimajor axis of the extrasolar planets found so far in Figs. 12.2 and 12.3, respectively.

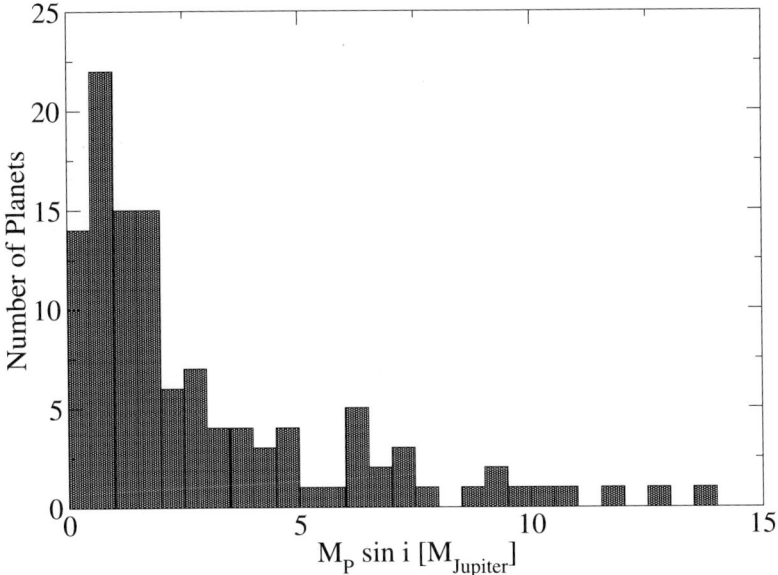

Fig. 12.2. Mass distribution of the extrasolar planets known by end of 2003 [16]

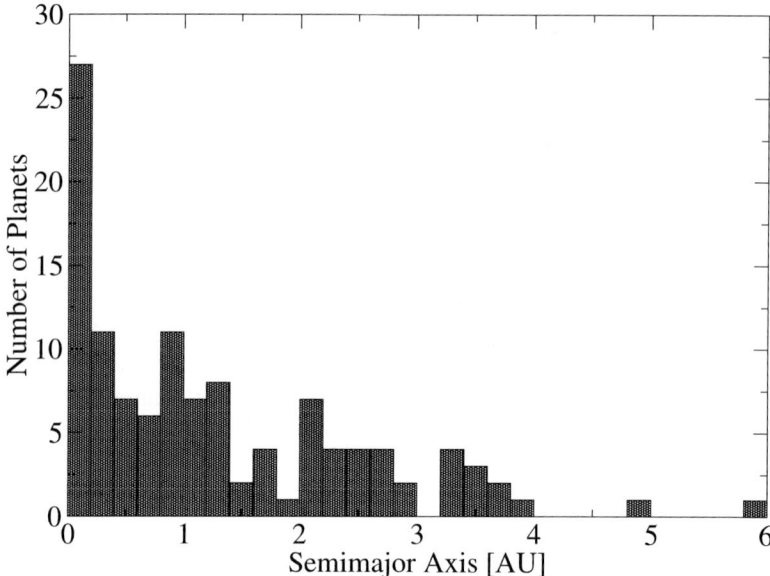

Fig. 12.3. Distribution of the semimajor axis of the extrasolar planets known by end of 2003 [16]

The mass of the known extrasolar planets ranges from that of Saturn up to more than ten Jupiter masses. A maximum is located at about 1 to 2 Jupiter masses and then the occurrence varies inversely with the mass. Their orbits are close to the star: about 30% have a semimajor axis smaller than 0.2 AU (Mercury: 0.387 AU, see Table 12.1), and about 50% have orbits within 1 AU. Assuming a synchronous rotation of these planets around their stars and, therefore, large temperature gradients in the atmosphere between day and night side, strong zonal winds of the order of $2\,\mathrm{km\,s^{-1}}$ may occur in these "hot Jupiters" compared with values of 100–$500\,\mathrm{m\,s^{-1}}$ observed in giant planets in our solar system [18]. On the other hand, Doppler velocity measurements require a full orbital period for the detection of an extrasolar planet so that the number of planets with larger orbits will increase with advancing observation time. The distribution of masses and semimajor axis as shown in Figs. 12.2 and 12.3 is, therefore, just a snapshot of the very limited knowledge that we have today about the properties of extrasolar planets. Especially, the occurrence of smaller, earth-like planets remains an open problem. The probability of planetary companions should also correlate with the properties and evolution of their stars. For instance, from planet-search surveys one can conclude that the occurrence of planets increases with increasing metallicity of the star [19].

12.4 Equation of State for Partially Ionized Plasmas

After we have given an overview of the Solar and extrasolar planets, we now focus on the "material properties" which are needed for a better understanding of their structure and evolution. In this section, we treat the EOS of hydrogen and helium and their mixtures for a wide range of densities and temperatures relevant for planetary interiors.

12.4.1 Dense Hydrogen and Helium

Various EOS were developed for hydrogen and helium. The standard source in this context is the Sesame EOS library for a great variety of materials, among them hydrogen and helium [20, 21]. Saumon and Chabrier [22] constructed a hydrogen EOS based on the free energy for a mixture of molecules, atoms, electrons, and ions including the processes of dissociation and ionization. Their EOS has found wide application in astrophysics [23] and is based, together with other work [24, 25], on the chemical picture. This approach treats elementary (electrons and ions) and composite particles (atoms, molecules) on the same footing. Their properties are modified by the medium compared with those of isolated particles which can be accounted for by effective Schrödinger equations for each species, considering many-particle effects such as dynamical screening, self-energy or Pauli blocking in a systematic way [26]. The concept of linear mixing of a neutral and a charged component has also been applied successfully to dense hydrogen [27, 28].

The physical picture starts from electrons and ions and is, therefore, well-suited for simulation techniques. For instance, quantum molecular dynamics (QMD) simulations within density functional theory [29, 30, 31, 32] and path-integral Monte Carlo (PIMC) simulations [33, 34] have been performed to study warm dense hydrogen. All theoretical results have to be checked against available experimental data. Isentropes and Hugoniot curves (see Sect. 12.4.5) for hydrogen and deuterium in the megabar pressure range can be determined by shock waves driven by gas guns [11], lasers [35], magnetically launched flyers [36], or chemical explosions [37]. The remarkable scatter of this data reflects the current controversy regarding the high-pressure EOS of warm dense hydrogen and, therefore, also the uncertainty with respect to models of planetary interiors. However, recent magnetically launched flyer-driven shock-wave experiments [36] indicate a less compressible behavior of hydrogen at high pressures than predicted earlier by laser-driven shock-wave experiments [35].

12.4.2 Free Energy

We will describe here the model of partially ionized plasmas (PIP) in more detail. This concept yieds the fully ionized plasma at high temperatures and/or densities and the atomic–molecular gas at low temperatures as limiting cases. The dissociated and ionized fractions have to be calculated self-consistently for given densities and temperatures by means of mass action laws, see [38]. The free energy of the system can be decomposed according to

$$F = F_0 + F^\pm + F^{pol},\qquad(12.4)$$

where F_0 is the contribution of the neutral (atomic-molecular) component, $F^\pm$ that of the charged (plasma) component, and F^{pol} describes the polarization interaction between them. We give expressions for these contributions in the next sections.

In order to treat planetary interiors at ultra-high pressures, a multicomponent systems with reactions (pressure dissociation and ionization) has to be treated, see [39, 40, 41]. The free energy $F(T,V,\{N_c\})$ for a system composed of various species c with a total particle number $N = \sum_c N_c$ determines other thermodynamic quantities such as the pressure P, the entropy S, the chemical potential μ_c of species c, and the internal energy U via standard thermodynamic relations:

$$P(T,V,\{N_c\}) = -\frac{\partial}{\partial V}F(T,V,\{N_c\}),\qquad(12.5)$$

$$S(T,V,\{N_c\}) = -\frac{\partial}{\partial T}F(T,V,\{N_c\}),\qquad(12.6)$$

$$\mu_c(T,V,\{N_c\}) = \frac{\partial}{\partial N_c}F(T,V,\{N_c\}),\qquad(12.7)$$

$$U(T,V,\{N_c\}) = F(T,V,\{N_c\}) + TS(T,V,\{N_c\}).\qquad(12.8)$$

For hydrogen–helium mixtures, the index c runs over neutral hydrogen atoms (H) and molecules (H$_2$) as well as helium atoms (He): $c = \{\text{H},\text{H}_2,\text{He}\}$. The dissociation of hydrogen molecules H$_2 \rightleftharpoons$ 2H is described by the equilibrium condition $\mu_{H_2} = 2\mu_H$. Ionization processes lead to further constraints: $\mu_{He} = \mu_{He^+} + \mu_e$ for He $\rightleftharpoons$ He$^+$ + e, $\mu_{He^+} = \mu_{He^{2+}} + \mu_e$ for He$^+ \rightleftharpoons$ He^{2+} + e, and $\mu_H = \mu_p + \mu_e$ for H $\rightleftharpoons$ p + e. Other species such as H$_2^+$ or H$^-$ but also further components as, e.g., H$_2$O, NH$_3$ or CH$_4$ can be considered. The neutral components (H, H$_2$, He) are treated within fluid variational theory (F^0). For the charged components (electrons and ions), a plasma EOS will be used ($F^\pm$). The polarization term between neutrals and charges F^{pol} is small and will be neglected here, see [38] for details.

12.4.3 Fluid Variational Theory

Fluid Variational Theory (FVT), initially developed to treat dense fluids [42], is a very efficient tool to determine the EOS of the neutral component. This method starts from the Gibbs–Bogolyubov inequality for the free energy F_0,

$$F_0 \leq F_{ref} + \langle \phi - \phi_{ref}\rangle_{ref},\qquad(12.9)$$

which is smaller than that of a known reference system F_{ref} plus the average of the difference of the real pair potential ϕ and the pair potential of the reference system ϕ_{ref}. The average is taken over the reference system. Considering noble gases or molecular fluids, a hard-sphere reference system is usually taken, and minimization of the free energy is performed with respect to the packing fraction η.

We take the expression for the free energy of a mixture of different hard spheres from [43]. The remaining correlation contribution F^{cor} is given by integrals over the effective pair potentials $\phi_{cd}(r)$ and the respective pair distribution functions $g_{cd}(r)$. The latter ones are calculated for the mixture of hard spheres within the Percus–Yevick approximation [44] and we have:

$$F^{cor} = \sum_c \frac{2\pi N_c^2}{V} \int_0^\infty dr\, r^2 \phi_{cc}(r) g_{cc}(r, \eta)$$

$$+ \sum_c \sum_{d \neq c} \frac{4\pi N_c N_d}{V} \int_0^\infty dr\, r^2 \phi_{cd}(r) g_{cd}(r, \eta)\,. \quad (12.10)$$

The total packing fraction $\eta = \sum_c \eta_c$ is given by the sum over all components with $\eta_c = \pi n_c d_c^3/6$, where $n_c = N_c/V$ is the number density and d_c the hard sphere diameter of species c. Effective pair potentials of the exponential–6 form are applied for the interactions between the various species, see [41, 42].

As a first result, the pressure dissociation of hydrogen molecules into atoms can be described within FVT by fulfilling self-consistently the condition of chemical equilibrium $\mu_{H_2} = 2\mu_H$ via (12.7). The correlation contributions to the chemical potentials lead to a lowering of the effective dissociation energy with increasing pressure (density). Isotherms of the respective dissociation degree are shown in Fig. 12.4. Pressure dissociation sets in at about 0.2–0.4 g/cm^{-3} and is more pronounced at lower temperatures. For higher temperatures, the steep increase is thermally smeared out. Good agreement with tight-binding molecular dynamics (TB-MD) simulations is found [45].

Similar calculations can be performed for other molecular gases such as nitrogen and oxygen or, even simpler due to the absence of dissociation, for noble gases. We have calculated the EOS for helium and combined the results with those for hydrogen. As an example, pressure isotherms of a hydrogen–helium mixture with a helium mass fraction of 15% as typical for giant planets are shown in Fig. 12.5 together with those for pure hydrogen [46]. The helium fraction leads to smaller pressures at the same density, especially at lower temperatures.

12.4.4 Plasma Component

An EOS for the plasma component can be derived within the efficient technique of Padé approximations [47]. These expressions are constructed by

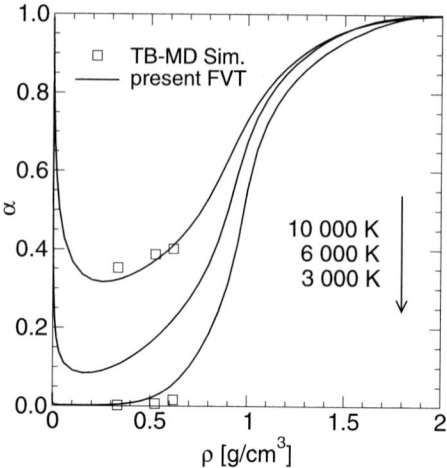

Fig. 12.4. Isotherms of the dissociation degree α in dense hydrogen fluid within FVT [40] compared with TB-MD simulations [45]

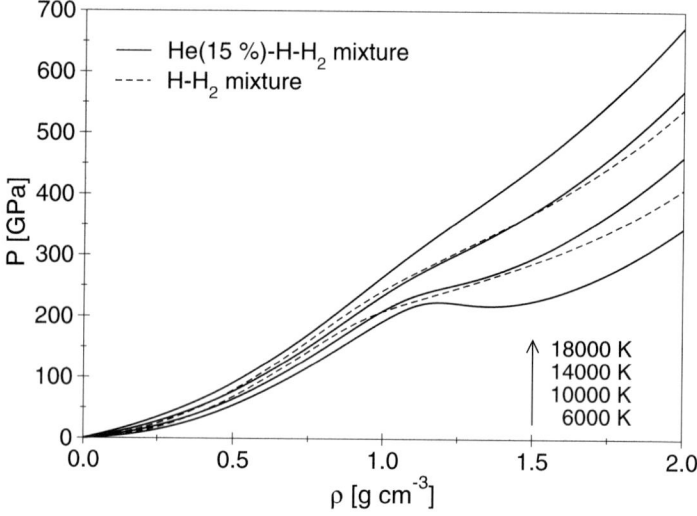

Fig. 12.5. Pressure isotherms for a He–H–H_2 mixture with a 15% He mass fraction (*solid lines*) and, for the lowest temperatures of 6 000 K and 10 000 K, also for hydrogen (*broken lines*) [46]

means of known results for the free energy in the limiting cases of low and high densities and interpolating in between. Extensive analytical expressions for the thermodynamic functions of fully ionized astrophysical plasmas were derived by Stolzmann and Blöcker [48].

Combining FVT with those expressions according to (12.4), the thermodynamic functions and the dissociation and ionization degree of partially ionized plasmas are derived. These calculations indicate that a phase instability, the so-called plasma phase transition (PPT), occurs in hydrogen at high pressures [49]. Other similar chemical models predict also a PPT with a coexistence line between a highly ionized and a weakly ionized phase which ends in a second critical point C_2 at about 15 000 K and 50 GPa [22, 50, 51], while again others show no instability in that region [21, 27]. Similarly, some first-principles simulations give no PPT at all [30, 33], while again others show precursors of an instability [52, 53, 54]. We show a comprehensive phase diagram of hydrogen as proposed by Ebeling et al. [51] in Fig. 12.6. Several tripel points Tr_i and the two critical points C_i are indicated. For an alternative phase diagram of hydrogen without a PPT, see [21].

The ultimate proof of the existence of a PPT has to be done experimentally but only dynamic methods are capable of reaching these extreme conditions. Up to now, no clear evidence for a PPT was found. Therefore, theoretical EOS data have to be checked carefully against available shock-wave experimental results.

12.4.5 Hugoniot Curves

Present shock-wave experiments are able to probe the EOS of materials up to several megabars. The Hugoniot curve connects all states $\{P, u, \varrho\}$ which can be reached by a single-shock experiment starting from initial conditions $\{P_0, u_0, \varrho_0\}$ (u: internal energy per mass). The respective Hugoniot equation [55]

$$u - u_0 = \frac{1}{2}(P + P_0)\left(\frac{1}{\varrho_0} - \frac{1}{\varrho}\right) \tag{12.11}$$

can be solved by using a theoretical EOS, for instance as derived from FVT.

We compare in Fig. 12.7 the Hugoniot curves for hydrogen, helium, and a He–H–H_2 mixture with 15% helium mass fraction. First we have to note that the compression reaches the value $4\varrho_0$ for atomic gases in the limit of ultra-strong shock waves, i.e. for $P \to \infty$ (ideal gas limit). The corresponding limiting value is $8\varrho_0$ for molecular gases if pressure dissociation is neglected and the vibrational and rotational levels are not shifted at high pressures. This behavior can be seen in Fig. 12.7 where we have performed model calculations for pure atomic hydrogen and helium neglecting ionization and pure molecular hydrogen neglecting dissociation.

Realistic models include dissociation and ionization as well as the temperature and density dependence of the vibrational and rotational levels [21] so that the compression is always $4\varrho_0$ in the limit of ultra-strong shock waves. Therefore, the maximum compression can be greater than $4\varrho_0$ at a certain pressure dependent on the actual dissociation–ionization process in the material. The Hugoniot curve for hydrogen thus interpolates between the pure

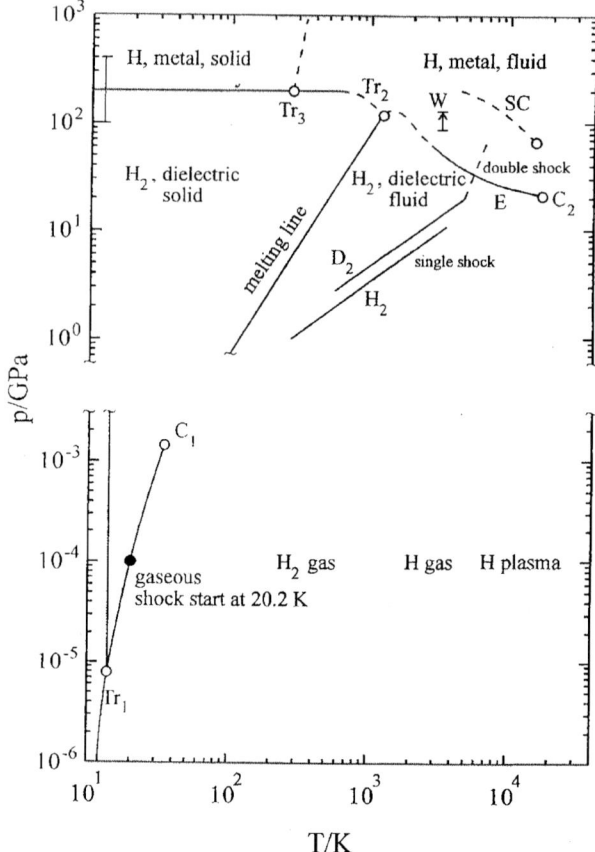

Fig. 12.6. Phase diagram of hydrogen as proposed in [51]. C_1: critical point of the liquid–gas phase transition, C_2: second critical point of the hypothetical PPT, Tr_n: possible triple points, W: location of the nonmetal-to-metal transition as found in gas gun experiments [11], SC: coexistence line and critical point of the PPT according to [22]. The paths of shock-wave experiments for H_2 and D_2 are indicated [27]

molecular case for lower pressures and the pure atomic one for the highest pressures. Our EOS yields a maximum compression of about $5\varrho_0$ for hydrogen at 100 GPa. The Hugoniot curve for the He–H–H_2 mixture is shifted to higher densities relative to that of the H–H_2 mixture because the helium fraction carries a higher mass, see also Fig. 12.5. A comparison of the present results with experimental Hugoniot curves gives reasonable agreement for hydrogen [40] and helium [41].

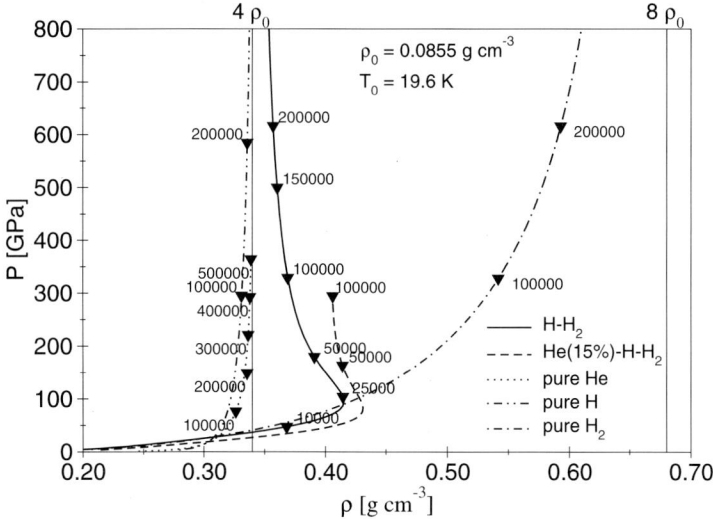

Fig. 12.7. Hugoniot curves for pure atomic helium and hydrogen, pure molecular hydrogen, a H–H$_2$ mixture with pressure dissociation, and a mixture of 15% mass fraction helium and 85% hydrogen considering again pressure dissociation. The same initial conditions $\varrho_0 = 0.0855\,\mathrm{g/cm^3}$ and $T_0 = 19.6\,\mathrm{K}$ were chosen for all substances. The maximum compression $4\varrho_0$ for atomic systems and $8\varrho_0$ for molecular systems is indicated by lines, and some temperatures are given along the Hugoniots [41]

12.5 Electrical and Thermal Conductivity

Besides the EOS, the electrical and thermal conductivity are further important "material properties" for models of planetary interiors. For their calculation we have to take into account ionization processes as described in Sect. 12.4.4. The transport coefficients of a partially ionized plasma can be determined within a general linear response theory valid for arbitrary degeneracy of the system; see [50, 56, 57] for details and also the contribution in this book (Chap. 6 in Part I). The Lorenz number L relates electrical (σ) and thermal conductivity (λ) according to

$$\lambda = L \left(\frac{k_B}{e}\right)^2 T\sigma \qquad (12.12)$$

and is used for a large domain of densities and temperatures to estimate the heat flow in giant planets. Isotherms for the electrical conductivity were calculated within the PIP model for hydrogen, see Fig. 12.8. Furthermore, we show in Fig. 12.9 the Lorenz number for hydrogen plasma at $15\,000\,\mathrm{K}$ as derived from the conductivities via (12.12).

As one can see from Fig. 12.8, this general linear response theory [50] is applicable for a large domain of densities and temperatures. The low-density

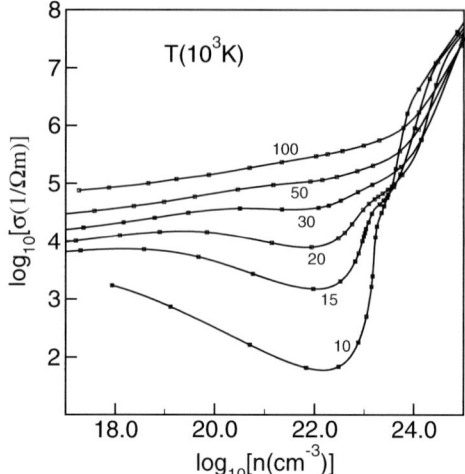

Fig. 12.8. Isotherms for the electrical conductivity of partially ionized hydrogen plasma as function of the total electron number density n [50]

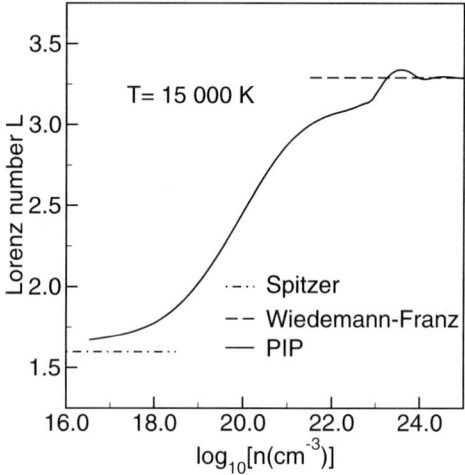

Fig. 12.9. Lorenz number for partially ionized hydrogen plasma (PIP) at 15 000 K as function of the total electron number density [50]. The low-density limit is given by the Spitzer value $L^{\mathrm{Sp}} = 1.5966$, the high-density limit by the Wiedemann–Franz law $L^{\mathrm{WF}} = 3.2897$.

behavior is given by the Spitzer results [58] while the high-density case is described by a Ziman–Faber formula [59]. The region of strong coupling in between is characterized by a nonmetal-to-metal transition, a steep increase of the electrical (and thermal) conductivity, which is caused by pressure ionization in our chemical model. This transition is very pronounced at low temperatures which are relevant for planetary interiors. Taking the Mott criterion which is strictly valid only for $T = 0\,\mathrm{K}$, the nonmetal-to-metal transition can be located at densities of about $1\,\mathrm{g\,cm^{-3}}$, in agreement with the experimental findings [11]. This means that the transition occurs at about 90% of the radius of Jupiter [12], i.e. much closer to the surface than previous estimates of about 75%. Besides the transport of free electrons, also hopping processes between weakly localized electrons become of importance in the transition range [60]. The location of this nonmetal-to-metal transition inside giant planets is important for the description of the heat and particle transport as well as for the generation of the planetary magnetic field.

The Lorenz number shown in Fig. 12.9 for a partially ionized hydrogen plasma at $15\,000\,\mathrm{K}$ varies only slightly between the Spitzer value $L^{Sp} = 1.5966$ for low densities and the original Wiedemann–Franz law $L^{WF} = 3.2897 = \pi^2/3$ for high densities. Thus, our calculations have shown that a density-dependent Wiedemann–Franz relation between electrical and thermal conductivity holds according to (12.12).

12.6 Conclusion

We have given here an overview of the planets in our solar system as well as of the extrasolar planets detected so far. For models of their interior, the thermophysical properties of hydrogen–helium mixtures are needed for extreme conditions, especially for high pressures in the megabar region and temperatures of several thousand Kelvin. We have presented results for the EOS and the conductivities within the chemical picture under those conditions. A central question is the location of the nonmetal-to-metal transition inside giant planets which occurs at about half a megabar and several thousand Kelvin [11, 12]. Comparison with available shock-wave experiments is performed.

The calculation of the EOS and of the conductivities for H–He mixtures as performed in [4, 22] is the aim of future work by combining the results for pure hydrogen and helium. This data will then be used in models of planetary interiors as in [7]. Several fundamental physical problems are closely related to this subject, see [5]. For instance, the phase diagram of matter at ultrahigh pressures is still unclear, see Fig. 12.6 and [21, 51]. The turn of the melting curve at high pressures is another central question in this context. A long-standing problem, the existence of the so far hypothetical PPT and of a second critical point as the end of its coexistence curve has to be proven or ruled out experimentally. The conditions under which hydrogen and helium are demixing have to be determined accurately [61]. This effect is one of the possible sources for the excess infrared luminosity of Jupiter.

The rapid progress in this interdisciplinary field between plasma physics, astrophysics, high-pressure physics and chemistry gives us a better and better understanding of the process of star formation and soon reliable probabilities for the occurrence of planetary companions. This is crucial for the ambitious goal of detecting earth-like planets in the near future and for the search for extraterrestrial life.

I thank David Blaschke, Werner Ebeling, Bastian Holst, Hauke Juranek, André Kietzmann, Sandra Kuhlbrodt, Nadine Nettelmann, Heidi Reinholz, Gerd Röpke, and Volker Schwarz for stimulating discussions, their help in preparing the lecture and this manuscript. I thank the organizers of the Summer School for the invitation and the WE–Heraeus Foundation for financial support.

References

1. J.N. Bahcall, M.H. Pinsonneault: Rev. Mod. Phys. **67**, 781 (1995)
2. R. Kippenhahn, A. Weigert: *Stellar Structure and Evolution* (Springer, Berlin Heidelberg New York 1990)
3. http://www.nineplanets.org
4. M.S. Marley, W.B. Hubbard: Icarus **73**, 536 (1988); W.B. Hubbard, M.S. Marley: Icarus **78**, 102 (1989)
5. D.J. Stevenson: J. Phys.–Condens. Matter **10**, 11227 (1998)
6. T. Guillot: Planet Space Sci. **47**, 1183 (1999)
7. T. Guillot: Science **286**, 72 (1999)
8. T.V. Gudkova, V.N. Zharkov: Planet Space Sci. **47**, 1201 (1999)
9. E.P. Wigner, H.B. Huntington: J. Chem. Phys. **3**, 764 (1935)
10. H.K. Mao, R.J. Hemley: Rev. Mod. Phys. **66**, 671 (1994)
11. S.T. Weir, A.C. Mitchell, W.J. Nellis: Phys. Rev. Lett.**76**, 1860 (1996)
12. W.J. Nellis: Planet Space Sci. **48**, 671 (2000)
13. http://www.public.asu.edu/~sciref/exoplnt.htm. There can be found a report on this topic by G.H. Bell: The search for extrasolar planets
14. J. Schneider, L.R. Doyle: Earth Moon Planets **71**, 153 (1995)
15. M. Mayor, D. Queloz: Nature **378**, 355 (1995)
16. http://www.obspm.fr/encycl/encycl.html for an overview and links to similar web sites
17. G.W. Marcy, R.P. Butler, D.A. Fisher et al: Properties of Extrasolar Planets. In: *Scientific Frontiers in Research on Extrasolar Planets, ASP Conference Series, vol CS-294* ed by D. Deming, S. Seager (Astronomical Society of the Pacific, San Francisco 2003) pp 1–16
18. T. Guillot, A.P. Showman: Astronom. Astrophys. **385**, 156 (2002); A.P. Showman, T. Guillot: ibid 166
19. D.A. Fisher, J.A. Valenti, G.W. Marcy: Spectral Analysis of Stars on Planet-Search Surveys. In: *Stars as Suns: Activity, Evolution and Planets, ASP Conference Series* ed by A.K. Dupree Astronomical Society of the Pacific, San Francisco 2004) in print
20. G.I. Kerley: A Theoretical Equation of State for Deuterium. Los Alamos Scientific Laboratory Report LA-4776, Los Alamos National Laboratories (1972)
21. G.I. Kerley: Equation of State for Hydrogen and Deuterium. Sandia Report SAND2003-3613, Sandia National Laboratories (2003)

22. D. Saumon, G. Chabrier, Phys. Rev. Lett. **62**, 2397 (1989); Phys. Rev. A **44**, 5122 (1991); ibid **46**, 2084 (1992)
23. D. Saumon, G. Chabrier, H.M. Van Horn: Astrophys. J. Suppl. Ser. **99**, 713 (1995)
24. A. Förster, T. Kahlbaum, W. Ebeling: Laser Part. Beams **10**, 253 (1992)
25. M. Schlanges, M. Bonitz, A. Tschttschjan: Contrib. Plasma Phys. **35**, 109 (1995)
26. W.D. Kraeft, D. Kremp, W. Ebeling et al: *Quantum Statistics of Charged Particle Systems* (Akademie-Verlag, Berlin 1986; Plenum, New York 1986)
27. N.C. Holmes, M. Ross, W.J. Nellis: Phys. Rev. B **52**, 15835 (1995)
28. M. Ross: Phys. Rev. B **58**, 669 (1998)
29. L.A. Collins, I. Kwon, J. Kress et al: Phys. Rev. E **52**, 6202 (1995)
30. T.J. Lenosky, S.R. Bickham, J.D. Kress et al: Phys. Rev. B **61**, 1 (2000)
31. G. Galli, R.Q. Hood, A.U. Hazi et al: Phys. Rev. B **61**, 909 (2000)
32. M.P. Desjarlais: Phys. Rev. B **68**, 064204 (2003)
33. B. Militzer, D. Ceperley: Phys. Rev. Lett. **85**, 1890 (2000)
34. B. Militzer, D. Ceperley, J.D. Kress, et al: Phys. Rev. Lett. **87**, 275502 (2001)
35. L.B. Da Silva, P. Celliers, G.W. Collins et al: Phys. Rev. Lett. **78**, 483 (1997)
36. M.D. Knudson, D.L. Hanson, J.E. Bailey et al: Phys. Rev. Lett. **87**, 225501 (2001); M.D. Knudson, D.L. Hanson, J.E. Bailey et al: Phys. Rev. Lett. **90**, 035505 (2003)
37. S.I. Belov, G.V. Boriskov, A.I. Bykov et al: JETP Lett. **76**, 433 (2002)
38. R. Redmer: Phys. Rep.-Rev. Sec. Phys. Lett. **282**, 35 (1997)
39. H. Juranek, R. Redmer: J. Chem. Phys. **112**, 3780 (2000)
40. H. Juranek, R. Redmer, Y. Rosenfeld: J. Chem. Phys. **117**, 1768 (2002)
41. R. Redmer, H. Juranek, S. Kuhlbrodt et al: Z. Phys. Chemie-Int. J. Res. Phys. Chem. Chem. Phys. **217**, 783 (2003)
42. M. Ross, F.H. Ree, D.A. Young: J. Chem. Phys. **79**, 1487 (1983)
43. G.A. Mansoori, N.F. Carnahan, K.E. Starling et al: J. Chem. Phys. **54**, 1523 (1971)
44. M. Shimoji: *Liquid Metals* (Academic Press, London 1977)
45. T.J. Lenosky, J.D. Kress, L.A. Collins, et al: Phys. Rev. E **60**, 1665 (1999)
46. V. Schwarz: Zustandsgleichung von Wasserstoff-Helium-Gemischen. Diploma Thesis, University of Rostock, Rostock (2003)
47. W. Ebeling, W. Richert: Phys. Status Solidi B-Basic Res. **128**, 467 (1985); Phys. Lett. A **108**, 80 (1985); Contrib. Plasma Phys. **25**, 1 (1985)
48. W. Stolzmann, T. Blöcker: Astronom. Astrophys. **361**, 1152 (2000)
49. D. Beule, W. Ebeling, A. Förster et al: Phys. Rev. B **59**, 14177 (1999); D. Beule, W. Ebeling, A. Förster et al: Phys. Rev. E **63**, 060202 (2001)
50. H. Reinholz, R. Redmer, S. Nagel: Phys. Rev. E **52**, 5368 (1995)
51. W. Ebeling, A. Förster, H. Hess et al: Plasma Phys. Contr. Fusion **38**, A31 (1996)
52. M. Knaup, P.-G. Reinhard, C. Toepffer: Contrib. Plasma Phys. **41**, 159 (2001); M. Knaup, P.-G. Reinhard, C. Toepffer et al: J. Phys. A–Math. Gen. **36**, 6165 (2003)
53. H. Xu, J.P. Hansen: Phys. Rev. E **60**, R9 (1999); Phys. Plasmas **9**, 21 (2002)
54. V.S. Filinov, V.E. Fortov, M. Bonitz et al: JETP Lett. **74**, 384 (2001)
55. Y.B. Zel'dovich, Y.P. Raizer: *Physics of Shock Waves and High-Temperature Hydrodynamic Phenomena* (Dover Publications, Mineola NY 2002)

56. R. Redmer: Phys. Rev. E **59**, 1073 (1999)
57. S. Kuhlbrodt, R. Redmer: Phys. Rev. E **62**, 7191 (2000)
58. L. Spitzer, R. Härm: Phys. Rev. **89**, 977 (1953)
59. J.M. Ziman: Philos. Mag.**6**, 1013 (1961)
60. R. Redmer, G. Röpke, S. Kuhlbrodt et al: Phys. Rev. B **63**, 233104 (2001); Contrib. Plasma Phys. **41**, 163 (2001)
61. O. Pfaffenzeller, D. Hohl, P. Ballone: Phys. Rev. Lett. **74**, 2599 (1995)

Part III

Methods and Applications

13 Plasma Diagnostics

H.-J. Kunze

Abstract. Measuring plasma quantities is a demanding challenge in experimental plasma physics. It requires detailed knowledge of basic plasma physics and related fields, such as atomic physics, aspects of thermodynamics and optics.

This chapter covers techniques in plasma diagnostics employing light scattering and emission spectroscopy and is complemented by Chaps. 2 and 5 in Part I. An overview on fluctuation diagnostics is given in Chap. 14 in Part III.

13.1 Introduction

Plasma diagnostics encompasses all methods and techniques employed for the determination of macroscopic as well as of microscopic properties and parameters of plasmas as a function of space and time. Since laboratory plasmas range from small-volume configurations of 10^{-16} m^3 (e.g. micro pinches) to large-volume systems of about 50 m^3 (magnetic fusion device JET), and density and temperature cover a range from 10^{14} to 10^{30} m^{-3} and from 10^3 K to several times 10^8 K (1 eV ... 10 keV), respectively, a large variety of different methods is used which originate in a number of fields of physics and applied physics. The breadth is enormous and challenges the experimentalists.

An impression of the spectrum of methods is conveyed, for example, by the monograph of Hutchinson [1], although it emphasizes subliminally diagnostics of fully ionized plasmas in contrast to the more recent book by Ovsyannikov [2] which focusses on methods for the investigation of low-temperature plasmas. For classical diagnostic techniques the books by Huddlestone and Leonard [3] and by Lochte–Holtgreven [4] are still valuable references. New schemes are constantly advanced and new applications become feasible by the development of new instruments. The biannual "Topical conference on high-temperature plasma diagnostics" with proceedings published in the journal "Review of Scientific Instruments" provides a forum for that. Other topical conferences concentrate on subfields like, for example, the application of lasers in plasma diagnostics.

This lecture will concentrate on the introduction to two fields – Thomson scattering of laser radiation and plasma emission spectroscopy. Both are selected because respective diagnostic methods are employed over nearly the full range of plasma parameters cited above.

13.2 Scattering of Laser Radiation by Plasma Electrons

13.2.1 Laser-aided Diagnostics

A laser beam traversing a plasma interacts with all its constituents, i.e. with electrons, ions, atoms and molecules. The geometry of such a set-up is shown in Fig. 13.1.

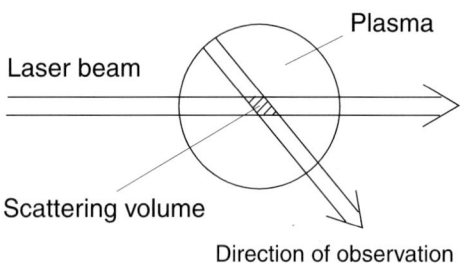

Fig. 13.1. Diagnostic geometry

Due to the tunability and the possible high power of lasers various interactions can be effective including possible heating of the plasma. Naturally these interactions modify the transmitted primary beam, which is utilized for several diagnostic techniques. It is clear that the information obtained represents an integral value along the path s of the laser beam through the plasma. Modification of the phase can be measured by interferometric methods and gives the integrated electron density $\int n_e \mathrm{d}s$. Magnetic fields cause a rotation of the plane of polarization of the laser by the Faraday effect yielding thus the quantity $\int n_e B \mathrm{d}s$. More details of both diagnostic methods are presented in Chap. 2 in Part I. Absorption leads to an attenuation of the intensity of the beam. It is strongest on transitions in atoms and ions and allows thus to measure the density $\int n_a \mathrm{d}s$ of the respective atomic species in the lower state of a transition. The availability of diode lasers accelerated the wide application of respective techniques especially to the diagnostics of low-temperature plasmas.

Interaction processes naturally also result in the re-emission of photons simply referred to as *scattering*. Laser induced fluorescence (LIF), Rayleigh scattering, Raman scattering, and coherent anti-Stokes Raman scattering (CARS) by atoms, ions and molecules are thus utilized. Scattering by the electrons known as Thomson scattering does not depend on atomic properties but leads to unique diagnostic possibilities which will be discussed in the following. One advantage of all scattering techniques is the possibility of *spatially resolved* information (Fig. 13.1): scattered radiation is received only from the hatched region in the plasma while radiation from the plasma is still being collected along the full line of sight. Muraoka and Maeda [5] treat all diagnostic methods utilizing lasers, whereas [6, 7, 8, 9] are devoted only to scattering by plasma electrons. Since its first application in 1963 Thomson scattering developed into one of the most powerful methods of plasma diagnosis.

13.2.2 Incoherent Thomson Scattering

Principles

The theory of scattering of electromagnetic radiation by plasma electrons is well understood and treated in all textbooks of modern physics. In the field of an electromagnetic wave focused into a plasma all charged particles are accelerated, oscillate and hence radiate like dipoles in all directions. This emitted radiation is the *scattered* wave. Because of their large mass the acceleration of the ions and thus their radiation is much less than that of the electrons, and so essentially only the electrons contribute. The total Thomson cross-section σ_{Th} for scattering from a single electron is given by

$$\sigma_{Th} = \frac{8}{3}\pi r_e^2 \cong \frac{2}{3} \times 10^{-24} \text{ cm}^2 , \qquad (13.1)$$

where $r_e = (1/4\pi\epsilon_0)(e^2/m_e c^2)$ is the classical electron radius. The very small magnitude of σ_{Th} reveals a serious problem. In order to detect sufficient photons of scattered radiation either long observation times are needed or high powers of the incident radiation. Radiation emitted by plasmas especially of high density imposes another constraint: the scattered radiation should be clearly above the level of plasma background radiation. Both requirements point to the use of *powerful lasers* as primary source of radiation. Their extreme directionality allowing good focusing for spatially resolved measurements and their narrow spectral bandwidth and purity are other welcome characteristics, which are made use of.

At low plasma densities with no correlations between the electrons they scatter independently of each other and the intensity contributions of all electrons simply add. The differential cross-section per unit volume is defined as the ratio of power scattered into a solid angle $d\Omega$ to incident power per unit area and is given by

$$\frac{d\sigma}{d\Omega} = n_e\, r_e^2 \sin^2\phi = \frac{3}{8\pi}\, n_e\, \sigma_{Th} \sin^2\phi \propto n_e , \qquad (13.2)$$

where ϕ is the angle between the direction of the electric field vector of the incident wave and the direction of observation. A measurement of the scattered radiation thus yields the *electron density* n_e.

In the plasma the electrons move and due to their small mass m_e the velocities are high, which results in considerable Doppler shifts of the scattered radiation. To be precise, the Doppler effect has to be considered twice: the moving electron experiences a Doppler shifted primary beam in its frame, and the radiation scattered by the moving electron is shifted again. If we denote the angular frequency and wave vector of the oncoming wave by $(\omega_o, \mathbf{k_o})$ and those of the scattered radiation by $(\omega_s, \mathbf{k_s})$, the resultant Doppler shift ω becomes

$$\omega = \omega_s - \omega_o = \mathbf{k} \cdot \mathbf{v}, \tag{13.3}$$

with

$$\mathbf{k} = \mathbf{k_s} - \mathbf{k_o}. \tag{13.4}$$

$\mathbf{k}$ is the scattering vector, and the triangular relation of the scattering geometry is illustrated in Fig. 13.2. The scattering angle θ is between the wave vectors of incident and scattered wave, $\theta = \angle(\mathbf{k_s}, \mathbf{k_o})$.

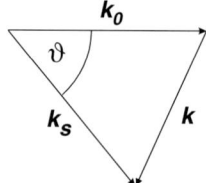

Fig. 13.2. Scattering geometry

Multiplication of (13.3) and (13.4) with $\hbar$ reveals that both equations represent nothing else but conservation of energy and linear momentum. As long as $k_s \simeq k_o$, the magnitude of the scattering vector $\mathbf{k}$ is given by

$$k \simeq 2k_o \sin\frac{\theta}{2} \simeq \frac{4\pi}{\lambda_o} \sin\frac{\theta}{2}. \tag{13.5}$$

Equation (13.3) shows that the resulting Doppler shift ω is according to the component v_k of the velocity in the direction of the scattering vector $\mathbf{k}$, and hence the spectrum of the scattered radiation from the ensemble of electrons mirrors the one-dimensional velocity distribution function $f_k(v_k)$ in the same direction. With (13.3) rewritten as

$$\frac{(\omega_s - \omega_o)}{k} = v_k, \tag{13.6}$$

the velocity distribution function transforms directly into the spectral profile of the scattered radiation:

$$f_k(v_k) \longrightarrow f_k\left(\frac{\omega_s - \omega_o}{k}\right). \tag{13.7}$$

It is this distribution function which is obtained experimentally. In an anisotropic plasma other directions may be probed by varying the scattering angle θ. The fraction of electrons dn_e giving the same Doppler shift is

$$\mathrm{d}n_e = n_e f_k(v_k)\mathrm{d}v_k = n_e \frac{1}{k} f_k\left(\frac{\omega_s - \omega_o}{k}\right) \mathrm{d}\omega, \quad (13.8)$$

and substituted into (13.2) leads to the differential scattering cross-section per unit frequency

$$\frac{\mathrm{d}^2\sigma}{\mathrm{d}\Omega_s \mathrm{d}\omega_s} = \frac{3}{8\pi} n_e \sigma_{Th} \sin^2\phi \frac{1}{k} f_k\left(\frac{\omega_s - \omega_o}{k}\right). \quad (13.9)$$

For a Maxwellian distribution the frequency spectrum is a Gaussian profile with a full width at half maximum (FWHM) in wavelength units

$$\Delta\lambda_{1/2} = 4\,\lambda_o \sin\frac{\theta}{2}\left(\frac{2k_B T_e}{m_e c^2}\ln 2\right)^{1/2}. \quad (13.10)$$

k_B is the Boltzmann constant and c is the velocity of light in vacuum. $\Delta\lambda_{1/2}$ yields directly the electron temperature T_e. In order to obtain an idea of typical widths, we consider the light of a ruby laser ($\lambda_o = 694.3$ nm) scattered at $\theta = 90°$: for $k_B T_e = 100$ eV, we have $\Delta\lambda_{1/2} = 32.4$ nm. Such large widths imply that usually low-resolution spectrometers suffice which, on the other hand, can have high a throughput to collect as much scattered light as possible.

With increasing electron temperature of the plasma relativistic effects have to be taken into account. The spectral profile changes because the oscillating electrons now preferentially radiate in the forward direction corresponding to an asymmetric dipole radiation pattern. This results in blue-shifted spectral profiles [10].

Scattering also changes with increasing density when correlations between the electrons start to become important; the total scattered power no longer is simply the sum of the intensity contributions from all electrons but now the coherent sum of the electric fields from scattering by the electrons has to be taken. A measure of the influence of correlations is k^{-1}, i.e. the distance over which the particle motion is sampled, in relation to the Debye length λ_D. As long as $k^{-1} \ll \lambda_D$ or $k\lambda_D \gg 1$, the electrons in the Debye sphere scatter incoherently. Quantitatively the effect of the correlations depends on the scattering parameter α which is defined as

$$\alpha = \frac{1}{k\lambda_D}. \quad (13.11)$$

In summary, incoherent Thomson scattering ($\alpha \ll 1$) yields the local electron density n_e and the one-dimensional electron velocity distribution function $f_k(v_k)$, respectively the electron temperature T_e.

Experimental Constraints and Techniques

As pointed out in the preceding section usually powerful lasers are needed in order to produce enough scattered photons. Where it is not possible to reach the goal of a reasonable signal-to noise ratio with a single laser pulse, repetitively pulsed lasers are employed and the photons are simply accumulated, which is naturally at the expense of time resolution. The quantum efficiency of the detectors is therefore important and has to be considered in addition to the throughput of the spectrographic system. Intensified charge coupled device (ICCD) cameras are now increasingly employed, which are gated and can record a full spectrum when mounted in the exit plane of the spectrograph. They also offer the possibility of simultaneous observations spatially resolved along the path of the laser beam in the plasma, when this path is imaged onto the entrance slit of a stigmatic spectrograph. Naturally, other schemes and detectors like photomultipliers and photodiodes are also in use. At very low densities one turns to photon counting techniques [11].

For the derivation of the electron density, the power of the primary laser beam and of the scattered radiation have to be determined, in principle. This is a rather difficult task, since in addition to the absolute sensitivity of the complete detection system also the volume in the plasma, from which scattered radiation is collected, must be known accurately. Fortunately, this can be avoided completely by utilizing Rayleigh or Raman scattering of gases, which are filled temporarily into the discharge vessel and for which the cross-sections are known. In this way even the scattering geometry remains identical in the "calibration".

The most serious problem of all scattering setups is *stray light* which reaches the detector and which is many orders of magnitude larger than the scattered radiation if no measures are taken. It is initially generated when the laser beam passes through the entrance and the exit windows of the plasma vessel. Windows at the Brewster angle, baffles in the path of the laser beam, the beam dump for the laser and the viewing dump providing a black background for the observation channel are standard remedies.

Other aspects to be considered are plasma background radiation and heating by the laser beam. *Heating* is due to absorption of the laser and occurs essentially by the process of inverse bremsstrahlung. At high temperatures it is usually negligible but must be taken into account when performing Thomson scattering on low-temperature plasmas. Plasmas emit continuum and line radiation which unavoidably is collected by the detection system along the full line of sight through the plasma. Since the *continuum* component [bremsstrahlung and recombination radiation, see subsequent (13.53)] scales with the density squared, it tends to be serious only at higher densities, where even the "noise"-denominator in the signal-to-noise ratio may be determined by this background.

One strives to stay clear of and even far away from *lines* by selecting a laser with a suitable wavelength. This can be a problem with low-temperature technical plasmas where many atomic and even molecular species can be present. In addition, fluorescence excited by the laser by absorption on the wings of molecular lines may even enhance the background dramatically.

We conclude the discussion of incoherent Thomson scattering by mentioning a scheme especially suitable for large-volume high-temperature fusion plasmas like in JET or ITER: by combining the LIDAR technique (an acronym for light detection and ranging) with Thomson backscattering electron density and temperature profiles are obtained [12]. Because the collection optics is simple the scheme is also well suited for fusion devices which are activated and become inaccessible. A short laser pulse (a pulse duration of 300 ps corresponds to a pulse length of 9 cm) traverses the plasma, and the backscattered radiation is simply recorded as function of time by a fast detection system. The position of the laser pulse within the plasma is known at any instant, and the time resolved spectra can be clearly assigned to respective spatial positions.

13.2.3 Collective Thomson Scattering

With increasing scattering parameter α correlations between the electrons modify spectral shape and magnitude of the scattered radiation. Adding the electric fields from all electrons the differential scattering cross-section takes on a form

$$\frac{d^2\sigma}{d\Omega_s d\omega_s} = r_e^2 \sin^2\phi \, n_e \, S(\mathbf{k},\omega) \,, \tag{13.12}$$

where the first two factors give again the well-known scattering characteristic of the single free electron, and $S(\mathbf{k},\omega)$ known as dynamic form factor contains the properties of the whole ensemble of electrons. In general, it is the time- and space-Fourier transform of the time-dependent electron density pair correlation function. $S(\mathbf{k},\omega)$ has been calculated by a number of authors, for plasmas with and without impurities, for stable and unstable plasmas. The most useful approach is still that of Salpeter [13], who gave $S(\mathbf{k},\omega)$ in an approximate form convenient for calculations.

Figure 13.3 illustrates the dynamic form factor as function of a normalized frequency x_i for various scattering parameters α; x_i is the frequency normalized to k times the mean thermal speed $v_{th,i}$ of the ions. It is obvious that for higher α three features develop, a narrow one of high intensity at the center displaying a width determined apparently by the mean velocity of the ions, and two side bands symmetrical to each other and shifted from the center by about the electron plasma frequency ω_{pe}. Hence the central part is called the ion feature and the side bands are the electron feature. In the approximation of Salpeter [13] the dynamic form factor can be written as

$$S(\mathbf{k},\omega) = S_e(\mathbf{k},\omega) + S_i(\mathbf{k},\omega) \,. \tag{13.13}$$

The electron feature corresponds to the electron term $S_e(\mathbf{k},\omega)$ and reflects scattering by those electrons which move uncorrelated or are correlated with the motion of other electrons. The ion feature given by $S_i(\mathbf{k},\omega)$ contains the scattering contributions of those electrons which are correlated with the

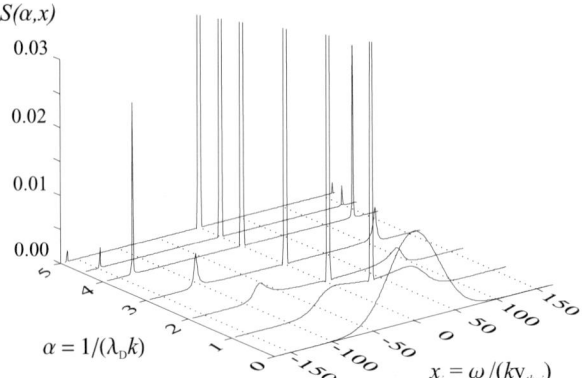

Fig. 13.3. Change of dynamic form factor $S(\alpha, x)$ with scattering parameter α, from [14]

motion of the ions. Hence, although scattering is by electrons, this part of the spectrum reflects the motion of the ions. Introducing two different scale lengths for the shifts of the electron and ion components

$$x_e = \frac{\omega}{k\, v_{th,e}} = \frac{\omega}{k\sqrt{(2k_B T_e)/m_e}} \quad \text{and} \quad x_i = \frac{\omega}{k\, v_{th,i}} = \frac{\omega}{k\sqrt{(2k_B T_i)/m_i}},$$
(13.14)

their respective dynamic form factors in the case of one ionic species in the plasma are of the same analytic form in the Salpeter approximation:

$$S_e(\mathbf{k}, \omega)\, d\omega = \Gamma_\alpha(x_e)\, dx_e \quad \text{and} \quad S_i(\mathbf{k}, \omega)\, d\omega = \frac{Z\alpha^4}{(1+\alpha^2)^2}\, \Gamma_\beta(x_i)\, dx_i,$$
(13.15)

with the Salpeter shape function

$$\Gamma_\alpha(x) = \frac{\exp(-x^2)}{|1 + \alpha^2\, W(x)|^2}.$$
(13.16)

$W(x)$ is the plasma dispersion function, and β_S^2 is given in the range $\beta_S^2 < 3.5$ by

$$\beta_S^2 = Z\, \frac{\alpha^2}{1+\alpha^2}\, \frac{T_e}{T_i}.$$
(13.17)

It is of interest to look also at the frequency integrated dynamic form factor

$$S_e(\mathbf{k}) = \frac{1}{1+\alpha^2} \quad \text{and} \quad S_i(\mathbf{k}) = \frac{Z\alpha^4}{(1+\alpha^2)\,[1+\alpha^2 + Z\alpha^2\,(T_e/T_i)]}.$$
(13.18)

The electron term $S_e(\mathbf{k})$ decreases strongly with α and the observation thus becomes difficult especially at high densities with high plasma background

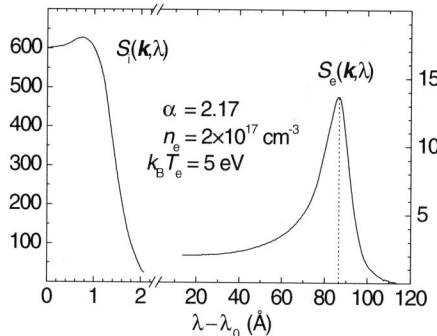

Fig. 13.4. Theoretical spectrum for $\alpha = 2.17$, $\lambda_0 = 694.3$ nm (ruby laser), and $\theta = 90°$, from [6]

radiation. Figure 13.4 illustrates a spectrum with $\alpha = 2.17$ for scattering from a hydrogen plasma at $\theta = 90°$ employing a ruby laser. The different scale length of both spectral features is clearly seen.

In the region $0.5 < \alpha < 2$ shape and position of the maximum of $S_e(\mathbf{k}, \omega)$ depend strongly on α and thus both n_e and T_e can be derived solely from the shape without absolute calibration of the radiation. On the other hand, in plasmas of low temperatures collisions can be important and modify the shape of the electron feature as given in the Salpeter approximation; they have to be taken into account.

The correlated motion of the electrons in the electron term reflects electron plasma waves (see Chap. 2 in Part I), and the position of the maximum of $S_e(\mathbf{k}, \omega)$ hence is given for large α by the Bohm–Gross dispersion relation

$$(\omega_s - \omega_o)^2 = \omega_{BG}^2 = \omega_p^2 + \frac{3k_B T_e}{m_e} k^2 . \tag{13.19}$$

For large α it approaches the plasma frequency ω_p.

The width of the ion feature $S_i(\mathbf{k}, \omega)$ is narrow, its magnitude increases with α and it is thus easily detected above the plasma background even at high plasma densities. It is displayed in Fig. 13.5 for various plasma conditions at $\alpha = 4$.

Even for $T_e = T_i$ it shows two maxima. They correspond to co- and counter-propagating ion acoustic waves obeying the following dispersion relation:

$$\begin{aligned}(\omega_s - \omega_o)^2 &= \left(\frac{Z k_B T_e}{m_i} \frac{\alpha^2}{1+\alpha^2} + \frac{3 k_B T_i}{m_i} \right) k^2 \\ &= \frac{k_B T_e}{m_i} \left(Z \frac{\alpha^2}{1+\alpha^2} + 3 \frac{T_i}{T_e} \right) k^2 \\ &\approx Z \frac{k_B T_e}{m_i} k^2 \quad \text{for large } \alpha \text{ and } Z \text{ and for } T_i \leq T_e . \end{aligned} \tag{13.20}$$

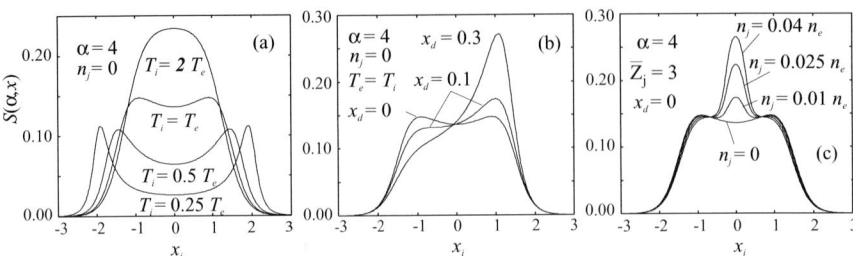

Fig. 13.5. Ion feature $S(\alpha, x)$ as function of the normalized frequency x_i for various conditions of a hydrogen plasma, from [14]

For electron temperatures $T_e > T_i$ (where Z is the charge of the ions) damping of the ion acoustic modes decreases and the maxima develop into two more or less well separated lines [Fig. 13.5(a)]. From the shape of the ion feature, T_i and T_e may be deduced. An exception are plasmas with large α, where the separation of the two "ion acoustic lines" becomes essentially a linear function of the product of ion charge Z and electron temperature T_e, (13.20). Respective methods are applied to the diagnostics of dense, inertial confinement fusion plasmas [15].

Drift velocities between electrons and ions influence the damping of the ion acoustic modes resulting in asymmetric profiles: electron Landau damping of the wave propagating in the same direction as the drift is decreased which results in an increase of the respective ion acoustic peak. Correspondingly, the drift leads to stronger damping of the counter-propagating wave and thus to a decrease of the scattered intensity. Figure 13.5(b) illustrates such cases for various values of the drift parameter x_d, which is defined as $x_d = (v_d/v_{th,e}) \cos \chi$, where χ is the angle between the drift velocity and the scattering vector **k**; the drift velocity v_d is taken as the shift of the electron velocity distribution function relative to that of the ions. Selecting different scattering geometries allows to probe the plasma accordingly [16]. In addition, a macroscopic motion of the plasma gives a shift of the ion feature as a whole, which is utilized for the diagnostics of respective velocities within the plasma as well.

Evans [17] extended Salpeter's theory to include small amounts of impurities. Figure 13.5(c) shows their effect on a hydrogen plasma for several impurity concentrations n_j: a narrow feature develops in the center of the profile. For small concentrations this impurity peak scales approximately with the product $n_j \overline{Z_j}^2$, where $\overline{Z_j}$ is the mean charge of the impurity species (j). If this mean charge can be derived from spectroscopic observations, the impurity peak thus yields the impurity concentration [18]. The width of the peak is given by the mean thermal speed of the ions. Finally, one has to be aware that even in the case of a pure high-density but low-temperature hydrogen plasma without impurities Rayleigh scattering by excited hydrogen atoms may occur on the wings of the strongly broadened H_α-line and is superposed onto the ion feature [19] complicating its interpretation.

For low-density plasmas the regime of collective Thomson scattering is reached either by observation at small scattering angles θ (forward scattering) and/or by employing long-wavelength lasers in the infrared or even powerful microwave radiation in the "mm" wavelength range from gyrotrons [20]. With these microwaves even backscattering is in the collective regime, and such methods are thus of special interest for the α-particle diagnostics of large fusion devices like JET and ITER; infrared lasers still require forward scattering, which is more difficult to perform [9].

13.2.4 X-ray Scattering

The most recent development extends scattering of radiation by plasma electrons to the x-ray regime and allows probing of the properties of solid state and super dense plasmas, from Fermi degenerate, to strongly coupled, to high temperature ideal gas plasmas [21]. Strong line radiation in the x-ray region from a plasma produced by powerful lasers focussed onto a solid was used as primary source. In this regime the energy of the incident photons is sufficiently high with respect to the rest energy of the electrons to yield a significant shift to the scattered radiation. The experiments were in the non-collective regime and both the elastic Rayleigh component and the broadened inelastic Compton scattering component were utilized: the spectral shape of the shifted Compton component provides the accurate measurement of the temperature, and the intensity ratio of inelastic to elastic scattering component gives the ionization balance.

13.3 Plasma Spectroscopy

13.3.1 Overview

All plasmas in the laboratory, in technical devices and in the universe emit electromagnetic radiation. The main task of emission spectroscopy is to analyze this radiation and to deduce as much information as possible on

- density n_e of the electrons,
- density n_z^i of the ions of species i in various charge states z,
- density n_o^i of neutral atoms and molecules, i.e. composition of the plasma,
- temperature of electrons T_e, ions T_i and neutral species T_o, or their velocity distribution functions $f_e(v), f_i(v), f_o(v)$,
- static, oscillating or turbulent electric fields,
- magnetic fields.

This is a formidable task and usually cannot be accomplished as desired for all plasmas existing in the wide range of parameters outlined in the introduction of this chapter. Experimental constraints on the one hand and a lack of sufficiently accurate theoretical models describing the emission from the huge number of atomic and ionic species on the other hand pose serious limitations. Standard spectroscopic techniques are naturally covered to some extent also in the books on plasma diagnostics [1, 2, 3, 4]. The main reference for this field, however, are the two monographs by Griem [22, 23], and a useful introduction to general spectroscopy is offered by [24]. Here, we will not discuss experimental details but concentrate on physical principles used in the analysis of the emitted radiation for the determination of plasma parameters.

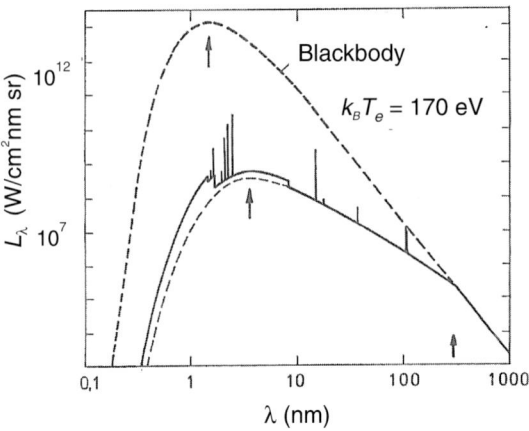

Fig. 13.6. Emission spectrum of a hydrogen plasma with impurities, $k_B T_e = 170$ eV

Figure 13.6 shows a typical spectrum emitted by a plasma, here of a temperature of 170 eV. We identify the following contributions:

– bremsstrahlung, a continuum radiation, is emitted when the electrons experience deflection in the field of the ions;
– recombination radiation, also a continuum but characterized by edges. It is emitted when electrons recombine with ions;
– line radiation corresponds to transitions between atomic energy levels.

The transitions are indicated in the schematic energy level diagram of a non-hydrogenic ion in Fig. 13.7 and are also termed free-free, free-bound and bound-bound transitions, respectively, transitions from doubly excited states included. With increasing size of the plasma and/or with increasing density the emission of a plasma characterized by the spectral radiance L_λ increases, emitted photons begin to be re-absorbed within the plasma (the

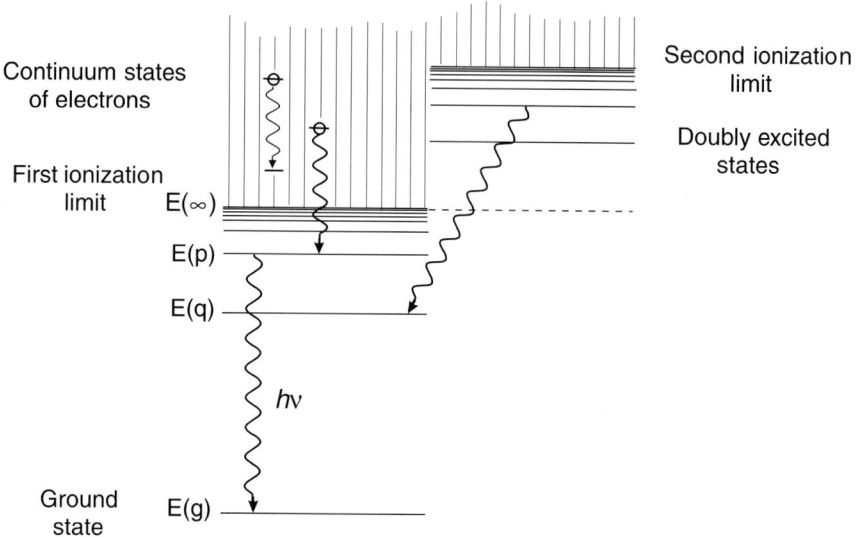

Fig. 13.7. Schematic energy level diagram

plasma becomes optically thick due to self-absorption), and lines first reach the limit of blackbody emission as given by Planck's radiation law for that temperature. Bremsstrahlung starts to touch the Planck curve first at long wavelengths.

The maximum of the bremsstrahlung emission is at the wavelength λ_{max} given by

$$\lambda_{max} k_B T_e = \frac{hc}{2}, \tag{13.21}$$

which may be written as

$$\frac{\lambda_{max}}{\text{nm}} \frac{k_B T_e}{\text{eV}} = 620, \tag{13.22}$$

i.e. it depends on the electron temperature $k_B T_e$. It is obvious that the emission of hot plasmas shifts to the x-ray region.

Most technical plasmas at low temperatures display a rich spectrum of lines from molecules, neutral atoms and ions of low charge. With increasing temperature the stages of ionization rapidly increase. For example, if we consider light or medium heavy elements of nuclear charge $6 \leq Z_n \leq 26$, the semi-empirical relation

$$\frac{k_B T_e}{\text{eV}} \geq 0.27 \, Z_n^{3.4} \tag{13.23}$$

gives a lower limit of the temperature range were at least 50% of the atoms are already completely stripped. Even "pure" hot plasmas of hydrogen or deuterium as they are studied for fusion contain light and heavy ions as *impurities*. They are unavoidable and enter the plasma by erosion of the walls. Although undesired by the plasma physicist, they are the spectroscopist's delight! Therefore, sometimes specific impurities are added intentionally for diagnostic purposes or for radiation cooling by the boundary region of the plasma.

The inherent shortcoming of all spectroscopic observations is the fact that radiation is collected along the full line of sight s in the direction of observation and any local information thus is lost. For optically thin plasmas, i.e. negligible absorption of the radiation within the plasma, the spectral radiance L_λ at the surface defined as spectral flux in watt per cm^2 per steradian sr and per wavelength interval 1 nm is related to the emission coefficient ϵ_λ, which characterizes the local emission in the plasma, by

$$L_\lambda = \int \epsilon_\lambda(s) \mathrm{d}s . \tag{13.24}$$

$\epsilon_\lambda \mathrm{d}V$ describes the local flux emitted from the volume element $\mathrm{d}V$ per spectral interval per steradian. In principle it would be possible to apply computer tomographic techniques to obtain local information but due to experimental constraints multi-directional observations are not feasible on nearly all plasma devices. With knowledge of the internal structure of a plasma, on the other hand, less observations are needed: for example, in case of axial symmetry one observation of the plasma cross-section in one direction suffices to deduce local emission coefficients via the procedure of Abel inversion [23]. One special case is known as particle beam diagnostics and is mostly applied to the boundary region of fusion devices: neutral atomic beams are injected and their emission is analyzed until they become ionized. Observation perpendicular to the beams thus gives directly the local information.

13.3.2 Charge State Distribution

Atomic species in a plasma are continuously ionized, they go successively through their ionization stages till they reach an equilibrium stage where further ionization is balanced by recombination. In principle, the equation of continuity has to be solved for the density n_z of each ionization stage z:

$$\frac{\partial n_z}{\partial t} + \nabla \cdot \mathbf{\Gamma_z} = Q_z , \tag{13.25}$$

$\mathbf{\Gamma_z}$ is the flux density due to diffusion and convection, in compressing or expanding plasmas respective changes have to be included. The source term Q_z couples all equations of continuity through ionization and recombination. The frequencies of atomic processes involving electrons are quantitatively characterized by the product of the respective cross-sections $\sigma(v)$ and the velocity v of the impacting electrons. In plasmas with Maxwellian velocity distribution these products $\sigma(v)v$ simply have to be averaged over the velocity distribution function $f_e(v)$ to yield the respective rate coefficients $\langle \sigma(v) v \rangle = \int \sigma(v) \, v \, f_e(v) dv$, which now are functions of the temperature.

Neglecting transport the continuity equations (13.25) may be written as rate equations

$$\frac{\partial n_z}{\partial t} = n_e (n_{z-1} S_{z-1} - n_z S_z) + n_e (n_{z+1} \alpha_{z+1} - n_z \alpha_z) , \qquad (13.26)$$

where S_z is the rate coefficient for ionization and α_z that for recombination. α_z includes radiative and dielectronic recombination [23]. With increasing electron density ionization from excited levels comes in and three-body recombination starts to become important so that both $\alpha_z = \alpha_z(n_e, T_e)$ and $S_z = S_z(n_e, T_e)$ are now also functions of the electron density in addition to the dependence on the temperature. Steady-state is reached for $(\partial n_z / \partial t) = 0$ which corresponds to

$$\frac{n_{z+1}}{n_z} = \frac{S_z}{\alpha_{z+1}} . \qquad (13.27)$$

At low densities this ratio is only a function of the temperature and well known as corona equilibrium in reference to the solar corona were such conditions prevail.

Figure 13.8 shows the relative ion stage distribution for neon; the Roman numeral is the spectroscopic notation of the ions and corresponds to $(z+1)$. For the diagnostician it is of interest that the heliumlike ion exists over the largest temperature range, e.g. NeIX in Fig. 13.8. This also holds for other ions with a closed shell, for example for neonlike ions. On the other hand, the mere existence of an ion in steady-state allows a rough estimate of $k_B T_e$. With increasing density the equilibrium is pushed to lower temperatures by the

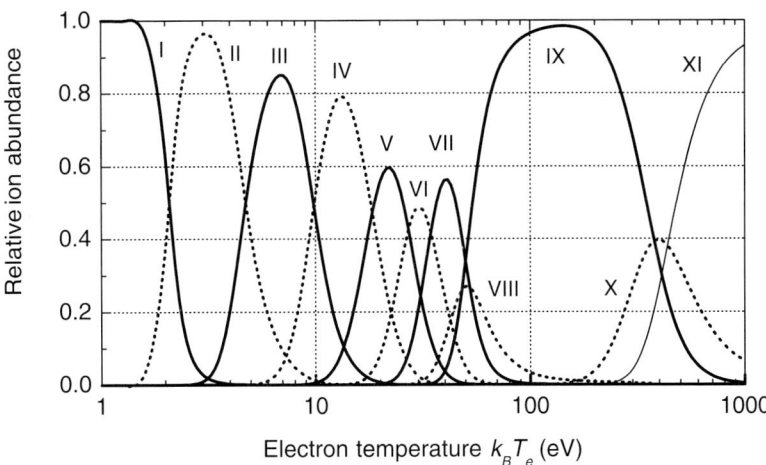

Fig. 13.8. Distribution of ionization stages of neon, from [25] as function of the electron temperature

increasing three-body recombination, and it approaches the Saha equilibrium given by

$$\frac{n_{z+1}n_e}{n_z} = 2\frac{U_{z+1}(T_e)}{U_z(T_e)}\left(\frac{m_e k_B T_e}{2\pi\hbar^2}\right)^{3/2}\exp\left[-\frac{E_z(\infty)}{k_B T_e}\right]. \qquad (13.28)$$

$U_z(T_e)$ is the partition function of the ion in charge state z, and in the ionization energy $E_z(\infty)$ a lowering due to the interaction of the ion with neighbouring charged particles has to be taken into account which becomes important at relatively high densities [22].

13.3.3 Line Emission

A line is emitted when an atomic species undergoes a transition from an upper level (p) of energy $E(p)$ to a lower level (q) of energy $E(q)$. The frequency of the lines is given by $\nu_{pq} = [E(p) - E(q)]/h$. The transition is *spontaneous*, i.e. the decay rate is simply proportional to the density $n_z(p)$ of the species in the upper state,

$$\left.\frac{dn_z(p)}{dt}\right|_{p\to q} = A(p\to q)\,n_z(p)\,, \qquad (13.29)$$

and the decay constant $A(p \to q)$ is a characteristic atomic constant for that specific transition and known as atomic transition probability or respective Einstein coefficient. With each transition a photon is emitted, and the emission coefficient ϵ_z of the line thus is given by

$$\epsilon_z(p\to q) = \frac{h\nu}{4\pi}A(p\to q)\,n_z(p)\,. \qquad (13.30)$$

This fundamental equation reveals that the population densities $n_z(p)$ of the excited states of the atomic species are the quantities obtained from measurements of the radiance of lines provided the transition probabilities are known. Hence, information on plasma parameters can only be deduced to that extent as they determine $n_z(p)$ directly and indirectly. Available transition probabilities are being critically evaluated by the National Institute of Standards and Technology (NIST) of the USA and are readily accessible via internet [26]. The rate of change of the population density of a level (p) is given in general by an equation of the type

$$\frac{dn_z(p)}{dt} = -n_z(p)A(p\to) - n_z(p)C(p\to) + R_r(\to p) + R_c(\to p)\,, \qquad (13.31)$$

where $A(p\to) = \sum_{r<p}A(p\to r)$ is the sum of all radiative transition probabilities from (p) to all lower levels (r), and $n_z(p)C(p\to)$ represents the sum of all collisional rates from that level to levels of the same atom, to levels of other atomic species and to the continuum. Correspondingly the sum of all

collisional rates into that level is $R_c(\to p)$, and the sum of all radiative rates from higher levels into (p) including radiative recombination but also photo-excitation by self-absorption is $R_r(\to p)$. Temperature and density of plasmas usually change on time scales that are long compared to intrinsic atomic relaxation times, so that a quasi-stationary population density, $\mathrm{d}n_z(p)/\mathrm{d}t = 0$, establishes itself on a time scale

$$\tau = \frac{1}{A(p \to) + C(p \to)}, \qquad (13.32)$$

and $n_z(p)$ practically reflects the existing plasma parameters at any instant. The longest relaxation time is $1/A(p \to)$. The emission coefficient of a line thus may be written as

$$\epsilon_z(p \to q) = \frac{h\nu}{4\pi} \frac{A(p \to q)}{A(p \to) + C(p \to)} \left[R_c(\to p) + R_r(\to p) \right]. \qquad (13.33)$$

A general solution of this equation is impossible, the number of participating processes is very large and many are not known with sufficient accuracy. Hence one usually limits the number of processes to the most important ones, which are:

- collisional excitation from the ground state by electrons,
- collisional excitation from other excited levels, especially from metastable or close-by levels,
- recombination (radiative, dielectronic, three-body),
- charge exchange collisions with neutrals and ions,
- radiative cascading from higher levels,
- self-absorption especially of resonance lines at high densities.

A number of steady-state models have been set-up for the atomic level populations in various ions and for various conditions. They are usually referred to as collisional-radiative (CR) models. Cross-sections and rate coefficients are found in the literature, for example in [27, 28, 29].

The population models simplify at low and high plasma densities. At *high plasma densities* collisional rates become much larger than radiative rates and $A(p \to)$ and $R_r(p \to)$ can be neglected in (13.33). In this limit the collisional rates between any two levels (p) and (r) become equal (principle of detailed balance), and because of their high velocity and large cross-sections electron collisions will usually establish this equilibrium first:

$$n_z(p) C(p \to) = R_c(\to p)$$
$$\to n_z(p) n_e X(p \to r) = n_z(r) n_e X(r \to p). \qquad (13.34)$$

$C(p \to r) = n_e X(p \to r)$, where $X(p \to r) = \langle \sigma(p \to r; v) v \rangle$ is the rate coefficient for electron collisional transitions from level (p) to level (r). With

$$g_z(p) X(p \to r) = g_z(r) X(r \to p) \exp\left[\frac{E(p) - E(r)}{k_B T_e}\right] \qquad (13.35)$$

we arrive at the Boltzmann distribution for the level population densities in the ion (z):

$$\frac{n_z(p)}{n_z(r)} = \frac{g_z(p)}{g_z(r)} \exp\left[-\frac{E(p) - E(r)}{k_B T_e}\right]. \tag{13.36}$$

$g_z(p)$ and $g_z(r)$ are the statistical weights of the levels. This condition in a plasma is known as *local thermodynamic equilibrium* (LTE). Summation over all levels gives the density n_z of the ionization stage and we may write

$$\frac{n_z(p)}{n_z} = \frac{g_z(p)}{U_z(T_e)} \exp\left[-\frac{E(p) - E(g)}{k_B T_e}\right] \tag{13.37}$$

with the partition function

$$U_z(T_e) = \sum_i g_z(i) \exp\left[-\frac{E(i) - E(g)}{k_B T_e}\right]. \tag{13.38}$$

Thus the measurement of the emission coefficient $\epsilon_z(p \to q)$ of a line gives the total density n_z of the ion provided the temperature and the partition function were known. On the other hand, the ratio of two lines is a sole function of the temperature since n_z and $U_z(T_e)$ cancel:

$$\frac{\epsilon_z(p \to q)}{\epsilon'_z(p' \to q')} = \frac{\nu}{\nu'} \frac{A(p \to q)}{A(p' \to q')} \frac{g_z(p)}{g_z(p')} \exp\left\{-\frac{[E(p) - E(p')]}{k_B T_e}\right\} = f(T_e). \tag{13.39}$$

For sufficient sensitivity one should choose $E(p) - E(p') > k_B T_e$. If this cannot be achieved within one ion, one may select two successive ionization stages:

$$\frac{\epsilon_{z+1}(p \to q)}{\epsilon'_z(p' \to q')} = \frac{\nu}{\nu'} \frac{A(p \to q)}{A(p' \to q')} \frac{n_{z+1}(p)}{n_z(p')}. \tag{13.40}$$

Both ionization stages are connected via the Saha–Boltzmann equation (13.28), and the ratio thus becomes a function of density and temperature:

$$\frac{\epsilon_{z+1}(p \to q)}{\epsilon'_z(p' \to q')} = \frac{f(T_e)}{n_e}. \tag{13.41}$$

The necessary condition of LTE certainly poses limitations and the validity of such an assumption must be checked. According to Griem [22] the electron induced collisional transition rate between the largest energy gap $[E(p) - E(q)]$ of the atomic species should be larger than the radiative decay rate by at least a factor of 10. This leads to the condition

$$\frac{n_e}{\text{cm}^{-3}} \geq 10^{14} \times \left(\frac{k_B T_e}{\text{eV}}\right)^{1/2} \left[\frac{E(p) - E(q)}{\text{eV}}\right]^3. \tag{13.42}$$

With decreasing electron density LTE seizes to exist but electron collisions will still dominate transitions between high-lying levels, since their spacing decreases towards the ionization limit. The lowest level, for which collisional transitions are faster than the radiative ones also by a factor of 10, is called *thermal limit*. Its principal quantum number n_{th} is given by [22]

$$n_{th} \approx 141 \times Z^{12/17} \left(\frac{n_e}{\text{cm}^{-3}}\right)^{-2/17} \left(\frac{k_B T_e}{\text{eV}}\right)^{1/17}. \qquad (13.43)$$

All levels above n_{th} are populated according to a Boltzmann distribution, and each level is connected to the ground state of the next ionization stage by the Saha–Boltzmann equation. One talks of *partial local thermodynamic equilibrium* PLTE.

At *low densities and weak radiation fields* radiative decay dominates collisional depopulation, and $C(p \rightarrow)$ and $R_r(p \rightarrow)$ can be neglected in (13.33). For higher accuracy maybe radiative cascading should be retained in $R_r(p \rightarrow)$. Because of the strong radiative decay compared to the collisional population, the steady-state population densities of the excited levels are very low, and in this limit practically all atomic and ionic species are in their respective ground state (g) and in a metastable level (m), if its radiative decay rate is low. Excitation thus occurs essentially by electron collisions from the ground state and to some extent from the metastable state, and we approximate

$$R_c(\rightarrow p) = n_z(g)\, n_e X(g \rightarrow p) + n_z(m)\, n_e X(m \rightarrow p). \qquad (13.44)$$

For low $n_z(m)$, the emission coefficient of a line reduces to:

$$\epsilon_z(p \rightarrow q) = \frac{h\nu}{4\pi} \frac{A(p \rightarrow q)}{A(p \rightarrow)} n_e X(g \rightarrow p)\, n_z(g). \qquad (13.45)$$

This low density limit is known as *coronal population model*. In order to deduce $n_z(g)$ from a measured emission coefficient,

- the radiative transition probabilities,
- the excitation function $X(T_e)$, which is a function of the electron temperature,
- and the electron density n_e

are needed.

In the ratio of two lines n_e and $n_z(g)$ cancel and it becomes a sole function of the temperature. Again for sufficient accuracy the energy separation of the upper levels should be large compared to $k_B T_e$. For emission from metastable levels of ions of low and medium nuclear charge Z_n collisional depopulation $n_e X(m \rightarrow)$ must be retained in the denominator of (13.45):

$$\epsilon_z(m \rightarrow q) = \frac{h\nu}{4\pi} \frac{A(m \rightarrow q)}{A(m \rightarrow) + n_e X(m \rightarrow)} n_e X(g \rightarrow m)\, n_z(g). \qquad (13.46)$$

This maintains a density dependence which is utilized, for example, in the ratio of resonance and intercombination line of heliumlike ions:

$$\frac{\epsilon_z(p \to g)}{\epsilon_z(m \to g)} = \frac{\nu_{pg}}{\nu_{mg}} \frac{X(g \to p)}{X(g \to m)} \left[1 + n_e \frac{X(m \to)}{A(m \to g)}\right]. \qquad (13.47)$$

This ratio thus is successfully used especially for the density determination in pinch and laser-produced plasmas, and for the experimentalist it is of advantage that both lines are neighboring.

Transitions from doubly excited states above the first ionization limit are characteristic features in the x-ray spectra of hydrogen- and heliumlike ions observed from solar flares and fusion plasmas, for example [30]. Since one electron is like a "spectator" to the other electron making the radiative transition, the wavelength is close to and on the long-wavelength side of the respective transition in the singly excited species of the next ionization stage, and the designation *satellite* for such lines is obvious. Doubly excited states are produced by dielectronic capture of an electron from the continuum and/or by excitation of an inner-shell electron. They offer thus the possibility to measure the electron temperature and the relative concentration of the two ionization stages.

Finally, injection of energetic neutral hydrogen beams usually is done for heating of fusion plasmas, but quite early also the diagnostic potential was recognized. Highly selective charge exchange of the hydrogen atoms even with completely ionized atoms takes place and allows the determination of the density of the respective ions in the plasma.

13.3.4 Line Profiles

Spectral lines do not have an infinitesimal spectral width but are broadened by several mechanisms: natural broadening, Doppler broadening and broadening under the influence of plasma particles. By introducing the line shape function $\mathcal{L}(\omega)$ the spectral emission coefficient is obtained by multiplying the previous equations with this function:

$$\epsilon_\omega(p \to q) = \epsilon(p \to q)\, \mathcal{L}(\omega) \quad \text{with} \quad \int_{-\infty}^{\infty} \mathcal{L}(\omega)\, d\omega = 1\,. \qquad (13.48)$$

The full width at half maximum (FWHM) is usually taken as measure of the line width, and we shall designate it by $\Delta\omega_{1/2}$.

Natural broadening is the consequence of the finite lifetime $\tau = 1/A(p \to)$ of all excited states of atomic systems: Heisenberg's uncertainty principle $\Delta E\, \tau = \hbar/2$ demands a corresponding energy spread $2\Delta E$ of both upper and lower level leading to a total spread of the emitted photons of $2[\Delta E(p) + \Delta E(q)] = \Delta \hbar \omega_{1/2}$. The line shape function is a Lorentzian with

$$\mathcal{L}(\omega) = \frac{1}{\pi} \frac{\Delta\omega_{1/2}/2}{(\omega - \omega_o)^2 + (\Delta\omega_{1/2}/2)^2}\,. \qquad (13.49)$$

Natural broadening is, however, negligibly small.

The motion of the emitting ions causes corresponding Doppler shifts, and for a Maxwellian velocity distribution this results in a Gaussian profile. Its FWHM is given by

$$\Delta\lambda_{1/2} = 2\,(2\ln 2)^{1/2}\,\lambda_0 \left(\frac{k_B T_i}{m_i c^2}\right)^{1/2} = 7.715 \times 10^{-5}\,\lambda_0 \left(\frac{k_B T_i/\mathrm{eV}}{m_i/\mathrm{u}}\right)^{1/2}, \quad (13.50)$$

where "u" is the atomic mass unit. Ion temperatures are thus readily deduced.

It is obvious that plasma particles perturb the emission of atomic species and thus reduce the lifetime of the states, which consequently leads to broadening of spectral lines and to a less pronounced shift. The total effect is known generally as *pressure broadening*. The perturbations are complex and manifold, and the relevant reference is the monograph by Griem [31]. In plasmas these perturbations are due to interactions of the radiating atom or ion with surrounding electrons and ions, if we exclude rather weakly ionized plasmas, in which interactions with neutral particles have to be accounted for, too. The charged particles produce electric fields in the vicinity of the radiator, and hence the resulting broadening is also called *Stark broadening*.

The interaction with the charged particles occurs on two time scales. Ions move slowly and their produced microfield may be considered constant during the relevant time interval. The respective Stark pattern of upper and lower level can thus be calculated using time dependent perturbation theory. The resulting broadening certainly reflects the distribution of the electric microfield. For isolated lines this ion broadening is essentially due to the quadratic Stark effect, lines of hydrogen and hydrogenic ions have the linear Stark effect resulting in much broader profiles with characteristic structures. Improved calculations replace this quasistatic approximation by procedures allowing for actual ion dynamics.

Fields by the fast moving electrons vary so rapidly that the process of interaction can be considered as collision. In this impact approximation, the line profile is Lorentzian with a width

$$\Delta\omega_{1/2} = n_e \langle v_e(\sigma_p + \sigma_q)\rangle\,, \quad (13.51)$$

where the average is over electron velocities and over cross-sections σ_p and σ_q of inelastic collisional transitions from all relevant levels. Also elastic collisions can contribute. For diagnostic applications it is important, that the widths are proportional to the electron density, and indeed line widths offer one possibility to measure n_e. Some ion broadening usually has to be accounted for also in lines mainly broadened by electron collisions, just as electron impact broadening modifies lines predominantly broadened by the linear Stark effect. The density scaling of these lines is different since their shape is essentially determined by the electric microfield. Thus we have, for example, for the P_α-line of HeII at 468.6 nm the semi-empirical relation [31]

$$\frac{n_e}{\mathrm{cm}^{-3}} = 3.31 \times 10^{17} \left(\frac{\Delta\lambda_{1/2}}{\mathrm{nm}}\right)^{1.21}. \quad (13.52)$$

Doppler and Stark broadening are usually present at the same time, and Gaussian and Lorentzian profile are convolved and yield a Voigt profile. On the other hand, if line profiles are measured with sufficient accuracy, Gaussian and Lorentzian component can be extracted. In the analysis of a line profile one has to check, of course, for self-absorption which sets in at high densities and causes broadening of the lines.

Finally, the electric microfield also mixes the wavefunctions of close-by levels with different parity, and a dipole-forbidden transition may start to appear at the side of an allowed transition. Its relative strength is a function of the electron density [31].

13.3.5 Continuum Radiation

We briefly consider the bremsstrahlung emission of hot plasmas (illustrated in Fig. 13.6) at long wavelengths, for example in the easily accessible visible spectral region. The emission coefficient of a pure hydrogen or deuterium plasma is of the form

$$\epsilon_\lambda \propto \frac{1}{\lambda^2} \frac{n_e^2}{(k_B T_e)^{1/2}}, \qquad (13.53)$$

i.e. the temperature dependence is weak and the dependence on the electron density is strong. This means that an approximate knowledge of T_e allows the determination of n_e. At short wavelengths in the x-ray region, on the other hand, the dependence on the temperature is strong and approximately given by

$$\lg \epsilon_\nu \simeq -\frac{1}{\ln 10} \frac{h\nu}{k_B T_e} + \text{const}. \qquad (13.54)$$

Now T_e is readily derived from a semi-logarithmic plot of the spectral emission coefficient versus the energy of the photons.

References

1. I.H. Hutchinson: *Principles of Plasma Diagnostics* (Cambridge University Press, Cambridge New York New Rochelle Melbourne Sydney 1987)
2. A.A. Ovsyannikov, ed: *Plasma Diagnostics*, (Cambridge International Science Publishing, Cambridge 2000)
3. R.H. Huddlestone, S.L. Leonard: *Plasma Diagnostic Techniques* (Academic Press, New York London 1965)
4. W. Lochte-Holtgreven: *Plasma Diagnostics* (North-Holland, Amsterdam 1968)
5. K. Muraoka, M. Maeda: *Laser Aided Diagnostics of Plasmas and Gases* (Institute of Physics Publishing, Bristol 2001)
6. H.-J. Kunze: The Laser as a Tool for Plasma Diagnostics. In: *Plasma Diagnostics* ed by W. Lochte-Holtgreven (North-Holland, Amsterdam 1968) pp. 550
7. D.E. Evans, J. Katzenstein: Rep. Prog. Phys. **32**, 207 (1969)

8. A.W. DeSilva, G.C. Goldenbaum: Methods of Experimental Physics **9**, ed by H.R. Griem, R.H. Lovberg (Academic Press, New York London 1970) pp. 61
9. J. Sheffield: *Plasma Scattering of Electromagnetic Radiation*, (Academic Press, New York, 1975)
10. O. Naito, H. Yoshida, T. Matoba: Physics of Fluids B **5**, 4256 (1993)
11. H.-J. Wesseling, B. Kronast: J. Phys. D: Appl. Phys. **29**, 1035 (1996)
12. H. Salzmann, J. Bundgaard, A. Gadd et al: Rev. Sci. Instrum. **59**, 1451 (1988)
13. E.E. Salpeter: Phys. Rev. **120**, 1528 (1960)
14. T. Wrubel: Stark-Breiten isolierter Linien in einem mittels ortsaufgelöster Thomson-Streuung diagnostizierten Plasma. PhD-thesis, Ruhr-University, Bochum, Germany (1999)
15. S.H. Glenzer, W.E. Alley, K.G. Eastabrook et al: Phys. Plasmas **6**, 2117 (1999)
16. T. Wrubel, S. Glenzer, S. Büscher et al: J. Atm. Terr. Phys. **58**, 1077 (1996)
17. D.E. Evans: Plasma Phys. **12**, 573 (1970)
18. A.W. DeSilva, T.J. Baig, I. Olivares et al: Phys. Fluids B **4**, 458 (1991)
19. S. Maurmann, H.-J. Kunze: Phys. Fluids **26**, 1630 (1983)
20. E.V. Suvorov, E. Holzhauer, W. Kasparek et al: Plasma Phys. Control. Fusion **39**, B337 (1997)
21. S.H. Glenzer, G. Gregori, F.J. Rogers et al: Phys. Plasmas **10**, 2433 (2003)
22. H.R. Griem: *Plasma Spectroscopy* (McGraw–Hill, New York San Francisco Toronto London 1964)
23. H.R. Griem: *Principles of Plasma Spectroscopy* (Cambridge University Press, Cambridge 1997)
24. A. Thorne, U. Litzén, S. Johannson: *Spectrophysics - Principles and Applications* (Springer, Berlin Heidelberg New York 1999)
25. Y. Ralchenko: private communication (National Institute of Standards and Technology 2004)
26. http://physics.nist.gov/PhysRefData/contents.html
27. V.G. Pal'chikov, V.P. Shevelko: *Reference Data on Multicharged Ions* (Springer, Berlin Heidelberg New York 1995)
28. R.K. Janev: *Atomic and Molecular Processes in Fusion Edge Plasmas* (Plenum Press, New York London 1995)
29. G.W.F. Drake: *Atomic, Molecular & Optical Physics Handbook* (AIP Press, Woodbury New York 1996)
30. G. Bertschinger, W. Biel, TEXTOR-94 Team et al: Physics Scripta **T83**, 132 (1999)
31. H.R. Griem: *Spectral Line Broadening by Plasmas* (Academic Press, New York London 1974)

14 Observation of Plasma Fluctuations

O. Grulke and T. Klinger

Abstract. The diagnostics of fluctuating plasma quantities and their evaluation is an important issue in the context of plasma waves, instabilities, and transport. This chapter introduces into the basic data analysis methods of fluctuations and gives a brief overview over the most important diagnostics for fluctuation measurements, which is strongly related to the chapters on waves in plasmas Chap. 2 in Part I and transport questions Chap. 9 in Part II.

14.1 Introduction

A plasma is never quiescent. Fluctuations ("fluctuosus" latin for "surging") are temporal deviations from the current state of the plasma. The most basic type of fluctuation is the thermal one, which is caused by the individual motions in the many particle system. This noise-type fluctuations are of less interest here and are covered by the thermodynamic description of plasmas (i.e. Klimontovich's many particle theory [1]). The collective behavior of a plasma can lead to a different class of plasma fluctuations, which are the subject of the present chapter.

Plasma fluctuations can be local or global, propagating or non-propagating, regular or turbulent. They play an important role in transport processes and are generally indicators for plasma instability. It is thus of greatest importance to observe and to characterize plasma fluctuations with the same care as steady-state parameters. Fluctuations can even been seen as quasi-particles (in a quantum-mechanical sense, defined by their frequencies and wave numbers) being an individual constituent of the plasma like electrons and ions. The interaction between the classical plasma particles and e.g. waves is then described by collisions with quasi-particles.

The physics of plasma fluctuations has a close relationship to the physics of plasma waves (see Chap. 2 in Part I), since plasma instabilities usually occur as waves, whose frequencies and wavenumbers determine fluctuation spectra and phase shifts. The fundamental difference between the two approaches is that plasma fluctuations are merely seen as the temporal variation of one or more plasma parameters. It is then the goal of the investigation to identify the origin of these fluctuations. In the physics of plasma waves, the waves' features are usually predefined and then propagation and dispersion properties are analyzed. Hence, the two concepts are fairly complementary but should be seen as just two different perspectives in the broad view of plasma dynamics.

This chapter gives an overview over measurement, analysis and some physics interpretation of plasma fluctuations. In Sect. 14.2, we introduce the most important basic concepts and notions, necessary to enter the review of statistical analysis tools. The different diagnostics used for fluctuation measurements are compiled in Sect. 14.3.

14.2 Basics

If a plasma fluctuates, this usually affects more than only one physical quantity. Instead, there is often a tight coupling between a number of fluctuating quantity, if not all. Mutual relationships like phase, amplitudes, correlations, coupling etc. can be very revealing in the search for the physical mechanisms. The common set of fluctuating quantities is compiled in Table 14.1.

Table 14.1. Plasma and field quantities observed in the laboratory to characterize fluctuations. A number of additional quantities can be derived from the listed ones

Class	Quantity	Symbol
plasma parameters	electron/ion density	n_e, n_i
	plasma potential	ϕ_p
	electron/ion temperature	T_e, T_i
electromagnetic fields	electric field	**E**
	magnetic field	**B**
particle flows	electron/ion velocities	$\mathbf{v}_e, \mathbf{v}_i$
	electron/ion fluxes	$\mathbf{\Gamma}_e, \mathbf{\Gamma}_i$
plasma kinetics	electron/ion distribution functions	f_e, f_i

All quantities are generally functions of space **r** and time t and one has to deal with fluctuation fields, either scalar-valued (e.g. density, potential) or vector-valued (e.g. electric field, velocity). As an example, we consider turbulent plasma density fluctuations[1] described as

$$n(\mathbf{r},t) = n_0(\mathbf{r}) + \tilde{n}(\mathbf{r},t) , \qquad (14.1)$$

where $n_0(\mathbf{r})$ is the average (mean) plasma density and $\tilde{n}(\mathbf{r},t)$ is the (zero mean) plasma density fluctuation. The ratio $\eta = \tilde{n}/n_0$ is called "fluctuation degree" and is usually found to be in the range of a few ten percent. Note

---

[1] For convenience we omit here the distinction between electron plasma density n_e and ion plasma density n_i. In most cases $n_e(\mathbf{r},t) = n_i(\mathbf{r},t) = n(\mathbf{r},t)$ holds.

that (14.1) is a separation ansatz for time and we have $n_0(\mathbf{r}) = \langle n(\mathbf{r},t) \rangle$ and $\tilde{n}(\mathbf{r},t) = n(\mathbf{r},t) - \langle n(\mathbf{r},t) \rangle$ with the averaging operator $\langle \cdot \rangle$ defined as

$$\langle x(\mathbf{r},t) \rangle = \int_{-\infty}^{\infty} x(\mathbf{r},t) P(x) \mathrm{d}x = \lim_{T \to \infty} \frac{1}{T} \int_{-T/2}^{+T/2} x(\mathbf{r},t) \mathrm{d}t , \qquad (14.2)$$

where $P(x)$ is the probability density function of x. In (14.2) the ergodic theorem was used to relate the ensemble average to the time average [2]. The use of ergodicity turns out to be extremely useful for practical purposes. The limit $T \to \infty$ is restricted, of course, by the time window of observation.

As mentioned above, the considerable interest in plasma fluctuations stems from their direct impact on particle and heat transport. Already in the early days of research on magnetic plasma confinement, it became clear that classical (diffusive) transport cannot explain the experimentally observed confinement times [3]. The additional ("anomalous") transport turned out to be caused by plasma fluctuations. In particular we obtain the particle fluxes [4]

$$\Gamma_E = \langle \tilde{E}_\Theta \tilde{n} \rangle / B_\phi , \qquad (14.3)$$

$$\Gamma_B = -\langle \tilde{j}_\parallel \tilde{B}_r \rangle / e B_\phi . \qquad (14.4)$$

Here, the fluctuating quantities are expressed in toroidal coordinates: $\tilde{E}_\Theta$ is the fluctuating poloidal electric field component, $\tilde{B}_r$ is the fluctuating radial magnetic field component, and $\tilde{j}_\parallel$ is the fluctuating current parallel to the magnetic field. The total anomalous particle flux is $\Gamma_{an} = \Gamma_e + \Gamma_B$. Similarly, for the anomalous heat flux one has

$$Q_{an} = \frac{3}{2} k_B n \langle \tilde{E}_\Theta \tilde{T}_\alpha \rangle / B_\phi + \frac{3}{2} k_B T_\alpha \langle \tilde{E}_\Theta \tilde{n} \rangle / B_\phi , \qquad (14.5)$$

where k_B is Boltzmann's constant. In (14.5) the effect of fluctuating magnetic field components was neglected. From (14.3)–(14.5) we see how important the phase relationship between the fluctuating quantities is: If a pair of multiplied and averaged fluctuating quantities has a random phase, the flux contribution vanishes. Conversely, any non-random phase relationship between each two fluctuating quantities leads to particle or heat flux. A closer look on transport phenomena in plasmas is taken in Chap. 9 in Part II.

Figure 14.1 shows raw data from density fluctuation measurements in a magnetized laboratory plasma [5]. The current flow of an electrostatic plasma probe (cf. Sect. 14.3) inserted into the edge of a cylindrical plasma column was recorded with a transient digitizer. The result is a "time series"

$$n_m = \{n_k : k = 1 \ldots m\} \quad \text{with} \quad n_k = \tilde{n}(\mathbf{r}_0, t_0 + k \Delta t) . \qquad (14.6)$$

Here $\mathbf{r}_0$ is the point of observation, t_0 is the start time of the measurement, Δt is the time interval between each two measurements (sampling rate) and

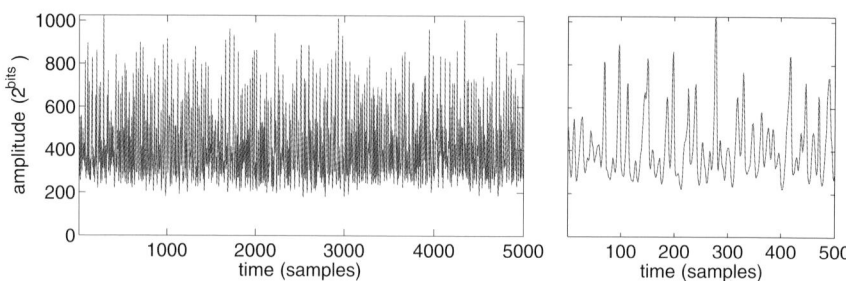

Fig. 14.1. *Left*: Raw data time series n_m ($m = 1\ldots 5\,000$) of turbulent density fluctuations measured with an electrostatic plasma probe. *Right*: The same but for the interval ($m = 1\ldots 5\,00$)

m is the length of the time series. The sampling rate defines the highest resolved frequency via the Nyquist limit $f_{Ny} = 1/2\Delta t$. Fluctuations with frequencies $f > f_{Ny}$ are undersampled and due to aliasing they spuriously occur in the lower frequency range. Care must be taken to limit the frequency range of fluctuations to the interval $[0, f_{Ny}]$ before digitizing (e.g. by the use of so-called anti-aliasing filters). The data is digitized at some point, and the vertical resolution is given by the number of bits of the analog-digital converter. In the present example, Fig. 14.1, we have 10 bit data sampled with $f_s = 500$ MHz which yields $\Delta t = 1/f_s = 2\,\mu$s.

Prior to a deeper analysis, the raw data must be preprocessed to introduce suitable scales and to meet assumptions made in the statistical framework. First, the fluctuation data should be zero mean. Further, it is useful to normalize fluctuation data to its standard deviation. This reads to be

$$n'_m = \{n'_t : t = 0\ldots m\Delta t\} \quad \text{with} \quad n'_t = \frac{\tilde{n}(\mathbf{r_0}, t_0 + t) - \mu(n)}{\sigma(n)}, \quad (14.7)$$

$$\mu(n) = \langle n(\mathbf{r_0}, t)\rangle \quad \text{and} \quad \sigma^2(n) = \langle [n - \mu(n)]^2 \rangle\,.$$

Figure 14.2 shows the same data as in Fig. 14.1 but preprocessed according to (14.8). Note that the density fluctuations are non-symmetric with respect to zero: positive excursions (up to 4σ) occur much more frequently than negative ones. One furthermore sees that the fluctuation signal is non-periodic with a fairly strong amplitude and phase modulation. To quantify these two visual impressions, we consider

1. the probability density function and
2. the power spectral density of the fluctuation signal.

Figure 14.3 shows the normalized amplitude histogram of the plasma density fluctuation time series. The histogram is obtained by partitioning the amplitude span into a discrete number of intervals (also called "bins") and to count the total number of data points falling into each interval. The histogram is

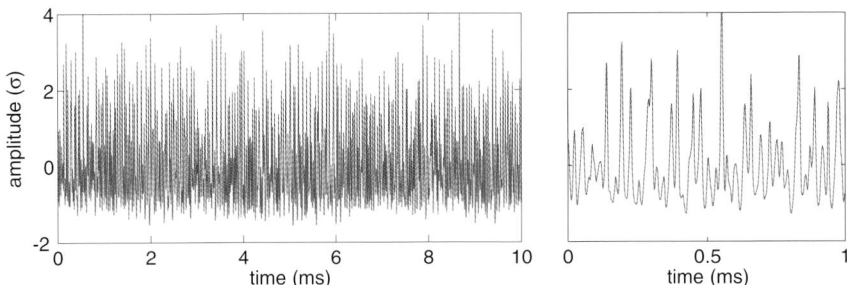

Fig. 14.2. *Left*: The same data as in Fig. 14.1 but normalized to standard deviation and with a time axis ($t = 0$–$10\,\mathrm{ms}$) in physical units. *Right*: The same but for the interval ($t = 0$–$1\,\mathrm{ms}$)

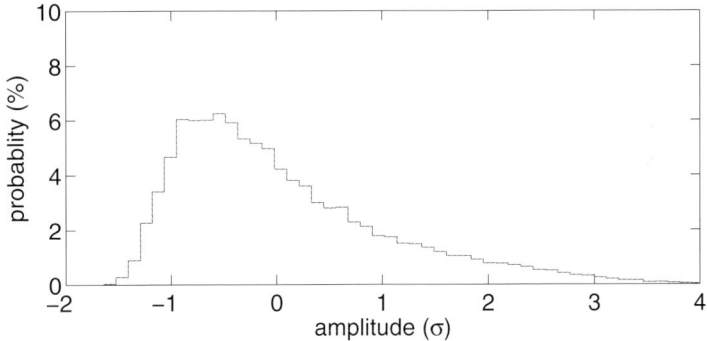

Fig. 14.3. Histogram of the density fluctuation time series

normalized to its sum to approximate the usual probability density normalization $\int P(x)\mathrm{d}x = 1$.

From Fig. 14.3 it becomes clear that the probability density function (PDF) of the plasma density fluctuations is far from being a Gaussian. The non-Gaussian shape can be measured by two higher-order moments of the PDF

$$S = \frac{\langle [n - \mu(n)]^3 \rangle}{\sigma^3(n)} \quad \text{and} \quad K = \frac{\langle [n - \mu(n)]^4 \rangle}{\sigma^4(n)} \,. \tag{14.8}$$

S is the skewness, K the kurtosis ('flatness') of the PDF. For a perfect Gaussian (normal distribution) $S = 0$ and $K = 3$ holds. For the PDF shown in Fig. 14.3 we obtain $S = 1.1$ (strong asymmetry) and $K = 3.9$ (more peaked than a Gaussian).

The power spectral density of the fluctuation signal can be obtained in a straightforward way by discrete Fourier transform of the time series. A problem is, that a fluctuation time series never meets the periodic boundary condition. This results in a feature known as "Gibbs phenomenon", which refers to the ripple effect in the partial sums of the Fourier series at a point

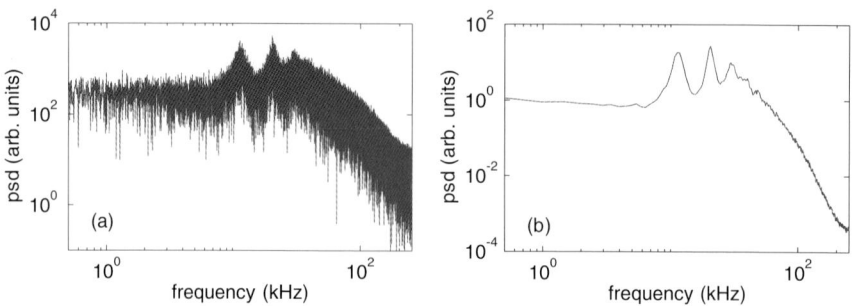

Fig. 14.4. Power spectral density of the fluctuations shown in Fig. 14.1. **(a)** Spectrum obtained from the simple modulus of the Fourier transform of the time series. **(b)** Spectrum obtained from Welch's averaged modified periodogram method

where the function is discontinuous [2]. These spurious side lobes can be seen in the power spectrum shown in Fig. 14.4 (a). The spectrum seems to have a significant noise level superimposed to the actual spectrum, which is however a numerical artifact only. The standard solution is simply to average a number of power spectra in Fourier space, which cancels out the (phase uncorrelated) Gibbs side lobes. For that, the time series (14.6) is split into M overlapping or non-overlapping "bins" (partitions of the time sequences) and the periodogram estimate is calculated as

$$S(\omega) = \frac{1}{M}\sum_{k=1}^{M} S_k(\omega) \quad \text{with} \quad S_k(\omega) = \frac{1}{N}\left|\sum_{l=1}^{N} \exp(i\omega l) n_l(k)\right|^2, \quad (14.9)$$

where $n_l(k)$ is the k-th partition of the time sequence (14.6). It is often appropriate to weight each partition $[n_1(k),\ldots,n_N(k)]$ by a window function $[w_1,\ldots,w_N]$. The corresponding periodogram then reads to be

$$S_k(\omega) = \frac{1}{N}\left|\sum_{l=1}^{N} \exp(i\omega l) w(l) n_l(k)\right|^2 \left|\left(\frac{1}{N}\sum_{l=1}^{N}|w(l)|^2\right)\right|. \quad (14.10)$$

The method was first proposed by Welch [6] and the result is shown in Fig. 14.4 (b). Spurious side lobes are not anymore visible and the fine structures of the spectrum are much better seen. An alternative way to calculate a power spectrum is the Blackman–Tukey method to calculate the Fourier transform of the autocorrelation function, which is of interest for measuring the decorrelation of a signal with time:

$$R(m) = \frac{1}{N}\sum_{k=0}^{N-m-1} n_{k+m} n_k. \quad (14.11)$$

The Blackman–Tukey method

$$S(\omega) = \frac{1}{M} \sum_{l=-M}^{M} \exp(i\omega l) R(l) \qquad (14.12)$$

has certain advantages and disadvantages and we refer the interested reader to the literature for a comprehensive discussion [7]. We finally note, that a robust estimate of the autocorrelation function can be obtained by Fourier transform of the power spectrum (14.9), (14.10) instead of the direct calculation (14.11).

The simple power spectrum provides useful information about how the signal energy is distributed over a certain spectral range. If two time series are available, from measurements of the same quantity at different points in space or from measurements of different quantities at the same point in space or even both, the phase information is of utmost importance. For the ensemble average (14.9) we use here the symbol $\langle \cdot \rangle$ which must not be confused with the real-space averaging (14.2) introduced earlier.

The most important quantities for two-point spectral analysis are summarized in Table 14.2.

Table 14.2. The most important spectral measures for processing two independent signals x and y. The symbols $\hat{x}$ and $\hat{y}$ stand for the Fourier transforms of the bins of the respective time series and $\langle \cdot \rangle$ for the ensemble average over the total number of bins

Quantity	Definition	Information
auto-power spectrum	$S_x(\omega) = \langle \hat{x}(\omega) \cdot \hat{x}^*(\omega) \rangle$	spectrum of x
auto-power spectrum	$S_x(\omega) = \langle \hat{y}(\omega) \cdot \hat{y}^*(\omega) \rangle$	spectrum of y
cross-power spectrum	$H_{xy}(\omega) = \langle \hat{x}(\omega) \cdot \hat{y}^*(\omega) \rangle$	phase, coherence
cross phase	$\Theta_{xy}(\omega) = \arg(H)$	phase angle
coherence spectrum	$\gamma_{xy}(\omega) = |H(\omega)|/[S_x(\omega) \cdot S_y(\omega)]^{1/2}$	phase coherence

If fluctuating quantities are measured at two different points in space, separated by the distance Δ, the wavelength can be measured by defining a local wavenumber $K(\omega) = \Theta(\omega)/\Delta$. Based on K, a number of useful estimates for the full wavenumber-frequency spectrum $S(K, \omega)$ and its moments can be derived [8]. Also the anomalous transport fluxes can be expressed in spectral quantities [9]. The average particle flux as caused by fluctuating electric field [cf. (14.3)] reads to be

$$\langle \tilde{E}_\Theta \tilde{n} \rangle / B_\phi = \frac{2}{B_\phi} \int_0^\infty k(\omega) |H_{n\phi}(\omega)| \sin[\alpha_{n\phi}(\omega)] \, d\omega , \qquad (14.13)$$

where the fluctuating space charge potential is related to the fluctuating electric field by $E_\Theta = -\partial \phi / \partial \Theta$. The phase of potential fluctuations minus the phase of density fluctuations at each frequency ω is denoted by $\alpha_{n\phi}(\omega)$. The sine of this phase difference decides about the magnitude of the particle flux caused by cross-correlated density-potential fluctuations.

A statistically more stable and more direct way to consider the space component is to do measurements at multiple points in space but simultaneously in time. This can be done, e.g., by sufficiently fast cameras [10] or by probe arrays [11]. Figure 14.5 shows a space-time diagram of density fluctuations caused by drift waves. The data was recorded with a 64-channel array of Langmuir probes (see below) arranged along a full circle around the cylindrical plasma column [5].

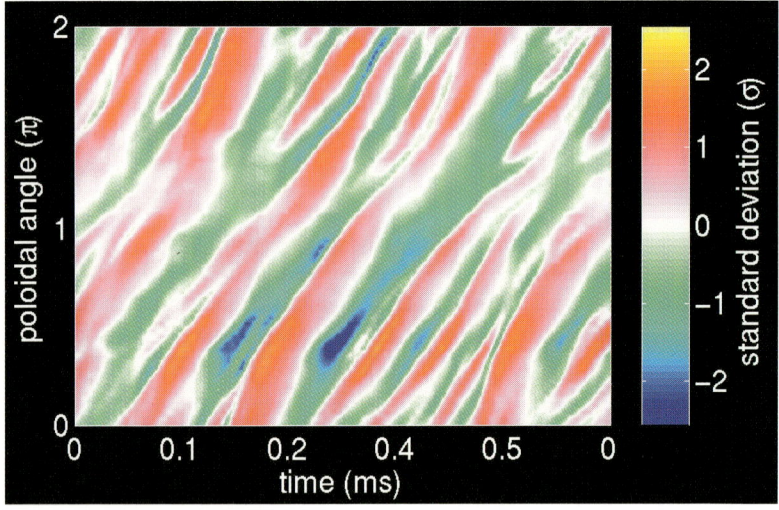

Fig. 14.5. Space-time diagram of turbulent drift waves observed with a Langmuir probe array in a laboratory experiment. The amplitude is normalized to the standard deviation. Note the different structure size in both time and space

The above discussed concepts for spectral analysis are extended in a straightforward manner for space-time data. The difficulties arising in local wavenumber analysis are avoided. For illustration, Fig. 14.6 shows the power spectrum (log scale) and the autocorrelation function of the space-time data shown in Fig. 14.5. It is seen the wide spread of the spectral power, centered roughly around the linear dispersion relation $\omega = k_\perp v_{e,d}$ with $v_{e,d} = c_s \rho_s / L_n$ the diamagnetic drift velocity [$L_n^{-1} = \partial_r \ln(n)$ is the inverse gradient length, $c_s = (k_B T_e / m_i)^{1/2}$ the ion sound velocity and $\rho_s = c_s / \omega_{ci}$ the drift scale]. The spectrum peaks around 10 kHz, indicating the remainings of an $m = 3$ mode structure. The periodic structure embedded into broad-band turbulence

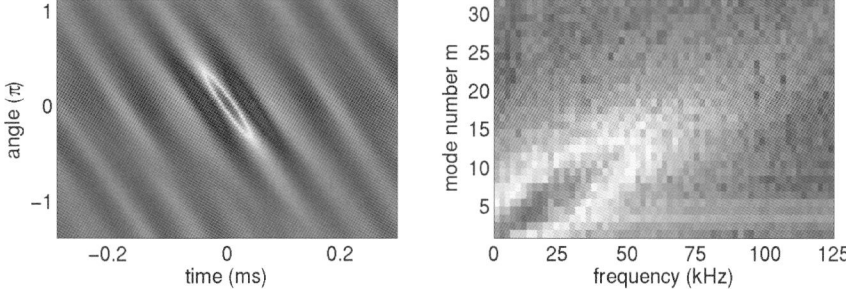

Fig. 14.6. Auto correlation function (*left*) and power spectral density (*right*) of the space-time data shown in Fig. 14.5. The color code ranges from blue (colored version) to red (low to high magnitude) [b/w version: high and low values correspond to deep gray; intermediate values to light gray. The maxima appear at about time ≈ 0 ms, angle ≈ 0 (*left*) and frequency ≈ 10 kHz, mode number ≈ 3 (*right*)]

is also observed in the autocorrelation function by the periodic "wings" l.h.s. and r.h.s. of $\tau = 0$ ms. The correlation function yields the average life time of turbulent structures $\tau_0 \simeq 10\,\mu$s and their average spatial extend $L_0 \simeq 0.2\pi r_0$ (with r_0 being the radial location of the fluctuations).

There is a wealth of further, partially advanced data analysis tools for plasma physics applications discussed in the literature. Even a brief summary of those is far beyond the scope of the present overview. To mention just the most important ones: Higher order (e.g. bi-) spectra are a tool to analyze three-wave coupling [12] and energy transfer between modes [13]. Conditional sampling is used to unveil large-scale turbulent structures embedded in broadband turbulence [14]. Wavelet analysis is an extremely useful instrument for studying transient phenomena and to make a spectral analysis on different consecutive scales [15]. It is worth noting that wavelets can also be taken as a basis set for bispectral analysis [16]. Non-linear phase space analysis (e.g. attractor dimensionality, Lyapunov exponents) is used to characterize complex dynamics which are characterized by only few degrees of freedom [17]. The space-time dynamics of low-dimensional systems was investigated and modeled on the basis of a bi-orthogonal decomposition analysis (also known as singular value decomposition) [18].

14.3 Fluctuation Diagnostics

Generally speaking, any plasma diagnostic with a certain time resolution can serve as a fluctuation diagnostic. It is, however, frequently a challenging task to meet the requirements given by the characteristic time-scale of the fluctuation, which is often in the 100 kHz–1 MHz range. Consequently, a sub-set of standard diagnostic tools could be developed to have a sufficiently high time-resolution to be of value for the measurement of fluctuating plasma quantities, without filtering out significant parts of the power spectrum. This set of tools is called "fluctuation diagnostics". We give here just a brief overview of the standard methods and their limitations. Table 14.3 compiles the standard diagnostics and indicates the measured fluctuating quantity. There is a certain redundancy between several diagnostic methods, but in general a multitude of methods must be used to obtain a fairly complete picture. The general development of diagnostic tools goes not only into the direction of higher time resolution and lower noise level but also in combining both time and space resolution. This is often called "imaging" and can be applied to many of the instruments discussed below.

We may briefly comment on the basic principle, prospects and limitations of each diagnostic tool listed in Table 14.3. For a more profound discussion we have to refer the reader to the literature.

Table 14.3. The most important fluctuation diagnostics. The dot indicates the measured fluctuating quantity

Diagnostics	n	T_e	f_e	T_i	f_i	ϕ_p	E	B	v_e	v_i	Γ_e	Γ_i
static Langmuir probe	•					•					•	•
swept Langmuir probe		•				•	•					
emissive probe						•	•					
Mach probe									•			
induction coil								•				
hall probe								•				
reflectometry	•						•					
electron cyclotron emission		•										
forward scattering	•											
beam emission spectroscopy	•											
heavy ion beam probe	•					•	•					
laser-induced fluorescence	•			•	•					•		

14.3.1 Invasive Fluctuation Diagnostics

Langmuir Probes

Langmuir probes are probably the first diagnostic instrument in plasma physics. The idea is simply to insert a (small) contactor into the plasma and to collect a current by appropriate biasing. If the probe is biased a potential sheath (Debye-sheath) forms around the probe. The current to the probe is determined by the charged particles that enter the sheath. The current-voltage $(I-V)$ characteristic of the probe depends sensitively on the electron density and electron temperature. If the probe is biased at potentials much smaller than the plasma potential the current to the probe is almost entirely carried by the plasma ions. This so-called *ion saturation current* reads:

$$j_{i,sat} = 0.61 n_e e \left(\frac{k_B T_e}{M}\right)^{1/2} \sim n_e T_e^{1/2} . \quad (14.14)$$

where n_e is the electron density, e the elementary charge, k_B the Boltzmann constant, T_e the electron temperature, and M the ion mass. For large positive probe potentials (larger than the plasma potential) the electrons can reach the probe due to their thermal motion and the ion current can be neglected. The current in this *electron saturation* regime is given by

$$j_{e,sat} = -\frac{1}{4} n_e e \left(\frac{8 k_B T_e}{\pi m_e}\right)^{1/2} . \quad (14.15)$$

For lower probe potentials the electron current is reduced, because only electrons with sufficiently high energy can contribute to the probe current. This reduced electron current is given by the electron saturation current multiplied with a Boltzmann factor

$$j_e = j_{e,sat} \exp\left[\frac{e(U - \Phi_p)}{k_B T_e}\right] , \quad (14.16)$$

where U is the probe potential and Φ_P is the plasma potential. The potential of the currentless probe, the so-called floating potential, is given by

$$\Phi_f = \Phi_p - \ln\left\{(1-\gamma)\left[\frac{m_i T_e}{2\pi m_e (T_e + T_i)}\right]^{1/2}\right\} \frac{k_B T_e}{e} = \Phi_P - \alpha(T_e, T_i) T_e . \quad (14.17)$$

Here, γ denotes the secondary electron emission coefficient of the probe material. The standard procedure is to measure the $I-V$ characteristic of the probe. By fitting an appropriate model function, the electron density, electron temperature, and the plasma potential is obtained. In low-temperature laboratory plasmas it is often assumed that electron and ion temperature fluctuations can be neglected. In that case one can measure at constant negative probe bias the ion saturation current and, more importantly, its fluctuations, which are directly proportional to the electron density and its fluctuations $j_{i,sat} \sim n_e$. Similarly, the floating potential fluctuations can be recorded, which under this assumption are proportional to plasma potential fluctuations ($\Phi_f \sim \Phi_p$). Floating probes are especially useful to measure electric field fluctuations in the plasma by using the floating potential difference of a double probe. The most important advantage of probes in fluctuation measurement is that they allow (under the above assumption) measurements of plasma density, potential, and electric field fluctuations with high spatial (mm range) and temporal (μs range) resolution. The localized measurement of plasma density and potential fluctuations provides a local estimate of the fluctuation induced particle flux by using a multi-tip probe. One probe tip measures density fluctuations $\tilde{n}$, two adjacent probes measure potential fluctuations, which provide an estimate of the local electric field fluctuation $\tilde{E}$ (perpendicular to the magnetic field B) at the position of the measured density fluctuation. With these information the local electrostatic particle flux

$$\Gamma = \frac{1}{B}\langle \tilde{E}\tilde{n}\rangle \qquad (14.18)$$

can be calculated. $\langle \ldots \rangle$ denotes the average over the fluctuation time series. If the assumption of non-existing temperature fluctuations must be relaxed, one faces the problem of recording the entire probe characteristic. In plasma turbulence typical fluctuation spectra show significant spectral power at frequencies up to 100 kHz and higher. In order to temporally resolve such spectra and to get a reasonable number of data points sufficient for further evaluation, the sweep frequency of the probe bias must be fast, much higher than the fluctuation spectrum.

The advantage of fast-swept probes is that simultaneously electron density, plasma potential, and electron temperature can be measured. Although the use of a fast swept Langmuir probe is a straightforward concept, the technical requirements are high to avoid parasitic capacitive currents, which cause hysteresis effects and therefore a distortion of the probe characteristic. In magnetic confinement devices it was shown by using fast-swept probes that temperature fluctuations, even in the edge of the plasma, cannot be neglected [20]. Similar to the local particle flux the use of multi-tip swept Langmuir probes provide electron temperature fluctuations $\tilde{T}_e$ and therefore a local estimate of the local energy flux is given by

$$Q = \frac{3}{2}k_B n \frac{\langle \tilde{E}\tilde{T}_e\rangle}{B} + \frac{3}{2}k_B T_e \frac{\langle \tilde{E}\tilde{n}\rangle}{B} \ . \qquad (14.19)$$

It must be noted that the above relations for the probe characteristics are valid for unmagnetized plasmas only, in which the spatial distribution of the current flowing to the probe is isotropic. The anisotropy introduced by a magnetic field complicates the situation considerably and no closed probe theory has been developed yet. Nevertheless, for special probe geometries Demidov et al. [21] developed a collisionless kinetic model to describe the $I - V$ characteristic of a probe in strongly magnetized plasmas.

Emissive Probe

An alternative technique for measuring the plasma potential is the use of emissive probes, first proposed by Langmuir [22] and further developed by several authors [23, 24, 25, 26]. The probe tip is heated to a temperature such that thermionic electron emission occurs. In contrast to the floating potential of a Langmuir probe, which is determined by the potential drop in the sheath, the idea of an emissive probe is to compensate for the sheath potential drop by electron emission. If the probe is biased to a potential less than the plasma potential Φ_p, the emitted electrons flow from the probe to the plasma. If the probe potential is larger than Φ_p, plasma electrons flow to the probe. The floating potential of the emissive probe reads:

$$\Phi_f = \Phi_p - \alpha T_e = \Phi_p - T_e \ln\left(\frac{j_{e,sat}}{j_{i,sat} + j_{em}}\right) , \qquad (14.20)$$

with j_{em} being the emission current of the probe. Due to this sudden change in the sign of the electron flow the probe current changes sign at the plasma potential. The plasma potential and its fluctuations are thus measured directly by the floating potential of the emissive probe to an accuracy on the order of the plasma electron temperature $(k_B T_e)/e$ [27]. In contrast to Langmuir probes, emissive probes measurements are fairly insensitive against electron and ion beams, probe geometry, and neutral pressure effects.

Figure 14.7 shows the characteristic of an emissive probe (blue curve, colored version). Without heating the usual characteristic of a Langmuir probe is obtained. If the probe is heated, the electron emission leads to increased current in the negative branch. The difference between the two characteristics is the emission current. It is clearly seen that the floating potential is found close to the plasma potential. Recently, emissive probes have been successfully used also in the edge plasma regions of high-temperature confinement devices and plasma potential as well as electric fluctuations could be deduced from the floating potential of an emissive probe [28].

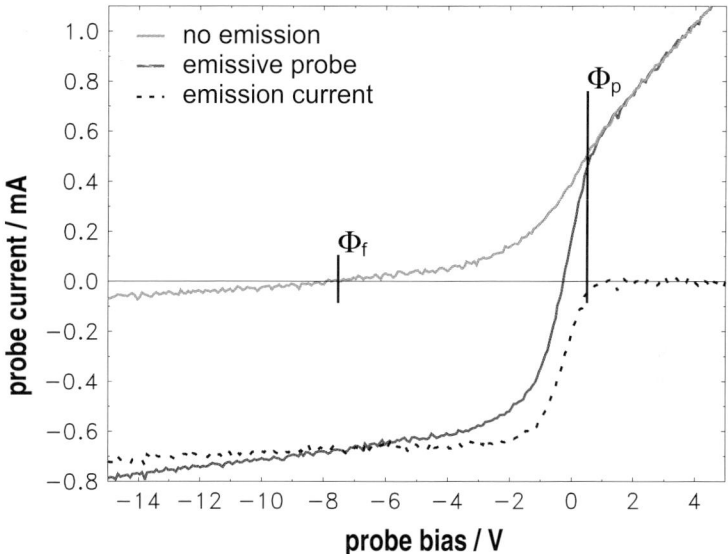

Fig. 14.7. $I - V$ characteristic of a cylindrical Langmuir probe (no emission) and an emissive probe together with the emission current. The plasma potential Φ_p and floating potential Φ_f of the Langmuir probe is indicated. Taken from [19]

Mach Probe

A Mach probe is essentially a directional Langmuir probe. The underlying principle is straightforward: In the presence of a mean plasma flow, a Langmuir probe biased in the ion saturation current regime collects more ions on the side directed upstream than on the downstream side. From a double sided insulated Langmuir probe the mean plasma flow velocity along the magnetic field is deduced from the ratio of the respective ion saturation currents j_u and j_d [29]

$$\begin{aligned}\frac{j_u}{j_d} &= \exp\left[\frac{(v_t+v_d)^2-(v_t-v_d)^2}{c_s^2}\right] \\ &= \exp\left(\frac{v_d/c_s}{c_s/4v_t}\right).\end{aligned} \quad (14.21)$$

Here, v_t denotes the ion thermal velocity, c_s the ion sound speed, and v_d is the plasma flow velocity. Equation (14.21) gives a reasonable estimate for the flow speed only if the ion temperature is on the same order or larger then the electron temperature $T_i/ZT_e \gg 1$ (Z is the ionization state) [30]. The straightforward approach is complicated by the calibration problem of Mach probes. In [30] it was noted that even in the strongly magnetized plasma case, the flow to the probe is not entirely given along the magnetic field. The collection is also partly perpendicular to the magnetic field due to ion orbit effects, which makes the ion collection three dimensional. In a strong magnetic field the relation

$$M = v_d/c_s \sim 0.4\ln(j_u/j_d) \quad (14.22)$$

gives a good estimate of the flow speed in units of the ion sound speed, i.e. the Mach number M [32]. Figure 14.8 shows a measurement of the plasma flow along the magnetic field in the scrape-off layer of the Alcator C-Mod tokamak for different line averaged densities. The results show that the parallel flows are predominantly determined by magnetic topology and not by scrape-off layer densities. In contrast to the normal Langmuir probe the interpretation of Mach probe data is getting more difficult in an unmagnetized plasma. With the help of self-consistent Particle-in-Cell simulations it was shown that ion orbit effects in an unmagnetized plasma leads to barely interpretable results [33, 34]. Against intuition the ion current downstream can be higher than upstream, which not only results in a wrong flow velocity magnitude but actually the flow direction cannot be recovered.

Induction Coil

Induction coils are used to measure magnetic fluctuations. The basic design is simple: It consists of a loop of n windings of copper wire enclosing an area

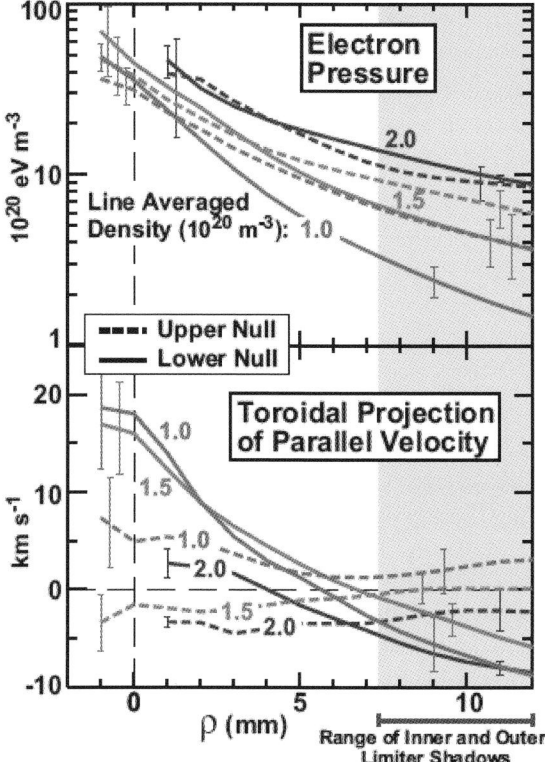

Fig. 14.8. Scrape-off layer densities and plasma flow velocities along the magnetic field as measured by Langmuir probes in the Alcator C-Mod tokamak. Taken from [31]

A. If the magnetic field is temporally varying a voltage $U_{ind} = -nA\partial B/\partial t$ is induced. The magnitude of the induced voltage U_{ind} depends on design parameters of the probe n, A and on the frequency of the magnetic field fluctuation. For a Fourier component of the magnetic fluctuations $B = \hat{B}\sin(\omega t)$ at frequency ω the induced voltage reads

$$U_{ind} = -nA\hat{B}\omega\cos(\omega t) \,. \tag{14.23}$$

For relatively high frequency fluctuations induction coils can be made relatively small and a fairly good spatial resolution is achieved. Induction coils are not only sensitive to magnetic fluctuations. The coil can couple capacitively also to plasma potential fluctuations and the resulting measured voltage cannot be distinguished between induced and capacitively coupled voltage. For reliable results induction coils require an effective capacitive pickup rejection scheme. One strategy is to construct a housing for the coil and its leads, which acts as a Faraday shield so that all coil components are shielded from the fluctuating plasma potential. This usually results in larger dimensions of the diagnostics. Another strategy makes use of the fact that the induced voltage changes sign when the probe is rotated by π but the capacitive voltage does not. At low frequencies $\omega \leq 10^6$ rad/s subtraction of the voltage signals at both leads of the coil by a differential amplifier extracts the induction voltage. The frequency limit is mainly given by the operation range of commercial differential amplifiers. For higher frequencies miniature center tapped transformers installed directly at the coil leads subtract the voltage signals. It has been shown that this concept is preferable and allows (after proper calibration) absolute magnetic fluctuation measurements [35], at least for higher frequencies.

Hall Probe

For low-frequency magnetic field fluctuations $\omega \leq 10^3$ rad/s induction coils are usually getting too big or the magnitude of the induced voltage is getting too small. Here, Hall sensors provide an alternative for the low frequency branch. Nowadays, they consist usually of a semiconductor (p or n type). In an external magnetic field (fluctuating or static) the Lorentz force leads to charge separation and thereby to a Hall voltage U_H given by $U_H = k_H I_H B \sin(\alpha)$, where α is the angle between B and the Hall current I_H, k_H is the Hall coefficient depending on the chosen semiconductor material. For fluctuating magnetic field the effective Hall constant $k \equiv k_H I_H \sin(\alpha)$ becomes frequency dependent $k(\omega)$. Furthermore, the finite time response of the Hall sensor to the magnetic field generally introduces a frequency dependent phase shift $\phi_H(\omega)$. The Hall voltage for a Fourier component of the magnetic fluctuation reads:

$$U_H(t,\omega) = k(\omega)\hat{B}\sin\left[\omega t + \phi_H(\omega)\right] . \qquad (14.24)$$

Today's commercial Hall sensors show an excellent frequency response up to $\omega = 10^3$ rad/s. However, operating Hall sensors inside the plasma requires active cooling to keep the temperature below $100°C$, the destruction temperature of the semiconductor. Additionally, the sensors must be electrically shielded to avoid capacitive pickup. Figure 14.9 shows an example of magnetic fluctuation measurements of a MHD instability (sawteeth precursors) in the TEXTOR tokamak using calibrated Hall sensors. The magnetic fluctuations at relatively low frequencies could be resolved.

14.3.2 Non-invasive Fluctuation Diagnostics

The probe methods briefly presented in the previous section provide high spatial and temporal resolution but are invasive diagnostics. In the following paragraphs non-invasive diagnostics are presented, which usually suffer less spatial resolution but have the advantage of probing the plasma without perturbation.

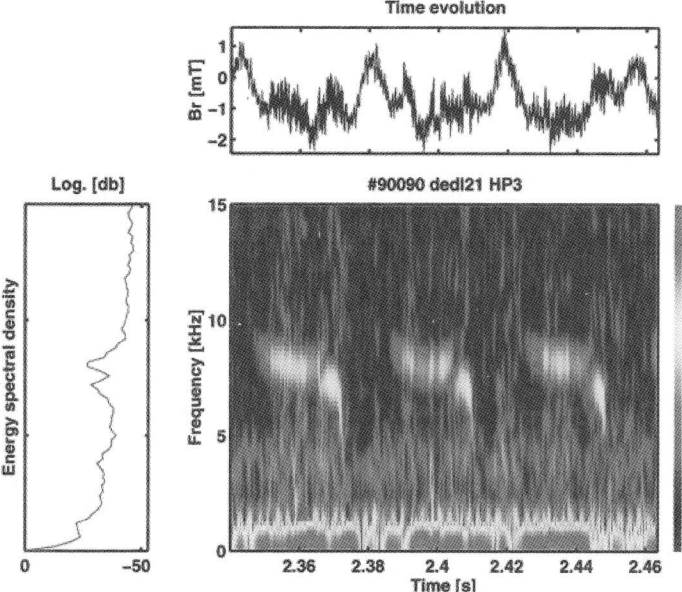

Fig. 14.9. Color-coded time-frequency plot of sawteeth precursors in the TEXTOR tokamak. Additionally, a time trace of the measured B signal (*top*) and the energy spectral density integrated over time (*left*) are shown. Taken from [36]

Reflectometry

Reflectometry belongs to the field of plasma microwave diagnostics. It is based on the radar principle: A wave with frequency ω is launched to the plasma and propagates through the plasma density gradient until the plasma density exceeds the cut-off density $n_{co}(\omega)$. At the cut-off the reflective index becomes purely imaginary and the wave is reflected back. The cut-off layer acts as a mirror for the incident wave. The similarities with a radar system are obvious. The distance to the cut-off position is determined by time delay τ measurements. The change of the phase with sign can be expressed as [37]

$$\frac{\partial \varphi}{\partial t} = \frac{4\pi}{c}\left(\frac{\partial f}{\partial t}\right)\int_{x=0}^{x_{co}} \mu(f,x)\mathrm{d}x + \frac{4\pi}{c} f \frac{\partial}{\partial t}\left(\int_{x=0}^{x_{co}} \mu(f,x)\mathrm{d}x\right). \quad (14.25)$$

In (14.25) f denotes the frequency of the incident wave, x_{co} is the position of the cut-off layer, and $\mu(f,x)$ is the index of refraction along the line of sight. Phase variations can results from two effects: The first term describes the distance to the reflecting layer $x_{co}(f)$ (optical path length). It is obtained by varying the frequency f, which gives $\partial \varphi/\partial t$. The second term in (14.25) describes phase changes due to spatial or temporal plasma density fluctuations. In contrast to standard radar systems, the index of refraction is not constant along the line of sight because the wave travels through a plasma density gradient before reaching the cut-off layer position. At fixed frequency, the phase changes are directly connected to density fluctuations and fluctuation spectra are obtained. The analysis is relatively simple if the index of refraction varies only along the line of sight and for fluctuations with wavelengths sufficiently long to make geometric optics applicable. In the case of multidimensional fluctuations with strong variations also perpendicular to the line of sight the spectrum becomes broader and the reflected wave field is a complicated interference pattern. For a detailed review see [38]. Current developments aim at microwave imaging reflectometry [39], which allows to resolve also multidimensional fluctuation patterns.

14.3.3 Electron Cyclotron Emission

Another microwave plasma diagnostic is electron cyclotron emission (see Chap. 2 in Part I), which provides measurements of electron temperature fluctuations in high-temperature plasmas, which are inaccessible with probes. The gyration of electrons around the magnetic field lines cause radiation at the cyclotron frequency given by

$$\omega_c = k\frac{eB}{m_e} \approx k\,B\,28\text{ GHz/T} \qquad k = 1, 2, 3, \ldots, \tag{14.26}$$

where k indicates the higher harmonics. The cyclotron frequency is magnetic field dependent. In toroidal confinement experiments, in which the magnetic field varies monotonic with radius, the frequency of cyclotron emission provides a spatial localization of the emission volume. If the optical thickness of the plasma is high the electron cyclotron radiation approaches black body radiation. Thus, using the Rayleigh–Jeans approximation the intensity of the radiation I is connected to the electron temperature T_e via

$$I(\omega) \sim \omega^2 k_B T_e \left[1 - \exp\left(-\tau\right)\right]. \tag{14.27}$$

Here, τ denotes the optical depth of the plasma.

Figure 14.10 (left) shows an advanced experimental setup of a two-dimensional ECE system as installed at the W7-AS stellarator [40]. It shows a poloidal array of a horn antenna-reflector system, which allows imaging of electron temperature fluctuations in a radial-poloidal plane. Reflectors (elliptical mirrors) are used to focus to an emission volume and a spatial resolution of 5 mm is achieved. In Fig. 14.10 (right) also the measured cross-spectral density of electron temperature fluctuations is shown. An elongation in the poloidal direction due to propagation of the electron temperature fluctuations in electron diamagnetic drift direction is clearly observed.

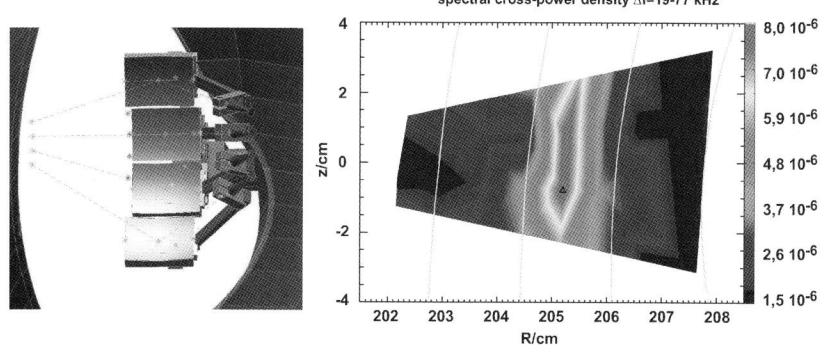

Fig. 14.10. Setup of a poloidal ECE system (*left*) showing the horn-reflector arrangement as installed at the W7-AS stellarator and a measurement of the radial-vertical cross-power spectral density of electron temperature fluctuations (*right*). Taken from [40]

14.3.4 Beam Emission Spectroscopy

Beam emission spectroscopy (BES) allows localized measurement of plasma density fluctuations by observing the light emitted from a neutral beam penetrating the plasma, which is excited by collisional processes (collisions with plasma ions or electrons) [41, 42] (see also Chap. 13 in Part III). The neutral beam species and the injection energy mainly determines the penetration depth. Light species as H, D, He can penetrate deeper in into the plasma and are used to study density fluctuations in high plasma density regions, whereas heavier species like Li are used to study the edge plasma regions with lower densities. For BES fluctuation measurement the emission intensity does generally not vary linearly with the plasma density and the details of the atomic processes must be considered by collisional radiative models . The experimental setup is as follows: a neutral particle beam of the appropriate particle species with energies in the keV range is radially injected into the plasma. The emission from the beam is viewed by an optical system under an angle to the beam. The emission volume is simply the intersection area between the beam and the focus of the collection optics. In toroidal magnetic field geometry the view is often tangential to the local magnetic field to get high radial resolution (in the mm range). The temporal resolution is determined by the bandwidth of the detection system and the signal-to-noise ratio due to photon noise, and the slowest atomic process. By beam extension a two-dimensional detection system BES can be used to image plasma density fluctuations in the beam plane [beam emission spectroscopy imaging (BESI)].

In Fig. 14.11 turbulent plasma density fluctuations in the edge and scrape-off layer plasma of the DIII-D tokamak as obtained by a BESI system [43] are shown. The temporal evolution and propagation of turbulent plasma density structures can be observed.

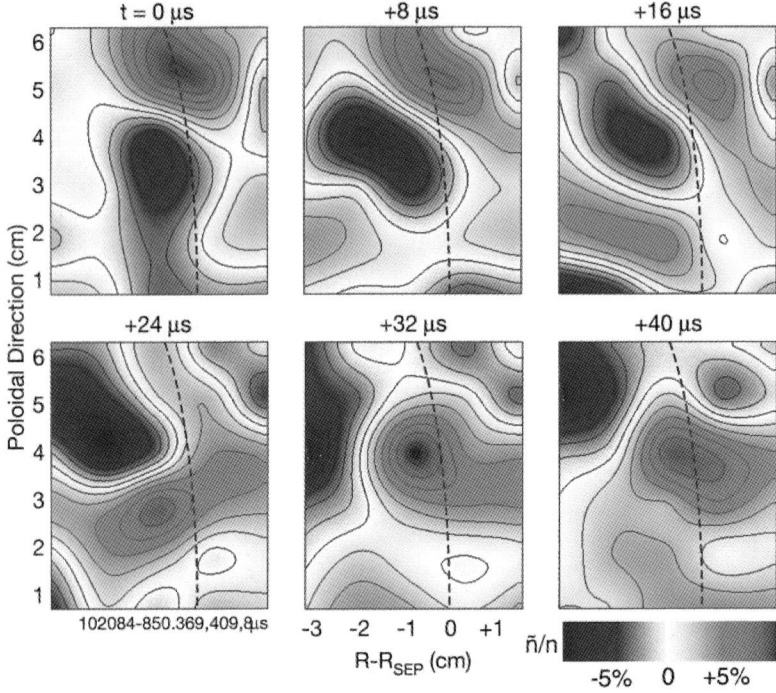

Fig. 14.11. Gray-scaled plot of turbulent density fluctuations as measured with a beam emission spectroscopy imaging system at the D-IIID tokamak. The dashed line represents the magnetic separatrix position. Taken from [43]

14.3.5 Heavy Ion Beam Probe

The heave ion beam probe (HIBP) is a different particle beam diagnostics, which is capable to measure plasma potential and plasma density and their fluctuations simultaneously. The basic principle is to inject a high energetic (in the range of 10 keV–4 MeV) heavy ion or neutral beam into the plasma (primary beam line). Independent of the ion species used the beam energy required to probe the plasma is proportional to the square of the magnetic field strength. A variety of ions can be chosen for the beam. Na, K, Li, Tl, Cs, Au are mostly used. As the primary beam ions or neutrals travel through the plasma they are further ionized by electrons (secondary beam line), which orbits are larger than the plasma radius (due to their high mass and high energy), and these secondaries are detected. The trajectory of the secondary beam line ions are complex and depend on the magnetic field components, which cause deflection of the beam. The detected beam intensity is a measure of the electron density because ionization and attenuation are insensitive to electron temperatures above ≈ 50 eV. The plasma potential and its fluctuations affect the secondary beam line energy and by resolving the energy of the secondary beam the plasma potential is obtained. Today's heavy ion beam probes can resolve fluctuation degrees down to 0.1% at a bandwidth of ≈250 kHz, the spatial and temporal resolutions are typically 1 cm and 1 μs, respectively.

14.3.6 Laser-induced Fluorescence

It is generally difficult to get access to the detailed dynamics of the plasma ions. In contrast to particle beam diagnostics, which provide information about the plasma dynamics by introducing an additional ion species to the plasma, laser-induced fluorescence (LIF) directly probes the plasma ions spectroscopically. The basic principle of LIF is to excite plasma ions from a (stable) ground state to a (short-lived) excited state with photons of a narrow bandwidth laser. The excited ions can loose the extra energy by emitting a photon – the fluorescence. Since the photon energy is quantized, the photon absorption can only occur at a certain laser frequency, which appears Doppler shifted if the ions are moving. Thus, the laser frequency selectively excites ions with the appropriate velocity

$$v_i = \frac{c(\omega - \omega_n)}{\omega_n}, \quad (14.28)$$

where c is the speed of light, ω_n is the frequency corresponding to the natural excitation energy (for the ions at rest), and ω is the actual laser frequency. By scanning the laser frequency and since the intensity of the fluorescence light is proportional to the number density of ions with the respective velocity, the full ion energy distribution function (IEDF) is obtained, thus the mean drift velocity of ions, their temperature, and deviations from thermal (Maxwell) distribution. Fluorescence also occurs naturally and one must discriminate the natural from the laser-induced light, which can be achieved by chopping the laser and using lock-in techniques. The same approach holds for measurements of perturbations of the IEDF due to fluctuating electric and magnetic fields. By recording time series of the fluorescence light intensity and the fluctuating quantity temporal resolution is achieved. For spatial resolution the laser and pick-up optics can be scanned over the spatial domain. A LIF probe was, e.g., used to measure the spatial ion velocity pattern corresponding to the propagation of an Alfvén wave [44]. From the ion velocities the electric fields of the Alfvén wave can be reconstructed. Another approach to obtain spatial resolution is to expand the laser beam to illuminate a spatial domain. In the Caltech's Encore tokamak a (powerful) laser was expanded into a sheet to illuminate a poloidal cross-section [45]. The fluorescence light was imaged to a microchannel plate and for a single time instant snapshots of the fluorescence light in the poloidal plane was obtained.

14.4 Concluding Remarks

The intention of this chapter was to give a brief overview over fluctuation diagnostics and analysis. This fields is strongly developing due to the importance of fluctuations in heating scenarios as well as in transport processes (see Chap. 9 in Part II). A rather comprehensive review of fluctuation diagnostics, particularly for fusion devices, can be found in [4] and the references therein.

References

1. D.R. Nicholson: *Introduction to Plasma Theory* (Wiley, New York 1983)
2. M.B. Priestley: *Spectral Analysis and Time Series*, 6th edn. (Academic, San Diego 1989)
3. J. Wesson: *Tokamaks*, 2nd edn. (Clarendon Press, Oxford 1997)
4. N. Bretz: Rev. Sci. Instrum. **68**, 2927 (1997)
5. T. Klinger, A. Latten, A. Piel et al: Plasma Phys. Contr. Fusion **39**, B145 (1997)
6. P.D. Welch: IEEE Trans. Audio Electroacoust. **AU-15**, 70 (1967)
7. L.R. Rabiner, B. Gold: *Theory and Application of Digital Signal Processing* (Prentice-Hall, New Jersey 1975)
8. J.M. Beall, Y.C. Kim, E.J. Powers: J. Appl. Phys. **53**, 3933 (1982)
9. E.J. Powers: Nucl. Fusion **14**, 749 (1974)
10. S.J. Zweben, R.J. Maqueda, D.P. Stotler et al: Nucl. Fusion **44**, 134 (2004)
11. A. Latten, T. Klinger, A. Piel et al: Rev. Sci. Instrum. **66**, 3254 (1995)
12. Y.C. Kim, J.M. Beall, E.J. Powers et al: Phys. Fluids **23**, 258 (1980)
13. C.P. Ritz, E.J. Powers: Physica D **20**, 320 (1986)
14. H. Johnsen, H.L. Pécseli, J. Trulsen: Phys. Fluids **30**, 2239 (1987)
15. B.B. Hubbard: *The World According to Wavelets: The Story of a Mathematical Technique in the Making* (A.K. Peters, Wellesley 1998)
16. B.P. van Milligan, C. Hidalgo, E. Sanchez et al: Rev. Sci. Instrum. **68**, 967 (1997)
17. H.D.I. Abarbanel: *Analysis of Observed Chaotic Data* (Springer, Berlin Heidelberg New York 1996)
18. N. Aubry, R. Guyonnet, R. Lima: J. Stat. Phys. **64**, 683 (1991)
19. K. Hansen: Experimentelle und numerische Untersuchungen zur Ausbreitung und Überlagerung HF-induzierter Dichtedepressionen. Ph.D. thesis, Christian-Albrechts-Universität, Kiel (1995)
20. R. Balbin, C. Hidalgo, M.A. Pedrosa et al: Rev. Sci. Instrum. **63**, 4605 (1992)
21. V.I. Demidov, S.V. Ratynskaia, R.J. Armstrong et al: Phys. Plasmas **6**, 350 (1999)
22. I. Langmuir: Phys. Rev. **33**, 954 (1929)
23. N. Hershkowitz, M.H. Cho: J. Vac. Sci. Technol. A **6**, 2054 (1988)
24. R.F. Kemp, J.M. Sellen: Rev. Sci. Instrum. **37**, 455 (1966)
25. E.Y. Wang, N. Hershkowitz, T. Intrator et al: Rev. Sci. Instrum. **57**, 2425 (1986)

26. D. Diebold, N. Hershkowitz, A.D. Bailey III et al: Rev. Sci. Instrum. **59**, 270 (1988)
27. M.Y. Ye, S. Takamura: Phys. Plasmas **7**, 3457 (2000)
28. P. Balan, R. Schrittwieser, C. Ioniŏ et al: Rev. Sci. Instrum. **74**, 1583 (2003)
29. M. Hudis, L.M. Lidsky: J. Appl. Phys. **41**, 5011 (1970)
30. I.H. Hutchinson: Phys. Plasmas **9**, 1832 (2002)
31. B. LaBombard, J.E. Rice, A.E. Hubbard et al: Nucl. Fusion **44**, 1047 (2004)
32. I.H. Hutchinson: Phys. Rev. A **37**, 4358 (1988)
33. I.H. Hutchinson: Plasma Phys. Contr. Fusion **44**, 1953 (2002)
34. I.H. Hutchinson: Plasma Phys. Contr. Fusion **45**, 1477 (2003)
35. C.M. Franck, O. Grulke, T. Klinger: Rev. Sci. Instrum. **73**, 3768 (2002)
36. I. Ďuran, J. Stöckel, G. Mank et al: Rev. Sci. Instrum. **73**, 3482 (2002)
37. H.J. Hartfuss, T. Geist, M. Hirsch: Plasma Phys. Contr. Fusion **39**, 1693 (1997)
38. E. Mazzucato: Rev. Sci. Instrum. **69**, 2201 (2003)
39. H. Park, C.C. Chang, B.H. Deng et al: Rev. Sci. Instrum. **74**, 4239 (2003)
40. S. Bäumel, G. Michel, H.J. Hartfuß et al: Rev. Sci. Instrum. **74**, 1441 (2003)
41. R.J. Fonck, P.A. Duperrex, S.F. Paul: Rev. Sci. Instrum. **61**, 3487 (1990)
42. R.D. Durst, R.J. Fonck, G. Cosby et al: Rev. Sci. Instrum. **63**, 4907 (1992)
43. G.R. McKee, C. Fenzi, R.J. Fonck et al: Rev. Sci. Instrum. **74**, 2014 (2003)
44. W. Gekelman, S. Vincena, N. Palmer et al: Plasma Phys. Contr. Fusion **42**, B15 (2000)
45. A.D. Bailey III, R.A. Stern, P.M. Bellan: Phys. Rev. Lett. **71**, 3123 (1993)

15 Research on Modern Gas Discharge Light Sources

M. Born and T. Markus

Abstract. This article gives an overview of today's gas discharge light sources and their application fields with focus on research aspects. In Sect. 15.1 of this chapter, an introduction to electric light sources, the lighting market and related research topics is outlined. Due to the complexity of the subject, we have focused on selected topics in the field of high intensity discharge (HID) lamps since these represent an essential part of modern lamp research. The working principle and light technical properties of HID lamps are described in Sect. 15.2. Physical and thermochemical modelling procedures and tools as well as experimental analysis are discussed in Sects. 15.3 and 15.4, respectively. These tools result in a detailed scientific insight into the complexity of real discharge lamps. In particular, analysis and modelling are the keys for further improvement and development of existing and new products.

15.1 Introduction to Light Sources

Electrical lighting became a commodity in our daily life since the invention and spread of incandescent lamps at the end of the 19th century. Today, electrical light sources based on incandescent, gas discharge and solid state lighting are used for numerous application areas [1, 2]. In this section, an overview of the lighting market and application fields of discharge lamps is given. The role of lamp research is an essential part with respect to further improvements of lamp performance such as light quality and high efficiency. In addition, environmentally friendly products, e.g. mercury-free lamps, become more important nowadays [3, 4].

15.1.1 The Lighting Market

In 2000 the world-wide lighting market consisted of about 14 billion lamps as shown in Fig. 15.1. The by far largest contribution arises from halogen and incandescent lamps followed by fluorescent lamps. Recently, energy saving fluorescent lamps, also known as compact fluorescent lamps, have been introduced to the market. A third block is represented by high pressure discharge lamps with various applications for general lighting, e.g. shop lighting, office lighting, city beautification and road lighting. Another market segment

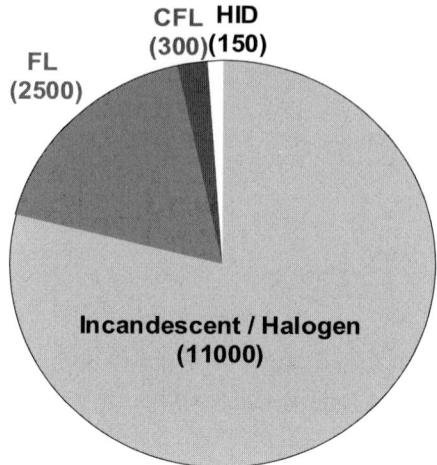

Fig. 15.1. World market in 2000 for electrical light sources in million pieces: incandescent or Halogen, high intensity discharge lamps (HID), fluorescent lamps (FL), compact fluorescent lamps (CFL)

includes lamps for special lighting purposes, such as projection lamps for beamers or automotive head-light lamps.

The lighting market has a volume of about 25 billion Euros (advertising lighting excluded) and is growing annually by some percent. Major players are Philips, Osram and General Electric, each having a market share of some 25%–30% followed by a number of other manufacturers like Ushio, Toshiba, Matsushita or Stanley. In the near future highly efficient white or colored light emitting diodes (LED) are expected to be applied substantially in existing and new lighting applications competing with incandescent lamps, for example. However, in this article we restrict ourselves to gas discharge lamps.

A major aspect of lamps is to efficiently convert electrical power into (visible) light. World-wide about 1% of the primary energy, i.e. about 10% of the electrical energy, is used for electrical lighting. This value corresponds to an energy of about 8×10^{11} kWh per year. For example, this energy would be delivered by about 90 power plants each having a power of 1 GW. Thus, energy saving aspects of electrical light sources have a significant environmental impact. Figure 15.2 shows the increase of luminous efficiency over the past decades for various lamp types being described in more detail in the next section. Besides efficiency, major improvements have also been achieved with respect to lamp life and color quality.

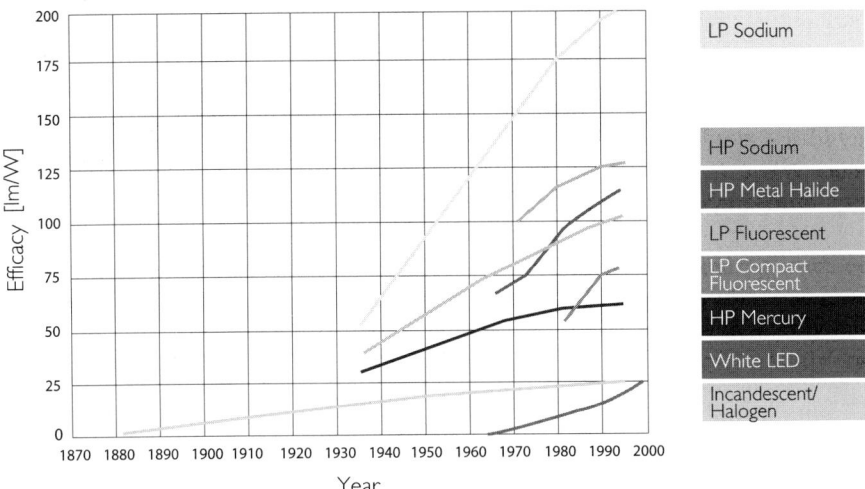

Fig. 15.2. Temporal development of luminous efficiency of various electrical light sources (LP=Low Pressure, HP=High Pressure). A luminous efficiency of 100 lm/W corresponds to an energy efficiency of about 30–35% in the visible wavelength region

15.1.2 Overview of Discharge Lamps and Applications

A gaseous discharge is obtained by driving an electric current through a gas, typically present between two electrodes (see Chap. 5 in Part I). Alternatively, electrode-less microwave excited or pulsed dielectric barrier discharges are also used as light sources. Many physical factors influence the properties of a gas discharge, the most important one is the type and partial pressure of the gas or gas mixtures. For light generation, two major types are distinguished: low pressure and high pressure discharge lamps (see Fig. 15.3).

In low pressure discharge lamps the gas pressure is typically less than 100 Pa. When igniting the plasma by means of applying a high voltage pulse to a starting gas, e.g. argon or xenon, the cold atoms, e.g. mercury at a partial pressure of 5 Pa, are excited by inelastic collisions with hot electrons. These have energies of typically more than 1 eV corresponding to an electron temperature of more than 10 000 K. Consequently, electrons and atoms are not in thermal equilibrium with each other.

In a low pressure mercury lamp, such as fluorescent or compact fluorescent lamp, ultraviolet (UV) photons are generated due to transitions of Hg atoms between excited state levels ($^1P_1$, $^3P_1$) and the ground state level ($^1S_0$). About 64% of the electric input power is converted into photons at a wavelength of 185 nm and 254 nm. The UV radiation is converted into visible light by means of a fluorescent powder or phosphor with a quantum efficiency close to unity. The composition of the fluorescent powder strongly determines the spectral power distribution and color of emitted light. As a result of the Stokes shift, the overall efficiency is only about 28% corresponding to a

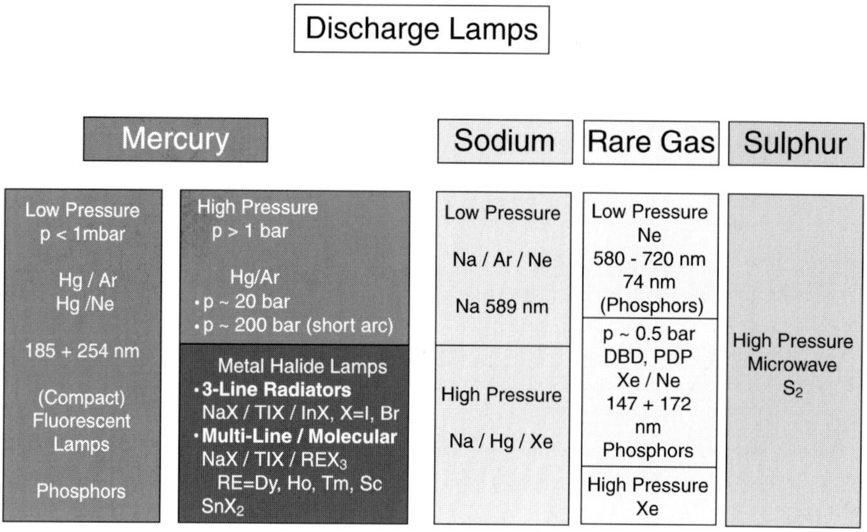

Discharge Lamps

Mercury | Sodium | Rare Gas | Sulphur

Mercury		Sodium	Rare Gas	Sulphur
Low Pressure p < 1mbar	**High Pressure** p > 1 bar	**Low Pressure** Na / Ar / Ne	**Low Pressure** Ne 580 - 720 nm 74 nm (Phosphors)	
Hg / Ar Hg /Ne	Hg/Ar • p ~ 20 bar • p ~ 200 bar (short arc)	Na 589 nm		
185 + 254 nm	Metal Halide Lamps • **3-Line Radiators** NaX / TlX / InX, X=I, Br • **Multi-Line / Molecular** NaX / TlX / REX$_3$ RE=Dy, Ho, Tm, Sc SnX$_2$	**High Pressure** Na / Hg / Xe	p ~ 0.5 bar DBD, PDP Xe / Ne 147 + 172 nm Phosphors	**High Pressure Microwave** S$_2$
(Compact) Fluorescent Lamps				
Phosphors			**High Pressure** Xe	

Fig. 15.3. Overview of types of gas discharge lamps used for lighting applications

luminous efficiency of about 100 lm/W. Fluorescent lamps have an electrical input power of up to 140 W and cover a broad range of white colors. Lamps are available in cylindrical or U-shaped tubes with a lifetime ranging from 5 000 to 25 000 hours.

Mercury based fluorescent and compact fluorescent lamps are typically used for indoor, outdoor and industrial lighting, e.g. in offices or factories. Besides mercury also low pressure sodium lamps are widely applied, e.g. for street or tunnel lighting. Here, sodium emits almost monochromatic yellow light at wavelengths of 589.0 nm and 589.6 nm originating from the Na-D lines. Due to the high melting point of Na, the lamps need to be operated at wall temperature of about 530 K. At these temperatures the highly reactive Na metal requires the usage of chemically resistive alumina (Al_2O_3) instead of quartz (SiO_2) as a wall material. Low pressure Na lamps exhibit the highest luminous efficiency of about 200 lm/W. On the other hand, color properties are poor which limits their application to the above mentioned areas.

Besides Hg and Na also other types of radiators are used in low pressure discharge lamps, e.g. Ne or Xe or mixtures of them. Red light from Ne is applied for automotive brake lights and advertisement lighting. Excimer radiation from excited Xe_2^* molecules is used in plasma displays and flat backlights, photocopier lamps and water purification. The UV light generated at wavelengths at 147 nm and 172 nm can either be directly used or is converted by luminescent materials. Such excimer discharges need to be operated in a pulsed mode with frequencies up to 100 kHz. The physical reason for ac operation is the accumulation of charged particles at the inner discharge wall forming an internal electrical field. Hence, during one pulse, the outside applied electrical field strength is finally compensated and the discharge extinguishes. The next pulse is applied with reversed polarity in order to remove the accumulated wall charges and to start a new cycle.

In high pressure discharge lamps – also known as high intensity discharge lamps (HID) the operating pressure is about several 10 bar or even up to several 100 bar. Under these conditions, electron and heavy particle temperatures are close to each other, typically in the range between 4 000...10 000 K. In high pressure Hg lamps the luminous efficiency is limited up to 60 lm/W, due to lack of transitions in the visible spectrum. In addition, thermal and infrared losses are significant [5]. High pressure mercury lamps are widely applied, e.g. for industrial and special lighting applications. Mercury emits radiation in the visible spectrum due to transitions between higher lying energy levels with the UV lines being suppressed by self-absorption. At very high pressures ($p > 200$ bar) – e.g. in projector lamps – also molecular radiation from Hg_2 molecules fills up the visible spectrum which results in improved spectra and color properties. In addition, higher pressure operation allows for point-like light sources due to the presence of large electrical field strengths. These arise from elastical scattering of electrons by mercury atoms and result in a low electrical conductivity of the plasma. Large electric fields (several 10 V/cm) are necessary in order to limit lamp currents at a given electrical lamp power, e.g. of more than 100 W. In addition, the introduction of a chemical tungsten transport cycle enables high electrode stability and a lamp life cycle of more than 5 000 h.

High pressure Na lamps emit a more white light than the low pressure ones because of spectral broadening of the Na-D lines. Due to corrosion of the hot Na vapor with quartz, polycrystalline alumina (Al_2O_3) is used as a wall material. Such lamps have high luminous efficiencies of up to 130 lm/W. Application areas are outdoor or horticultural lighting, for example.

An important variant of high pressure mercury lamps are so called metal halide lamps. Here, Hg is still used as a buffer gas in order to adjust sufficiently large electrical field strengths. The radiation is emitted from various other metal atoms and molecules with low excitation energies. In multi-line radiators rare-earths and associated elements like Dy, Ho, Tm or Sc and Na are added to the lamp filling. Argon or xenon are used as a starting gas. In order to establish sufficiently large vapor pressures, elements are dosed as metal halides, e.g. NaI or DyI_3, which are much more volatile than the pure metals. As a result of the large number of transitions in the visible spectrum such plasmas emit multi-line radiation with high luminous efficiencies of up to 100 lm/W.

Another type of a metal halide lamp is based upon molecular radiation originating from SnI_2, for example. Such molecules produce a quasi-continuous spectrum in the visible wavelength range. Finally, a third class – so called three-band radiators – emit line radiation from NaI, TlI and InI in the yellow, green and blue region, respectively.

Metal halide lamps are used for a variety of indoor and outdoor applications, such as shop lighting or office lighting. At a power range between 35...1000 W they offer luminous efficiencies up to 100 lm/W with excellent color quality and long lamp life of about 20 000 h.

15.1.3 Aspects of Lamp Research

In gas discharge lamps physical and thermochemical analysis and modelling is the basis for a continuous improvement of lamp performance and the development of new products. Physical and chemical properties of the lamps are strongly related to the overall energy balance and life time behavior. One aim of modelling is to gain a detailed knowledge about spectral radiation emission and energy loss mechanisms. Here, thermal heat management of wall and electrode materials is a major topic. Modelling tools are used to predict optimal lamp design parameters such as choice of discharge tube (burner) and electrode materials and their geometries, lamp fillings and operation mode. Complex chemical reactions such as corrosion and transport properties of lamp fillings, wall and electrode materials and interactions between these components need to be studied in order to realize stable lamp performance over lamp life. Here, the influence of impurities like water or oxygen is essential with respect to suitable lamp processing and high quality production on a large scale. Combining physical and thermochemical modelling tools with data from experimental analysis, e.g. from plasma spectroscopy, pyrometry or Knudsen Effusion Mass Spectrometry (KEMS), finally results in an optimized design of real lamps (see also Sect. 15.4.1).

15.2 High Intensity Discharge Lamps

15.2.1 Construction and Working Principle

In contrast to low pressure mercury lamps, high intensity discharge lamps, i.e. mercury, sodium or metal halide lamps, directly emit radiation in the visible wavelength region operated at pressures of up to 200 bar. This is possible since higher lying energy levels are efficiently excited in the plasma which is close to local thermal equilibrium (LTE) (see also Chap. 13 in Part III). Plasma temperatures are in the range between 4 000–10 000 K. Due to thermal losses and absorption of infrared radiation the discharge vessel heats up to temperatures of about 1 000–1 400 K. Recrystallization of quartz (SiO_2) limits its application for HID lamps to temperatures below 1 300 K. At higher temperatures, ceramic wall materials like poly-crystalline alumina (PCA, Al_2O_3) are used. In Fig. 15.4 typical wall and plasma temperature distributions of a metal halide discharge lamp with ceramic envelope are shown. The lamp is operated at an electrical input power of 75 W. Values of the inner diameter,

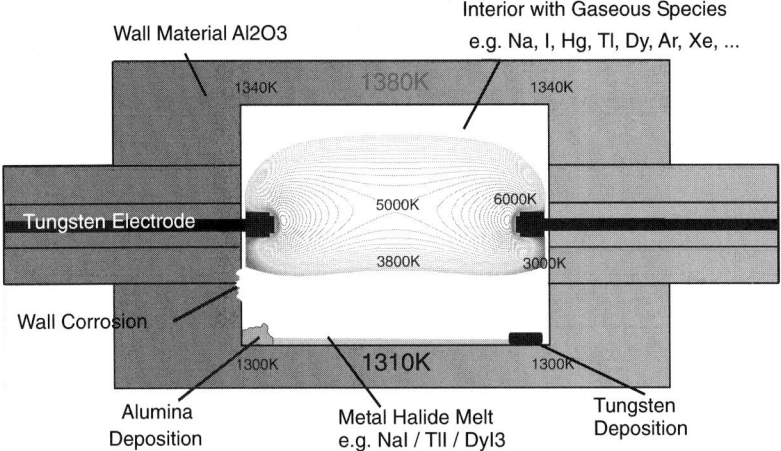

Fig. 15.4. Wall and plasma temperature distributions in a HID lamp with ceramic envelope

electrode distance and wall thickness are 6 mm, 8 mm and 0.8 mm, respectively. The electrode is typically formed as a rod with a diameter of a few 100 μm, depending on the electrical power of the lamp. Electrode tips are made of tungsten since they must withstand temperatures of up to 3500 K. The middle part is made e.g. of molybdenum in order to effectively conduct thermal heat to the end construction. In PCA a special feed-through made of Niobium is necessary which matches the thermal expansion coefficients of the electrode and wall material over a wide temperature range (300–1 400 K). Vacuum tight sealing of the lamps is achieved by use of special frits consisting, for example, of a mixture of Al_2O_3, SiO_2 and Dy_2O_3, depending on the lamp filling.

When igniting the lamp the argon arc heats up the discharge vessel resulting in an enhanced evaporation of filling species such as Hg, NaI, DyI_3. The mercury atoms effectively scatter the electrons elastically due to a high momentum transfer cross section and high partial pressure of Hg at several bars. Consequently, the electrical field strength and lamp voltage rises resulting in an increase of electrical input power at a constant current. Thus, wall and plasma temperature further increase with time until a steady state situation is reached, where the electrical input power is balanced by radiation emission and non-radiative losses such as conductive, convective and electrode losses.

As illustrated in Fig. 15.4, key parameters of a robust thermal burner design are the coldest and hottest spot temperatures. The former defines the position of the liquid salt and must be as high as possible in order to effectively evaporate the filled species. Due to convective heat flow, the hottest spot is usually located at the upper central part of the vessel. Here, the thermal load of the discharge wall is largest and corrosion or recrystallization (quartz) may be critically. Consequently, the hottest spot temperature must be as low as possible. Optimized thermal burner design is a central subject of lamp modelling as discussed in Sect. 15.3.

Looking to the plasma, atomic, molecular and charged species establish a complex chemical equilibrium over the parabolic-like temperature profile. In addition, gaseous and condensed phases may react with the discharge wall or electrode material resulting in a formation of corrosion products (e.g. aluminates) or etching of the wall. Another effect is the transport of species like tungsten causing wall blackening, for example. These processes strongly limit lamp maintenance and life. A detailed knowledge of the chemical equilibrium is a central topic of modern lamp research. For example, gaseous densities of species must be known for energy balance analysis or calculation of spectral radiation emission power. High temperature chemistry analysis of discharge lamps is discussed in Sects. 15.3 and 15.4.

15.2.2 Light Technical Properties

Metal halide discharge lamps offer high luminous efficiencies with excellent color quality and long lamp life. A spectrum of a 75 W PCA lamp filled with a mixture of NaI/TlI/DyI$_3$/Hg/Ar is shown in Fig. 15.5. Main contributions arise from multi-line radiation of Dy, the spectral broadened Na-D transition and atomic line radiation from Tl and Hg. In addition, molecular (DyI) and continuous recombination radiation are observed, the latter resulting in infrared losses. About 60% of the electrical input power is converted into radiation, about one third being emitted in the visible. This value corresponds to a

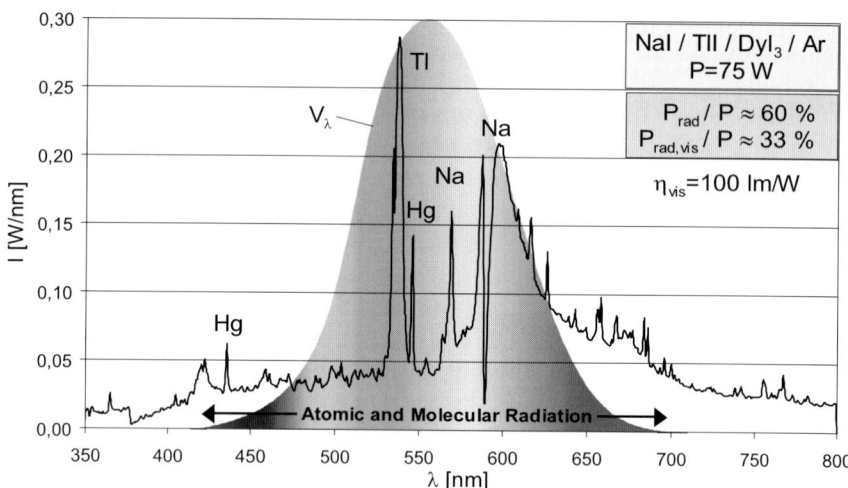

Fig. 15.5. Spectral intensities of a 75 W PCA HID lamp with indicated filling. V_λ is the normalized eye sensitivity curve

luminous efficiency of about 100 lm/W. Metal halide lamps are widely used for shop- and accent-lighting applications, for example.

15.3 Modelling of High Intensity Discharge Lamps

The relevance of lamp research has already been pointed out in Sect. 15.1. In the following sections modelling tools are described which are used for optimization of thermal burner design, characterization of the plasma arc and evaluation of thermochemical properties. A central assumption is to have a local thermal and chemical equilibrium of the plasma which is almost the case for HID lamps. However, large temperature and particle density gradients result in complex transport phenomena, like diffusion or radiative transfer. In addition, gravity causes convection effects which must be considered in three dimensional geometry.

Besides, physical and chemical modelling must take into account a lot of uncertainties or lack of input data, e.g. transition probabilities, cross-sections for particle collisions, enthalpies and entropies of formation, impurity species and levels etc.. However, estimations and measurements of input parameters often help to overcome these gaps. In this sense the procedures described below result in quantitative lamp data in a predictive way.

15.3.1 Physical Modelling

We describe the plasma of a dc operated HID lamp by the following stationary and local equations:

$$\text{energy balance}: \sigma E^2 = U_{rad} - \nabla \cdot (\kappa \nabla T) + \rho c_p \mathbf{v} \cdot \nabla T , \quad (15.1)$$

$$\text{momentum balance}: \rho(\mathbf{v} \cdot \nabla)\mathbf{v} = \rho \mathbf{g} - \nabla p_{total} + \nabla \cdot \underline{\underline{\tau}} , \quad (15.2)$$

$$\text{viscosity tensor}: \tau_{ij} = \eta \left[\frac{\partial v_j}{\partial x_i} + \frac{\partial v_i}{\partial x_j} - \frac{2}{3}\delta_{ij}(\nabla \cdot \mathbf{v}) \right] , \quad (15.3)$$

$$\text{Ohm's law}: \mathbf{j} = \sigma \mathbf{E} , \quad (15.4)$$

$$\text{conservation of charge}: \nabla \cdot \mathbf{j} = 0 , \quad (15.5)$$

$$\text{conservation of mass}: \nabla \cdot (\rho \mathbf{v}) = 0 , \quad (15.6)$$

$$\text{equation of state}: p_{total} = \frac{\rho}{M_{carrier}} RT . \quad (15.7)$$

Equation 15.1 describes the conversion of electrical input power density ($\mathbf{E}$: electric field strength, σ: electrical conductivity of the plasma) into total net radiation power density U_{rad} and thermal and convective losses.

Thermal losses are caused by temperature gradients with κ being the total thermal plasma conductivity. According to (15.2), convection losses are driven by gravity (**g**: constant of gravity on Earth). They are described by treating the plasma as a liquid having a macroscopic flow velocity **v**, specific heat c_p and mass density ρ. In case of HID lamps, the flow is assumed to be laminar having a certain viscosity η, see (15.3). In addition, Ohm's law and the continuity of charge and mass must be fulfilled [(15.4), (15.5) and (15.6)]. Mass density ρ and total plasma pressure p_{total} are related to the carrier mass $M_{carrier}$ which represents the majority of species, e.g. mercury [see (15.7)].

This set of equations must be solved simultaneously in three dimensional geometry under the following boundary conditions:

$$\text{total electrical input power}: P = \int_V \sigma E^2 \, dV, \quad (15.8)$$

$$\text{zero flow velocity at inner wall}: \mathbf{v}|_{r=R} = 0, \quad (15.9)$$

energy conservation
$$\text{at electrodes and discharge wall}: \nabla \cdot (\kappa_{e,w} \nabla T) = 0, \quad (15.10)$$

$$\text{heat flux at outer discharge wall}: q_h = \epsilon(T_w)\,\sigma_{\text{SB}}(T_w^4 - T_{ref}^4)$$
$$+ h\,(T_w - T_{ref}). \quad (15.11)$$

Instead of choosing a fixed total current, (15.8) allows for solutions at a given electrical input power. Equation (15.9) describes the property of a laminar flow at the inner side of the discharge vessel ($r = R$, R: inner radius of discharge vessel). The heat flux of the plasma entering the electrode (e) or discharge wall (w) results in temperature gradients according to (15.10). The heat at the outer surface of the discharge wall, q_h, is related to its temperature via the Stefan-Boltzmann law (see (15.11)[1]). Here, ϵ is the thermal emissivity of the wall material, being dependent on temperature. Finally, the lamp may be operated in a specific environment at an ambient temperature T_{ref}. Additional convective losses at the outer surface are represented by the heat transfer coefficient h. This fact must be taken into account if the burner is not operated in an evacuated outer bulb, e.g. in an ambient atmosphere, which results in a redistribution of burner temperatures.

Solving the above equations involves the knowledge of the transport coefficients σ and κ. In HID lamps the electrical conductivity σ is mainly dominated by (elastical) collisions between electrons and neutral atoms (index $e0$)

---

[1] Stefan–Boltzmann constant: $\sigma_{\text{SB}} \approx 5.67 \times 10^{-8}\,\text{W}\,\text{m}^{-2}\,\text{K}^{-2}$

$$\frac{1}{\sigma} = \frac{1}{\sigma_{e0}} + \frac{1}{\sigma_{ei}} \approx \frac{1}{\sigma_{e0}}, \qquad (15.12)$$

$$\sigma_{e0}(T) = \frac{e^2 n_e(T)}{m_e \sum_j n_j(T) \langle q_j(T) v_e(T) \rangle}. \qquad (15.13)$$

Collisions between electron and ions (index ei) are less significant due to the low degree of ionization in the order of a few percent where n_j is the neutral density of species j, q_j the momentum transfer cross-section, v_e the mean thermal velocity of electrons and m_e the electron mass.

From (15.13) it is obvious that collisions between electrons and the buffer gas, e.g. mercury, dominate due to its large neutral density. In addition, mercury has a large value of q. The brackets indicated in (15.13) correspond to the average of q_j over the electron velocity distribution.

The thermal conductivity κ contributes of a translational, chemical and radiative contribution as illustrated in Fig. 15.6

$$\kappa = \kappa_{trans} + \kappa_{chem} + \kappa_{rad}. \qquad (15.14)$$

At lower temperatures a peak of κ_{chem} reflects the dissociation of molecules whereas ionization of species takes place at higher temperatures. Absorption and re-emission of radiation dominates the total thermal conductivity at temperatures of more than 3 000 K. This fact underlines the importance of including radiation transport calculations into the model, for example by means of Monte-Carlo techniques. Radiation transport in HID lamps is a very complex subject since such simulations involve a detailed knowledge

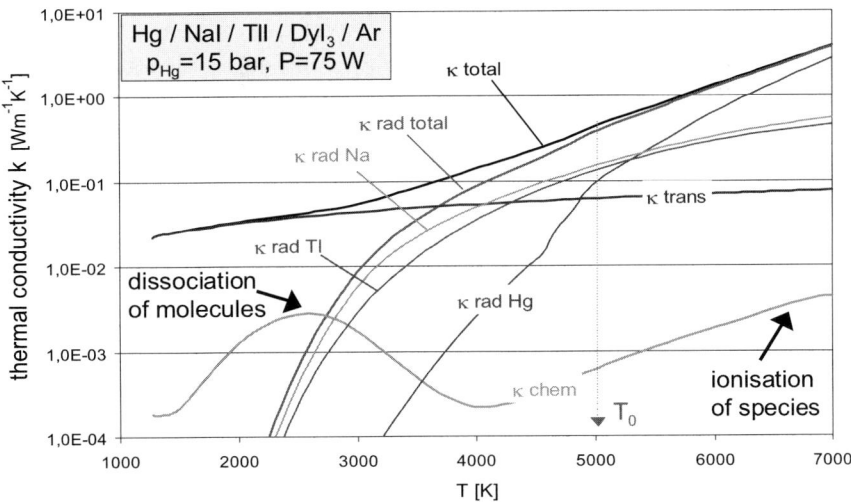

Fig. 15.6. Contributions to thermal conductivity in a rare-earth metal halide lamp. T_0 indicates the central plasma temperature

of spectral emission and absorption parameters. For example, spectral line broadening due to resonance, van-der-Waals and Stark effects [4] must be taken into account at least for the most dominant species. These are linked to the (local) net emitted radiation power density U_{rad} of (15.1) which is equal to the difference of local emission and radiation power density. For the net absorption coefficient of a transition i we find

$$a_i(\lambda, T) = \frac{\lambda_{i,0}^4}{8\pi c} A_{i,21} \frac{g_{i,2}}{Z_i(T)} n_{0,i} \exp\left(-\frac{E_{i,2}}{k_B T}\right)$$

$$\times \left[\exp\left(\frac{hc}{\lambda_{i,0} k_B T}\right) - 1\right] P_i(\lambda), \qquad (15.15)$$

$$Z_i(T) = \sum_i g_{i,j} \exp\left(-\frac{E_{i,j}}{k_B T}\right), \qquad (15.16)$$

where $\lambda_{i,0}$ is the central wavelength, $A_{i,21}$ the transition probability, $g_{i,21}$ the statistical weight, $n_{0,i}$ the neutral density of radiating species, $E_{i,j}$ the energy level, and Z_i is the partition function. The indices 2, 1 indicate the upper and lower energy level, respectively.

The normalized spectral line shape is assumed to be Lorentzian (pressure broadening)

$$P_i(\lambda) = \frac{\Delta\lambda_i/2}{\pi} \frac{1}{(\lambda - \lambda_{i,0} - \delta\lambda_i)^2 + (\Delta\lambda_i/2)^2} \qquad (15.17)$$

with $\Delta\lambda_i$ is the spectral line width (FWHM) and $\delta\lambda_i$ the spectral shift of central wavelength.

The set of (15.1)–(15.11) is solved numerically using a finite-element code, e.g. *FIDAP* [6]. *FIDAP* is a commercially available software package for solving the equations of fluid flow and heat transfer [7]. Therefore, the discharge volume is divided into several thousands of cells, each of them representing a region of local thermal equilibrium (see Fig. 15.7).

The input parameters of the numerical tool are lamp and electrode geometry, wall and electrode material parameters, lamp filling and electrical input power. From this data, electrical and thermal conductivity, viscosities and specific heats, emissivities and radiation emission are calculated as functions of temperature. Particle densities are evaluated using a commercial tool such as *FactSage* [8] (see also Sect. 15.3.2).

Output of the model are local data of plasma, wall and electrode temperatures, particle densities, spectra and electrical field strengths. The calculations are performed iteratively, since after each iteration source and sink terms of (15.1) vary as a consequence of radiation transport. In addition, after each iteration temperature profiles and thus wall temperatures change which requires a new calculation of particle densities, radiation fields and related quantities.

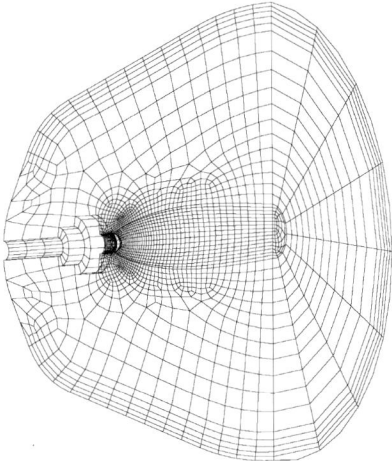

Fig. 15.7. Typical grid of a HID discharge volume used within the finite element code *FIDAP*

Figure 15.8 shows a result of a simulation for a metal halide lamp with the grid taken from Fig. 15.7. Due to symmetry only a quarter of a lamp needs to be considered. Using a state of the art 3 GHz Pentium processor, a full lamp simulation takes several hours of computing time. This value may be exceeded significantly, depending on the speed of convergency, number of nodes, complexity of the filling, gradients in temperature and densities etc. Note that in Fig. 15.8 the effect of arc bending due to convection is pronounced.

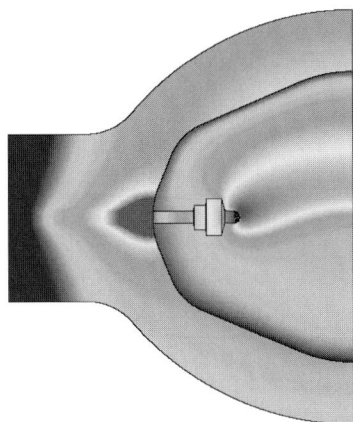

Fig. 15.8. Temperature distributions calculated with the finite element code package *FIDAP*. The colors represents wall, electrode and plasma temperatures in the ranges between 1 000 K–1 300 K, 800–3 000 K and 1 300–5 500 K, respectively

15.3.2 Thermochemical Modelling

Partial densities of gaseous species need to be known for the solution of the energy balance. On the other hand, thermochemical modelling leads to a detailed insight of the chemical interactions. They originate from evaporation of condensed phases as well as from corrosion reactions, such as from electrode and wall materials. Besides, interaction of the liquid phases of the lamp filling with gaseous species must be taken into account.

The simulation of the chemical equilibrium – being assumed in HID lamps – is performed using the *FactSage* code [8]. This tool is used for thermodynamic calculations based upon a minimization of the Gibbs free energy. Complex equilibrium simulations are possible for systems with many components over a broad range of temperatures and pressures. In this section an introduction into thermochemical basics with application to lamp research is given.

In general, the fundament of all thermodynamic simulations is to minimize the Gibbs free energy of the system under consideration. Hence, the first step is to express the Gibbs free energy for a chemical reaction

$$\sum_i \nu_i A_i \Longleftrightarrow \sum_f \nu_f A_f \qquad (15.18)$$

with A_i denoting the educts, A_f the products formed and $\nu_{i,f}$ the stoichiometric coefficients. The Gibbs free energy is a function of pressure, temperature and stoichiometric coefficients:

$$G = G(p, T, \nu_1 \cdots \nu_n) . \qquad (15.19)$$

The derivative of G (15.19) can be written in its differential form

$$\mathrm{d}G = V \mathrm{d}p - S \mathrm{d}T + \sum_i \mu_i \mathrm{d}\nu_i . \qquad (15.20)$$

At a minimum of G its derivative vanishes: $\mathrm{d}G = 0$. In chemical equilibrium we assume $\mathrm{d}p = 0$, $\mathrm{d}T = 0$ so that G remains as a function of the chemical potentials μ_i

$$G = \sum_i \mu_i \nu_i . \qquad (15.21)$$

The chemical potentials are calculated as a function of temperature and pressure according to

$$\mu_i(T, p_i) = \mu_i^0(T) + RT \ln \frac{p_i}{p_0} \qquad (15.22)$$

with p_i for the partial pressures of species i, $p_0 = 10^5$ Pa for the reference pressure and $R = 8.314$ J mol^{-1} K^{-1} for the gas constant. μ_i^0 is the chemical potential at reference pressure.

Combination of equation 15.21 and 15.22 leads to

$$\sum_i \left[\mu_i^0(T) + RT \ln\left(\frac{p_i}{p_0}\right)\right] \nu_i = \sum_f \left[\mu_f^0(T) + RT \ln\left(\frac{p_f}{p_0}\right)\right] \nu_f . \quad (15.23)$$

Equation (15.23) results in the well-known law of mass action, see (15.27). Experimental determination of enthalpies and entropies from partial pressure measurements are performed as described in Sect. 15.4.

In order to determine the minimum of G the chemical potential functions need to be known. They are related to the changes of enthalpy ΔH_r^0 and entropy ΔS_r^0 of the system in the reference state

$$\sum_i \mu_i \nu_i = \Delta G_r^0 = \Delta H_r^0 - T\Delta S_r^0 . \quad (15.24)$$

If available, values for ΔH_r^0 and ΔS_r^0 are taken from literature or may be derived from calculations involving heat capacity functions $[c_p = c_p(T)]$. Unfortunately, in most cases such data are unknown and need to be determined by means of thermochemical measurements. These comprise calorimetric methods or the technique of Knudsen Effusion Mass Spectrometry (KEMS). The latter is explained in the following section. For more details concerning interrelations of thermodynamic properties, the reader is refereed to the literature [9].

Commercially available thermodynamic simulation tools are based upon different numerical procedures for minimizing G assuming chemical equilibrium. As an example, the software package *FactSage* consists of a database, a numerical simulation tool and a data assessment module. User defined databases may be implemented. Typically, HID lamp modelling involves a large number of various species and substances resulting in complex partial pressure distributions. As mentioned in the previous section, such data are used as input parameters for physical modelling of lamp plasmas.

In Fig. 15.9 an example of such a calculation is given for a rare-earth metal halide lamp filling. The composition of the salt mixture serves as an input parameter assuming an equimolar dose between NaI(s) and DyI$_3$(s). *FactSage* computes the partial pressures of gaseous species being in equilibrium with the condensed phases as a function of indicated temperatures. Each value corresponds to an independent simulation run. Parameters of species formed during vaporization are taken from the implemented database.

In Fig. 15.9 the heterocomplex NaDyI$_4$ is formed during evaporation of NaI(s) and DyI$_3$(s) according to the reaction

$$\text{NaI(g)} + \text{DyI}_3(\text{g}) \rightleftharpoons \text{NaDyI}_4(\text{g}) . \quad (15.25)$$

Obviously, this heterocomplex is dominating with respect to partial pressure. Consequently, Dy and Na plasma densities are enhanced as compared to those of the pure phases (DyI$_3$ and NaI) by formation of NaDyI$_4$ [10]. The

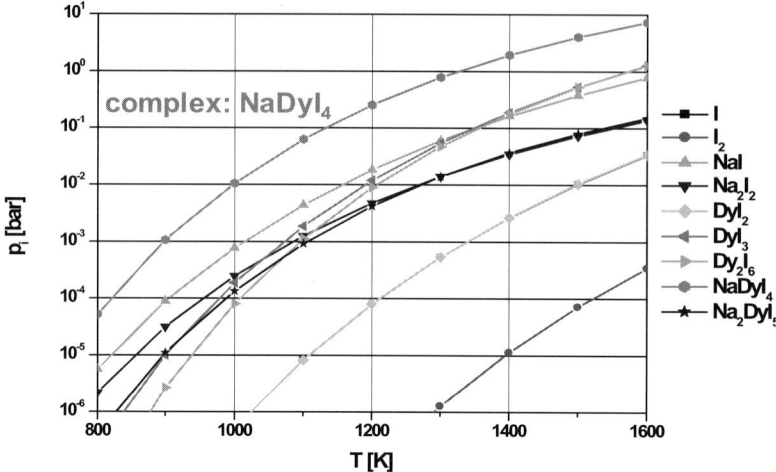

Fig. 15.9. Calculation of partial pressures of species in a NaI/DyI_3 mixture using *FactSage*

result is an increase of spectral radiation power at improved color properties. In the following section we discuss the experimental determination of thermochemical input data needed for the model.

15.4 Thermochemical Experiments

15.4.1 Knudsen Effusion Mass Spectrometry (KEMS)

Particle densities of relevant species which are present in the plasma of high intensity discharge lamps are unknown in most cases and have to be determined experimentally. Partial pressures and thermodynamic input data like enthalpies and entropies of formation can be derived by elucidation of the vaporization processes. Here, high temperature mass spectrometry is the most versatile method for such studies. The interaction between species in the liquid and gaseous phase is considered. Thermodynamic data of the condensed phase are obtained from the partial pressures of the evaporating species. From such analysis chemical reactions, e.g. corrosion and transport properties of species in the lamp burner are derived.

Below we describe the determination of thermodynamic data for gaseous species and condensed phases by the study of the vaporization processes. The method of choice is Knudsen effusion mass spectrometry which has been developed in the early 1990's [11]. Since then, vaporization studies have been carried out with KEMS for almost all groups of inorganic substances such as, for example, for borides, carbides, fullerens, nitrates, sulphates, halides, metals, alloys, oxides, glasses and ceramics. The KEMS technique is mainly characterized by the following two features:

- Identification of the equilibrium forming gaseous species of the vapor in the Knudsen cell.
- Partial pressure determination of the equilibrium species generally in the range between 10^{-8} Pa and 10 Pa at up to temperatures of more than 2 800 K.

The aim of KEMS is to determine the temperature dependency of thermodynamic data, e.g. enthalpies and entropies of formation, vaporization and dissociation. From partial pressure measurements, equilibrium constants of various chemical reactions are derived.

Figure 15.10 shows the principle of a magnetic type sector field mass spectrometer with a Knudsen cell assembly. The Knudsen cell, which comprises the filling mixture under investigation, can be heated up to temperatures of more than 3 300 K. Mechanisms are resistance heating up to 1 000 K and electron bombardment for higher temperatures. The temperature inside the Knudsen cell is measured by means of an optical pyrometer or thermocouple. Electronic control allows for a constant temperature inside the cell.

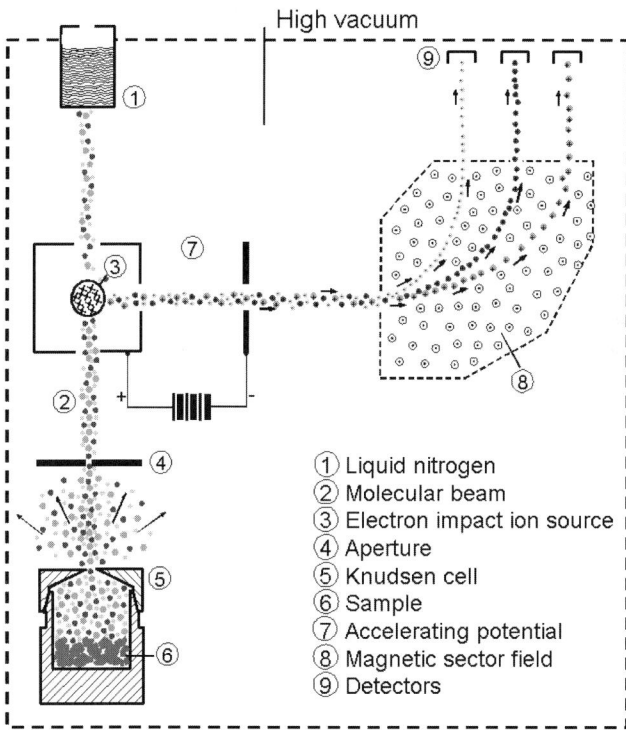

Fig. 15.10. Schematic representation of a Knudsen cell mass spectrometer system

Partial pressures and thermochemical data are computed from these quantities for the identified vapor species. In order to keep a chemical equilibrium inside the Knudsen cell, the evaporated species pass an effusion orifice with a typical diameter of 0.1...1 mm. As indicated in Fig. 15.10 a molecular beam representing the equilibrium vapor inside the Knudsen cell is formed by the effusing species. This particle beam crosses the shutter valve and enters an electron impact ion source. Ions and fragments are formed by electron bombardment. The charged species are accelerated in an electric potential and enter the magnetic sector field. Here, they are separated according to their mass to charge ratios. As detector systems for the ion species electron multiplier and a Faraday cup are used. A problem is to distinguish between species formed in the effusing vapor and the residual gas present in the ion source chamber. This background noise is reduced by means of a movable shutter, which is placed between the effusion orifice and the ion source.

Another difficulty is caused by the production of various fragments due to electron bombardment. The origin of these fragments needs to be known in detail in order to trace back the densities of the species of interest. Such fragmentation patterns and the degree of fragmentation strongly depend on the electron energy, typically in the order of 20...40 eV, the molecular structure and strength of chemical bonding.

An example of fragmentation for DyI_3 is schematically drawn in Fig. 15.11. For the determination of partial pressures and thermochemical data it is important to assign the ions observed in the mass spectrum to their neutral precursors qualitatively as well as quantitatively. Due to the complexity of the vapor composition and the fragmentation this assignment often is difficult.

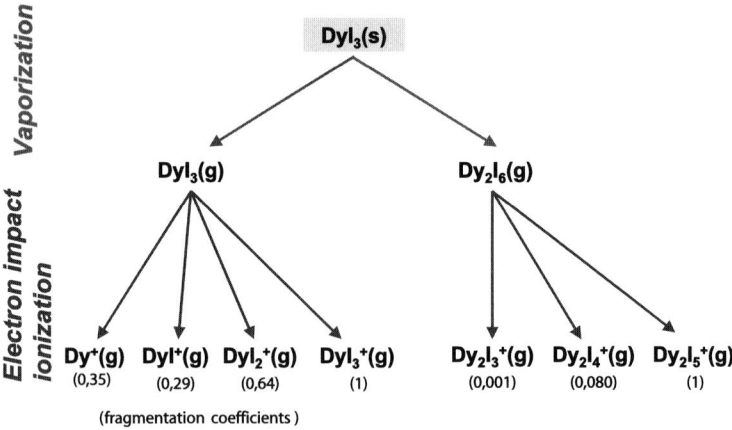

Fig. 15.11. Fragmentation scheme for the vaporization of $DyI_3(s)$. The fragmentation coefficients correspond to the intensity ratio of an ion species i and related ion j with maximum intensity

From the example shown in Fig. 15.11 it can be concluded that seven different ionic species are identified in the mass spectrum which are assigned to two neutral precursors in the vapor: the monomer $DyI_3(g)$ and the dimer $Dy_2I_6(g)$. The latter are present in the vapor over solid DyI_3 (s). From the measured temperature dependence of the ion intensities, partial pressures of neutral molecules are determined according to

$$p_i = k \frac{1}{\sigma_i} T \sum \frac{100}{\gamma_{i,j} A_{i,j}} I_{i,j}^+ = k \frac{1}{\sigma_i} \frac{100}{\gamma_i A_i} I_i^+ T , \qquad (15.26)$$

where k is a pressure calibration constant and σ_i denotes the ionization cross section of the ion i.

The calibration constant k is specific to each measurement including geometrical factors of the cell and the ion source arrangement. $A_{i,j}$, $\gamma_{i,j}$ and $I_{i,j}$ are the isotopic abundances in percent, the multiplier gains and the intensities of the ions j which are generated by the ionization of the neutral species i, respectively, $\gamma_{i,j}$ the amplification factor of the detector. Pressure calibration takes into account deviations of non-monochromatical energy distributions of the effusing molecular beam. In addition, transmission losses of the ion source and the analyzer are included.

The accuracy of determined partial pressures critically depends on the knowledge of the ionization cross-sections σ_i as a function of ionization energies. For example, computational data of ionization cross-sections can be found in [12, 13].

From the partial pressures of the molecules thermodynamic data such as enthalpies and entropies of formation and reactions are calculated. For this purpose the following equations are considered

$$\Delta_r G_T^0 = -RT \ln K_p^0 \qquad (15.27)$$

and

$$\Delta_r G_T^0 = \Delta_r H_T^0 - T \Delta_r S_T^0 , \qquad (15.28)$$

where R is the gas constant ($8.314 \, \text{J mol}^{-1} \text{K}^{-1}$). $\Delta_r G_T^0$ and $\Delta_r S_T^0$ are the changes of Gibbs free energy and entropy, respectively. $\Delta_r H_T^0$ is the change of reaction enthalpy. K_p^0 is the well-known equilibrium constant of the reaction under consideration

$$K_p^0 = \prod_j \left(\frac{p_j}{p^0} \right)^{\nu_j} , \qquad (15.29)$$

where ν_j are the stoichiometric coefficients of the reaction partners j and p^0 denotes the reference pressure, typically 10^5 Pa. Combining (15.27) and (15.28) yields the so called 2$^\text{nd}$ law of thermodynamics

$$\ln K_p^0 = \frac{A}{T} + B , \qquad (15.30)$$

where
$$A = -\Delta_r H_T^0/R, \qquad B = \Delta_r S_T^0/R . \qquad (15.31)$$

From the temperature dependence of K_p [see (15.30)] the thermodynamic parameters $\Delta_r H_T^0$ and $\Delta_r S_T^0$ are derived according (15.31) by means of a least square fit procedure. From the measured temperature T the values for enthalpy and entropy changes are evaluated for standard conditions ($T = 298$ K) by using the specific heat functions $c_p^0(T)$ of all reactands [9].

In addition to the 2nd law, changes of enthalpy and entropy are evaluated independently according to the 3rd law method

$$\Delta_r H_T^0 = -T[R \ln K_p^0 - \Delta_r S_T^0] . \qquad (15.32)$$

Thus, enthalpy changes for the reaction at $T = 298$ K follow from

$$\Delta_r H_{298}^0 = -T \left[R \ln K_p^0 + \Delta_r \left(\frac{H_T^0 - H_{298}^0}{T} \right) - \Delta_r S_T^0 \right], \qquad (15.33)$$

$$= -T \left[R \ln K_p^0 + \Delta_r \left(\frac{G_T^0 - H_{298}^0}{T} \right) \right] . \qquad (15.34)$$

For computation of (15.33) and (15.34) the values of $\Delta_r \left[(G_T^0 - H_{298}^0)/T \right]$ need to be known. These functions are tabulated for many compounds in different phases and at different temperatures [14, 15, 16].

For gaseous species the terms $\Delta_r \left[(G_T^0 - H_{298}^0)/T \right]$ can be computed from molecular parameters by the rules of statistical thermodynamics [17, 18]. In addition to the determination of thermodynamic data for homogeneous and heterogeneous equilibria relevant thermodynamic properties of condensed phases can be evaluated from gas phase studies, such as the enthalpies of formation of solid compounds [19, 20, 21].

Figure 15.12 shows the temperature dependence of the equilibrium constant of the sublimation reaction of DyI_3(s) together with derived thermodynamic data. Calculations are based upon the second law method.

15.4.2 Corrosion Analysis

In burner vessels of HID lamps strong corrosion effects are observed during operation time. To understand the origin of wall and electrode attack, which take place simultaneously, it is necessary to separate the different phenomena. One approach to analyze the corrosion of the wall material is to perform annealing experiments which are described in this section as well as selected results.

As an example, Fig. 15.13 shows a cross section of a PCA burner which has been operated for about 9 000 h. The quadrangle shows the shape of the burner just after manufacturing. As compared with Fig. 15.4, alumina is depleted in the electrode region and at the inside bottom of the vessel. Here, the molten salt is located, typically. The Al_2O_3 is chemically transported

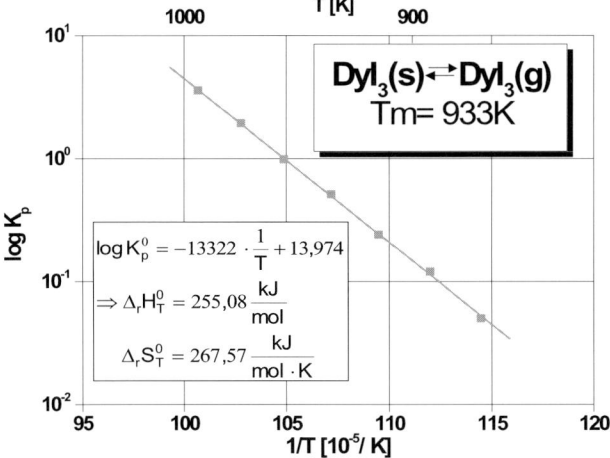

Fig. 15.12. Temperature dependence of the equilibrium constant K_p^0 for the reaction $DyI_3(s) \Leftrightarrow DyI_3(g)$

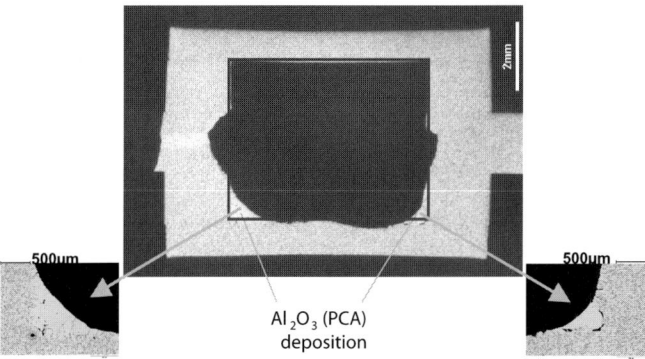

Fig. 15.13. Cross section of a corroded discharge vessel in horizontal burning position after 9 000 h of operation

to the colder parts of the discharge vessel resulting in a deposition at the indicated edges.

In order to analyze the corrosion and transport mechanisms of the alumina wall material and consequences on the gas phase properties annealing experiments are carried out. They allow for a separation of corrosion phenomena from additional overlapping effects, e.g. tungsten corrosion. For this purpose, electrode-less vessels are annealed in an argon atmosphere under isothermal or temperature gradient conditions. The latter results in a transport of species from hot to colder parts like in real lamps. In contrast, isothermal annealing experiments yield basic thermochemical data at a well defined temperature, such as solubilities or phase formation parameters for the species under investigation.

Quartz or PCA vessels are filled with the salt mixtures of interest. They contain plain quartz or PCA plates which are removed after a specified annealing time. Typical analysis of the plates concentrates on their surface characterization by Scanning Electron Microscopy (SEM) and X-ray diffraction (XRD). The result is the identification of stable corrosion phases, e.g. aluminates. In addition, the remaining vessels are studied by means of chemical analysis in order to characterize the residual salt mixtures, e.g. amounts and composition of species. Aluminum is frequently dissolved in the salt as a result of corrosive interactions between the wall material and lamp fillings.

Studies under isothermal conditions are carried out at a temperature of $1475\,\text{K} \pm 5\,\text{K}$ (PCA) or $1325\,\text{K} \pm 5\,\text{K}$ (quartz). Linear temperature gradient experiments are performed at $1450\,\text{K}{-}1600\,\text{K}$ (PCA) or $1125\,\text{K}{-}1325\,\text{K}$ (quartz) in a specially designed gradient furnace. For both types of experiments, the heated chambers of the furnaces must be gas tight and need to be continuously flushed with argon in order to inhibit oxygen migration through the wall. This effect preferably takes place in PCA at higher temperatures.

An example of such an annealing experiment with applied temperature gradient is given in Fig. 15.14. PCA ampoules are filled with a mixture of NaI and DyI_3. After $1000\,\text{h}$, mixed oxides such as $Dy_3Al_5O_{12}(s)$ and $DyAlO_3(s)$ are identified as corrosion products being accumulated at the cold part of the vessel. Deposited Al_2O_3 originates from hot parts where a depletion of wall material is observed. It is concluded that Al_2O_3 is chemically transported by a transport agent. The latter is identified as AlI_3 which is a reaction product of the wall material with the salt filling.

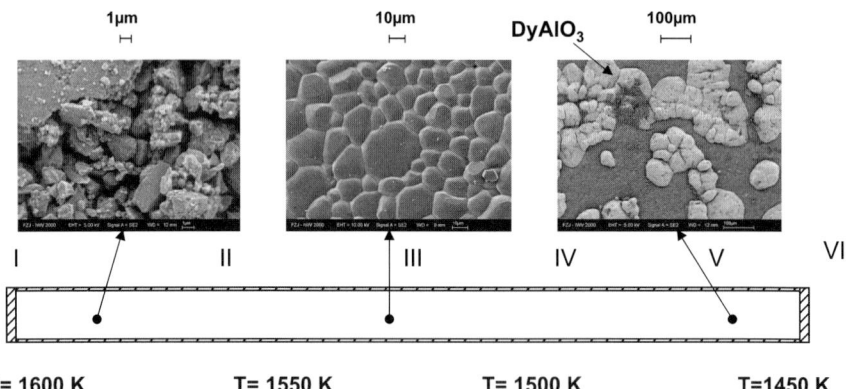

Fig. 15.14. SEM investigation of a PCA vessel after annealing in a temperature gradient between $1450\,\text{K}{-}1600\,\text{K}$ after $1000\,\text{h}$

From systematic investigations using different salt mixture compositions and filling amounts it is deduced, that chemical transport of alumina in PCA vessels is initiated by condensed phases [22].

In contrast, for quartz corrosion attack takes place solely via the gas phase. This effect is shown in Fig. 15.15. As indicated, two different transport cycles of silica via the transport agents SiI_2 and SiI_4 lead to depletion, transport and deposition of SiO_2. For more detail, the reader is referred to the respective literature, e.g. [23, 24].

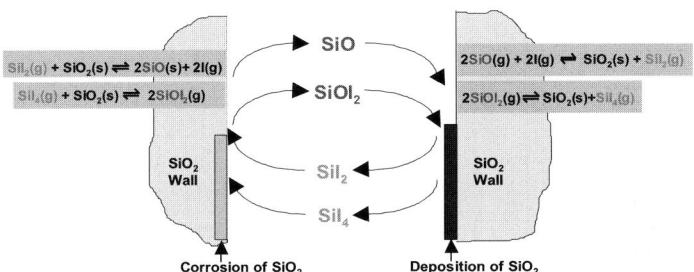

Fig. 15.15. Transport of SiO_2 via the gas phase in a quartz vessel. Two different cycles lead to depletion, transport and deposition of quartz according to indicated reactions

15.5 Conclusions

Research and development of modern gas-discharge light sources is an essential element of the lighting industry. Major improvements of lamp performance and technical data such as efficiencies, color properties and lamp life are achieved by systematic and detailed analysis of physical and thermochemical properties. In HID lamps, representing one of today's most efficient light sources, outstanding thermal parameters and material properties must be combined. Therefore, predictive lamp models are needed in order to further improve lamp characteristics especially with respect to operation at physical and chemical limits. Experiments and simulation codes are used in order to reach this aim.

Three-dimensional plasma modelling using e.g. the *FIDAP* code package, results in a detailed knowledge of plasma temperature distributions, heat fluxes to the discharge wall and electrodes, electrical field strengths and radiation properties. A key input for such tools is a precise description of the chemical interactions between lamp materials and fillings. Central parameters are the particle densities in the gaseous phase. They are determined using thermochemical simulation tools such as *FactSage*. Input data like enthalpies and entropies of formation are derived from thermochemical experiments.

The most versatile method is KEMS based upon the measurement of particle densities. From such experiments, reliable databases for HID lamps are build up and further developed. In addition a detailed understanding of transport processes and corrosion phenomena in real lamps is achieved.

In future, this kind of approach must be further intensively applied to technologically driven lamp research and development. New requirements and specifications as well as environmental aspects need to be considered, such as improved efficiencies, compact lamp design and the usage of green materials.

References

1. J.R. Coaton, A.M. Marsden: *Lamps and Lighting* (Arnold and Contributors, London Sydney New York 1997)
2. W. Elenbaas: *Light Sources* (Crane, Russek & Company Inc., New York 1972)
3. M. Born, T. Jüstel: Physik Journal **2**, 43 (2003)
4. M. Born: J. Phys. D–Appl. Phys. **34**, 909 (2001)
5. W. Elenbaas: *The High Pressure Mercury Vapour Discharge* (Publishing Company, Amsterdam 1951)
6. H. Giese: Theoretische Untersuchungen zur Konvektion in Quecksilber-Hochdruckgasentladungslampen. ISBN: 3-86073-573-X, Dissertationsschrift, Verlag der Augustinus Buchhandlung, Aachen (1997)
7. *FIDAP Theory Manual* (Fluent Inc., Lebanon New Hampshire 1998)
8. C.W. Bale, P. Chartrand, S.A. Degterov et al: Calphad **26**, 189 (2002)
9. P.W. Atkins, C.A. Trapp, M.P. Cady et al: *Atkins' Physical Chemistry*, 7th edn (Oxford University Press, Oxford 2001)
10. K. Hilpert, U. Niemann: Thermochim. Acta **299**, 177 (1997)
11. P.G. Wahlbeck: High Temp. Sci. **21**, 189 (1986)
12. J.W. Otvos, D.P. Stevenson: J. Am. Chem. Soc. **78**, 546 (1956)
13. J.B. Mann: Recent Developments in Mass Spectroscopy. In: *Proceedings of the International Conference on Mass Spectrometry* ed by K. Ogata, T. Hayakawa (University of Tokyo Press, Tokyo 1970) pp 814–819
14. I. Barin: *Thermochemical Data of Pure Substances*, 3rd edn (VCH Verlagsgesellschaft mbH, Weinheim 1995)
15. O. Knacke, O. Kubaschewski: *Thermochemical Properties of Inorganic Substances* (Springer, Berlin Heidelberg New York 1993)
16. D.R. Stull, H. Prophet: *JANAF Thermochemical Tables 2, NSRDS-NBS 37* (National Bureau of Standards, Washington D.C. 1971)
17. C.W. Bauschlicher, H. Partridge: J. Chem. Phys. **109**, 4707 (1998)
18. D.J. Frurip, C. Chatillon, M. Blander: J. Phys. Chem. **86**, 647 (1982)
19. R. Odoj, K. Hilpert: Z. Phys. Chem. Neue Folge **102**, 191 (1976)
20. K. Hilpert: Z. Metallk. **75**, 70 (1984)
21. K. Hilpert: Ber. Bunsenges. Phys. Chem. **88**, 37 (1984)
22. T. Markus: Thermochemische Untersuchungen zur Hochtemperaturkorrosion von polykristallinem Aluminiumoxid (PCA) durch Metallhalogenide. Berichte des Forschungszentrums Jülich Juel–3955 (ISSN 0944-2952), Forschungszentrum Jülich, Jülich, Germany (2002)

16 Computational Plasma Physics   427

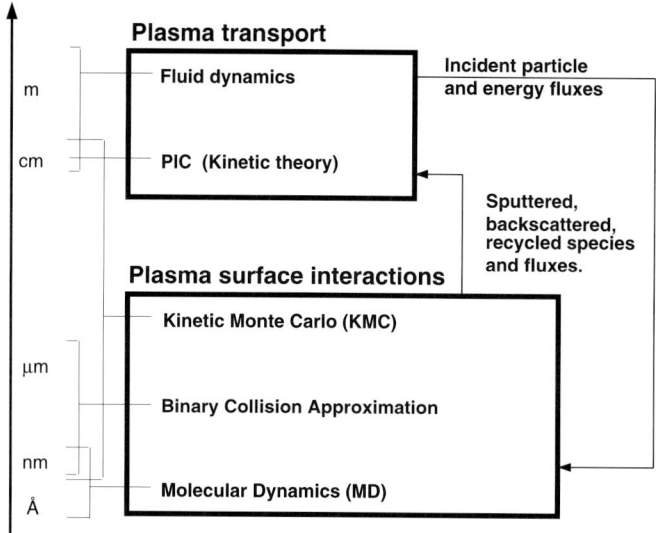

Fig. 16.1. The different length scales and methods used for plasma edge modelling

in amorphous materials) are analyzed with Monte Carlo methods (kinetic Monte Carlo with input from molecular dynamics or experiment).

The plasma description again has different levels of complexity. A full kinetic description (including ions, electrons, neutrals and their collisions) is possible for some low temperature plasmas (e.g. electron cyclotron resonance heated methane plasmas) and for qualitative studies of edge plasma effects in fusion edge plasmas. Here, the limitations are given by the fact that the Debye length and the plasma frequency have to be resolved.

For the study of the physics of the edge of magnetically confined plasmas (2D tokamaks, tokamaks with ergodic perturbations, 3D stellarators) fluid codes are used for understanding the complex physics in such devices. Depending on the geometrical complexity (2D tokamaks, 3D stellarators) and on the additional effect of ergodicity, different numerical methods (finite volume, finite difference and Monte Carlo methods) are used.

The plasma surface interactions influence the plasma transport through sputtered, backscattered and recycled particles and fluxes. On the other hand, the incident particle and energy fluxes determine the plasma surface interaction.

The different codes not only describe different time and spatial scales, but also different parts of the plasma edge: codes describing plasma wall interaction processes and codes resolving the sheath in front of a wall are models for the near-wall physics, whereas the plasma transport fluid codes try to cover the whole scrape-off layer (SOL) replacing the near-wall physics by effective boundary conditions.

A typical example involving all the different scales mentioned before, is the study of carbon as a wall material. The release of hydrocarbons from the walls of a fusion device [2, 3, 4] due to the influx of hydrogen ions and/or neutrals create, after break-up in the plasma, a source of carbon ions for the plasma. However, the hydrocarbons also tend to form co-deposited layers far away from the plasma like in pump ducts. This results from additional transport through neutrals and/or low temperature plasmas in the periphery. These layers pose a severe safety problem for any reactor because they trap tritium. Therefore, understanding of such processes and plasmas is critical for fusion reactor design and will strongly influence the choice of the wall material.

This chapter describes codes developed and used in one group to study many aspects of plasma edge physics.

16.2.1 Models

Plasma Wall Interaction

Molecular Dynamics

A molecular dynamics code HCParcas (developed by K. Nordlund) is used to study the transport of interstitials in a graphite crystal. This code uses the semi-empirical Brenner potential [5] for treating the hydrogen and carbon system and the Nordlund interlayer term [6] to simulate a graphite crystal. The interstitial trajectories (over a sample size of 100 Å for 100 picoseconds) are analyzed to obtain input parameters for our KMC code [7]. A first study of the diffusion of hydrogen in porous graphite showed [8] Levy flight type behavior. The diffusion process proceeds via vacancy jumps towards neighboring atoms. These jumps are thermally activated processes with jump frequencies ω determined by $\omega = \omega_0 \cdot \exp(\frac{-\Delta E}{k_B T})$, where ω_0 is the jump attempt frequency and ΔE is the activation energy for this process. There exist two different channels of diffusion for hydrogen isotope interstitials in graphite crystallites: one is a high frequency, high migration energy channel which matches the graphite phonon frequencies, and the other is a low frequency, low migration energy channel which shows a $1/\sqrt{m}$ mass dependence for the jump attempt frequencies [8].

Binary Collision Code

A computationally much less expensive technique is also successfully used for studies of the interaction of particles with (homogeneous) materials [9]. Here, only two-particle interactions are taken into account. A successful application of this code is the description of physical sputtering of surfaces including dynamical changes of the composition [10, 11, 12]. However, due to its simplification it fails as soon as chemical processes get important [2].

Kinetic Monte Carlo

The KMC code DiG (Diffusion in Graphite) is being developed to treat hydrogen transport in graphite. It is designed to use the information from MD or from experiments to study the transport and interactions of hydrogen as it diffuses in a realistic porous graphite structure. The advantage of using a KMC scheme is that it allows us to model multiple scale lengths in time in an efficient way using the scheme described in [13]. It models graphite crystals ranging from 100 Å across to graphite granules of a few microns, and with time scales ranging from picoseconds to seconds (depending on the graphite temperature and the trap energies). Experimental results for diffusion in graphite were matched in the trapping/de-trapping dominant regime. It was shown that the diffusion coefficient depends on the structure of the graphite used (void sizes) and the trapping mechanism [8].

Plasma Modelling

Kinetic PIC

Microscopic models of plasmas are conceptually easy: one has to solve the equations of motion and, self-consistently, the resulting fields, which again influence the particle motion. However, it is impossible to solve such a system directly due to the large number of particles. Therefore, a so-called particle-in-cell (PIC) method is used [14, 15]. Here, we deal with "super-particles", which are collections of thousands of real particles. Since their charge mass ratio is the same as for normal particles they obey the same equation of motion as real particles. Since charges are shielded on the Debye length scale, the plasma parameter n (see also Chap. 1 in Part I) can be appropriately described by means of such super particles. This allows one to compute electric and magnetic fields only at grid points separated by about one Debye length and thus considerably reduce computational time. A general multi-species electrostatic PIC code was developed including all collisions between electrons, ions and neutrals. This code was successfully applied to the study of low temperature methane plasmas [16] (which are model systems for chemical sputtering studies), capacitive rf discharges [17] and complex plasmas [18].

Plasmas which, in addition to electrons, ions and neutrals, also contain microscopic particles of nanometer–micrometer size are called dusty (complex) plasmas (Chap. 11 in Part II). The dust particles in such plasmas gain an electric charge, the sign and magnitude of which depends on the balance between different charging processes. The absorption of electron and ion fluxes, thermo-, photo- and secondary electron emissions are the most typical mechanisms of particle charging in complex plasmas. Such charged micro-particles substantially change plasma behavior and are responsible for the unusual properties of complex plasmas. In a capacitive rf discharge the gravitational force acting on the particles can be equilibrated by the electrostatic force acting in front of the lower electrode due to a strong repulsive electric field in the rf sheath. In this case particles are trapped in the discharge and form a cloud levitating above the lower electrode. The dust particles interact with each other through the repulsive Coulomb potential, screened by the plasma electrons and ions. In the case of strong electrostatic coupling, i.e. when the energy of the inter-particle interaction is large compared to the particle thermal energy, particles self-assemble into ordered structures, known as plasma crystals. Due to the large mass of the dust particles the characteristic relaxation time for the plasma crystals is usually of the order of seconds, making such structures easy to observe with ordinary video-observation techniques. The inter-particle distance in dusty plasma crystals is usually of the order of a fraction of millimeter, so that it is possible to observe such structures even with the naked eye. The plasma crystals represent a bridge connecting the atomic or molecular scale of matter with the macroscopic scale of a dusty particle system, giving a unique possibility to observe processes in the condensed matter on the kinetic level.

We have studied the formation of dust structures in a capacitive coupled rf discharge using a self-consistent particle simulation. For this purpose we have utilized the PIC code with a Monte Carlo collisions (MCC) package resolving three spatial dimensions and three velocity components [16]. The dust particles were introduced into the model as an additional charged species, using the cloud-in-cell weighting formalism [14], so that no finite size effects for dust particles were considered. In addition to the electrostatic force the gravitational and neutral gas friction forces were also considered for the dust particles.

In Fig. 16.2 we present the dust structure equilibrated over the lower electrode of a capacitive rf discharge.

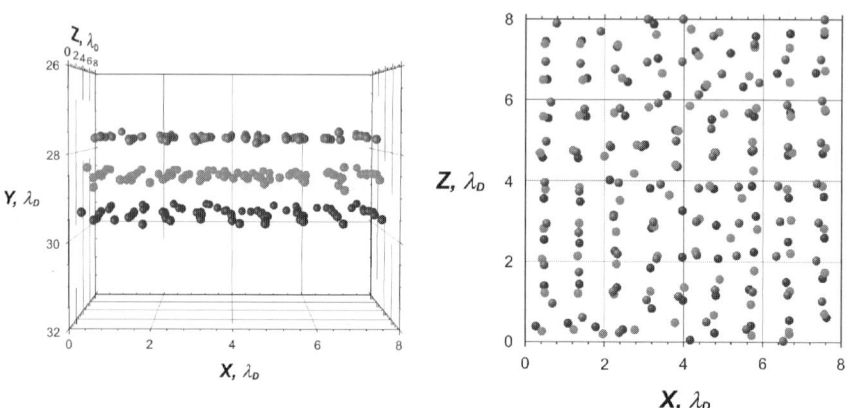

Fig. 16.2. *Side* (*left*) and *top* (*right*) view of the 3D plasma crystal

We can see that the particles are divided in three horizontal layers with a separation of about one Debye length (for convenience we highlighted the layers with different gray-scales). When looking on the dust structure from above (Fig. 16.2) we can note that particles tend to form "triads", as particles belonging to three different layers are aligned vertically. This type of alignment is caused by the polarization of the ion flow in the sheath region. Thus the dust formation in Fig. 16.2 shows a quasi-two-dimensional structure of vertically aligned horizontal layers, each with a similar structure.

Application of the PIC method for turbulence modelling will be discussed in Sect. 16.3.

2D Fluid Code (B2)

The kinetic models also allow one to formulate effective boundary conditions at the plasma sheath, which is needed for multi-fluid models [19]. These models use the fluid equations derived as moments from the kinetic equations, including sinks and sources (like radiation or heating). Closing of the system is obtained by the proper transport coefficients [20, 21]. The applicability of these models is limited to collisional regimes, for which the all mean-free paths are smaller than the relevant system lengths, namely the temperature gradient length and the connection length along the field lines (see also Chap. 7 in Part II). Applying the fluid model can be problematic in front of the divertor where temperature gradients can become steep, resulting in non-local heat conduction. Further problems can arise on closed field lines near the separatrix, where for high temperatures the heat conduction of electrons has to be modified according to kinetic flux limits for periodic systems. Then, (gyro-kinetic) models are necessary for a complete and correct description. In addition, the effect of plasma turbulence has to be introduced by parametrization of the radial anomalous fluxes either by empirical fitting [22] or from ab-initio models [23]. The neutrals, which are quite important for many edge effects, are modelled either by fluid models are through Monte Carlo codes [24]. Numerically, a generalized finite volume scheme for mixed conduction–convection is used, where each individual equation (continuity equations, parallel momentum equations, electron and ion internal energy equations, potential equation) is solved by iterating through this coupled set of equations until convergence is obtained [25, 26]. This code package has been successfully applied to many tokamaks, especially to ASDEX Upgrade, e.g. for the optimization of the divertor [27]. The latest development was the completion of the physics regarding inclusion of drifts and currents, where the observed radial electric fields and flow patterns close to the separatrix were studied [28].

3D Fluid Code (BoRiS)

An extension of this finite volume scheme is necessary for stellarators, which are intrinsically 3D. BoRiS is a 3D SOL transport code under development which is designed to simultaneously solve a system of partial differential equations (PDE) in three dimensions. Although developed in the framework of 3D edge modelling for the new W7-X stellarator, the code development follows a more general concept to allow for different applications. The main characteristics of BoRiS are:

1. finite volume method,
2. use of generalized (magnetic) coordinates,
3. general interpolation for mixed convection–diffusion problems and
4. Newton's method.

These features are strongly influenced by the experiences derived from edge physics modelling with model-validated 2D codes like B2-Eirene [25] and UEDGE [29]. The use of magnetic (Boozer) coordinates s, θ, ϕ allows for standard discretization methods with higher-order schemes to describe a complex 3D geometry (see Fig. 16.3). Therefore, the existence of intact flux surfaces is a prerequisite of this concept.

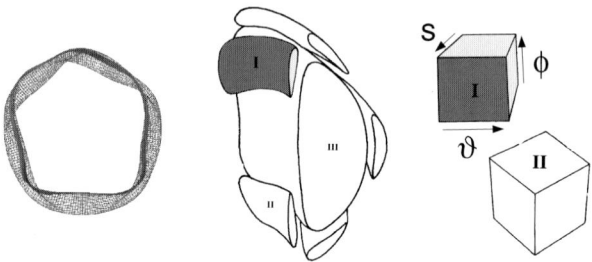

Fig. 16.3. A set of seven sub-grids forms the complete W7-X grid (*five islands, core and outer SOL*)

With Newton's method, BoRiS utilizes a variety of sophisticated sparse matrix solvers (sparse direct, sparse iterative and matrix-free iterative) and preconditioners [30]. In order to deal with large systems, a parallel version of the code was developed.

BoRiS was initially tested on the coupled Laplace equations for electron and ion temperatures in a W7-X island flux tube [31]. Several complexity levels (1D, 2D, 3D) were benchmarked against UEDGE [32] and a 3D Monte Carlo code [33]. In addition, a complete Navier–Stokes neutral fluid model was also compared with simplified models [34].

Transport Codes for Ergodic Configurations

A further complication appears as soon as ergodic configurations are studied. Here, like in **TEXTOR-DED** [35], a special set of coils is installed to give an additional control of the magnetic topology in the periphery. Indeed, the stronger the perturbation, the larger the islands on the resonance surfaces, until they overlap and build a so-called ergodic layer: the behavior of a field line becomes stochastic (two originally neighboring field lines diverge from each other exponentially with a Kolmogorov characteristic length L_K), and we obtain a nonzero projection of the strong parallel heat transport onto the radial direction and, therefore, a flattening of the temperature profile (see also Chap. 9 in Part II).

Configurations of this kind are very general: they contain intact magnetic surfaces, islands, ergodic and open field lines, and are, in that respect, similar to the edge region of stellarators.

The main problem of the transport modelling in such mixed regions is as follows: the coordinate system must be aligned to the unstructured magnetic field. For realistic plasma edge conditions, the ratio between $\chi_\parallel$ and $\chi_\perp$ can be up to eight orders of magnitude, and numerical diffusion becomes a severe problem.

If we choose a coordinate system in which the magnetic field has only one nonzero component, we guarantee the separation of the parallel and perpendicular fluxes. The question is how to build such a coordinate system. We begin by choosing a surface (called the reference cut) which intersects all field lines of interest and then draw some reasonable (say, Cartesian) mesh on this cut. We then trace field lines through the mesh lines. The surfaces we obtain are our coordinate surfaces. The metric of this system can be obtained by field line tracing. Indeed, the real space coordinates are linked with these magnetic coordinates by field line tracing, and we can extend this procedure to calculate the relevant transformation matrix. These Clebsch coordinates have a remarkable advantage: they can be used locally. By this, we mean that more than one reference cut (and, therefore, multiple coordinate systems) can be used, in order to keep the scope of a single system well below the Kolmogorov length. The price for such flexibility is clear: the system is non-periodic, i.e. the coordinate surfaces of neighboring systems overlap arbitrarily at the interface between the two systems. Also we have a full metric tensor.

If we try to solve the problem by interpolation, we induce numerical diffusion of the same nature as before: a contribution of the parallel flux to the perpendicular fluxes. For instance, if one were to pass a beam (a δ-function) through such a system, one would obtain a response in all corners of the cell at the interface, and then on the next interface it would spread further. There are two possible ways of dealing with this phenomenon: The first is to optimize the mesh on the reference cut, i.e. to produce it by field line tracing itself. This idea is good for the field lines starting and ending on the wall, or for closed field lines.

Any other field line can be treated as "almost closed" as long as the following criterion is fulfilled (see Fig. 16.4): $\Delta \ll L_\parallel \sqrt{\chi_\perp/\chi_\parallel}$, where Δ is the

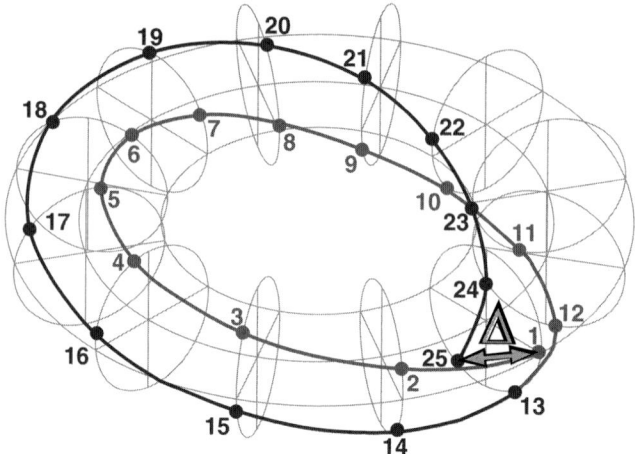

Fig. 16.4. Grid construction criterion for the finite difference code

excursion between the start and end points on the cut and $L_\parallel \approx 2\pi R N$ is the distance between the same points measured along the field line with $N \approx 100$ toroidal turns. We select only those field lines which come within $\Delta = 1\,\mathrm{mm}$ of closing on themselves. These field lines are used in building the mesh. This minimizes the induced numerical diffusion to about $10^{-4}\,\mathrm{m^2\,s^{-1}}$. The penalty one pays for this method is an unstructured mesh [36]. The local system is a moving patch covering three neighboring reference cuts simultaneously, because one must be able to write and discretize the divergence of gradients within its scope.

For the parallel terms, we use only the three points in our scheme located along the same field line at its intersections with the reference cuts. The rest is more or less conventional work with an unstructured mesh: the perpendicular problem is quasi-isotropic, and the choice of the numerical method is uncritical. Normally, one uses so-called Delaunay triangulation to build the mesh – a method commonly used in finite elements. First successful studies were done for W7-X cases [37].

The second idea is to switch to a Monte Carlo method with an appropriate mapping technique [38, 39].

16.3 Turbulence

In tokamak devices the observed transport is much higher than the neoclassical transport (transport caused by collisions in a toroidal confinement system) thus it must be due to transport caused by turbulence (anomalous transport). Observations suggest that ion temperature gradient (ITG) driven instabilities and the subsequent turbulence are the main cause for anomalous transport in the plasma core of tokamaks (see also Chap. 8 in Part II).

For stellarator devices the situation is less clear. In most classical stellarator experiments the transport was dominated by neoclassical transport making it impossible to observe the anomalous contribution. For the newly designed advanced stellarators (e.g. Wendelstein 7-X, currently under construction in Greifswald) the neoclassical transport has been greatly reduced so that the anomalous transport may become detectable.

16.3.1 Gyro-kinetic Theory

Gyro-kinetic theory for ions is the standard theory used for describing the core (where temperature and density are high). It is a simplification of full kinetic theory where the Vlasov equation

$$\frac{\mathrm{d}f}{\mathrm{d}t} = \frac{\partial f}{\partial t} + \mathbf{v} \cdot \nabla_r f + \frac{q}{M}(\mathbf{E} + \mathbf{v} \times \mathbf{B}) \cdot \nabla_v f = 0 \qquad (16.1)$$

(q, M: ion charge, mass; $\mathbf{E}, \mathbf{B}$: electric, magnetic field) describes the time evolution of a six dimensional distribution function $f(\mathbf{r}, \mathbf{v})$ in phase space. Solving this equation numerically should be avoided since it is very costly; thus approximations are required which are more feasible for numerical simulations (see also Chap. 8 in Part II).

For the general case the electric and magnetic field ($\mathbf{E}, \mathbf{B}$) are determined by Maxwell's equations where the density and current are obtained by taking velocity moments of f. In the following we will only look at the electrostatic case where a fixed magnetic field is prescribed externally and only the electric field $\mathbf{E} = -\nabla \phi$ is allowed to change.

The motion of a charged particle in a magnetic field can be decomposed into a fast gyration with the gyro-frequency $\Omega_i := (qB)(M)$ along a circle – called the gyro-ring – with radius $\rho := v_\perp / \Omega_i$ centered at the slowly drifting gyro-center $\mathbf{R}$. The position of the particle can thus be written as $\mathbf{x} = \mathbf{R} + \boldsymbol{\rho}$ with $\boldsymbol{\rho} = \rho[\cos(\alpha)\mathbf{e}_{\perp 1} + \sin(\alpha)\mathbf{e}_{\perp 2}]$ and $\mathbf{e}_{\perp 1,2}$ two base vectors perpendicular to $\mathbf{B}$. The velocity $\mathbf{v}$ can then be decomposed into components parallel ($v_\parallel$) and perpendicular ($v_\perp$) to the magnetic field and the gyro-phase α.

If the characteristic frequency ω of the processes one wants to describe is much smaller than Ω_i – this is the case for turbulence and ITG instabilities – it is possible (under assumptions on the space variation of the quantities) to average over the fast gyro-motion. Due to the averaging procedure the gyro-phase α disappears and the result is the gyro-kinetic equation [40] now describing the evolution of a distribution function $f(\mathbf{R}, v_\parallel, \mu)$ in a five dimensional phase space spanned by the gyro-center position $\mathbf{R}$ and two velocity components (here the magnetic moment per unit mass $\mu := v_\perp^2/(2B)$ is used instead of $v_\perp$):

$$\frac{\mathrm{d}f}{\mathrm{d}t} = \frac{\partial f}{\partial t} + \dot{\mathbf{R}} \cdot \nabla f + \dot{v}_\parallel \frac{\partial f}{\partial v_\parallel} + \dot{\mu} \frac{\partial f}{\partial \mu} = 0 \qquad (16.2)$$

with the equations of motion for the gyro-center (for the reason of simplicity we assume a vacuum magnetic field in the following)

$$\dot{R} = v_\| \mathbf{b} + \frac{1}{B} \mathbf{b} \times \nabla \langle \phi \rangle + \frac{\mu B + v_\|^2}{B\Omega_i} \mathbf{b} \times \nabla B \,, \tag{16.3}$$

$$\dot{v}_\| = -\mu \mathbf{b} \cdot \nabla B - \frac{q}{M} \left(\mathbf{b} + \frac{v_\|}{B\Omega_i} \mathbf{b} \times \nabla B \right) \cdot \nabla \langle \phi \rangle \,, \tag{16.4}$$

$$\dot{\mu} = 0 \,, \tag{16.5}$$

($\mathbf{b} := \mathbf{B}/B$). Equation (16.3) gives the motion of the gyro-center as a parallel motion plus $\mathbf{E} \times \mathbf{B}$-drift and grad-B/curvature-drift; the first term in the equation for $\dot{v}_\|$ shows the mirror effect; finally, μ is a constant of motion (see also Chap. 1 in Part I).

The equations of motion do not depend on the electrostatic potential directly but on its average over the gyro-ring described by the gyro-averaging operator $\langle . \rangle$

$$\langle \phi \rangle (\mathbf{R}) := \frac{1}{2\pi} \int_0^{2\pi} \phi(\mathbf{R} + \rho) \, d\alpha \,. \tag{16.6}$$

The space density n_i of ions is given by the gyro-averaged ion density $\langle n_i \rangle$ defined as

$$\langle n_i \rangle (\mathbf{x}) := \int f(\mathbf{R}, v_\|, \mu) \, \delta(\mathbf{R} + \rho - \mathbf{x}) \, d\mathbf{R} \, d\mathbf{v} \tag{16.7}$$

plus a correction term called the polarization density (n_0 denotes the equilibrium ion density and $\nabla_\perp$ the gradient perpendicular to the magnetic field)

$$n_i = \langle n_i \rangle + \nabla_\perp \cdot \left(\frac{n_0}{B\Omega_i} \nabla_\perp \phi \right) \,. \tag{16.8}$$

Since the gyro-radius of the electrons is much smaller than that of the ions the former can be described by a simple approximation called adiabatic electrons: $n_e = n_0 + (en_0)/(k_B T_e)\phi$. Together with charge neutrality $n_e = n_i$ this leads to the final Helmholtz equation determining ϕ

$$\frac{en_0}{k_B T_e} \phi - \nabla_\perp \cdot \left(\frac{n_0}{B\Omega_i} \nabla_\perp \phi \right) = \langle n_i \rangle - n_0 \,. \tag{16.9}$$

Equations (16.2)–(16.7) and (16.9) constitute a closed system for ϕ and f.

Gyro-kinetic theory may seem complicated but can be understood intuitively by a simple physical picture: Due to the time-scale separation between gyration and drift motion the ion can be replaced by a ring of charge with radius ρ centered around $\mathbf{R}$. f now becomes the distribution function of these rings located in space at the position $\mathbf{R}$. Equations (16.3)–(16.5) describe the movement of such a ring in five dimensional space. As a consequence the electric field does not act at the center of the ring $\mathbf{R}$ but on the ring itself; thus to get its action it must be integrated along the ring what is expressed by (16.6). Equation (16.7) can be understood by noting that the charge at one specific space point is the sum of the charge from all the rings with different ρ passing through it. Finally, the polarization density can only be understood by deriving the gyro-kinetic theory rigorously using the theory of Lie transformations [40].

16.3.2 The PIC Method

The gyro-kinetic equation can be solved as a partial differential equation on a five dimensional grid or by particle in cell (PIC) simulations. Since the PIC method combined with a δf approach has proven extremely powerful we restrict ourselves to this approach. Also we concentrate – as an example – on the line of PIC codes originating from the CRPP at the EPFL Lausanne. We do not want to delve into the details and problems of linear/nonlinear numerical implementation, geometry or extensions for non-adiabatic electrons and electromagnetic effects (see e.g. [41, 42, 43, 44, 45, 46]) but merely present some of the basic methods used in a modern PIC code.

The fluctuation amplitude of ITG turbulence in a plasma is much smaller than the magnitude of the equilibrium quantities. For a simulation using particles this means that most of the particles get wasted by representing the equilibrium and only a small number is left over for the description of the fluctuations. Since the numerical noise depends on the particle number N as $N^{-1/2}$ this leads to a bad statistical behavior that can be overcome by employing the δf approach [47, 48, 49]; here the time dependent distribution function f is split into a fluctuating part δf and a time–independent part f_0 representing the equilibrium: $f(t) = f_0 + \delta f(t)$. If now f_0 is given analytically (usually it is assumed to be a Maxwellian distribution) all the particles can be used to represent the fluctuation δf resulting in highly improved statistics.

Using this ansatz in (16.2) one gets $\mathrm{d}\delta f/\mathrm{d}t = S$ (S is a source term resulting from the application of $\mathrm{d}/\mathrm{d}t$ on f_0). This PDE (partial differential equations) can be solved formally using the method of characteristics: Given an initial function $\delta f(t=0)$ integration of the ODE (ordinary differential equations) $\mathrm{d}\delta f/\mathrm{d}t = S$ along the characteristics determined by (16.3)–(16.5) gives the general solution $\delta f(t)$.

A gyro-kinetic PIC code rests on four procedures subsequently executed at each time step:

1. Particle pushing: Integrating the equations of motion for each particle in a given potential.
2. Gyro-averaging: The gyro-average $\langle\phi\rangle$ of the electrostatic potential ϕ needs to be calculated by numerically approximating the integral in (16.6).
3. Charge assignment: the potential equation (16.9) is discretized on a grid but the particles are distributed irregularly; thus a prescription must be provided specifying how to generate from the particles (and their weights) a density defined on the grid (in gyro-kinetics one has also to take

into account that one does not deal with point particles but with gyro-rings). Charge assignment can thus be regarded as the link connecting the Lagrangian (particles) with the Eulerian part (grid).

4. Potential solver: given the gyro-averaged density on the grid the Helmholtz equation must be solved in order to obtain the potential.

Introducing N marker particles each one carrying a weight w_p the quantity δf is discretized by writing it as a sum of delta functions

$$\delta f = \sum_{p=1}^{N} \frac{1}{J} w_p(t)\, \delta[\mathbf{R} - \mathbf{R}_p(t)]\, \delta[v_\| - v_{\|p}(t)]\, \delta[\mu - \mu_p(t)] \qquad (16.10)$$

($J = 2\pi B$ is the phase space Jacobian) resulting in the equation

$$\dot{w}_p = S|_{\mathbf{R}_p, v_{\|p}, \mu_p} \qquad (16.11)$$

for the weights.

For the numerical integration of (16.3)–(16.5) and (16.11) with e.g. a Runge–Kutta– or a Predictor–Corrector–method $\nabla \langle \phi \rangle$ must be known at each time step. In order to compute this one uses $-\nabla \langle \phi \rangle \approx \langle \mathbf{E} \rangle$ and approximates the integral by an average over N_a points ρ_j evenly spaced on the gyro-ring:

$$\langle \mathbf{E} \rangle = \frac{1}{N_a} \sum_{j=1}^{N_a} \mathbf{E}(\mathbf{R} + \rho_j) \,. \qquad (16.12)$$

ϕ is obtained by solving the Helmholtz equation (16.9) discretized on a grid; also the right hand side $\langle n_i \rangle$ must be computed by the charge assignment process. An elegant way that combines the charge assignment and the discretization of the Helmholtz equation uses B-splines as finite elements [41]. The potential is discretized by writing it as a sum over the finite element basis

$$\phi(\mathbf{x}) = \sum_\nu \phi_\nu \Lambda_\nu(\mathbf{x}) \qquad (16.13)$$

where Λ_ν is a tensor product B-spline of order k. B-splines are continuous non-negative functions with finite support piecewise defined by polynomials of order k. Their most interesting property here is that B-splines provide a partition of unity, i.e. $\sum_\nu \Lambda_\nu(\mathbf{x}) = 1$ [50]; this guarantees the conservation of charge during charge assignment.

Using (16.13) and the particle discretization (16.10) in (16.9) one can derive the following matrix equation for the coefficient vector ϕ_ν

$$\sum_{\nu'} A_{\nu\nu'} \phi_{\nu'} = n_\nu \qquad (16.14)$$

with the matrix

$$A_{\nu\nu'} = \int \left(\frac{n_0}{B\Omega_i} \nabla_\perp \Lambda_\nu \cdot \nabla_\perp \Lambda_{\nu'} + \frac{en_0}{k_B T_e} \Lambda_\nu \Lambda_{\nu'} \right) d\mathbf{x} \quad (16.15)$$

that is the same at each time step and thus can be precomputed.

The charge assignment process defines the vector n_ν

$$n_\nu = \sum_{p=1}^{N} w_p \frac{1}{2\pi} \int_0^{2\pi} \Lambda_\nu(\mathbf{R}_p + \rho_p) d\alpha \quad (16.16)$$

where the last integral is again discretized by an average over N_a points on the gyro-ring [analogously to (16.12)].

This completes the short description of the ingredients needed for a numerical solution of the gyro-kinetic equation. We now demonstrate the application by a linear and a nonlinear example.

If $\delta f \ll f_0$ is assumed then the gyro-kinetic equation can be linearized. The EUTERPE code [44] solves this equation globally in full three dimensional geometry; thus it can be used to investigate stellarator configurations (e.g. Wendelstein 7-X [51]).

As a simple example the result of a calculation for a circular tokamak configuration is shown in Fig. 16.5 (the ion temperature gradient driving the instability is chosen such that it is maximal close to half the minor radius of the configuration). In this simulation one follows the time evolution of small initial fluctuations imposed onto an equilibrium plasma. In the beginning ($t \lesssim 4 \times 10^3 \, \Omega^{-1}$) the plasma oscillates randomly but for $t > 4 \times 10^3 \, \Omega^{-1}$ a coherently oscillating mode is established and grows exponentially in time. The slope of the bold line then gives the growth-rate while the frequency of the mode can be obtained from the thin solid dashed curve. The mode

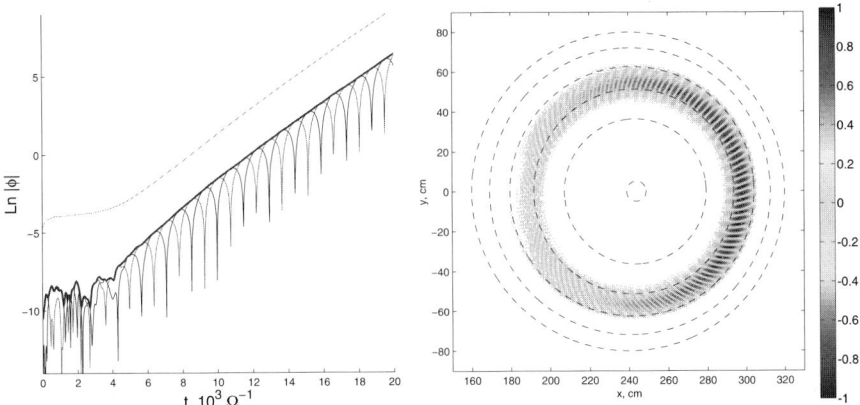

Fig. 16.5. Linear ITG mode in a tokamak. *Left*: Time traces of $Re(\phi), Im(\phi), |\phi|$ (*red, black, blue*; colors in online version) and field energy (*green*). *Right*: $Re(\phi)$ (color coded) in the poloidal plane (courtesy of V. Kornilov)

structure at the end of the simulation is presented in Fig. 16.5, right (since the configuration is toroidally symmetric only one poloidal cut is depicted). The structure shown rotates poloidally but its envelope always has its maximum at the right hand side (corresponding to the low field side of a tokamak); thus it is a typical toroidal ballooning mode. This kind of mode is characterized by small scale structures perpendicular to the magnetic field and long structures nearly following the field lines; this reflects the high anisotropy of the system caused by the magnetic field.

Linear calculations are important in many ways, e.g., to identify parameter regions where the plasma becomes unstable. However, for the calculation of anomalous transport caused by turbulence, nonlinear calculations are necessary. Figure 16.6 presents results from the TORB code [42, 43] which solves the nonlinear global gyro-kinetic equation for a straight cylinder geometry (θ-pinch). The left plot shows the ion heat flux as a function of time (in this simulation the formation of zonal flows, known to strongly suppress the turbulence, was artificially inhibited). The simulation starts from small initial random noise. During the linear phase the perturbations grow exponentially until $t \approx 1.5 \times 10^{-4}$ s. At this time the system saturates due to the nonlinearity but the transport increases because the turbulence spreads radially. After the flux reaches its maximum the transport induced flattening of the temperature profile leads to its slow decay. This is connected to the absence of an external source of energy maintaining the temperature profile. The space structure of the turbulence at $t = 3 \times 10^{-4}$ s (Fig. 16.6, right) is again characterized by very different length scales perpendicular and parallel to the magnetic field.

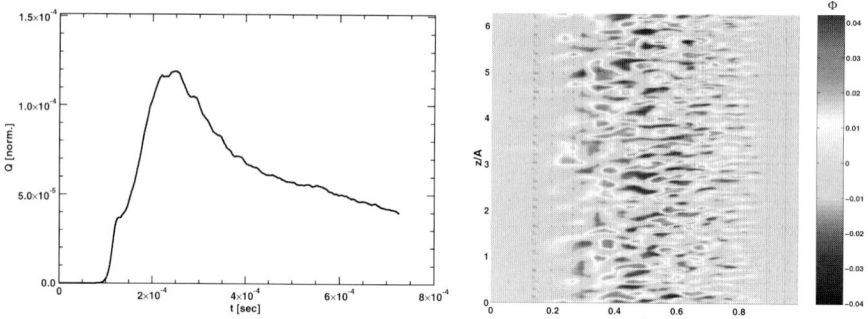

Fig. 16.6. ITG turbulence in a θ–pinch. *Left*: Time trace of the (normalized) ion heat flux. *Right*: Potential fluctuations (color coded in online version) in the $r - z$ plane of the cylinder (the magnetic field points in the z direction; also note that z has been scaled with $A = 10$) (courtesy of S. Sorge)

16.4 Outlook

Plasma physics is a typical example of a discipline which is necessarily strongly linked to computational physics. Better understanding is directly connected with the development of better computational models based on improved theory and validated on experiments. As in many other fields in physics, the complex physics requires complex numerical tools to be developed and used. The transfer and coupling of the different length and time scales poses a specific problem, where multi-scale techniques and strategies have to be applied. The ultimate goal of an ab-initio model including all the different scales in one numerical model remains a real challenge, which should be able to be achieved within the next ten years.

16.5 Seminars

You can find example exercises for this contribution at
http://www.rzg.mpg.de/~stel/Heraeus/uebungen.html.

References

1. C.K. Birdsall: IEEE Trans. Plasma Sci. **19**, 65 (1991)
2. J. Roth: J. Nucl. Mater. **266–269**, 51 (1999)
3. J. Küppers: Surf. Sci. Rep. **22**, 249 (1995)
4. E. Salonen, K. Nordlund, J. Keinonen et al: Phys. Rev. B **63**, 195415 (2001)
5. D.W. Brenner: Phys. Rev. B **42**, 9458 (1990)
6. K. Nordlund, J. Keinonen, T. Mattila: Phys. Rev. Lett. **77**, 699 (1996)
7. M. Warrier, R. Schneider, E. Salonen et al: Physica Scripta **T108**, 85 (2004)
8. M. Warrier, R. Schneider, E. Salonen et al: Contrib. Plasma Phys. **44**, 307 (2004)
9. W. Eckstein: *Computer simulation of ion-solid interactions* (Springer, Berlin Heidelberg New York 1991)
10. W. Eckstein, C. García-Rosales, J. Roth et al: Sputtering. IPP-Report 9/82, Max-Planck-Institut für Plasmaphysik, Garching, Germany (1993)
11. W. Eckstein: Sputtering, reflection and range values for plasma edge codes. IPP-Report 9/117, Max-Planck-Institut für Plasmaphysik, Garching, Germany (1998)
12. W. Eckstein: Calculated sputtering, reflection and range values. IPP-Report 9/132, Max-Planck-Institut für Plasmaphysik, Garching, Germany (2002)
13. A.B. Bortz, M.H. Kalos, J.L. Lebowitz: J. Comp. Phys. **17**, 10 (1975)
14. C.K. Birdsall, A.B. Langdon: *Plasma physics via computer simulation* (McGraw–Hill, New York 1985)
15. R. Hockney, J. Eastwood: *Computer simulation using particles* (McGraw–Hill, New York 1981)
16. K.V. Matyash, R. Schneider, A. Bergmann et al: J. Nucl. Mater. **313–316**, 434 (2003)

17. K.V. Matyash, R. Schneider: Contrib. Plasma Phys. to be published
18. K.V. Matyash, R. Schneider: Contrib. Plasma Phys. **44**, 157 (2004)
19. R. Chodura: Phys. Plasmas **25**, 1628 (1982)
20. S.I. Braginskii: Transport Processes in Plasmas. In: *Reviews of Plasma Physics*, vol 1, ed by M.A. Leontovich (Consultants Bureau, New York 1965) pp 205–311
21. V.M. Zhdanov: Plasma Phys. Control. Fusion **44**, 2283 (2002) (re-edition of a Russian document, first published in 1982)
22. D.P. Coster, J.W. Kim, G. Haas et al: Contrib. Plasma Phys. **40**, 334 (2000)
23. X.Q. Xu, R.H. Cohen, T.D. Rognlien et al: Phys. Plasmas **7**, 1951 (2000)
24. D. Reiter: The EIRENE Code, Version: Jan. 92 Users Manual. Jül-Report 2599, KFA Jülich, Jülich (1992)
25. R. Schneider, D.P. Coster, B.J. Braams et al: Contrib. Plasma Phys. **40**, 328 (2000); V.A. Rozhansky, S. Voskoboynikov, E. Kovaltsova et al: ibid 423
26. B.J. Braams: A Multi-Fluid Code for Simulation of the Edge Plasma in Tokamaks. NET Report No. 68, EUR–FU/XII–80/87/68, Commission of the European Communities Directorate General XII - Fusion Programme, Brussels (1987)
27. R. Schneider, H.-S. Bosch, J. Neuhauser et al: J. Nucl. Mater. **241–243**, 701 (1997); R. Schneider, H.-S. Bosch, D.P. Coster et al: J. Nucl. Mater. **266–269**, 175 (1999)
28. V.A. Rozhansky, E. Kaveeva, S. Voskoboynikov et al: Contrib. Plasma Phys. **42**, 230 (2002); V.A. Rozhansky, E. Kaveeva, S. Voskoboynikov et al: J. Nucl. Mater. **313–316**, 1141 (2003); V.A. Rozhansky, E. Kaveeva, S. Voskoboynikov et al: Nucl. Fusion **42**, 1110 (2002)
29. T.D. Rognlien, J.L. Milovich, M.E. Rensink et al.: J. Nucl. Mater. **196–198**, 347 (1992)
30. M. Borchardt, J. Riemann, R. Schneider: Numerics in BoRiS. In: *Parallel Computational Fluid Dynamics*, ed by K. Matsuno, A. Ecer, J. Periaux, N. Satofuka and P. Fox, (Elsevier, Amsterdam 2003) pp 459–465
31. M. Borchardt, J.S. Riemann, R. Schneider et al: J. Nucl. Mater. **290–293**, 546 (2001)
32. J.S. Riemann, M. Borchardt, R. Schneider et al: J. Nucl. Mater. **313–316**, 1030 (2003)
33. A.M. Runov, S. Kasilov, J.S. Riemann et al: Contrib. Plasma Phys. **42**, 169 (2002)
34. J.S. Riemann, M. Borchardt, R. Schneider et al: Navier–Stokes neutral and plasma fluid modelling in 3D. In: *Europhysics Conference Abstracts ECA27A*, ed by R. Koch and S. Lebedev (European Physical Society, Geneva 2003) p P-2.151
35. K.H. Finken, G.H. Wolf: Fusion Eng. Des. **37**, 337 (1997); K.H. Finken: ibid 379; K.H. Finken, G. van Oost: ibid 411; K.H. Finken: ibid 445
36. A.M. Runov, S. Kasilov, D. Reiter et al: J. Nucl. Mater. **313–316**, 1292 (2003)
37. N.A. McTaggart: Contrib. Plasma Phys. **44**, 31 (2004)
38. A.M. Runov, D. Reiter, S.V. Kasilov et al: Phys. Plasmas **8**, 916 (2001)
39. A.M. Runov, S.V. Kasilov, N. McTaggart et al: Nucl. Fusion **44**, S74 (2004)
40. T.S. Hahm: Physics of Fluids **31**, 2670 (1988)
41. M. Fivaz, S. Brunner, G. de Ridder et al: Comp. Phys. Comm. **111**, 27 (1998)
42. T.M. Tran, K. Appert, M. Fivaz et al: Global Gyrokinetic Simulation of Ion-Temperature-Gradient-Driven Instabilities using Particles. In: *Proceedings of*

the Joint Varenna-Lausanne International Workshop Theory of Fusion Plasmas 1998, ISPP-18, ed by J.W. Connor, E. Sindoni, J. Vaclavik (Societa Italiana di Fisica, Bologna 1999) pp 45–58
43. R. Hatzky, T.M. Tran, A. Könies et al: Phys. Plasmas **9**, 898 (2002)
44. G. Jost, T.M. Tran, W.A. Cooper et al: Phys. Plasmas **8**, 3321 (2001)
45. S. Sorge and R. Hatzky: Plasma Phys. Control. Fusion **44**, 2471 (2002)
46. S. Sorge: Plasma Phys. Control. Fusion **46**, 535 (2004)
47. M. Kotschenreuther: Bulletin of the American Physical Society **34**, 2107 (1988)
48. S.E. Parker and W.W. Lee: Phys. Fluids B **5**, 77 (1993)
49. A.Y. Aydemir: Phys. Plasmas **1**, 822 (1994)
50. C. de Boor: *A practical guide to splines* (Springer, Berlin Heidelberg New York 2001)
51. V. Kornilov, R. Kleiber, R. Hatzky et al: Phys. Plasmas **11**, 3196 (2004)

17 Nuclear Fusion

H.-S. Bosch

Abstract. Nuclear fusion is one of the most important physics effects as it is the basis of power production and synthesis of nuclei in stars. The basic physics of this process is described. Based on this knowledge, the idea comes up to use this process also for power production on Earth. The physics of the possible nuclear reaction as well as the required physics parameters are described. From the power balance one receives a criterion for the fusion triple product, opening up two possible ways to achieve fusion conditions, i.e. magnetic and inertial confinement. The status of magnetic confinement fusion as well as the plans for the next step, ITER are described. Finally, an interesting physics principle to enhance fusion reactions at room temperature, i.e. muon-catalyzed fusion, is discussed.

17.1 Introduction

Nuclear fusion is one of the two possible reaction types than can occur for atomic nuclei, nuclear fission being the second type which is more familiar for the public. This chapter will give a short introduction into the physics of nuclear reactions, then explain in detail nuclear fusion and its role in the universe and possibly on Earth.

All nuclear reactions are based on differences in the nuclear binding energy. Figure 17.1 shows the nuclear binding energy per nucleon (proton or neutron) as a function of the nuclear mass for the nuclei bound strongest. Such graphs have been derived from measurements of the masses of the nuclei already in the beginning of the twentieth century, when it was observed that the masses of nuclei are always just a little smaller than some multiple of the hydrogen mass. After the discovery of the nuclear constituents, the proton and neutron, it became finally evident why nuclear masses should roughly be multiples of the mass of the protons and neutrons which constitute the nucleus and that the mass difference corresponds to the nuclear binding energy according to Einstein's energy-mass relation $\Delta E = \Delta m\, c^2$.

An explanation of the coarse structure shown in Fig. 17.1 was given by Carl Friedrich von Weizsäcker in 1935 with the liquid-drop model. Based on the very limited range of the strong nuclear force he concluded that each nucleon just influences its nearest neighbors. The binding energy per nucleon would thus be constant. The smaller binding energies for smaller nuclei are then due to the relatively large surface-to-volume ratio: The nucleons at the

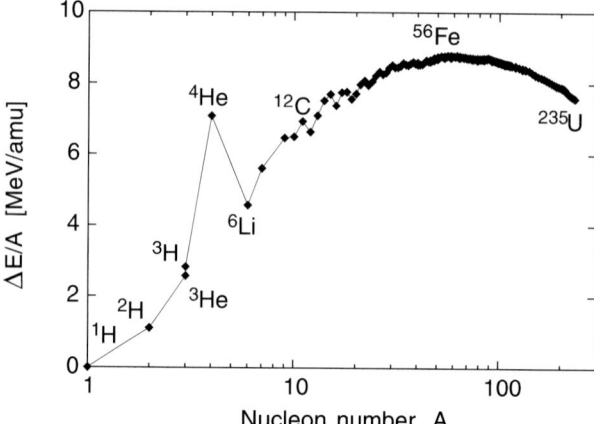

Fig. 17.1. Nuclear binding energy per nucleon as a function of the nucleon number A

surface have missing partners and thus their contribution to the total binding energy of the nucleus is reduced. The decrease of binding energy per nucleon for nuclei beyond $A \approx 60$ is due to the repulsive Coulomb force of the large number of positive protons.

The very fine structure in Fig. 17.1 can, however, not be explained by the liquid-drop model. It is due to quantum mechanical effects, i.e. at certain so-called "magic" proton and neutron numbers the nucleus formed is a very stable configuration. This is roughly comparable to the stable electron configurations of the noble gases, for which electron shells are completed. The first magic number is 2, which is manifested as a most remarkable example of a local maximum in Fig. 17.1, i.e. the helium nucleus with 2 protons and 2 neutrons, i.e. A = 4.

Looking into the binding energies shown in Fig. 17.1 with regard to their possible use for energy production, one notes immediately that the energy release per nucleon is of the order of 1 MeV ($= 10^6$ eV) for fission reactions and in the order of a few MeV for fusion reactions. This is 6–7 orders of magnitude above typical energy releases in chemical reactions, which explains the effectiveness and potential hazard of nuclear power. Also, this figure shows clearly that there are two ways of gaining nuclear energy, i.e. setting free the binding energy of nuclei:

1. **transforming (i.e. splitting) heavy nuclei into medium-size nuclei.** Fission of uranium-235 or other fissile nuclei constitutes the basis of nuclear weapons as well as of energy production in nuclear power stations. Nuclear fission is not treated in this course but is mentioned here to indicate the physics relation of the different types of nuclear reactions.
2. **fusion of light nuclei into heavier ones.** Especially fusion of hydrogen isotopes into stable helium offers the highest energy release per mass unit.

This is the basis of energy production in stars as well as of thermonuclear weapons. Doing this, however, in a controlled manner has been the goal of fusion research for about 40 years. Many of the lectures in this course are related to this goal.

Nuclear reactions are governed by the strong nuclear force acting over very small distances in the order of the radius of the nuclei (i.e. strong interaction), but for reactions between two nuclei, i.e. two positively charged partners, other forces enter into the picture. For distances above a few fermi (i.e. 10^{15} m) the repulsive Coulomb force between the nuclei becomes dominant. The potential energy of two nuclei as a function the distance between the nuclei is shown in Fig. 17.2. The depth of the deep well at small radii is determined by the binding energy, discussed before, while the barrier at a few fermi is given by the Coulomb potential of $Z_1 Z_2 e^2 / 4\pi\epsilon_0 r_m$ which is much smaller, but still poses a principal problem. For alpha-particle decay (where a ^{4}He nucleus separates itself out of a positive nucleus) as well as for fusion of lighter nuclei, this diagram demands a particle energy of the order of 500 keV, and this would make fusion processes almost impossible. However, in the beginning of the twentieth century, is was known that α-particle decay

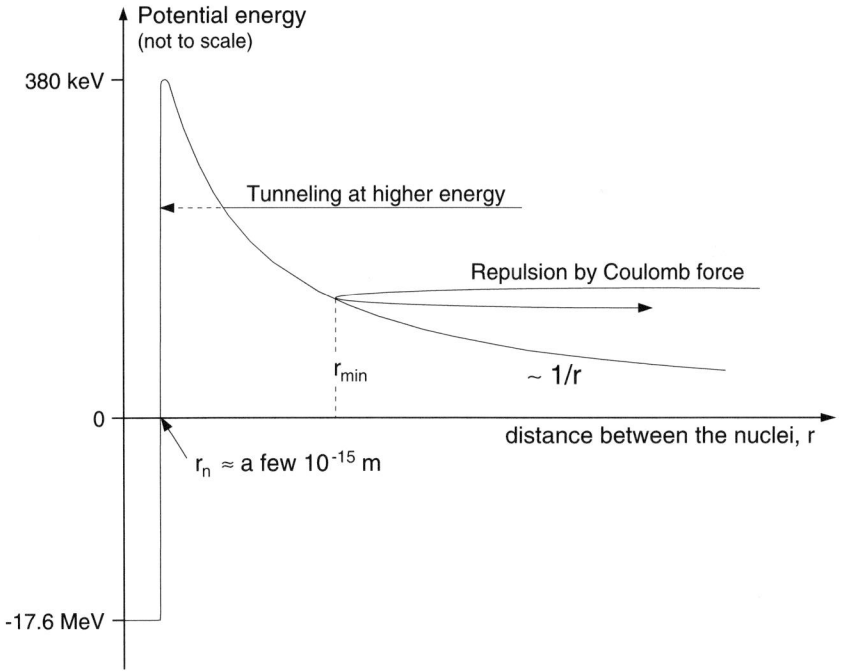

Fig. 17.2. Schematic representation of the potential energy U of two nuclei as a function of their distance. The energies quoted are for a system of D ($=^2$H) and T ($=^3$H)

occurs at room temperature, although the energy is not sufficient to overcome the barrier. Only in 1928 Gamov [1] explained this by the tunnelling effect, which in turn also allows fusion reactions to occur at temperatures far below the Coulomb barrier: Due to quantum mechanical effects the minimum distance between the two nuclei is not a fixed value determined by the energy in this two-particle system (as it is indicated schematically by the repulsion in Fig. 17.2), but there is a finite probability for the nuclei to get closer, and eventually "tunnel" through the Coulomb barrier, as indicated by the dotted line in Fig. 17.2. In terms of wave functions, the amplitude is not zero for $r \leq r_{min}$, but it is finite and decays slowly for smaller radii. Therefore it can still be finite for $r \leq r_n$, i.e. the particles have a possibility to approach close enough for a fusion reaction to occur. This tunnelling probability is a strong function of the relative velocity v of the reacting particles with charge Z_1, Z_2,

$$P_{tunneling} \sim \exp\left(-\frac{2\pi Z_1 Z_2 e^2}{\hbar\, v}\right). \tag{17.1}$$

This equation shows that reaction partners with small mass and small charge Z are preferred, and that the reaction probability increases strongly with the relative velocity v of the reacting nuclei.

In the light of the above discussion it becomes clear why fission energy has been much more readily obtained than fusion energy: fission is triggered by capture of thermal neutrons, e.g.

$$^{235}\text{U} + n_{thermal} \longrightarrow {}^{89}\text{Kr} + {}^{144}\text{Ba} + 3\,\text{n} + 174\,\text{MeV}, \tag{17.2}$$

where no force prevents the neutron from entering the uranium nucleus at room temperature and from causing a fission reaction. Using a nomenclature from chemistry, one could state that the activation energy for fusion reactions is extremely high, while for fission it is close to zero, which on the other hand is also the source of problems.

17.2 Energy Production in the Sun

Though still a somewhat exotic topic on Earth, nuclear fusion is one of the most important phenomena in the universe. Not only is it the process by which all the elements (at least for $A \leq 60$) were created after the big bang and are still created in stars, it is also the energy source of all stars [2]. An excellent review on stellar nucleosynthesis is given in [3], and a recent review on cross sections of fusion reactions occurring in the Sun, is presented in [4].

On the Sun, the main fusion reactions are the following:

$$\text{p} + \text{p} \longrightarrow \text{D} + \text{e}^+ + \nu_e \tag{17.3}$$

$$\text{D} + \text{p} \longrightarrow {}^3\text{He} + \gamma \tag{17.4}$$

$$^3\text{He} + {}^3\text{He} \longrightarrow {}^4\text{He} + 2\,\text{p} \tag{17.5}$$

where p denotes a proton, D a deuteron, a heavy hydrogen isotope with one proton and one neutron, ^{3}He, ^{4}He are helium isotopes, γ stands for a high-energy photon, e$^+$ for a positron (anti-electron), and ν_e for an electron neutrino. This reaction chain accounts for about 85% of the energy production from the pp-reaction, the rest being from two more seldom reaction branches. Theses are important at temperatures above about 1 keV, and produce $^7_4$Be, $^7_3$Li, $^8_5$B and $^8_4$Be, which decays into 2 α-particles. Also in these reactions neutrinos are produced, however with a higher kinetic energy than those from the reactions mentioned above.

A very important feature of the energy production on the Sun is the need for the weak interaction which transforms protons to neutrons (β^+-decay), in the first of the above listed reactions. All weak interaction processes involve the emission of neutrinos (thereby keeping the lepton number constant). As all reactions involving the weak interaction have a very low reaction cross section, the rates of such reactions are very low.

Since neutrinos react only to the weak interaction, they are the only product from the fusion reactions that can leave the core of the Sun directly and therefore the only probe we have to investigate directly the energy production processes in stars. For the same reason – neutrinos ignore both the strong and electromagnetic interactions – it is extremely hard to detect them experimentally. It was not till 1992 that the European GALLEX collaboration detected the low energy solar neutrinos from the main energy-producing reactions (pp-chain) listed above. This was the first experimental validation of theoretical models of energy production in the stars [5].

For much higher temperatures ($T \geq 2\,\text{keV}$), i.e. in stars with a higher mass than the Sun, fusion of four protons to $^4_2$He can also occur in a catalytic process based on the involvement of ^{12}C. In this so-called Bethe–Weizsäcker cycle, shown in Fig. 17.3, oxygen, nitrogen and ^{13}C are only present in intermediate stages, and the net reaction again is

$$4p \rightarrow {}^4\text{He} + 2e^+ + 2\nu_e \ . \tag{17.6}$$

17.3 Fusion on Earth

The fundamentals of fusion physics are described in great detail in Chap. 4 in Part I, Chap. 7 in Part II and in [6].

For energy production on Earth the weak interaction has to be avoided since it would lead to unacceptably small reaction rates. The Sun (and all other stars) compensate for the low reaction rates mentioned above by a large number of reactants (i.e. by their tremendous volume), leading to huge energy production. For a terrestrial fusion reactor, however, this is not a viable solution.

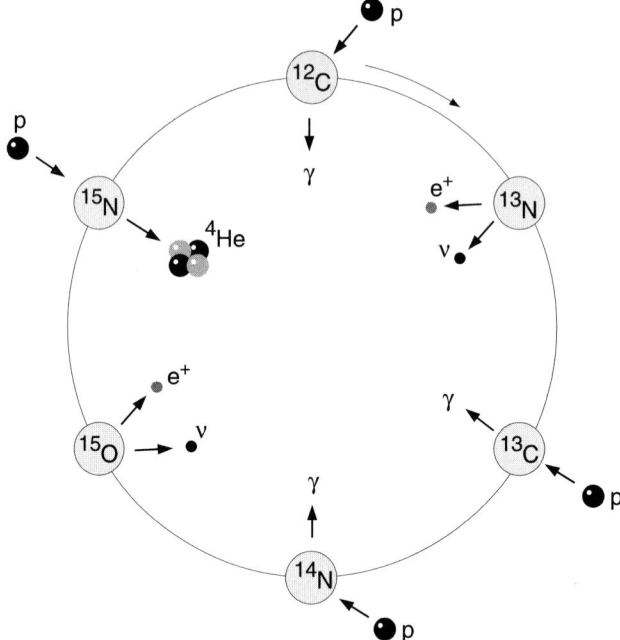

Fig. 17.3. Schematic of the Bethe–Weizsäcker cycle. Reaction products are shown inside the cycle, with the reactants outside the cycle. The C-, N- and O-atoms have a catalytic role only

Possible candidates for the use of nuclear fusion on Earth are therefore the following reactions ($T = {}^3_1\text{H}$ denoting tritium, the heaviest hydrogen isotope with two neutrons):

$$D + D \longrightarrow {}^3\text{He} + n + 3.27\,\text{MeV} \quad (50\%) \quad (17.7)$$
$$\text{or} \quad T + p + 4.03\,\text{MeV} \quad (50\%) \quad (17.8)$$
$$D + {}^3\text{He} \longrightarrow {}^4\text{He} + p + 18.35\,\text{MeV} \quad (17.9)$$
$$D + T \longrightarrow {}^4\text{He} + n + 17.59\,\text{MeV} \quad (17.10)$$
$$p + {}^{11}B \longrightarrow 3\,{}^4\text{He} + 8.7\,\text{MeV} \quad (17.11)$$

The kinetic energy of the reactants is much lower than the energy gained in the reaction, and therefore can be neglected in first order for the energy of the reaction products. Due to energy and momentum conservation the distribution of the reaction energy onto the two product particles is inverse proportional to their mass, i.e. $E_1/E_2 = m_2/m_1$. Hence, in the D-T reaction, e.g. the α-particle has an energy of 3.54 MeV, and the neutron has 14.05 MeV. The first four reactions (for which the cross sections are shown in Fig. 17.4) can be summarized as

$$3\,D \longrightarrow {}^4\text{He} + p + n + 21.6\,\text{MeV} \quad (17.12)$$

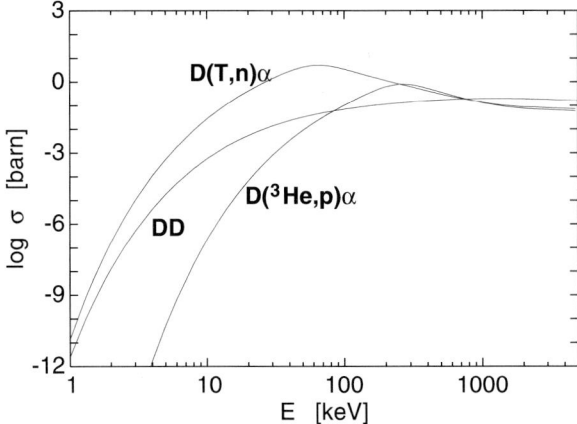

Fig. 17.4. Measured cross-sections for different fusion reactions as a function of the center of mass energy [7]. The curve labelled DD shows the sum of the cross sections for both branches of this reaction. Nuclear reaction cross sections are measured in barn, where 1 barn = 10^{-28} m^2. The notation (**a**(**b**,**c**)**d**) means that the projectile (**a**) hits a target (**b**) and the reaction produce a light particle (**c**) and the nucleus (**d**). DD indicates that this curve shows the cross section for the sum of the two reactions (17.7) and (17.8)

and therefore rely on deuterium as fuel only. Since the weight fraction of deuterium in water is 3.3×10^{-5}, the energy content of water is about 11.5 GJ per liter, which is about 350 times larger than the chemical energy density of gasoline. This demonstrates the huge potential afforded by nuclear fusion as an energy source.

All reaction cross sections in Fig. 17.4 show a steep increase with the relative energy, as discussed before, but the D-T reaction (17.10) has by far the largest cross-section at the lowest energies. This makes the D-T fusion process the most promising candidate for an energy-producing system.

The special role of D-T reactions becomes clear from the energy levels of the unstable ^{5}He nucleus shown in Fig. 17.5. It has an excited state just 64 keV above the sum of the masses of deuterium and tritium. The D-T fusion cross-section reaches its maximum at this energy difference, due to this resonance-like reaction mechanism. The D-D reactions (Fig. 17.5, left part) do not show such resonances, and their cross sections are solely governed by the tunnelling probability, showing a smooth increase without any maximum, while the D-^{3}He reaction also has a resonance at about 270 keV in the ^{5}Li system.

To be a candidate for an energy-producing system, the fusion fuel for this reaction has to be sufficiently abundant. As mentioned earlier, deuterium occurs with a weight fraction of 3.3×10^{-5} in water. Given the water of the

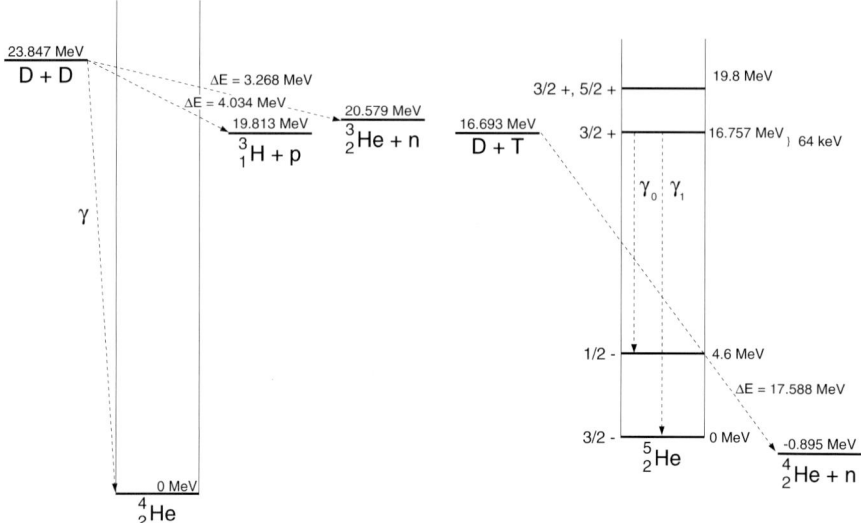

Fig. 17.5. Energy diagram of the ^{4}He nucleus (*left*), and of the unstable ^{5}He nucleus (*right*), where the resonance at 16.757 MeV dominates the reaction cross section. The numbers left of the level scheme indicate the J-value and parity of the respective level

oceans, the static energy range is larger than the time the Sun will continue to burn (a few billion years).

Tritium, however, is an unstable radioactive isotope. It decays to

$$\text{T} \longrightarrow {}^3\text{He} + e^- + \bar{\nu}_e \tag{17.13}$$

with a half-life of 12.3 years. Note the previously mentioned neutrino production of this β^- decay. Owing to the unstable character of tritium no significant amounts exist naturally, but tritium can be produced with nuclear reactions of the neutrons from the D-T reaction and lithium:

$$n + {}^6\text{Li} \longrightarrow {}^4\text{He} + \text{T} \quad + 4.8\,\text{MeV} \tag{17.14}$$

$$n + {}^7\text{Li} \longrightarrow {}^4\text{He} + \text{T} + n' - 2.5\,\text{MeV} \tag{17.15}$$

The reaction with ^{7}Li is particularly important because it does not consume a neutron and therefore allows that each fusion-produced neutron creates even more than one new tritium nucleus. As inevitably some neutrons will escape from the breeding blanket, this reaction nevertheless opens up the possibility for self-sufficient tritium production in a fusion reactor.

The ultimate fusion fuel will thus be deuterium and lithium. The latter is also very abundant and widespread in the Earth's crust and even ocean water contains an average concentration of about 0.15 ppm (1 ppm = parts per million). Table 17.1 summarizes the estimated world resources of various energy-relevant materials (see also Chap. 18 in Part III).

Table 17.1. Estimated world energy resources. The figures are only indicative, being dependent on prices and subject to uncertainty because of incomplete exploration

Present world annual primary energy consumption	3×10^{11} GJ	
Resources	**in GJ**	**in years**
Coal	10^{14}	300 years
Oil	1.2×10^{13}	40 years
Natural Gas	1.4×10^{13}	50 years
^{235}U (fission reactors)	10^{13}	30 years
^{238}U and ^{232}Th (breader reactors)	10^{16}	30000 years
Lithium (D-T fusion reactors) :		
on land	10^{16}	30000 years
in oceans	10^{19}	3×10^7 years

17.4 Conditions for Nuclear Fusion

As discussed before, for a fusion reaction to occur, the two nuclei have to "touch" each other since the range of the nuclear force is of the order of the dimensions of the nuclei. The repulsive Coulomb force counteracts all attempts to bring them close together. This is what the difficult research on nuclear fusion is all about: how can the two reaction partners be brought into contact?

The seemingly simplest approach to realizing the fusion reactions would be to accelerate the reactants to about 100 keV and bring them to collision. This, however, does not lead to a positive energy balance, since the elastic Coulomb scattering occurring simultaneously, has a much larger cross-section as compared with the fusion cross-section. Thus the two particle beams would just scatter and diverge after one interaction (see also Chap. 7 in Part II).

A way of overcoming this problem is to confine a thermalised mixture of deuterium and tritium particles at energies of about 10 keV. Since the average energy of particles at a certain temperature is about $k_B T$, where k_B is the Boltzmann constant, temperatures are often given in energy units of electron volt (1 eV $\widehat{=}$ 1.16×10^4 K). At energies of 10 keV the hydrogen atoms are completely ionized and form a plasma of charged ions and electrons. The basic physics of plasmas is discussed in Chap. 1 in Part I. For now it should suffice to observe that in a plasma the particles thermalize as a result of many Coulomb scattering processes and thus entail a Maxwellian velocity distribution:

$$f(v) = n \left(\frac{m}{2\pi k_B T}\right)^{3/2} \exp\left(-\frac{mv^2}{2 \, k_B T}\right) , \qquad (17.16)$$

where f is the number of particles in the velocity interval between v and $v + dv$, n is the spatial density of particles, m is their mass, and $k_B T$ is their temperature.

The reaction rate per unit volume R can be written as

$$R = n_D\, n_T\, \langle \sigma v \rangle \tag{17.17}$$

with v now being the relative particle velocity and $\langle \sigma v \rangle$ being the reaction parameter, i.e. the average of the product of cross-section times velocity.

Calculation of the reaction parameter (rate coefficient) requires integration over the distribution function of deuterium and tritium. After some numerical transformations one obtains

$$\langle \sigma v \rangle = \frac{4}{(2\pi m_r)^{1/2}(k_B T)^{3/2}} \int \sigma(E)\, E\, \exp\left(-\frac{E}{k_B T}\right)\, dE\,, \tag{17.18}$$

where m_r is the reduced mass, and E denotes here the kinetic energy in the center of mass coordinate system. Figure 17.6 shows the reaction parameter for some important fusion reactions. At the temperatures of interest the nuclear reactions come predominantly from the tail of the distribution. This is illustrated in Fig. 17.7, where the integrand of (17.18) is plotted versus E together with the two factors $\sigma(E)$ and $E\,\exp(-E/k_B T)$ for a D-T plasma at a temperature of 10 keV.

Thermalization is thus not just a way of handling the large cross-sections for elastic Coulomb scattering, it also considerably increases the reaction rate compared to beam experiments with single particle energies.

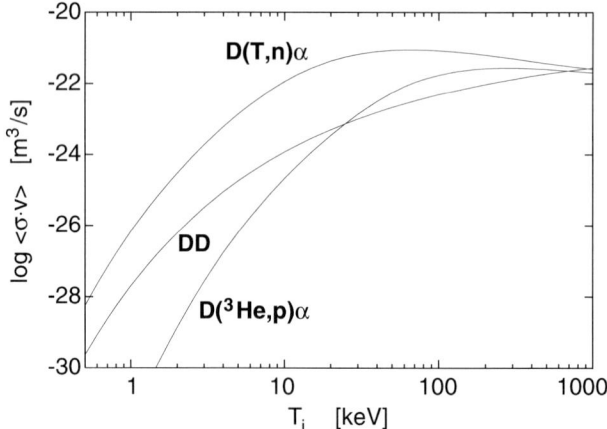

Fig. 17.6. Reaction parameter $\langle \sigma v \rangle$ as a function of ion temperature T_i for different fusion reactions [7]

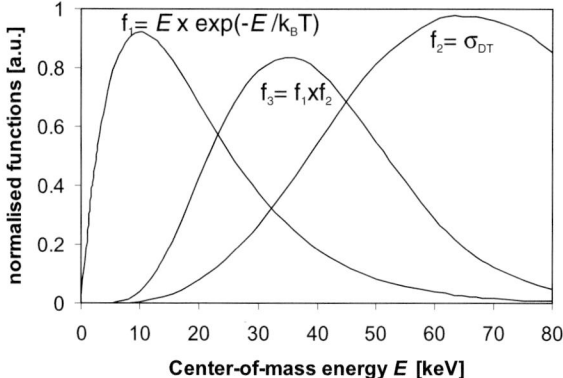

Fig. 17.7. Graph of integrand of the reaction parameter equation and of its two factors $\sigma(E)$ and $E\exp(-E/k_BT)$ versus the center of mass energy E for a D-T plasma at T = 10 keV

17.5 Power Balances

In the following the condition which a thermal D-T plasma has to satisfy to serve as an energy producing system is investigated. Historically, in 1957 John D. Lawson deduced a criterion for a positive energy balance using quantities such as the thermal cycle efficiency η of power reactors [8]. Today the approach has changed slightly to a more physics oriented condition: the aim is an ignited plasma where all energy losses are compensated by the α-particles from the fusion reactions, which transfer their energy of 3.5 MeV to the plasma while slowing down. The neutrons cannot be confined and leave the plasma without interaction.

Transport processes such as diffusion, convection, charge exchange and others are empirically described by an energy confinement time τ_E leading to the power loss term (per volume unit) $3n_e k_B T/\tau_E$ with $3n_e k_B T$ as the inner thermal plasma energy (n_e is the electron density). Note that this is twice the ideal gas value since every hydrogen atom is split into two particles (electron and nucleus). Another loss mechanism is bremsstrahlung, which becomes particularly important at high temperatures and impurity concentrations. The power loss due to bremsstrahlung can be written as

$$P_{\text{bremsstrahlung}} = c_1 n_e^2 Z_{eff} (k_B T)^{1/2} \quad (17.19)$$

with c_1 being the Bremsstrahlung constant ($c_1 = 5.4 \times 10^{-37}$ W m^3 keV$^{-1/2}$), and Z_{eff} the effective charge of the plasma, including all (impurity) species: $Z_{eff} = \Sigma_z n_z Z^2/n$.

The energy balance now reads

$$\left(\frac{n}{2}\right)^2 \langle\sigma v\rangle E_\alpha = 3nk_BT/\tau_E + c_1 n^2 Z_{eff}(k_BT)^{1/2} \quad (17.20)$$

and this can be rewritten to the ignition condition

$$n\tau_E = \frac{12\, k_BT}{\langle\sigma v\rangle E_\alpha - 4c_1 Z_{eff}(k_BT)^{1/2}}, \quad (17.21)$$

where E_α is the energy of the α-particle, 3.54 MeV. This equation shows that the product of the particle density and energy confinement time is only a function of the plasma temperature, with a minimum at about 13 keV. In the range of 10 keV the reaction parameter $\langle\sigma v\rangle$ is roughly proportional to T^2, which motivated the definition of the so-called fusion product

$$n\tau_E T = \frac{12\, k_BT^2}{\langle\sigma v\rangle E_\alpha - 4c_1 Z_{eff}(k_BT)^{1/2}} \quad (17.22)$$

which has a flat minimum of about 35×10^{20} keV s/m^3 around 10 keV. The fusion product dictates the strategy for developing fusion power as an energy producing system: One has to attain temperatures of around 10 keV (about 100 million K) and achieve the required density and energy confinement time simultaneously. There are two distinct approaches:

1. In **Magnetic Confinement Fusion:** the energy confinement time is maximized. The hot plasma is confined by strong magnetic fields leading to maximum densities of about 1.5×10^{20} m^{-3}, which is 2×10^5 times smaller than the atom density of a gas under normal conditions. With these densities, the energy confinement time required is in the range of 2–4 seconds. Magnetic confinement of plasmas is the main line in fusion research and is discussed in detail in Chapt. 7 in Part II.
2. In **Inertial Confinement Fusion:** the density of the plasma is maximized. This can be done by strong, symmetric heating of a small D-T pellet. The heating can be done with lasers or particle beams and leads to ablation of some material causing implosion due to momentum conservation. It is clear that the energy confinement time is extremely short in this concept as it is given by the time required for the particles to leave the hot implosion center, i.e. the plasma sound speed. Since it is the mass inertia which causes the finiteness of this time, this approach to fusion is called "inertial fusion". The density required is about 1000 times the density of liquid D-T; the pressure in the implosion center reaches (at temperatures of 10 keV) that in the center of the Sun. The physics of this approach is well described in [9, 10], and a detailed description of the actual status in this area can be found in [11, 12]. Inertial fusion allows studying the very interesting physics of hot dense matter under extreme conditions and also offers the possibility to study in the laboratory conditions which are quite

similar to those in thermonuclear explosions. The efficiency of the implosion drivers (presently mostly lasers are used for this) and their repetition rate, the necessary accuracy of the targets and their position in the target chamber as well as the frequency of the pellet production, however, pose technical obstacles that make it seem unlikely that the first fusion power plants will be based upon this technique.

Figure 17.8 illustrates the progress of nuclear fusion research with magnetic confinement in approaching the required $n\tau_E T$ condition. Today a factor seven is missing for ignition, whereas in the mid-sixties the best experiments fell short of the required conditions by more than five orders of magnitude. However, it has to be kept in mind that achieving ignited plasmas is not sufficient for building fusion reactors. In addition, this plasma state has to be maintained stably for very long times to allow continuous energy production. One of the most difficult problems will be the interaction of the edge plasma with the surrounding structures and the removal of helium ash. Consequently, edge plasmas and plasma-wall interaction constitute an important research topic. Other important topics are related to plasma heating and fuelling.

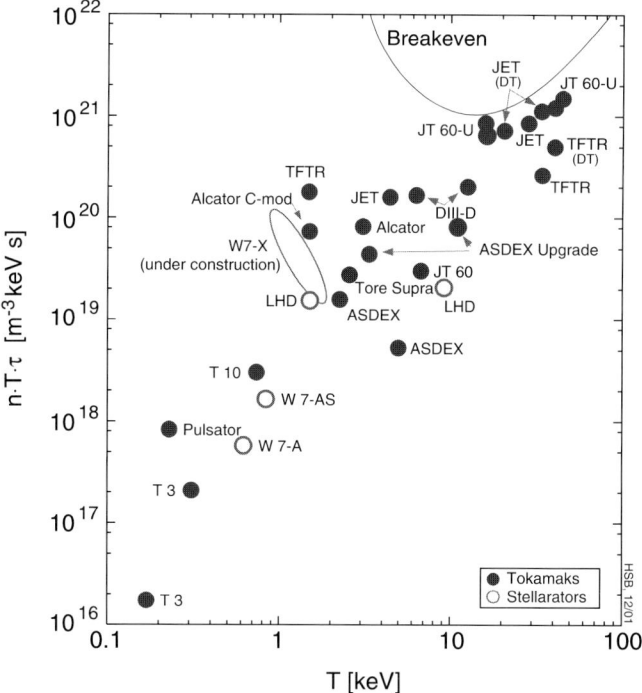

Fig. 17.8. Diagram of $nT\tau_E$ values versus T obtained in different magnetically confined fusion plasma experiments

17.6 Development of a Fusion Power Plant

As can be seen in Fig. 17.8, fusion research is close to achieving a power producing plasma. This is the goal of ITER (latin: the way), a tokamak experiment that has been developed since 1984 in a worldwide collaboration between Europe, Russia, Japan and the US. The ITER torus will have a large radius of 6.2 m, a minor radius of 2 m and a height of 6.8 m [13]. It will be equipped with 18 superconduction toroidal field coils and the goal is to produce a fusion power of about 500 MW for pulse lengths of up to 400 seconds. In summer 2001 a final design report was published and presently political discussions are in progress on if and where ITER should be constructed. If such a decision is taken soon, ITER could become operational in about 2016/7.

While ITER is planned to produce ten times the power that is needed to produce and to heat the plasma, this is not yet sufficient for a power plant. This next step after ITER will be taken by a device often generically dubbed DEMO, that presently exists only as a very general sketch. It will have to settle the transition from physics experiments like ITER to a commercial fusion power plant. In parallel to the physics investigations and the technical development in ITER, an accompanying fusion technology programm is required to develop low-activation structural materials and other technologies for DEMO. The basic geometry of fusion reactors using magnetically confined plasmas will be a torus (ring). The hot plasma is surrounded by the first wall and blanket. The latter is filled with lithium to produce the tritium and the majority of thermal energy of the plant is delivered here by neutron moderation. A shield is provided behind the blanket to stop the neutrons not captured by the blanket in order to reduce the heat and radiation loads to the cold structures of the superconducting magnets. The application of superconduction is mandatory for fusion reactors to obtain a positive energy balance. If the development of fusion power plants runs according to the present roadmap, DEMO could for the first time produce electrical power in about 2036, and the first commercial power plant could start operation in about 2050.

17.7 Muon-catalyzed Fusion

At the end of this basic chapter on nuclear fusion which finally aims for a new type of power plants, we want to dwell shortly on a more exotic way to produce nuclear fusion.

The issue in achieving nuclear fusion is to overcome the repulsive Coulomb forces of light nuclei. The "conventional" approach discussed above is strong heating to reach the ignition conditions as was described. Another, more exotic, way is to screen the electric charge by replacing the electron in the hydrogen atom with a muon [14]. This idea was first proposed by A.D. Sacharov in 1948. A muon is an elementary particle, a so-called lepton ("light" particle), with the same properties as an electron, the only difference being that its mass is 200 times larger. The muon is unstable and decays with a half-life of 1.5×10^{-6} seconds:

$$\mu^- \longrightarrow e^- + \bar{\nu}_e + \nu_\mu \,. \tag{17.23}$$

Muons can be produced in accelerators which provide collisions between accelerated protons and some other material. In these collisions many pions are produced. A pion is the lightest elementary particle participating in the strong interaction (so-called hadrons). The three types of pions (π^-, π^+ and π^0) are produced in equal amounts. A negative muon is formed from the decay of a negative pion:

$$\pi^- \longrightarrow \mu^- + \bar{\nu}_\mu \,. \tag{17.24}$$

In a D-T mixture the muon slows down very fast ($\approx 10^{-9}$ s) and forms a $D\mu$ or $T\mu$ atom with a small Bohr radius of

$$a_\mu = a_e \, m_e/m_\mu \approx 2.5 \times 10^{-13} \text{ m} \,. \tag{17.25}$$

This reduced atomic radius is the key point for possible catalyzed nuclear fusion: If a $D\mu T$ molecule is formed, it takes just 10^{-12} s until quantum mechanical tunnelling triggers a fusion reaction:

$$D\mu T \longrightarrow {}^4\text{He} + n + \mu^- \quad + 17.6 \text{ MeV} \,. \tag{17.26}$$

There are two limiting factors in this approach: the first is the time needed to form a $D\mu T$ molecule. This time can be influenced by some resonance mechanisms. The second limitation comes from the 0.6% probability that the muon will stick to the helium atom after the fusion reaction and will thus be lost for catalyzing more fusion reactions during its lifetime. Up to now, experiments obtained fusion rates of more than 100 reactions per muon. For a pure energy producing system, however, this rate is not sufficient, since about 5 GeV ($= 5 \times 10^9$ eV) is needed to produce one negative muon. Taking also the efficiency for converting fusion power into electrical power into account, however, about 3000 fusion reactions per muon would be required [10]. This means, that muon-catalyzed fusion is a very interesting principle to enhance fusion reactions at room temperature, and deserves further research, but it is not the direct route to a commercial fusion power plant.

References

1. G. Gamov: Z. Phys. **51**, 204 (1928)
2. A.C. Phillips: *The Physics of Stars*, 2nd edn (Wiley, Manchester 1999)
3. E.M. Burbidge, G.R. Burbidge, W.A. Fowler et al: Rev. Mod. Phys. **29**, 547 (1957)

4. E.G. Adelberger, S.M. Austin, J.N. Bahcall et al: Rev. Mod. Phys. **70**, 1265 (1998)
5. GALLEX Collaboration, P. Anselmann, W. Hampel et al: Phys. Lett. B **285**, 376 (1992)
6. M. Kaufmann: *Plasmaphysik und Fusionsforschung* (B.G. Teubner, Stuttgart 2003)
7. H.-S. Bosch, G.M. Hale: Nucl. Fusion **32**, 611 (1992)
8. J.D. Lawson: Proc. Phys. Soc. B **70**, 6 (1967)
9. W.J. Hogan, R. Bangerter, G.L. Kulcinski: Physics Today **Sep**, 42 (1992)
10. S. Atzeni, J. Meyer-ter-Vehn: *The Physics of Inertial Fusion* (Clarendon Press, Oxford 2004)
11. J.D. Lindl, R.L. McCrory, E.M. Campbell: Physics Today **Sep**, 32 (1992)
12. J.D. Lindl, P. Amendt, R.L. Berger et al: Phys. Plasmas **11**, 339 (2004)
13. R. Aymar: Fus. Eng. Design **55**, 107 (2001)
14. S.E. Jones: Nature **231**, 127 (1986)

18 The Possible Role of Nuclear Fusion in the 21st Century

T. Hamacher

Abstract. The possible role of fusion as a future energy source is examined. These considerations have to link physical issues and socio-economic aspects.

18.1 Introduction

The 19th and 20th centuries saw an unprecedented increase in population, economic activity and energy consumption. The distance people travelled – walking excluded – increased from 1–40 km per day [1] in France between 1900 and 1990. The primary energy consumption increased worldwide tenfold. Life expectancy increased not only in industrialized countries, but in the whole world working hours decreased leaving considerable time to leisure activities. Technological innovations were a major driver for these developments and energy and energy conversion technologies played a crucial role. This can well be exemplified by the innovation waves, first the steam turbine utilizing coal, marking the starting point of industrialization, then the introduction of electricity in the second half of the 19th century, which diffused deep into industrial and household applications and then in the beginning of the 20th century the revolution of the individual mobility by cars using oil. But even in the last decade of the 20th century, which is often considered as the beginning of the information society, the energy demand increased in the European Union by more than ten percent, while the increase in population was less then four percent.

Figure 18.1 shows the increase in global primary energy consumption in the last 150 years. A steady increase was mainly managed by a diversification of the energy sources. First coal was added to the traditional biomass, than oil and late in the 20th century natural gas started to play an important role. In the complete 20th century a total of $14\,620 \times 10^{18}$ J ($=14\,620$ EJ) of primary energy were consumed, most of it supplied by coal, followed by oil and natural gas. Hydro power was present from the very beginning of industrialisation. In some world regions e.g. South America hydropower covers large fraction of the electricity system. Nuclear energy is present only in a selected number of countries, but as exemplified in France, nuclear has the potential to dominate the electricity sector completely. Nevertheless is the absolute contribution of

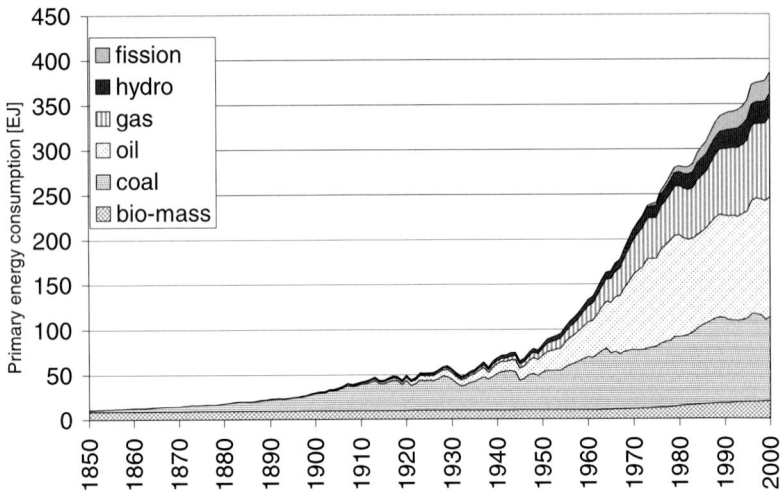

Fig. 18.1. The development of the global energy consumption in the last 150 years [1, 2]

nuclear to the total energy supply of the 20th century with 2.6% marginal. It takes decades before a new technology shows impact on the global system.

Figure 18.2 emphasizes another very important fact, the global energy system is still in its fossil area and will stay there at least three to four more decades.

There is another important issue: The community of the states has emphasized in numerous resolutions and summits that the concept of *sustainable development* should become the guideline for future international and national politics. The term *sustainable development* has suffered by numerous and partly contradictory definitions. Still it is obvious that the energy system poses the biggest challenge to *sustainable development* by its very nature. While all material cycles can at least in principle be closed energy can always only be used once and is for ever degraded by its use. If there is a sector which deserves special attention in respect to sustainable development, then it is the energy sector.

18.2 The Challenges

Fossil energy carriers need to be replaced for various reasons beginning from local air pollution over climate change issues up-to rather dangerous geopolitical frictions. The range of the future development is huge. Cultural, political, economic structures that might evolve can not be foreseen. Pictures of the future range from a prosperous globalized world, with vanishing disparities

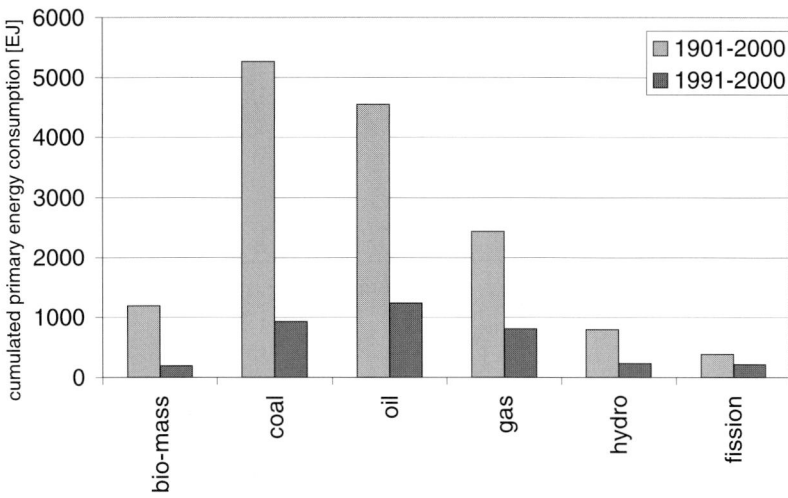

Fig. 18.2. Contributions to the energy supply in the last 100 and last 10 years. Fossil fuels dominated not only the whole century but also the last decade. The impact of new renewable technologies like wind or solar are invisible on the scale depicted

between world regions, up to a very heterogeneous regionalised world, with very different development patterns. Discussions of the future need to capture these uncertainties, either be allowing for very different development patterns or by identifying extreme directions of the development that might pose the most demanding challenge. This does of course not mean that an insurance for every risk can be developed, but certainly it is necessary to prepare for a rich family of possible developments. The willingness of a society to invest in the future and even a future which goes beyond the living generations should be equivalent to the burdens these generations will inherit. As an example: a good fraction of the CO_2 emissions of the year 2004 will still be in the atmosphere in 2100. Therefore it is well justified that people in 2004 invest in technologies which come only to a real application 2100.

18.2.1 Energy Demand and Lifestyle

Life styles and consumer habits are not constants over time, in contrast they are subject to constant change. A very prominent example is the change of consumer behaviour in the US in the years between 1909 and 1929. The expenses for private consumption nearly tripled in these years. This change went along with new family values, with a new role of women and so forth. The energy demand of a society correlates with the life style. Distance and mode of transport, kind and size of houses, food consumption, fraction of meat etc.

If we consider behaviours and life styles being open to political guidance and influence than it seems simple to reduce the energy demand. Life style concepts like the one proposed and practised by M. Ghandi would only require a small fraction of the energy demand observed in OECD countries. Actually a large fraction of the human population is forced to be "happy" with a considerable low per capita energy consumption. But as can be observed nearly everywhere in the world, affluent people consume energy. The disparity does not only sustain between people of different countries, it also sustains within countries. Inhabitants of the European community fly in average not more than 756 km per by plane within Europe per year [3]. Certainly affluent Europeans fly a multiple of this.

Some simple estimate should visualize what range of future energy demand might evolve in the next hundred years. Assuming that in 2100 9 billion people populate the Earth and that the per capita demand is like that of an average EU or Japanese citizen of the year 2000. This would lead to a fourfold increase in the global energy consumption. If we assume the per capita energy demand of a US citizen in the year 2000 the increase would be even eightfold. An eightfold increase corresponds to the increase in energy consumption in the last hundred years. Later it will become more obvious what it will mean to realise an increase of that order of magnitude.

18.2.2 Efficient Use of Energy and Energy Saving

An increase in energy efficiency is necessary to keep the energy growth rates moderate, but from all historic trends it seems unfeasible to reverse the trend so strongly that the whole increase in economic activity could be managed without an increase in energy consumption.

18.2.3 Energy Resources

The discussion and debate about energy resources is as old as industrialization. The most prominent example is the book "Limits to growth" [4] in the early seventies of the 20th century, which claimed that energy resources will become scarce within the next 30 years. Several institutes work out regularly surveys of the most prominent energy resources, prominent examples are the US Geological Survey in the US and the Bundesanstalt für Geowissenschaften und Rohstoffe in Germany. Exact use of terminology is important in this context. Simply speaking the term "reserve" and "resource" is distinguished. Reserve means those stocks of energy carriers which are well proven and can be mined economically with existing technology. Resources are those stocks which are only proven with less certainty and which can either not be mined with current technology or not at competitive costs. The major findings can be summarized as follows. Carbon resources are still huge, especially if unconventional resources like hydrates are counted. But production of conventional oil will peak within the next fifty years. Especially the later is an important message. Thus oil, driving especially an ever increasing transport sector, needs to be replaced.

Beside the "non renewable" energy sources a number of so called renewable energy sources exist. These are the flows of energy in the environment like solar radiation and wind energy. The actual size of these flows is tremendous, they outweigh the human energy demand by orders of magnitude.

18.2.4 Geopolitical Frictions

Oil and gas are not only limited in absolute amount, their most abundant appearances is concentrated at a few places in the world, especially in the Middle East, the Caspian sea and Russia. The import dependence of the industrialized countries is expected to increase strongly in the next decades. The EU Green Book [5] expects an increase from 50–70% import dependence for energy in the next 20 years. Given that in the same time period countries like China and India will also import large amounts of oil and gas and so will the US, it is likely that the already unstable region of the Middle East will become even more unstable and more subject to foreign influence [6].

18.2.5 Environmental Damages

The shear number of humans and the largely increased technological potential of men have changed the qualitative impact of humans on the global environment. Until the beginning of the industrial revolution men had already the capability to change a region completely, but on a global scale the impacts were marginal. Now men starts to alter the global environment [7]. The most prominent example is the CO_2 emission, which will if, no major contra measures are taken, lead to at least to a doubling of the CO_2 concentration in the global atmosphere. This will change the global climate drastically at rates of change, which were hardly ever observed in climate history. Another example are the material flows. The movement of materials by human activities comes now close to the material cycles observed on nature by erosion and so forth.

If concentrations of CO_2 should not exceed levels of more than 550 ppm, which seems to be a realistic target, than not more than (870–990 GtC) should be emitted in the time period between 1991–2100. The current level of annual emissions is 6.4 GtC. This means that actually only a small fraction of the expected increase in energy demand can be satisfied by fossil fuels. In India alone an increase of a factor ten in electricity demand is expected in the next 100 years (see Fig. 18.3). Heavy investment in coal technologies is prohibited in such a scenario anyhow, due to the high specific CO_2 emissions of coal, if not CO_2 sequestration becomes an established technology.

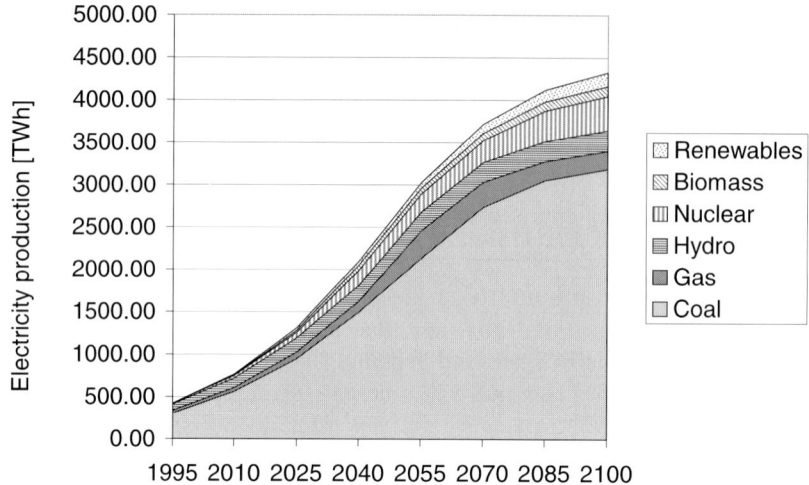

Fig. 18.3. Development of the Indian electricity supply, coal would be the mainstay of the supply even in the year 2100

18.2.6 Possible Supply Options

The supply options should be an adequate answer to the mentioned challenges, but they have of course also to be competitive from an economic point of view.

Fossil fuels are limited by two major constraints:

1. only a limited amount of fossil fuels is available and
2. unlimited emissions of CO_2 of the carbon resources would lead to dramatic changes of the global climate.

Both issues were already mentioned. Only if unconventional fossil fuel resources like methane hydrates can be mined and only if carbon can be sequestered in an economic and safe fashion, only then fossil fuels will have a chance to contribute substantially to a sustainable energy system even beyond the 21st century. Ideas how such an carbon-free fossil energy system could look like do exist and they are followed rather seriously especially in the US.

Harvesting energy streams in the environment is another option, which is widely discussed. Solar, wind, hydro, biomass, geothermal and ocean power seem to offer on a first look an inexhaustible stream of power and energy. Just the solar radiation reaching the Earth surface is more than 6000 times bigger than the annual primary energy demand. Still the power density, the geographical variation, the daily, seasonal and stochastic variations pose severe problems to mine these energies properly and economically. A simple estimate should visualize the order of magnitude of the problem. The end-energy use in

Germany accounted in 2000 for 9195 PJ. If we assume that this would solely be satisfied by direct solar radiation a space of roughly $140\,\text{m}^2$ per person of active collector space would be necessary. The area covered by collectors would be as the area covered by roads and other transport infrastructure.

Nuclear energy as applied today faces in very general terms three major challenges:

1. resources with high uranium concentrations are limited,
2. major accidents although rather unlikely can not be excluded deterministically and
3. the radioactive waste passes burdens to future generations.

In principle all of these challenges can be solved by technical means, although a complete system which solves all problems only exists on paper and needs decades of intense research and development (R&D). A number of the industrialized countries did form a new nuclear initiative to develop these technologies, which is called Generation IV.

Without going into any detail, it can be concluded that none of the options is completely ready and applicable by now and that none of the problems is without major obstacles and identified problems.

18.3 Characteristics of Nuclear Fusion as Power Source

The basics of nuclear fusion are described in this volume elsewhere and need not to be repeated here (see also Chap. 17 in Part III).

18.3.1 Overall Design of a Fusion Power Plant

Various features of a fusion plant such as steam turbine and current generator will be the same as in conventional nuclear or fossil-fuelled power plants. A flow chart of the energy and material flows in a fusion plant is shown in Fig. 18.4.

The fuel – deuterium and tritium – is injected into the plasma in the form of a frozen pellet so that it will penetrate deeply into the center. The neutrons leave the plasma and are stopped in the so-called blankets which are modules surrounding the plasma. The neutrons deposit all their kinetic energy as heat in the blanket. The blankets also contain lithium in order to breed fresh supplies of tritium via a nuclear reaction. The "ash" of the fusion reaction – helium – is removed via the divertor. This is the section of the containing vessel where the particles leaving the plasma hit the outer wall. The outer magnetic field lines of the tokamak are especially shaped so that they intersect the wall at special places, namely the divertor plates. Only a small fraction of the fuel is "burnt" so that deuterium and tritium are also found in the "exhaust" and can be re-cycled. The tritium produced in the

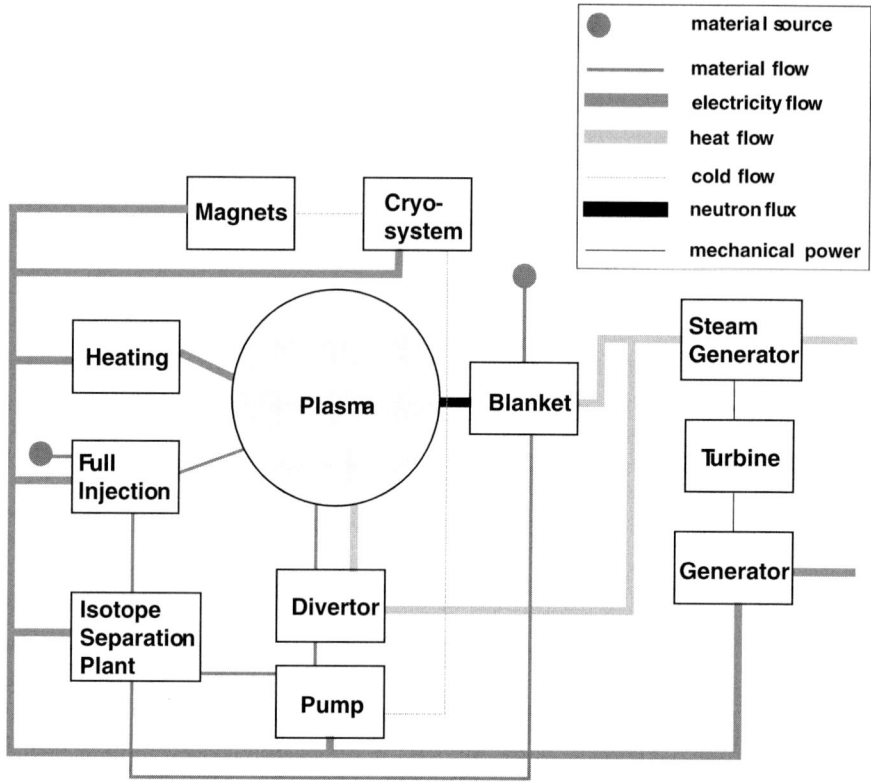

Fig. 18.4. Flow chart of a future fusion reactor

blankets is extracted with a flushing gas – most likely helium – and delivered to the fuel cycle.

The heat produced in the blanket and the divertor is transported via water or helium to the steam generator and used to produce electricity to feed to the grid. A small fraction is used to supply electricity to the various components in the plant itself. Electrical power is required mainly for the cryo-system which produces low temperature helium for the super-conducting magnets, the current in the magnets, the current drive and the plasma heating systems.

The reactor core is arranged in different layers like an onion. The inner region is the plasma, surrounded by first wall and blanket. All this is contained in the vacuum vessel. Outside the vacuum vessel are the coils for the magnetic field. Since the magnets operate at very low temperatures (superconductors), the whole core is inside a cryostat. Two different magnetic confinement schemes are under development: the tokamak and the stellarator. Beside the magnetic confinement schemes so called inertial confinement schemes are exist.

18.3.2 Resources

One of the main motivations from the very beginning of fusion research has been that fusion can be considered as a practically unlimited source of energy. The argument is based on the abundance of the fusion fuels – lithium and deuterium – and the very small quantities required [8]. A 1 GW electric (GWe) fusion power plant would require annually 110 kg deuterium and 380 kg lithium consumption.

Deuterium is a hydrogen isotope. In terrestrial hydrogen sources, such as sea water, deuterium makes up one part in 6 700. Given the above annual consumption rates it can be shown that fusion could continue to supply energy for many millions of years. The oceans have a total mass of 1.4×10^{21} kg and therefore contain 4.6×10^{16} kg of deuterium; moreover, there is already a mature technology for extracting the deuterium. One of the main applications is the production of heavy water for heavy water-moderated fission reactors. Existing plants can produce up to $250 \, \text{t} \, \text{a}^{-1}$ of heavy water which means a production of $50 \, \text{t} \, \text{a}^{-1}$ of deuterium. This would be enough to supply deuterium for 500 fusion plants each with 1 GWe capacity. Obviously deuterium supply places no burden on the extensive use of fusion. What about tritium? As we have mentioned above, tritium, also a hydrogen isotope, will be bred from lithium using the high flux of fusion neutrons. Lithium is found in nature in two different isotopes ^{6}Li (7.4%) and ^{7}Li (92.6%). The two nuclear reactions

$$^6\text{Li} + \text{n} \rightarrow \text{T} + ^4\text{He} + 4.8 \, \text{MeV} \tag{18.1}$$

$$^7\text{Li} + \text{n} \rightarrow \text{T} + ^4\text{He} + \text{n} - 2.5 \, \text{MeV} \tag{18.2}$$

are relevant. Since the second reaction is endothermic only neutrons with an energy higher than the threshold can initiate this process. In most blanket concepts the reaction with ^{6}Li dominates, but in order to reach a breeding ratio exceeding unity the ^{7}Li content might be essential.

Lithium can be found in:

- **salt brines**: in concentrations ranging from 0.015–0.2%,
- **minerals**: spodumene, petalite, eucryptotite, amblygonite, lepidolite, where concentration varies between 0.6–2.1% and
- **sea water**: the concentration in sea water is $0.173 \, \text{mg} \, \text{l}^{-1}$ (Li+).

The land-based reserves are given in Table 18.1 according to two different sources.

While the annual consumption of lithium in a fusion plant is low, the lithium inventories in the blankets are much larger [9, 10]. At least a couple of hundred tons of lithium are necessary to build a blanket. It is expected that most of the lithium can be recovered and re-used, although radioactive impurities such as tritium will complicate the handling. No detailed concept for recovering lithium has been developed so far. The lithium supply is, however, a minor problem in the context of the construction of the whole plant:

Table 18.1. Land reserves of lithium

Material	Present Production	Reserve [18]	Reserve Base [18]	Reserve [19]
Lithium	15 000 t	34 000 000 t	9 400 000 t	1 106 000 t

lithium can be purchased today for around 17 € kg^{-1} and the blanket containing 146 t of lithium needs to be replaced five times in the life of a fusion plant, which would amount to only 12 M€. Beside the land-based resources there is a total amount of 2.24×10^{11} t lithium in sea water. Techniques to extract lithium from sea water have already been investigated [11]. The associated energy consumption has also been investigated [12]. The ultimate lithium resources in sea water are thus practically unlimited. Besides fuel numerous other materials will be necessary in order to construct and operate a fusion power plant [9, 13]. The energy necessary to produce, transport and manufacture all the materials to build a fusion plant add up, in a conservative model, to 3.15 TWh [14, 15]. The energy pay back time, the time necessary for the plant to deliver the same amount of energy necessary for its construction, is roughly half a year and thus comparable with conventional power plants.

18.3.3 Environmental and Safety Characteristics, External Costs

A fusion power plant is a nuclear device with large inventories of radioactive materials. The safe confinement of these inventories and the minimization of releases during normal operation, possible accidents, decommissioning and storage of waste are major objectives in the fusion power plant design. Besides tritium the other source of the radioactivity in the plant is the intense flux of fusion neutrons penetrating into the material surrounding the plasma and causing "activation".

Three confinement barriers are foreseen: vacuum vessel, cryostat and outer building. Small fractions of the radioactive materials are released during normal operation. The amounts depend strongly on design characteristics such as cooling medium, choice of structural materials and blanket design. The expected doses to the public stay well below internationally recommended limits [16].

Detailed accident analyses have been performed within the framework of system studies [17] and in even more detail for ITER [18]. Although ITER is not in all aspects comparable to a later power plant many of the characteristics are similar. Different methods (bottom-up and top-down) are applied to guarantee a complete list of the accident sequences. Reactivity excursions are for several reasons not possible in a fusion power plant. Therefore, the most severe accidents are all related to failures of the cooling system. These failures can be caused either by power failures or ruptures in cooling pipes or both. As an example of one of the most severe accident sequences, a total loss of coolant accident, should be described. Shortly after the accident the fusion reaction will come to a halt. This happens because the walls surrounding the plasma are no longer cooled and their temperature increases. Impurities, evaporated from the hot walls, enter the plasma. The larger impurity content in the plasma disturbs its energy balance and more energy is radiated, thus cooling down the plasma. Fusion reactions are extinguished. With no more fusion reactions, only the decay heat of the activation products in the structural materials and the blanket produce heat. Detailed calculations show that the heat produced will be dissipated by heat radiation to the inner walls of the cryostat. Temperatures in the structural materials will stay well below the melting temperature and keep the confinement barriers intact. During such an accident sequence not more than 1 PBq of tritium would be released. Doses for the population would stay in the range of 1 mSv [17, 19], which is of the same order as the exposition due to natural radiation, like cosmic rays.

As a worst case scenario it was assumed that the complete vulnerable tritium inventory (roughly 1 kg) of the fusion plant is released at ground level. The initiator of such an accident could only be very energetic outside events such as an aeroplane crash on the plant. Even if the worst weather conditions are assumed, only a very small area, most likely within the perimeter of the site (around 1–4 km^2), would have to be evacuated [19].

All the radioactive material produced in a fusion plant is neutron-induced. A detailed analysis of the amount and composition of the fusion power waste was performed in [17, 19]. The time evolution of the radiotoxicity of the waste is shown in Fig. 18.5. The plant model assumed is based on available materials. The picture shows a rapid decrease in radiotoxicity once the plant is shut down. The time evolution of the fusion waste is compared with the time evolution of the waste from a pressurised water reactor (PWR) fission plant and with the radiotoxicity of ash in a coal-fired power plant. The radiotoxicity of the waste of fission plants hardly changes on the time scale of a few hundred years and stays at a high level. Fusion approaches rapidly the radiotoxicity of the coal ash. It is a fair conclusion to say that the radiotoxicity of fusion waste does not place a major burden on future generations.

The impact on the population is rather low. Doses below $60\,\mu\text{Sv}\,\text{a}^{-1}$ are expected in the case the fusion waste stored in typical waste repositories like Konrad in Germany or SFR or SFL in Sweden. The value represents a rather conservative estimate.

Comparisons between competing technologies on an economic basis is mainly based on cost arguments. Comparison on the basis of environmental performance or safety issues is often more interesting. It is tempting therefore to look for a scale which also covers these aspects. One promising approach in this direction is the concept of "external costs" or "externalities" [20]. All the damage and problems not contributing to the market price are reflected in the

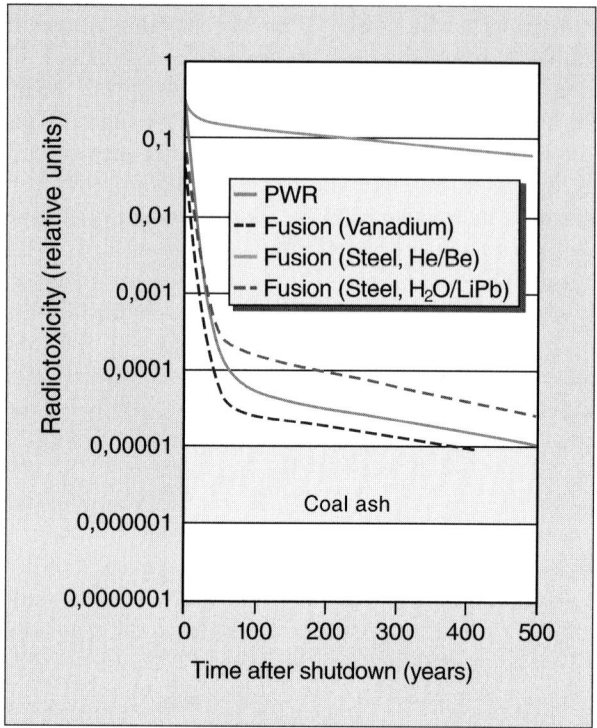

Fig. 18.5. Development of radiotoxicity for a fusion plant, a fission plant and the ash of a coal plant. It is assumed that all the plants produce the same quantity of electricity. The volume of coal ash is of course 2–3 orders of magnitude greater than that of fusion or fission waste

"external costs" which are normally borne by society as a whole. Examples of externalities are the damages to public health, to agriculture or to the ecosystem.

A methodology for the assessment of the environmental externalities of the fusion fuel cycle has been developed within the ExternE project [21]. The method used is a bottom up, site specific and marginal approach, i.e. it considers extra effects due to a new activity at the site studied. Quantification of impacts is achieved through damage functions or impact pathway analysis. The whole fuel and life cycle of the plant is considered.

The hypothetical plant under investigation is sited at Lauffen in Germany on the river Neckar. Two different fusion plant models are considered. Most characteristics of these models are taken from the European fusion safety study SEAFP [17]. The first model utilises martensitic steel for the structural materials and helium as coolant (Model 3, see Fig. 18.6). The second model has a water-cooled blanket and martensitic steel as structural material

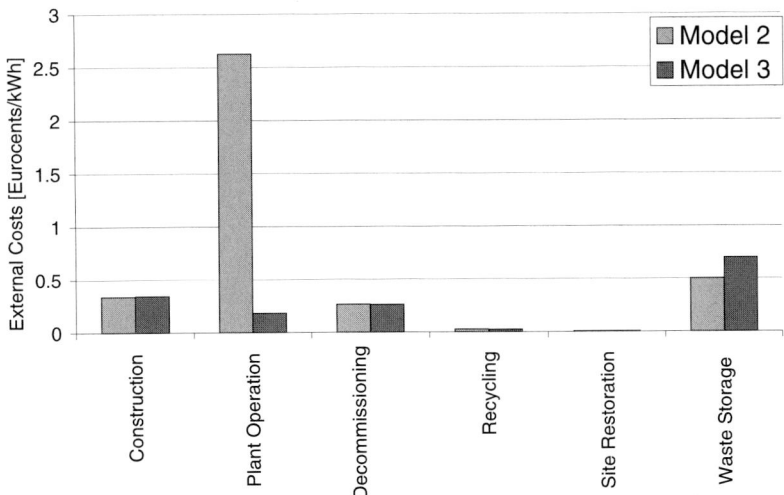

Fig. 18.6. External costs of fusion [22]

(Model 2). The parts of the plant not included in the above-mentioned study are taken from the ITER design and from data for a fission plant.

The results shown in Fig. 18.6 indicate that the external costs of fusion do not exceed those of renewable energy sources [22]. A major factor in the external costs of plant Model 2 are the ^{14}C isotopes released during normal operation which enter the world-wide carbon cycle. Nevertheless, the individual doses related to these emissions are orders of magnitude below the natural background radiation. For all models a considerable fraction of the external costs is due to material manufacturing, occupational accidents during construction and decommissioning.

18.3.4 Economic Consideration

Basis for the following cost estimates of fusion power is a plant of 1 GWe capacity based on the tokamak concept. Conceptually the plant can be divided up between the fusion core – the heat source – and the rest consisting of turbines, generators, switchboards. The assumptions in the underlying physics and technology seem well within the reach of current achievements. If progress in fusion technology is faster, it might of course lead to considerably lower costs.

Most of the components of the fusion power core are unique for fusion. The basis for the cost estimates of these components is

- existing experience with operating fusion experiments,
- the experience with designing ITER [23] and
- numerous system studies.

The ITER experience is of particular importance because it combines system studies and real manufacturing experience. As mentioned earlier, part of the ITER activities to date have been the design, construction and testing of central components of the experiment. The following discussion is based on [24, 25, 26].

Magnets make up 30% of the investment costs of the fusion core for a prototype and another big item are the buildings. The rest splits up into numerous items. Blanket and divertor make up 14% and 3%, respectively, although these items will have to be replaced regularly. The divertor will be replaced every second year, the blanket every fifth year. Two possible technological developments should be mentioned which might lead in the long run to cost reductions. The magnetic field has to balance the pressure of the plasma (see Chap. 7 in Part II). For specific physical reasons, however, the magnetic pressure needs to be much higher than the plasma pressure in current installations. Progress in plasma physics could reduce this ratio in future and thus reduce the size and cost of the magnets. Also a lower replacement frequency of blanket and divertor due to the development of advanced materials might lead to a further reduction.

Cost of electricity (COE) is the sum of the capital costs for the fusion core (39%) and the rest of the plant (23%), the costs for the replacement of divertor and blanket during operation (30%), fuel, operation, maintenance and decommissioning (8%). An annual load factor of 75%, an operating lifetime of 30 years and an interest rate (corrected for inflation) of 5% are assumed. The investment costs for DEMO, which is expected to be the first experimental demonstration fusion power plant, are expected to be roughly $10\,000\,€\,kW^{-1}$ (1995) [24] giving an expected COE of $165\,m€\,kWh^{-1}$.

Collective construction and operation experience are expected to lead to considerable cost reduction due to accumulated learning processes [27]. Learning curves describe the correlation between the cost reductions and the cumulated installed capacity. The slope of the curve – the so-called progress ratio – gives the cost reduction for a doubling of the capacity. A progress ratio of 0.8 is assumed for the novel components in the fusion core. This ratio is well within the values generally experienced in industry; possible physics progress is also included.

Further cost reductions can be achieved by scaling up the plant size or by siting two or more plants at the same site. When fusion is a mature and proven technology in 2100, costs are expected to be in the range described in Table 18.2.

Studies performed in the US and in Japan arrive at even lower investment and electricity costs [28]. The underlying assumptions do not violate any physical principles but assume considerable progress in technology.

Table 18.2. Estimated cost of energy (COE) for fusion power plants

Plant Capacity (GW)	Number Plants	Study	COE (cents kWh^{-1})
1.0	1	Knight [26]	9.6
1.0	1	Knight [26]	7.1
1.0	1	Gilli [24]	8.7
1.5	2	Gilli [24]	6.7

18.4 The Possible Role of Fusion in a Future Energy System

18.4.1 The Global Dimension

What is the possible impact of fusion on future energy systems? What role could fusion play to mitigate greenhouse gas (GHG) emissions? First, a general answer can be given which reflects well-known patterns of technological change. (For a review see [1].) Technological change is described by two phases, the first being that of invention. In case of fusion, invention would be the point in time when the first commercial power plant goes into operation. The second phase, in which numerous power plants would be constructed in many different places, is represented by the time of diffusion. It usually follows very general patterns, which can be described by an S-shaped curve, starting with a smooth increase in market share, followed by a robust growth and finally a smooth approach to a saturation level.

The "market" share of different primary energy sources in the past 150 years has always developed according to this pattern. In the nineteenth century wood was replaced by coal. In the first half of the 20th century oil started to replace coal and now natural gas begins to replace oil. Extrapolation of the current trend would mean that gas would become the most important primary energy carrier in the first half of the 21st century [29]. This would mean that fusion can only hold a considerable market share by the end of this century since the invention phase is expected to happen around 2050. Therefore fusion can not play a role as greenhouse gas mitigation technology before that time. Second, it means that even without further incentives the primary energy carrier natural gas, which has a specific lower CO_2 emission than coal and oil and which can be converted at least to electricity with very high efficiencies (nearly 60% today, roughly 70% in the foreseeable future), would in any case lead to a specific reduction of greenhouse gas emissions. In comparison with coal this combined advantage would produce roughly a factor of three lower CO_2 emissions per kWh delivered. If all coal-fired plants were to be replaced by very efficient gas-fired plants the electricity demand could triple without increase in emissions. Third, the time when the share of natural gas will pass its maximum roughly coincides with the "invention" (the technological and economic proof of principle) of fusion.

Another very important point is of course the future development of energy usage and, in particular, the electricity demand. Scenarios made by the International Institute of Applied System Analysis (IIASA) and the World Energy Council (WEC) describe various possible paths into the future [30]. Of these scenarios labelled A, B and C, A is a high growth scenario, B an average growth scenario and C an ecologically driven scenario. Even in the C scenario electricity consumption will increase considerably even after 2050, leaving enough space for fusion, even without replacing older technologies. It must be noted that, given the long lead-time, alternative low-GHG electricity generating techniques might compete for the same potential market as fusion. While predicting winners or losers is obviously a very long shot, continued R&D is an absolute necessity for all of them.

18.4.2 Fusion in Western Europe

In the framework of socio-economic studies on fusion (SERF), which have been conducted by the European Commission and the Fusion Associations, a study was carried out on the possible impact of fusion on the future West-European energy market, on the assumption that fusion is commercially available in the year 2050. The scenario horizon is based on the complete 21st century. The scenarios were performed with the programme package MARKAL [31]. MARKAL is a bottom-up technology energy model. All energy flows and transformation processes are represented in the model. The dynamics of the future development is described by a cost-minimisation process. All costs, investment, variable and fixed costs are summed in one objective function. Future costs are discounted. Details of the analysis can be found in [24].

Two different scenarios were explored which differ in the discount rates, level of energy demand, availability of fossil fuels and energy price projections.

1. The first scenario is called **Market Drive**: interest rates on power generation investments are 8%, interest rates on end-use investments are higher. 15% of the world resources of fossil fuels are available to Western Europe and a rapid increase in the oil price is expected.
2. The second scenario is called **Rational Perspective**: discount rates are 5% across the whole energy sector, but only 10.5% of the world fossil fuel resources are available to Western Europe. The oil price increases more slowly.

Energy demand is higher in scenario Market Drive. Both scenarios assume that the capacity of nuclear fission never exceeds the current level. Fission is expected to phase out at 2100.

The demand for energy increases in both scenarios. In Market Drive it more than doubles in relation to the 1990 value and in Rationale Perspective it increases by more than 50%. Steady increases in efficiency keep the overall primary energy demand roughly constant over the whole scenario horizon. The demand for electrical energy increases in both scenarios roughly by a factor two.

The development of energy supply and conversion technologies, especially further progress in economic performance and efficiencies, is based upon detailed assessments of the literature and on the studies by Fusion Associations. The increase in efficiency or the decrease in costs are time-dependent. A detailed description of the supply technologies can be found in [32]. Another important point is the future development of fuel prices. An increase in the oil price to $25\,\$\,\text{bbl}^{-1}$ (Rational Perspective) or $29.5\,\$\,\text{bbl}^{-1}$ (Market Drive) in 2100 is expected. This prices based on estimates to recover unconventional oil resources and should not be compound to the daily market prices. The gas price is strongly tied to the oil price. The price for hard coal is assumed to be flat over the whole period investigated. In both scenarios neither new renewables nor fusion will win considerable market shares until the year 2100. Fossil fuels remain the most important primary energy sources. Two shifts in the use of fossil fuels can be identified. The use of gas increases considerably until the middle of the 21st century when the easily accessible natural gas reserves are exhausted and its price has substantially increased. Coal will then win again a market share and advance to the most important primary energy carrier at the end of the 21st century. The picture changes drastically, however, if future CO_2 emissions are to be restricted in order to reduce the risk of climate changes. These cases are constructed in such a way that the global emissions would lead in the long term to a stabilization of the CO_2 concentration in the atmosphere. Different values for the stabilization concentration are assumed. Western Europe would be allowed to produce 10% of the total global emissions. The time-dependent allowed emissions are constraints in the energy model. If these constraints are applied to the scenarios, the energy mix changes considerable. The share of the electricity supply technologies in 2100 is shown in Fig. 18.7. Fusion and new renewables such as wind and solar win considerable market shares. The conclusion can be summarized as follows: fusion can win shares in the electricity market if

1. the further use of fission is limited and
2. if greenhouse gas emissions are constrained.

Similar studies have been performed in Japan [33] and the US [34].

18.4.3 Fusion in India

The development of the global energy system will strongly depend on the development in Asia and particularly in China and India. Both countries cover roughly one third of the world population. In the last decades both countries showed a very dynamic development of their economies with quite considerable growth rates. If this development is sustained both countries

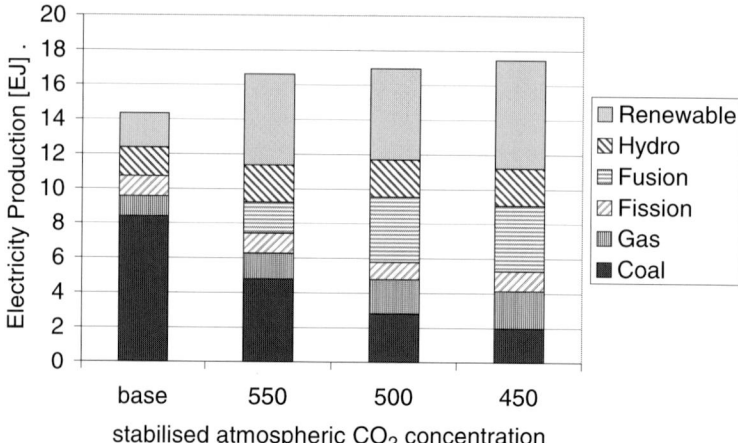

Fig. 18.7. The possible role of fusion in 2100 in the European electricity market [24]. Four different cases were calculated, which correspond to more strict regulations of CO_2 emissions

might become the biggest national economies in the world with huge local markets.

The dynamics in economic development is reflected in the development of the energy system. Both countries did undergo a substantial increase in primary energy consumption. In India the consumption increased by more than a factor of six in the last 35 years (see Fig. 18.8).

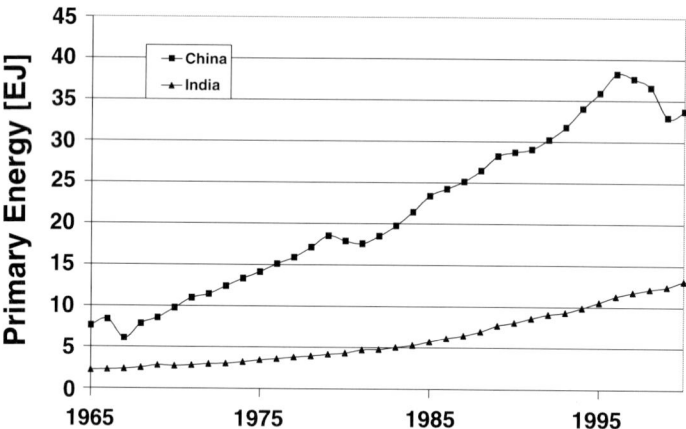

Fig. 18.8. The increase in primary energy consumption in India and China [2]. The increase in China since 1965 was more than fourfold, the increase in India more than sixfold

Special emphasis is given to the problem of global warming and the possibilities to restrict carbon emissions by implementing new technologies like nuclear fusion. The central software tool behind the investigations is again the MARKAL model generator as in the European case.

India had a per capita energy demand as low as 0.243 toe adult a^{-1} 1994. For comparison the per capita consumption of an average German was 4.097 toe adult a^{-1} and the one of an US citizen was 7.905 toe adult a^{-1} in 1994 [35]. The demand for goods and services will increase considerable. Although the demand increases, the per capita consumption level will stay even in 2100 below the average values of the industrialised countries in 2000.

The increase in demand is estimated with extrapolations which were calibrated by the historic development. The considerable increase in goods and services leads to a similar increase in electricity demand from roughly 500 TWh in 2000 to more than 4 000 TWh in 2100. For comparison the global electricity demand was 13 719 TWh in 2000 [36]. The per capita demand of electricity in India in 2100 would still only be half of the demand in Germany in the year 2000. This indicates that the estimations are certainly not unrealistic.

The only "cheap" energy resource, which is amply available in India, is coal. Coal prices are well competitive with world market prices, although the quality of the Indian coal is quite low. Indian and global coal resources are big enough to supply a sustained coal demand even after the year 2100. For the base case no major intervention from politics are assumed. The results of the MARKAL calculations are depicted in Fig. 18.3.

This energy system could by no means be considered sustainable. In particular it would lead to huge carbon emissions. The carbon emissions from the power sector account for roughly 45% of the overall emissions.

In the above base case fusion or other renewable energy sources played no or only a minor role. This picture changes dramatically if the amount of CO_2 emissions is drastically restricted. The assumption here is that the CO_2 concentration in the atmosphere should be stabilized at a level of 550 ppm, which is roughly twice the preindustrial level of 280 ppm. This can only be achieved if the global CO_2 emissions in the time period 1990–2100 do not exceed 980 GtC. The macro economic model calculations indicate that India would be allowed to make 7.5% of these emissions.

If this constraint is applied to the model the energy mix changes considerable. The role of coal diminishes, while a mixture of gas, renewable and nuclear technologies cover most of the supply. Fusion makes only an inroad to the system, if the capacity of fission is restricted exogenously. Under this assumptions fusion would cover 9.9% of the total electricity production or 429 TWh with 67 GW of installed fusion capacity (see Fig. 18.9).

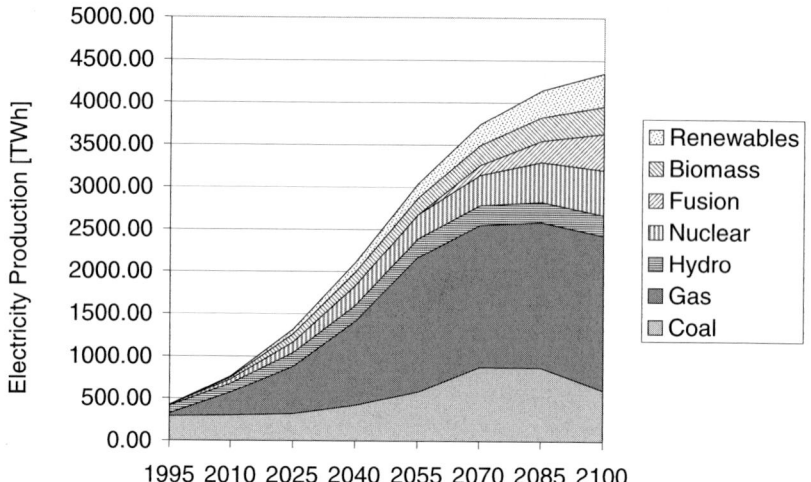

Fig. 18.9. The development of the Indian electricity supply system under very strict CO_2 emission constraints

18.5 Conclusion and Outlook

Considerable challenges are ahead of us to master the energy supply of the future without major political ruptures or environmental damages. The characteristics of fusion are well compatible with the goals of sustainable development: especially no greenhouse gas emissions during operation and no long-lived radioactive waste. This makes fusion to one of the most promising options for the future energy system. This system is expected to be dominated by fossil fuels well into the 21st century. Still environmental concerns and resource scarcity will drive the system away from the widespread use of fossil fuels. In two detailed case studies one for Europe and one for India the dynamics of the future development were investigated. Both studies showed similar patterns. In case no bounds on CO_2 emissions exist, coal will play a dominant role again at the end of the 21st century. With CO_2 emission restrictions a zoo of new technologies enters the market. Renewable technologies will replace hand in hand with fusion fossil capacities. The replacement will not be complete at the end of the 21st century. Still fusion seems capable to win considerable market shares already till then, roughly 30% in Europe and 10% in India in the electricity sector. Fusion would be by then one of the corner stones of the future energy system.

References

1. A. Grübler: *Technology and Global Change* (Cambridge University Press, Cambridge 1998)
2. BP: *Statistical Review of US Energy 2001* (BP Distribution Service, Bournemouth 2001)
3. European Commission Directorate-General for Energy and Transport: *EU Energy and Transport in Figures: Statistical Pocket Book 2003* (European Union, Brussels 2003)
4. D.H. Meadows, D.L. Meadows, J. Randers et al: *Limits to Growth* (Potomak Associates, New York 1972)
5. European Commission: *Green paper – Towards a European strategy for the security of energy supply* (European Communities: Official Publications of the European Communities, Luxembourg 2001)
6. International Energy Agency: *World Energy Investment Outlook – 2003 Insights* (IEA, Paris 2003)
7. H.J. Schellnhuber, V. Wenzel: *Earth System Analysis* (Springer, Berlin 1999)
8. S. Glasstone, R.H. Lovberg: *Controlled Thermonuclear Reactions* (Van Nostrand Corp., Princeton 1960)
9. AEA Technology National Environmental Technology Centre: *Appraisal of Resources and Costs of Critical Chemical Elements Used in Fusion*, internal report: AEA/CS/ZJ/16401091/P01 (AEA, Culham 1994)
10. J. Raeder, K. Borrass, R. Bünde et al: *Kontrollierte Kernfusion* (Teubner Studienbücher: Physik, Stuttgart 1980)
11. Y. Miyai, H. Kanoh, Q. Feng et al: Reports of the Shikoku National Industrial Research Institute **28**, 2217-14 (1996)
12. K. Tokimatsu: Quantitative Analysis of Economy and Environmental Adaptability of Tokamak Fusion Power Reactors. PhD Thesis, University of Tokyo, Tokyo (1998)
13. Battelle Northwest Laboratory: *Materials Availability for Fusion Power Plant Construction*, internal report: BNWL-2016, UC-20 (Battelle Northwest report series, Richland 1976)
14. R. Bünde: Nucl. Eng. Des. Fusion **3** 1 (1985)
15. L. Schleisner: *Socio-Economic Research on Fusion, SERF: 1997-98. Macro Task E2: External Costs and Benefits. Sub Task : Life Cycle assessment of a fusion power plant*, internal report R2.1 (RisøNational Laboratory, Roskilde 1998)
16. R.H. Clarke, F.A. Fry, J.W. Stather et al: Document of the NRPB **4**, 1 (1993)
17. J. Raeder, I. Cook, F.H. Morgenstern et al: *Safety and Environmental Assessment of Fusion Power*, report of the SEAFP project EURFUBRU XII-217/95 (European Commission, Brussels 1995)
18. FDR Safety Assessment. In: *Technical Basis for the ITER Final Design Report, Cost Review and Safety Analysis (FDR)*, ITER EDA Documentation Series No. 16, Chap. 4 (IAEA, Vienna 1998)
19. I. Cook, G. Marbach, L. Di Pace et al: *Safety and Environmental Impact of Fusion*, EUR(01) CCE-FU/FTC 8/5, EFDA-S-RE-1 (EFDA, Garching 2001)
20. A.C. Pigou: *Wealth and Welfare* (Macmillian, London 1912)
21. ExternE 1995: *ExternE: Externalities of Energy*, Vol. 1, EUR 16520 EN (European Commission, Directorate-General XII, Science Research and Development: Office for Official Publications of the European Communities, Luxembourg 1995)

22. R. Sáez: *1999 Socio-economic Research in Fusion SERF 1997-98: Externalities of the Fusion Fuel Cycle. Final Report* (Coleccion Documentos CIEMAT, Madrid 1999)
23. S. Barabaschi, C. Berke, F. Fuster Jaume et al: *Fusion programme evaluation 1996*, EUR 17521 (Office for Official Publications of the European Communities, Luxembourg 1997)
24. P. Lako, J.R. Ybema, A.J. Seebregts et al: *Long term scenarios and the role of fusion power*, Report ECN BS: ECN-C–98-095 (ECN Policy Studies, Petten 1998)
25. T.C. Hender, P.J. Knight, I. Cook: Fusion Technol. **30**, 1605 (1996)
26. P.J. Knight, S.C. Donovan: *Calculations with SUPERCODE for SERF Task E1*, Progress Report (UKAEA Fusion, Culham 1998)
27. J. Edmonds, H.M. Pitcher, D. Barns et al: *Modelling Future Greenhouse gas emissions: The second generation model description, in Modeling global Change* (United Nations University Press, Tokyo 1993); J. Edmonds, J.A. Reilly: Ener. Econ. **5**, 74 (1983)
28. F. Najmabadi, R.W. Conn, P.I.H. Cooke et al: *The ARIES-I Tokamak Fusion Reactor Study – The Final Report*, UCLA report UCLA-PPG-1323 (UCLA, San Diego 1991); J.G. Delene: Fusion Technol. **26**, 1105 (1994)
29. J.H. Ausubel, A. Grübler, N. Nakicenovic et al: Climate Change **12**, 245 (1988)
30. N. Nakicenovic, A. Grübler, A. McDonald: *Global Energy Perspectives* (Cambridge University Press, Cambridge 1998)
31. P. Lako, J.R. Ybema, A.J. Seebregts: *The Long-Term Potential of Fusion Power in Western Europe*, Report ECN BS: ECN-C–98-071 (ECN Policy Studies, Petten 1998)
32. P. Lako, A.J. Seebregts: *Characterisation of Power Generation Options for the 21st Century*, Report ECN BS: ECN-C–98-085 (ECN Policy Studies, Petten 1998)
33. K. Tokimatsu, J. Fujino, Y. Asaoka et al: Studies of Nuclear Fusion Energy Potential Based on Long-term World Energy and Environment Model. In: *Proceedings of the 18th IAEA Fusion Energy Conference, Sorrento, Italy, 4.–10. October 2000*, paper IAEA-CN-77/SEP/03 (IAEA, Vienna 2001)
34. J. Edmonds: private communication (2004)
35. *Der Fischer Weltalmanach* (Fischer Taschenbuch Verlag GmbH, Frankfurt am Main 1996)
36. http://www.eia.doe.gov/emeu/iea/table62.html

Abbreviations

ACTEX	**act**ivity **ex**pansion	
APGL	**a**tmospheric **p**ressure **g**low discharge	
ASDEX Upgrade	**A**xial **S**ymmetric **D**ivertor **EX**periment Upgrade	fusion device (IPP)
AU	**a**stronomical **u**nit	
B2	**B**raams **2** (developer)	2D Fluid code
B2-Eirene	B2 and Eirene	2D Fluid code
bbc	**b**ody **c**entered **c**ubic	
BBGKY	**B**orn, **B**oguljubov, **G**reen, **K**irkwood, **Y**von-theory	
BES	**b**eam **e**mission **s**pectroscopy	
BESI	**b**eam **e**mission **s**pectroscopy **i**maging	
BoRiS	**Bo**rchardt **Ri**emann **S**chneider (developers)	3D fluid code
CARS	**c**oherent **a**nti-Stokes **R**aman **s**cattering	
CCD	**c**harge **c**oupled **d**evice	
CCP	**c**apacitively **c**oupled **p**lasmas	
CFL	**c**ompact **f**luorescent **l**amp	
CHS	**C**ompact **H**elical **S**ystem	fusion device (NIFS)
CME	**c**oronal **m**ass **e**jection	
COE	**c**ost **o**f **e**lectricity	
COREX	**Co**operative **R**esonance Cone **Ex**periment	
CR	**c**ollisional-**r**adiative (model)	
CRPP	**C**entre de **R**echerches en **P**hysique des **P**lasmas	EPFL
DALF3	**d**rift **Alf**vén turbulence code **3**D	
DAW	**d**ust-**a**coustic **w**ave	
DBD	**d**ielectric **b**arrier **d**ischarge	
DEMO	**demo**nstration electricity-generating power plant	

DEOS	Department of Earth Observation and Space Systems	
DESY	Deutsches Elektronen Synchrotron	
DFT	density functional theory	
DIAW	dust-ion-acoustic waves	
DiG	diffusion in Graphite	
DIII-D	Doublet III-D	fusion device (GA)
DKES	drift kinetic equation solver	numerical code
DNS	direct numerical simulations	
DOS	density of states	
DLW	dust lattice wave	
ECE	electron cyclotron emission spectroscopy	
ECR	electron cyclotron resonance	
ECRH	electron cyclotron resonance heating	
EEDF	electron energy distribution function	
ELM	edge localized mode	
EOS	equation of state	
EPFL	Ecole Polytechnique Federale de Lausanne (Switzerland)	
ETG	electron temperature gradient	
FactSage	combines FACT-Win and Chemsage	code, commercial tool
fcc	face-centered cubic	
FEL	free electron laser	
FIDAP	Fluid Dynamics Analysis Package	software package
FIDF	full ion distribution function	
FL	fluorescent lamp	
FLOP	floating point operation	
FOM	Fundamenteel Onderzoek der Materie	Netherlands
FVT	fluid variational theory	
FWHM	full width half maximum	
GA	General Atomics	San Diego (USA)
GHG	greenhouse gas	
GtC	Gigatonnes of carbon	
GWe	GW electric	
HFS	high field side	
HIBP	heavy ion beam probe	
HID	high intensity discharge	
HMDSO	hexamethyldisiloxane	
HP	high pressure	
IAW	ion-acoustic waves	
ICCD	intensified charge coupled device	
ICP	inductively coupled plasmas	

ICRH	ion cyclotron resonance heating	
IIASA	International Institute of Applied System Analysis	
IMS	Intelligent Maintenance Systems	
IPP	Max–Planck–Institut für Plasmaphysik	Garching, Greifswald (Germany)
ISS95	International Sterallator Scaling 95	
ITER	formerly interpreted to stand for International Thermonuclear Experimental Reactor	projected fusion device, latin: the way
ITG	ion temperature gradient	
JET	Joint European Torus	fusion device, Culham, GB
KAM theorem	theorem of Kolmogorov, Arnold and Moser	
KEMS	Knudsen Effusion Mass Spectrometry	
KMC	kinetic Monte Carlo	
LASNEX		Los Alamos ICF design code
LED	light emitting diode	
LES	large eddy simulation	
LFS	low field side	
LHD	Large Helical Device	fusion device (NIFS)
LIDAR	light detection and ranging	
LIF	laser-induced fluorescence	
LP	low pressure	
LLNL	Lawrence Livermore National Laboratory	
LTE	local thermodynamic equilibrium	
MAGPIE	mega-ampere generator for plasma implosion experiments	
MARKAL	MARKet Allocation	linear programming model
MC	Monte-Carlo method	numerical method
MCC	Monte Carlo collisions	
MD	molecular dynamics	
MF	melamine formaldehyde	
MHD	magnetohydrodynamics	
NBI	neutral beam injection	
NIF	National Ignition Facility	LLNL, USA
NIFS	National Institute for Fusion Science	Toki (Japan)
NIST	National Institute of Standards and Technology	Gaithersburg, USA

OCP	one component plasma	
ODE	ordinary differential equation	
OH	ohmic heating	
OML	orbital motion limit	
PCA	poly-crystalline alumina	
PDE	partial differential equation	
PDF	probability density function	
pe	polyethylene	
PIC	particle-in-cell	simulation technique
PIMC	path-integral Monte Carlo	simulation technique
PIP	partially ionized plasmas	
PLTE	partial local thermodynamic equilibrium	
PPT	plasma phase transition	
P.S.	Pfirsch-Schlüter	
ps	polystrene	
PTE	partial thermodynamic equilibrium	
PWR	pressurised water reactor	
QEOS	quotidian EOS	
QMD	quantum molecular dynamics	
RABER	Radio Beacon on Rocket	
R&D	research and development	
rf	radio frequency	electromagnetic waves
RPA	random phase approximation	
RTP	Rijnhuizen Tokamak Project	fusion device (FOM)
SEAFP	Safety and Environmental Assessment of Fusion Power	
SEM	scanning electron microscopy	
SERF	Socio-Economic Research on Fusion	
SOL	scrape-off layer	
STP	standard temperature and pressure	
TB-MD	tight-binding molecular dynamics	
TEC	total electron content	
TEM	trapped particle modes	
TEXTOR	Tokamak Experiment for Technology Oriented Research	fusion device, Forschungszentrum Jülich, Germany
TEXTOR-DED	TEXTOR - dynamic ergodic divertor	

TEXT-U	**TEX**as **T**okamak-**U**pgrade	fusion device, U-Texas at Austin (USA)
TF	**t**oroidal **f**ield	
TJ-II	second upgrade of **T**orus **J**EN (former name of Spanish energy agency; now CIEMAT)	fusion device, Madrid, Spain
TORB	**t**heta-pinch **orb**it code	gyro-kinetic global nonlinear code
UEDGE	**u**niversal **Edge** code	2D fluid transport code
UV	**u**ltra**v**iolet	
VUV	**v**acuum **u**ltra**v**iolet	
W7-AS	**W**endelstein7-**A**dvanced **S**tellarators	fusion device (IPP)
W7-X	**W**endelstein7-X	fusion device (IPP) (under construction)
WEC	**W**orld **E**nergy **C**ouncil	
XRD	**X**-**r**ay **d**iffraction	

Index

$1/\nu$ regime   239, 242

absorption coefficient   410
adiabatic
   electrons   194
   invariant   18, 19, 144, 145
   response   191, 194
advective derivative   55
Alfvén velocity   66, 84
Alfvén waves   *see* waves
ambipolar diffusion   101, 104, 233, 303
ambipolarity   241, 242
Ampere's law   62
amplitude histogram   378
anomalous transport   *see* transport
ASDEX Upgrade   75, 76, 153, 154, 431, 457
aspect ratio   152
atomic relaxation times   367

B-spline   438
banana orbit   167, 227, 228, 237
BBGKY-hierarchy   *see* kinetic equation
BES   *see* plasma diagnostics
β *see* plasma beta
Bethe–Weizsäcker cycle   *see* CNO cycle
Blackman–Tukey method   380
Bohm
   criterion   106, 107
   diffusion coefficient   169, 199
   velocity   106, 319
Boltzmann
   distribution   368
   equation   19, 78, 96, 128, 222
   relation   251, 257, 385
bootstrap current   156, 169, **222**

bounce motion   18, 227, 228, 238
breathing mode   280
bremsstrahlung   130, 215, 356, 363, 455
Brillouin density   279
Brownian motion   165, 320, 325

canonical momentum   144
capacitively coupled plasmas   105
CCP   *see* capacitively coupled plasmas
center-of-mass mode   280
charge state distribution   364
Child-Langmuir law   107
CHS   42, 244, 263
circulation   247
Clebsch coordinates   433
cluster decomposition   122
clusters   124, 125, 324
CNO cycle   331, 450
CO_2 emission   465, 475, 480
COE   *see* cost of electricity
collision time   *see also* collisions   10, 12, 13
collisional radiative model   367, 393
collisionality   238
collisions
   charge transfer   96
   Coulomb   8, 10, 11, 78, 138, 148, 304
   electron–electron   13
   electron–ion   13
   ion–electron   13
   ion–ion   13
conduction–convection problems   426
conductivity   127, 343
confinement   213
   H-mode   169, 216
   L-mode   169, 216
   scaling laws   169, 216
continuity equation   54, 62

490 Index

Corona model 369
coronal loop 88
correlation 250
cost of electricity 474
Coulomb
  barrier 448
  cross section 12
  crystal 269, 294
Coulomb logarithm 12, 232
coupling parameter 6, 117, 276, 309
cross section
  differential 357
  Thomson 353
cross-correlation function 250
curvature drift *see* drifts
cut-off **40**
  density 26, 391
  frequency 26
  wavelength 30
cyclotron frequency 16, 140, 273
cyclotron heating *see* heating

D-T reaction *see* fusion
DAW *see* waves
Debye length 7, 77, 278, 355
degeneracy parameter 117
Delaunay triangulation 434
DEMO 458, 474
diamagnetism 16, 140, 162, 233
DIAW *see* waves
dielectric tensor 23
diffusion 223
  banana 227, 238
  classical 165
  neoclassical 166
  random walk diffusion coefficient 224
diocotron waves *see* waves
discharge
  arc 95, 97, **102**
  corona 97, **102**
  dc 97, 104, 107
  dielectric barrier 103
  glow 95, 99–101, 104
    atmospheric pressure 103
  micro 102, 103
  microwave 110
  mircowave 97, **104**, 105
  overview **97**

rf 97, **104**, 105, 107–110, 113, 115, 305, 429
Townsend 100
dispersion relation *see* corresponding waves
disruptions 91
divertor 162, 467, 474
DLW *see* waves
Doppler cooling 276, 277
Doppler shift 354
drift ordering 198
drift parameter 70
drift surface 149, 237
drift waves *see* waves
drifts
  ∇B 17, 142, 149, 150, 436
  $\mathbf{E} \times \mathbf{B}$ 16, 141, 249, 250
  curvature 17, 143, 230, 235, 436
  diamagnetic 233, 241
  general force 16, 141
  gravitational 17
Druyvesteyn method 112
dual cascade 248
dust charge 301, 307
dust charging 298, 300, 302
dust plasma frequency 313
dust resonance 306, 308
dynamic form factor 357
dynamical screening 123
dynamo effect 70, 189

$\mathbf{E} \times \mathbf{B}$ velocity 61
ECE *see* plasma diagnostics
ECR *see* electron cyclotron resonance
ECRH *see* plasma heating
eddy mitosis 174
edge localised mode 91
EEDF *see* electron energy distribution function
Ehrenfest theorem 269
Einstein's energy-mass relation 445
electric breakdown 98, 99, 102–104
electric probes *see* plasma diagnostics
electrical conductivity 15, 23, 51, 82, 85, 117, 118, 122, 127, 130, 233, 235, 333, 345, 403, 408
electron attachment 96
electron cyclotron frequency 40

electron energy distribution function 96
electron root 243
electron volt (eV) 4
electron–diamagnetic direction 253
elementary processe
    excitation 96
elementary processes 96
    charge exchange 367
    excitation 367, 369
    recombination 367
    self-absorption 367
ELM *see* edge localised mode
Elsässer variables 188
emissive probes 386
energy
    consumption 461
    resources 464
energy confinement time 137, 214, 215
enstrophy 175, 183, 247
EOS *see* equation of state
Epstein friction coefficient 304
equation of motion 15, 22
equation of state 118, 125, 127, 128, 331, 333, **337**, 338, 339, 341, 342
equilibrium
    corona 365
    in toroidal geometries 149
    local thermodynamic 57, 58, 102
    local thermodynamic (LTE) 95, 368, 404
    partial local thermodynamic (PLTE) 369
    partial thermodynamic (PTE) 95
    Saha 366
    thermodynamic 51, 57, 79, 139
ergodic magnetic fields 90, 150, 427, 432
ETG *see* instability
eV 4
extrasolar planets 334

F-layer 40, 46
Faraday effect 41, 352
Fermi distribution 124
Fermi energy 6, 117
Feynman diagram 119
Fick's law 165, 215
field line tension 68, 82, 83

finite clusters 324–326
floating point operations 425
floating potential 300
FLOP *see* floating point operations
fluctuations 19, 32, 164, 195, 245, 249, 251, 254
fluid equations *see* MHD equations
fluid variational theory 338
Fokker–Planck equation 78, 166
forces on dust particles 302
    electric force 302
    gravity 302
    ion drag 303
    neutral drag 304
    nonreciprocal attraction 311
    theromophoresis 304
Fourier notation 31
free energy 338
fusion **445**
    D-T reaction 450, 451, 454
    impurity radiation 364
    inertial 360, 445, 456
    muon-catalyzed 458
    power balance 455, 456
    power plant 458, 467, 469–471
        economic considerations 473
        safety 470
    reactions 448, 450, 454
    reactor 138, 156, 428, 449, 452, 458
    resources 469
    socio-economic stuides 476
    triple product 137
FVT *see* fluid variational theory
FWHM *see* spectral lines

Gibbs free energy 412
Gibbs phenomenon 379
Gibbs–Bogolyubov inequality 338
Green's function 119
group velocity 24
guiding centre 140, 150, 166, 229
gyro radius 16, *see* Larmor radius
gyro-averaging 436
gyro-Bohm scaling 169
gyro-kinetic ordering 198
gyro-kinetic theory 435, 437
gyro-motion 140

H-mode *see* confinement

492   Index

Hall sensors   390
Hall term   193
Hartree–Fock approximation   120
heavy ion beam probe   *see* plasma diagnostics
helical magnetic field lines   156
helically trapped particles   229–231, 239
heliotron *see also* stellarator   41, 217
Helmholtz equation   436, 438
HIBP   *see* plasma diagnostics
HID   *see* high intensity discharge lamps
high intensity discharge lamps   399, 400, 403–405, 408
   chemical modelling   413
   corrosion analysis   418, 419
   modelling   407
   radiation transport   409
   spectral intensity   406
high-field side (HFS)   225
Hugoniot curve   127, 341, 343

ICP   *see* inductively coupled plasmas
ICRH   *see* plasma heating
ideal plasma   *see* plasma
ideal two fluid model   51
impedance probe   *see* plasma diagnostics
inductively coupled plasmas   105, 110
inertial confinement   456
infrared absorption   114
instability
   beam–plasma   35, 36
   electron-temperature gradient (ETG)   254
   interchange   87, 253
   ion-temperature gradient (ITG)   168, 254, 434, 435, 437, 439
   MHD   86
   Rayleigh–Taylor   46
   trapped electron modes (TEM)   254
interferometry   *see* plasma diagnostics
inverse cascade   175
inward pinch   221
ion acoustic velocity   33
ion cyclotron frequency   40
ion energy distribution function   107
ion root   243

ion trap   269
   cooling   278
   Paul trap   270, 271, **271**, 278, 287, 289
   ion clouds   275
   potential   270, 271
   stability   272
   Penning trap   270, **273**, 274, 279, 282
   collective effects   284
   ion crystals   291
   loading   275
   plasma   280
   potential   270
ionisation   95
   associative   96
   collsional   96
   Penning effect   96
   photo   96
ionosondes   *see* plasma diagnostics
ionosphere   40, 46
island   *see* magnetic islands
ITER   42, 154, 361, 458, 470, 473, 474
ITG   *see* instability

JET   153, 154, 351, 357, 361, 457
Jupiter   117, 297, 331–333, 335, 336

K41 theory   176, 247
KEMS   *see* Knudsen effusion mass spectrometry
kinetic equation   19, 78, 80, 96, 166, 431
   BBGKY-hierarchy   78
   drift-kinetic equation   166
kink instability   87
Knudsen effusion mass spectrometry   404, 413–415, 422
Kolmogorov length   433
kurtosis   379

L-mode   *see* confinement
L-wave   *see* waves
Lüst-Schlüter-Grad-Shafranov-equation   154
Landau damping   78, 79, 87, 253, 360
Landau length   8
Langmuir oscillations   9

Langmuir probe *see* plasma diagnostics
Laplacian pressure method 188
Larmor radius 16, 138
laser manipulation of dust 307, 311, 319–321
Lawson parameter 137
LHD 41, 170, 217, 457
LIDAR 357
LIF *see* plasma diagnostics
linear response theory 127
local thermal equilibrium 404
Lorentz force 144
Lorenz number 343
loss cone 147, 150, 168
low-field side (LFS) 225
lower hybrid resonance 44
LTE *see* equilibrium
Lundquist number 73, 88

Mach cone
  compressional 320
  in Saturn's rings 322
  shear 322
Mach probe *see* plasma diagnostics
Mach-Zehnder interferometer 27
magnetic axis 150
magnetic flux tube 64
magnetic islands 89, 90, 150
magnetic mirror 143, 225
magnetic moment 18, 145
  gyrating particles 140
magnetic pressure 65, 82
magnetic pressure waves 66
magnetic pumping 240
magnetic surface 150
magnetic tension 65
magnetohydrodynamics *see also* MHD 51
  adiabatic pressure 62
  flux conservation 62
  force equation 62
  force free equilibrium 65
  kinematic equation 62
  time scales 68
  validity 69
magnetron frequency 284
Maxwell's equations 22
MD *see* modelling

mean free path 10
  at magnetic fusion conditions 138
MHD *see also* magnetohydrodynamics
  equations 82
  equilibrium 82, 160, 193
  force equation 60
  kinematic equation 61
microgravity 305
mirror machine 147
mixing length 251
modelling
  2D fluid 431
  3D fluid 431
  discharge lamps 407
  kinetic Monte Carlo 429
  kinetic PIC 429
  molecular dynamics (MD) 279, 426–428
  Monte Carlo methods 119, 409, 427, 429–432, 434
  particle–in–cell (PIC) 427, 430, 437
  plasma edge 427
  radiation transport 409
  thermochemical 412
  turbulence 434
Mott effect 122, 345

Navier–Stokes equation 246, 248, 432
NBI *see* plasma heating
neoclassical transport *see* transport
Noether's theorem 144
ν regime 244
nuclear binding energy 446
Nyquist limit 378

Ohmic heating *see* plasma heating
OML *see* orbital motion limit
orbital motion limit 111, 112, 298

particle confinement time 215
particle–in–cell *see* modelling
partition function 124, 366, 368, 410
Paschen law 100
passing particles 226, 234, 236, 237
passive scalar 175, 192
Paul trap *see* ion trap
PCA *see* poly-crystalline alumina
PDF *see* probability density function

494 Index

Penning effect 96
Penning trap *see* ion trap
Pfirsch–Schlüter current 161, 165, 235, 236
Pfirsch–Schlüter transport *see* transport
phase velocity 24
PIC *see* modelling
plasma
  beta 70, 77, 162, 170
    low-β 70, 175, 187, 190
  crystal 279, 287, 289, 291, 294, 310, 430
    phase transition 312
  degenerated 6
  diagnostics *see* plasma diagnostics
  edge 91, 169, 219, 257, 426, 427, 431–433
  frequency 9, 10, 25, 32, 279, 313, 318
  heating 356
    electron cyclotron resonance heating (ECRH) 105, 215
    ion cyclotron resonance heating (ICRH) 215
    neutral beam injection (NBI) 215
    Ohmic 105, 215
  ideal 5, 6, 117, 161, 164
  low temperature 95
  non-thermal 95
  oscillations 9
  parameter 8
  polarisation 141
  quasi-neutrality 7, 9, 10, 13, 14, 33, 60, 69, 81, 235
  reactive 114
  relativistic 6
  sheath 105, **106**, 107–110, 114
  surface interaction 427
  waves *see* waves
plasma diagnostics
  beam emission spectroscopy (BES) 393
  density measurements for dense plasmas 30
  double probes 112
  electric probes 110–112
  electron cyclotron emission 41, 384, 392

  electron cyclotron emission (ECE) 41, 90
  emissive probe 386
  emissive probes 387
  Hall probe 390
  heavy ion beam probe (HIBP) 244, 259, 384, 394
  impedance probe 46
  induction coil 388
  interferometry 26–29, 352
  ionosondes 44
  Langmuir probe **110**, 384, 387
    electron saturation current 385
    floating potential 385
    ion saturation current 384
  laser aided diagnostics **352**
  laser induced fluorescence (LIF) 33, 279, 395
  Mach probe 388
  plasma oscillation method 37
  probe arrays 382
  radio beacon technique 29
  reflectometry 391
  resonance cone 46, 48
  Thomson scattering 118, 126, 220, 352, 353
    collective 357
    incoherent 353
    LIDAR 357
plasma phase transition 341
plasma surface
  interaction 96, 114
  transition 95, **106**
plateau regime 238
Poisson's equation 7, 32, 78, 107, 313
poloidal coordinate 151
poloidal Larmor radius 226
poly-crystalline alumina 404–406, 420
power spectral density 379
PPT *see* plasma phase transition
pressure dissociation 339
probability density function 179, 224, 258, 378, 379
PTE *see* equilibrium

quasi-neutrality *see* plasma

radial electric field 229
radiation transport 410

radio beacon technique   *see* plasma diagnostics
Raman scattering   356
random walk   223
rate coefficient   454
rate equations   365
Rayleigh scattering   356
recombination   96, 356
reconnection   73, 85, 86, 88, 89, 190
reflectometry   *see* plasma diagnostics
resistive decay time   73
resistive MHD   72
resonance cone   *see* plasma diagnostics
Reynolds number   175, 177, 178, 246
Reynolds stress   189, 263
rf discharge   *see* discharge
rotational transform   149, 150, 156

safety factor   *see* rotational transform
Saha's equation   5, 126, 368, 369
Salpeter
    approximation   357
    shape function   358
scaling laws   168
scanning electron microscopy   420
scattering   352
    cross-section   355
    parameter   355
    vector   354
    X-ray   361
scrape-off layer   388, 427
secondary electron emission   96, 98, 99, 101, 102, 104, 106
self-energy   119, 120
SEM   *see* scanning electron microscopy
Shafranov shift   154, 161, 162, 165
shock waves   337, 341
skewness   379
SOL   *see* scrape-off layer
solar system   332
Sonin plot   111
space-time data   382
spectral emission coefficient   364
spectral function   120
spectral lines
    Doppler broadening   370
    Doppler line shape function   370
    full width half maximum (FWHM)   355, 370, 410
    Gaussian   371
    Lorentzian line shape function   370, 410
    natural broadening   370
    pressure broadening   371
    Stark broadening   371, 410
    van-der-Waals broadening   410
    Voigt line shape function   372
spectral radiance   364
Spitzer resistivity   164, 300
Stark broadening   *see* spectral lines
stellarator   **156**, 158
    modular   159
    optimisation   161
Stix parameter   39
stray light   356
streamer   102, 103, 253
strongly coupled plasmas   117
strongly coupled systems
    one component plasma (OCP)   310
    Yukawa systems   310
sustainable development   462

τ_E *see* energy confinement time
tearing mode   90
TEC   *see* total electron content
TEM   *see* instability
thermal limit   369
thermodynamic equilibrium   *see* equilibrium
θ–pinch   440
Thomas–Fermi model   118
Thomson scattering   *see* plasma diagnostics
three-body collision   96
three-wave coupling   173, 174, 176, 181, 182, 186, 189, 192, 207
tokamak   75, 76, 90, 91, **151**
    coil system   152
    equilibrium   153
Tore Supra   152, 457
toroidal confinement   148
toroidal coordinate   151
toroidal resonance   230, **236**, 242
torsatron *see also* stellarator   158, 159, 228, 244, 259
total electron content   30

Townsend mechanism 98, 102
  first coefficient α 98, 99
  second Townsend coefficient β 98
  third Townsend coefficient γ 98
  Townsend discharge *see* discharge
transport 164, **213**
  anomalous 168, 213, 245, 377, 381, 434, 435
  coefficient in partially ionized plasmas 127
  heat flux 218
  ITG *see* instability
  matrix 221
  neoclassical 166, 231, 238, 242
  particle flux 220
  Pfirsch–Schlüter 234, 238
  timescale 91
  timescales 92
trapped particles 226
tunnelling probability 451
turbulence
  drift wave 191
  electromagnetic 245
  electrostatic 245, 257
  entropy 192
  helicity conservation 189
  high Reynolds number regime 175, 178, 191, 194, 199, 200, 203, 206
  inertial range 177
  Kolmogorov spectrum 177
  polarization current 196
  saturated state 179
turbulence modelling 434
two-point spectral analysis 381

upper hybrid resonance 44

Virial theorem 151
Vlasov equation 19, 78
void 305
vortex tube stretching 184

vorticity 175, 176, 183, 192, 195, 196, 199–201, 247, 248, 256

warm dense matter 331
waves **21**
  Alfvén 67, 83, 189, 395
    fast 69
    slow 69
  diocotron wave 280
  drift 191, 194, 210, 252, 253
  dust acoustic (DAW) 313
  dust ion-acoustic (DIAW) 33, 34, 316
  dust lattice (DLW) 313, 318–320, 323
  dust-acoustic (DAW) 313–315
  electron beam driven 35
  electron plasma 359
  electrostatic 32
  ion-acoustic 31–33, 359
  L-wave 39, 40
  longitudinal **31**
  magnetic pressure 65
  magnetized plasma **37**
  normal modes 23
  O-mode 43, 46
  R-wave 39–42, 105
  transverse 25
  unmagnetized plasma 24
  whistler 42
  X-mode 43, 46
Welch's method 380
Wendelstein 7-A 158
Wendelstein 7-AS 160, 171, 242
Wendelstein 7-X (W7-X) 162, 163, 217, 435, 439, 459
Wiener–Khintchine theorem 250
Wigner–Seitz radius 6, 117, 277

X-ray diffraction 420
XRD *see* X-ray diffraction